21st Australasian Weeds Conference 2018

Weed Biosecurity – Protecting Our Future

Sydney, Australia
9-13 September 2018

Editors:

Stephen Johnson
Leslie Weston
Hanwen Wu
Bruce Auld

ISBN: 978-1-5108-8847-0

Printed from e-media with permission by:

Curran Associates, Inc.
57 Morehouse Lane
Red Hook, NY 12571

Some format issues inherent in the e-media version may also appear in this print version.

Printed by Curran Associates, Inc. (2019)

For permission requests, please contact Council of Australasian Weed Societies Inc. (CAWS) at the address below.

Council of Australasian Weed Societies Inc. (CAWS)
Massey University, Private Bag
11-222 Palmerston North 4442
New Zealand

Phone: +61 6 350 4926
Fax: +61 8 9821 1028

www.caws.org.au

Additional copies of this publication are available from:

Curran Associates, Inc.
57 Morehouse Lane
Red Hook, NY 12571 USA
Phone: 845-758-0400
Fax: 845-758-2633
Email: curran@proceedings.com
Web: www.proceedings.com

21st Australasian Weeds Conference

21AWC

SYDNEY 2018 *WEED BIOSECURITY ~ PROTECTING OUR FUTURE*

PROCEEDINGS

Editors: Stephen Johnson, Leslie Weston, Hanwen Wu and Bruce Auld

9–13 September 2018

Sydney, New South Wales

The Weed Society of NSW

Council of Australasian Weed Societies Inc.

Cataloguing-in-Publication entry is available from the National Library of Australia
http://catalogue.nla.gov.au/

Citation
Please cite this publication as:
Proceedings of the 21st Australasian Weeds Conference, 9–13 September 2018, Sydney, New South Wales, eds S. Johnson, L. Weston, H. Wu and B. Auld. (Weed Society of New South Wales).

Proceedings designed and typeset by R.G. and F.J. Richardson, Meredith, Victoria
www.weedinfo.com.au

Council of Australasian Weed Societies Inc.

The Weed Society of NSW

Committees

Organising Committee

Kim Hignell, Chair
John Cameron
Hillary Cherry
Maria Edmonds
Stephen Johnson
Deirdre Lemerle
Rex Stanton
Michael Walsh
Leslie Weston
Hanwen Wu

Proceedings Editors

Stephen Johnson
Leslie Weston
Hanwen Wu

Coordinating Editor

Bruce Auld

Program Editor

Hillary Cherry

Field Trip Committee

Maria Edmonds
Jillian Macintyre
Nicola Dixon

This is a refereed conference proceedings and all published papers have been peer reviewed.

21st Australasian Weeds Conference
21AWC
SYDNEY 2018 WEED BIOSECURITY ~ PROTECTING OUR FUTURE

Sponsors and Supporters

The Organising Committee sincerely thanks our generous sponsors

Major Partner:

Grains Research and Development Corporation

The grains industry plays a vital role in Australia's economy, comprising 24 percent of total agricultural exports. The Grains Research and Development Corporation (GRDC) supports the industry through investing in research, development and extension (RD&E) to create enduring profitability for Australian grain growers. The GRDC invests over $192 million in around 900 research, development and extension (RD&E) projects to directly benefit growers across a broad range of research areas. Within their carefully balanced portfolio is a range of investments, from long-term, high risk, 'blue sky' research to short-term, outcome-focused applied research at the local level.

GRDC™
GRAINS RESEARCH & DEVELOPMENT CORPORATION

Platinum Partner:

About BASF's Crop Protection division

BASF
We create chemistry

With a rapidly growing population, the world is increasingly dependent on our ability to develop and maintain sustainable agriculture and healthy environments. BASF's Crop Protection division works with farmers, agricultural professionals, pest management experts and others to help make this possible. With their cooperation, BASF is able to sustain an active R&D pipeline, an innovative portfolio of products and services, and teams of experts in the lab and in the field to support customers in making their businesses succeed. For more information, please visit us at http://agro.basf.com.au/ or on any of our social media channels.

Gold Partners:

NSW Government – Department of Primary Industries

NSW Government – Local Land Services

Silver Partners:

AgriFutures Biocontrol of Weeds

Centre for Invasive Species Solutions

CORTEVA agriscience – Agriculture Division of DowDuPont

OCP Organic Crop Protectants – Crop Health Specialists

MACSPRED Australia

Dinner Partner:

BAYER

Welcome Reception Partner:

WEED smart – every weed every seed every farm every year

Contents

DAY 1: MONDAY 10 SEPTEMBER

PLENARY PRESENTATIONS

Herbicide resistance in Canada: Biggest threats, best management, and lessons learnt
Hugh J. Beckie 1

National priority list of exotic environmental pests and diseases: an update
Sandra Parsons, Jessica Evans, Bo Raphael and Marwan El Hassan 4

HERBICIDE RESISTANCE

Next generation testing of herbicide resistance: Rapid detection, multiple herbicides, a single assay
Aaron L. Preston and James E. Pratley 5

Multiple herbicide resistance in winter grass (*Poa annua*) in Australia
Peter Boutsalis, Jenna Malone and Christopher Preston 10

Eradication or mitigation? How to manage patches of glyphosate resistant *Echinochloa colona* L.
Jeff A. Werth, David F. Thornby, Michelle D. Keenan, James P. Hereward and Bhagirath S. Chauhan 11

Control of thiocarbamate resistant annual ryegrass (*Lolium rigidum*) in lentils (*Lens culinaris*) in southern Australia
David J. Brunton, Peter Boutsalis, Gurjeet Gill and Christopher Preston 12

Glyphosate resistant common sowthistle (*Sonchus oleraceus* L.)
Jenna M. Malone, Tony Cook, Mahima Krishnan and Christopher Preston 15

Extent of herbicide resistant common sowthistle (*Sonchus oleraceus*) in southern Australia
Alicia Blyth Merriam, Peter Boutsalis, Jenna Malone, Gurjeet Gill and Christopher Preston 16

Genome sequencing of glyphosate resistant fleabane (*Conyza bonariensis*) and sowthistle (*Sonchus oleraceus*)
James P. Hereward, Jeff A. Werth, David F. Thornby, Michelle Keenan, Bhagirath Singh Chauhan and Gimme H. Walter 20

Mechanisms of resistance to penoxsulam and quinclorac in *Echinochloa crus-galli*
Yongfeng Li, Xia Yang, Mingchao Dong, Zichang Zhang, Tao Gu, Jingjing Cao and Qin Yu 21

Understanding the mechanisms of 2,4-D resistance in *Sonchus oleraceus* (common sowthistle)
Mahima Krishnan, Tijana Petrovic, Alicia Merriam, Geetha Velappan and Christopher Preston 22

BAS684H a new tool for selective control of multi-resistant ryegrass
Roberto Busi and Stephen Powles 23

Target-site and metabolic resistance mechanisms to dinitroaniline herbicides in a *Lolium rigidum* population
Jinyi Chen, Heping Han, Danica Goggin, Qin Yu and Stephen Powles 24

A novel *psbA* mutation (Phe-274-Val) confers resistance to PSII herbicides in wild radish (*Raphanus raphanistrum*)
Huan Lu, Qin Yu, Heping Han, Mechelle J. Owen and Stephen B. Powles 25

Metabolic herbicide resistance gene discovery in wild oat (*Avena fatua*)
Qiong Peng, Qin Yu, Heping Han, Lianyang Bai and Stephen Powles 26

WEEDS OF CROPS AND PASTURES – ECOLOGY AND BIOLOGY

Abutilon theophrasti – its biology and management in New Zealand
Trevor K. James and Heidi M. Pene 27

Germination ecology of two Australian populations of African turnip weed (*Sisymbrium thellungii* L.)
Gulshan Mahajan, Amar Matloob, Michael Walsh and Bhagirath S. Chauhan 32

Response of common sowthistle, *Sonchus oleraceus* L. to simulated herbivory
Nagalingam Kumaran, Michelle A. Rafter, Kylie B. Ireland, Gavin Hunter, Louise Morin and S. Raghu 33

Allelopathic potential of pasture legume species and selected winter crops on barnyard grass and rice
Jhoana Opena, Hanwen Wu, Sergio Moroni, Deirdre Lemerle and James Pratley 34

Inter-specific variations in seed germination biology and seedling emergence of bladder ketmia (*Hibiscus* spp.)
Md Asaduzzaman and Eric Koetz 35

Predicting the mass emergence of *Vulpia bromoides*
Claire A. Dowsett, Christopher E. Buddenhagen, Richard J. Chynoweth and Trevor K. James 36

Fact or Fiction: Do glyphosate resistant cropping systems reduce crop or soil health?
Martin M. Williams II, Stephen O. Duke, Agnes M. Rimando, Jude E. Maul, Krishna N. Reddy and James V. Cizdziel 42

Allelopathic interference of annual ryegrass (*Lolium rigidum*) on lucerne nodulation
H.M. Zubair, J.E. Pratley, G.A. Sandral, A.W. Humphries, S.G. Nielsen and A. Price 43

Seed germination ecology of *Sesbania cannibana* (Retz.) Poir.
Nadeem Iqbal, Sudeesh Manalil, Bhagirath S. Chauhan and Steve Adkins 47

WEEDS OF CROPS AND PASTURES – STRATEGIC MANAGEMENT

Developing a framework to prioritise plant biosecurity threats to New Zealand's dairy industry
David Hodges, Trevor James and Petra Muellner 48

Exploring and extending integrated weed management opportunities in the Australian vegetable industry
Michael Coleman, Christine Fyfe, Sita Tiwari Pokhrel, Graham Marshall, Brian Sindel and Paul Kristiansen 49

Developing new options for weed control in Australian cotton systems
James Neilsen, Meredith Conaty and Kate Lang 54

An integrated approach to managing Noogoora burr in maize crops
Shane R. Hona, Trevor K. James and Andrew Blayney 55

BUILDING CAPACITY TO MANAGE WEEDS

From little things big things grow – A review of behaviour change initiatives for weeds management in NSW
Birgitte Verbeek, Elissa van Oosterhout and Wendy Gibney 60

Building capacity for weed management in the Lao PDR
Deirdre Lemerle, John Burley, Sarah Hain, Brendon Bangma, Nicholas Pain, Sengphet Phantavong, Kylie Ireland, Madaline Healey and Lester Burgess 64

Floating Landcare: Bringing a bit of WOW to weedy hotspots on the Hawkesbury Estuary and NSW Central Coast waterways
Rebecca Mooy and Tegan Burton 68

Using the Atlas of Living Australia to support weed biological control
Greg Lefoe, Raelene Kwong and Kerinne Harvey 69

Practical best practice management of opuntioid cacti – from desk to dirt
Shauna Potter, Matthew R. Sheehan, Kate Blood and Kay Bailey 70

STRATEGY – ASSESSING WEED RISK

Introducing the Plant Sure Ornamental Plant Decision Support Tool
Michelle Leishman, Nola Hancock, Victoria Graham, Vanessa Adams, Tim Maher and Anthony Manea 71

Refining the environmental weed risk assessment for non-indigenous plants which may have agricultural potential
Christine J. Munday, Geoff Moore, Rachel Whitsed, Margaret Byrne and Clinton Revell 72

WRASP: A spatial weed risk assessment tool reveals critical variationsin weed risks
Darren J. Kriticos, Josef Beautrais and Mike B. Dodd 77

A national weed management decision-support system
Graeme W. Bourdôt, Darren J. Kriticos and Michael B. Dodd 82

Futures in risk assessment: Challenges in preventing and mitigating weed threats
Stephen B. Johnson, Belinda Mitterdorfer, Dorjee, Tony Buckmaster and Paul O. Downey 86

STRATEGIC PLANNING

Let's get strategic: Weed management for conservation in Western Australia
Kellie Passeretto 87

Buffel grass in South Australia: Progress and future directions
Ross Meffin, Ellen Ryan-Colton, John L. Read, Troy Bowman and Michaela Heinson 91

An ounce of prevention: An accreditation scheme for ornamental plant industries
Hillary Cherry, Stephen B. Johnson and Des Boorman 96

A new approach to managing invasive pests at the local government level
Craig Magnussen 100

Helping secure the future
Our programs and partnerships grow farm productivity and healthy environments and we play a vital role in helping to protect against pests, diseases and environmental threats. We connect our customers with the best services, advice and networks.
Local Land Services offers a range of specialist services, tailored for each community, industry and environment. We understand the needs of the communities we work with and we are well placed to deliver services that address local needs.
Our team is here when it matters, on the ground helping you with knowledge, networks and local experience.
NSW GOVERNMENT
Local Land Services
1300 795 299 www.lls.nsw.gov.au

Heritage or threat, conserve or destroy? The heritage plant management dilemma in protected landscapes
David C. Havell and Angela L. Scott 101

The changing weedscape: Policy and management implications of unknown knowns
Paul O. Downey and Stephen B. Johnson 102

STRATEGIC MANAGEMENT APPLICATIONS

Lake Benmore islands wilding conifer control
Pete Caldwell 103

Managing WoNS for better drinking water
Dennis J. Gannaway and Tim Odgers 104

Containment of Western Cape bridal creeper in peri-urban Adelaide – an eleven-year review
Henry Rutherford, Michaela Heinson, Susan Ivory and Phil Elson 107

WEED ERADICATION

The research behind New Zealand's aquatic weed eradication programmes
Paul D. Champion and Deborah E. Hofstra 108

10 years after Victoria's major Mexican feather grass incursion: Assessing eradication methods and germination behaviour
Nicholas P. Moran, Alexei D. Rowles and Angela B. Constantine 109

Pushing 37 weed species towards extinction in Queensland
Steve Csurhes 115

The effects of cyclones on a tropical weed eradication program
Simon J. Brooks and Mick Jeffery 119

The eradication of weed species in New Zealand's National Interest Pest Responses Programme
Frances G. Velvin 124

DAY 2: TUESDAY 11 SEPTEMBER

PLENARY SESSION
NEW TECHNOLOGIES – USING REMOTE DETECTION TO IMPROVE WEED BIOSECURITY

Machine vision systems for robotic weed sensing in field environments
Cheryl L. McCarthy, Steven R. Rees and Craig P. Baillie 125

Needle in a haystack – detecting hawkweeds using drones
Mark Hamilton, Robert Matthews and Jo Caldwell 126

Multispectral imagery and weed biocontrol: A useful pairing?
A.J. McConnachie, P. Jones, R. Shilpakar and P. Dawson 131

GRAINS RESEARCH AND DEVELOPMENT CORPORATION – GWAC SYMPOSIUM

What's over the horizon for weed management in Australia – science outcomes to benefit Australian growers from the Bayer/GRDC partnership
Marco Busch 132

Research priorities for weed suppression by crops in Australia
Michael Widderick, Deirdre Lemerle, Cameron Taylor, Chris Johansen, Abul Hashem, Vikki Osten, Tony Cook, Kerry Bell and Andrew Storrie 133

Managing clethodim-resistant annual ryegrass (*Lolium rigidum*) in canola with herbicides and competition
Samuel G.L. Kleemann, Gurjeet S. Gill and Christopher Preston 137

Reduction in sowthistle biomass and seed production through manipulation of pulse population and row spacing
Adam B. McKiernan, Greg Harvey, Linda Heuke, Michael Walsh and Michael Widderick 138

Glyphosate and 2,4-D amine resistance in common sowthistle (*Sonchus oleraceus*) and fleabane (*Conyza bonariensis*) in the Northern Grain growing region of Australia
Adam Jalaludin, Michael Widderick, John Broster and Michael Walsh 139

Microwave: A novel approach to tackle herbicide resistance in no-till farming systems
Muhammad J. Khan, Graham I. Brodie and Dorin Gupta 140

Comparing the weed seed collection efficacy of stripper and draper type harvester fronts during harvest
Annie Rayner, John Broster, Alison Chambers and Michael Walsh 144

The extent of herbicide resistance in ryegrass and wild oats in New South Wales and Queensland
John C. Broster, Allison J Chambers, Adam Jalaludin, Michael J. Widderick and Michael J. Walsh 145

Metabolic profiling for benzoxazinoids in weed suppressive and early vigour wheat genotypes
James M. Mwendwa, Paul A. Weston, Inge Fomsgaard, Bente B. Laursen, William B. Brown, Hanwen Wu, Jane C. Quinn, Jeffrey D. Weidenhamer and Leslie A. Weston 146

Opportunities for weed management using virtual fencing technology
Rick Llewellyn, Danila Marini, Marta Monjardino, Sue Belson, Jackie Ouzman and Caroline Lee 147

Ecology of threehorn bedstraw (*Galium tricornutum*): Implications for management and harvest weed seed control
Samuel G.L. Kleemann and Gurjeet S. Gill 148

Lime improves control of wild radish and annual ryegrass in acid soils of Western Australia
Abul Hashem and Catherine Borger 153

Developing biological control options for the integrated management of *Conyza bonariensis* and *Sonchus oleraceus*
L. Morin, K. Nagalingam, K.B. Ireland, V. Lesieur, M. Ollivier, M. Rafter, G.C. Hunter, T. Thomann, M. Jourdan, A. Sheppard, J.F. Martin, M.S. Tixier, M.D. Vitorino and S. Raghu 157

Recruitment and fecundity of annual ryegrass, great brome grass, barley grass, doublegee and sowthistle
Catherine P.D. Borger and Abul Hashem 158

Impact of chaff lining/tramlining on the seed persistence and emergence of weeds
Annie van der Meulen, Annie Rayner, Michael Widderick, Adam Jalaludin, John Broster, Alison Chambers, Linda Heuke, Shona Robilliard and Michael Walsh 163

Newly-developed RIM models for integrated management of brome and barley grass
Marta Monjardino and Rick Llewellyn 164

Technical manuals and training on the soil behaviour of pre-emergent herbicides and modes of action (MOA's) of post-emergent herbicides
Mark Congreve and John Cameron 169

NEW SURVEILLANCE TECHNOLOGIES

How distinguishable are hawkweeds from co-occurring vegetation species in the eye of hyperspectral sensor?
Hsing-Chung Chang, Kerrie Tomkins, Rene Heim, Will Farebrother, Chad Ajamian, Mark Hamilton and Hillary Cherry 175

Use of GIS technology in weed management
Paras Acharya 176

Unmanned aerial vehicles used to control giant reed (*Arundo donax* L.)
Philip Milling 177

Collector app mapping to assess effectiveness of invasive grass control
Steve Taylor, Jenny Conolly and Renee Brawata 181

Weed surveillance using search theory
Oscar Cacho, Dane Panetta, Hillary Cherry, Susie Hester, Sue Bower and Hank Bower 187

FINDING AND MAPPING WEEDS – MAPPING, SURVEILLANCE AND CONTROL

The state (and territory!) of data collection on national priority weeds
Bo Raphael, Jessica Evans, Katherina Ng, Luke Maloney and Sandra Parsons 192

The Victorian Weed Spotters: Recruitment and training of citizen scientists
Zachariah Munakamwe, Angela Constantine and Catherine McInerney 193

Weeds at the early stage of invasion: pilot project evaluation
Kate Blood and Bec James 197

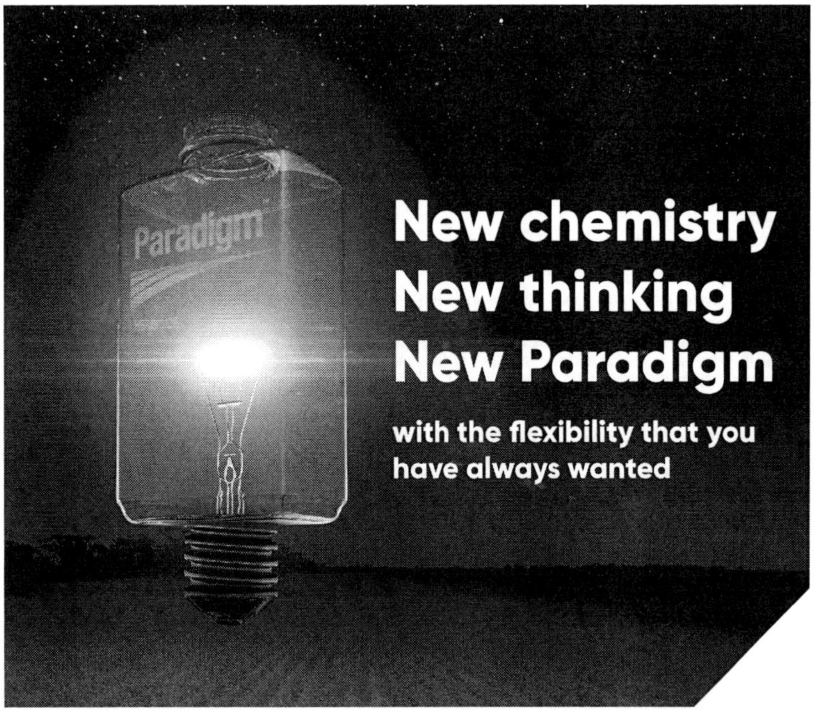

First incursion, weed policy implementation and eradication of *Boerhavia erecta* (Nyctaginaceae) in northern Australia
John O. Westaway, Bert Lukitsch and Roni Opden 200

Iris pseudacorus L. (Irideae, Iridaceae) in South Africa: Feasibility of eradication
Thulisile P. Jaca, Reshnee Lalla and Menzi Nxumalo 201

Frogbit (*Limnobium laevigatum*) detection and eradication in New South Wales
Charlie Mifsud and Terry Inkson 202

BIOLOGICAL CONTROL

Australian weed biocontrol: A look at the past, present and future
Kerinne Harvey and Andrew McConnachie 203

Impacts of a pre-dispersal seed predator on achene production in the aquatic macrophyte, *Sagittaria platyphylla*
Raelene M. Kwong, Jean Louis Sagliocco, Nathan E. Harms and Julie G. Nachtrieb 204

Project summary of biocontrol agent *Eueupithecia cisplatensis* for *Parkinsonia aculeata* in Queensland by Queensland DAF
Kelli Pukallus, Joshua Nicholls, Centaine Ferris and Judy Clark 211

Progress towards biological control of bellyache bush and prickly acacia
Dianne B.J. Taylor, Boyang Shi, Jason Callander and Kunjithapatham Dhileepan 215

Overwintering habitat and the survival of the biocontrol beetle, *Cassida rubiginosa*
Michael G. Cripps, Sarah D. Jackman, Chikako van Koten, Cor J. Vink and Michael B. Dodd 216

Molecular analysis of ecological interactions for optimising biocontrol of the invasive weed *Sonchus oleraceus* L. (Asteraceae) in Australia
M. Ollivier, V. Lesieur, M. Jourdan, T. Thomann, S. Raghu, L. Morin, A. Sheppard, M.S. Tixier and J.F. Martin 220

Biological control of *Sonchus oleraceus*: what is known, what is new and what is still missing?
V. Lesieur, T. Thomann, M. Jourdan, M. Ollivier, A. Sheppard, J.F. Martin, M.S. Tixier, L. Morin and S. Raghu 221

Matching expectations and reality: Potential impact of biological control on flaxleaf fleabane
Michelle A. Rafter, Kerri Moore, Nagalingam Kumaran, Gavin Hunter, Kylie B. Ireland, Louise Morin and S. Raghu 222

Genomics, lineages and phenotypes: Investigations into the stability of the *Puccinia chondrillina–Chondrilla juncea* pathosystem in Australia
Gavin C. Hunter, Kylie B. Ireland, Mireille Jourdan, Jim Cullen and Louise Morin 223

Boxthorn rust (*Puccinia rapipes*): a potential biological control agent of African boxthorn (*Lycium ferrocissimum*)
Kylie B. Ireland, Gavin Hunter, Alan Wood and Louise Morin 224

Water influences the impact of a native parasitic plant ('snotty gobble') on gorse
Robert M. Cirocco, José M. Facelli and Jennifer R. Watling 225

Promising new biological control agents for Queensland
Elizabeth L. Snow, Peter Jones, Natasha Riding and Michael Day 226

Biological control of mistflower (*Ageratina riparia*) with a fungal agent facilitates native vegetation regeneration
Ben Gooden, Louise Morin, Andrew White, Melissa Piper, Jamie Fagg, Kris French, Alice Yeates and Shon Schooler 230

Biocontrol post-release: could earlier failures lead to overlooked successes?
Iain D. Paterson, Zezethu Mnqeta and Paul O. Downey 231

Biocontrol research for environmental weeds management in NSW
Louise Morin, Andrew McConnachie and Peter Turner 232

Weed biocontrol in temperate Australia – a new manual for best management practice
Paul R. Sullivan 233

DAY 3: WEDNESDAY 12 SEPTEMBER

PLENARY SESSION: WEED FUTURES

Drivers of change and adaptation tools for reducing the impact of invasive plants under a changing climate
Michelle Leishman 234

Future weeds RD&E investment under a collaborative national biosecurity approach
Andreas Glanznig and Richard J. Price 235

CHEMISTRY INNOVATIONS

Trifludimoxazin: A new PPO inhibitor that controls PPO resistant weed biotypes
G.R. Armel, I. Francis, R. Nielson, M. Witschel and R. Liebl 236

Trifludimoxazin: A new novel herbicide for grass and broadleaf weed control in Australian winter cereals
Ian Francis, Marco Montagna, Russell Ison and Gavin Heard 237

Cinmethylin – a new herbicide for cereals: Mode of action and biokinetic properties
Ruth Campe, Eva Hollenbach, Lara Kämmerer, Janneke Hendriks, Wolfgang Höffken, Helmut Kraus, Jens Lerchl, Thomas Mietzner, Matthias Witschel, Johannes Hutzler, Laurent Picard and Klaus Kreuz 238

Agixa™ herbicide – a new treatment for selective weed control in drill-sown rice
Gregory S. Wells, Malcolm Taylor and Colin Plater 239

Innovations for weed management – From new herbicide modes of action to novel technologies for weed management
Stefan Tresch 242

Topramezone (Frequency® Herbicide), a new flexible tool for weed control in winter cereals across Australia
Ian Francis, Marco Montagna and Russell Ison 243

A new mode of action (MOA) pre-emergence herbicide for the control of annual ryegrass (*Lolium rigidum* Gaud.) and other monocotyledon weeds in cereals in Australia
Ian Francis, Marco Montagna, Russell Ison and Gavin Heard 244

ForageMax™ herbicide – a new treatment for selective weed control in graze and grain canola
Gregory S. Wells and Colin Plater 245

Encapsulated dry herbicides: A novel approach for control of trees
Ken C. Goulter, Victor J. Galea and Peter Riikonen 247

INNOVATIVE CHEMISTRY APPLICATION

Stopping the invasion of *Persicaria chinensis* (Chinese knotweed) – a herbicide solution
Kim Hignell 251

Drive-by shooting: Increasing weed treatment speeds using a skattergun
Nathan March and Wayne Vogler 257

A magic bullet for the Northern Australian neem nightmare?
Jessica Miller and Louise Beames 261

Herbicidal control of environmental weeds from sub-Antarctic Macquarie Island
Brian M. Sindel, Waqas Zahid, Paul E. Kristiansen, Susan C. Wilson, Justine D. Shaw and Laura K. Williams 265

COMMUNITY ENGAGEMENT AND ENVIRONMENTAL WEED MANAGEMENT

More than a Chore: The pleasures of weed management in high amenity rural landscapes
Nicholas Gill 266

Buffel grass management in Indigenous communities
Troy A.J. Bowman 267

Restoring coastal ecosystems for the Great Barrier Reef and the RAMSAR wetlands of Bowling Green Bay
Scott Fry 271

Managing bitou and boneseed on Long Reef Headland
Jillian Macintyre 272

Weed management is not quite 'Bush Regeneration' – An opinion
Peter Harper and Nimal Chandrasena 273

WEEDS IN NATURAL SYSTEMS (WINS) HOW DO WEEDS WIN?

Effects of soil nitrogen reduction on competition between *Phragmites australis* and *Melaleuca ericifolia*
Md N. Uddin, Randall W. Robinson and Takashi Asaeda 280

The invasive *Parthenium hysterophorus* L. has limited impact on soil chemistry and enzyme activities but influences above and below ground biodiversity
Olusegun O. Osunkoya, Boyang Shi, Kerri Moore and Kunjithapatham Dhileepan 281

Weeds in Natural Systems (WINS): Weighing up the importance of different invasion mechanisms
Kristine French 285

CAWS INTERACTIVE FORUM

Global learnings for research impact in weed management in Australia
Rachel Melland 286

WEEDS OF CROPS AND PASTURES – APPLIED MANAGEMENT

Emergence patterns and herbicide control of prickly lettuce (*Lactuca serriola* L.)
Hanwen Wu, Adam Shephard and Michael Hopwood 290

The effect of integrating defoliation with herbicides on barley grass survival in a legume pasture
Jane E. Kelly, William Brown, John Broster, Paul A. Weston, Wayne Robinson, Jane Quinn and Leslie A. Weston 295

Performance of INTERCOM for predicting rice-barnyard grass interference in dry seeded rice systems
Tahir Hussain Awan and Bhagirath Singh Chauhan 300

Effect of high temperature and soil moisture on seed germination of four summer weeds
Sudheesh Manalil and Bhagirath S. Chauhan 301

GENETICS IMPROVING WEED MANAGEMENT

Specificity of LAMP for genetic diagnostics of Chilean needle grass and serrated tussock
David Gopurenko, Aisuo Wang and Hanwen Wu 302

Molecular identification of *Solanum elaeagnifolium* in Australia using DNA barcoding, a solution for better management
Xiaocheng Zhu, David Gopurenko, Laurence A.R. Haegi and Hanwen Wu 307

Identification of eight *Panicum* species in Riverina region of NSW using DNA sequence analysis
Yuchi Chen, Xiaocheng Zhu, David E. Albrecht, Panayiotis Loukopoulos, Jane C. Quinn and Leslie A. Weston 312

Screening of gene regions for genetic diversity in global parthenium weed (*Parthenium hysterophorus* L.) populations
Ali A. Bajwa, Xiaocheng Zhu, Bhagirath S. Chauhan, Steve W. Adkins and Leslie A. Weston 318

THINKING OUTSIDE THE SQUARE

A conversation starter: Standardising terminology and definitions for weeds status in Australia – What do you think?
Chris Brodie, Peter Lang and Michelle Waycott 322

The challenges of herbicide resistance in non-agricultural weed management systems
Andrew Storrie 323

Using detection dogs to assist weed eradication in conservation areas
Hillary Cherry 327

POSTER PRESENTATIONS

Evolved resistance to glyphosate in barley grass (*Hordeum* spp.) in South Australia
Patricia Adu-Yeboah, Jenna Malone, Benjamin Fleet, Samuel Kleeman, Gurjeet Gill and Christopher Preston 328

Longevity of blady grass (*Imperata cylindrica*) seeds
Husham A. Alqaderi, Brian M. Sindel, Paul E. Kristiansen and Ralph D.B. Whalley 329

Efficacy of some selective herbicides against broad leaved weeds of wheat crop grown under moisture deficit conditions of Pakistan
Muhammad Asad, Muhammad Mudassar, Zahid Mahmood and Muhammad Rasheed 333

Effect of parthenium weed on maize yield at different competition durations in Ethiopia
Ali A. Bajwa, Tamado Tana, Bhagirath S. Chauhan and Steve W. Adkins 334

The effects of parthenium weed density on yield attributes and yield of maize in Ethiopia
Ali A. Bajwa, Tamado Tana, Bhagirath S. Chauhan and Steve W. Adkins 338

Emergence of great brome grass and barley grass
Catherine P.D. Borger and Abul Hashem 341

Management of *Chromolaena odorata* in the Douglas Shire
Simon J. Brooks and Peter Logan 345

Progress in the eradication of *Mikania micrantha* from Australia
Simon J. Brooks and Mick Jeffery 350

The effects of burial depth and water stress on Melastome weed seeds
Simon J. Brooks, Rose K. Easton and Kirsty L. Gough 354

The NSW Biological Control Taskforce – collaborating to maintain a functional biocontrol agent delivery pipeline
Troy R.F. Brown 359

Geospatial assessment to predict potential African olive (*Olea europaea* spp. *cuspidata*) infestation sites in remnant vegetation communities
Doug Campbell 360

Causing a stink: *Paederia foetida* found on mainland Australia
Joel Daniels and Stephen B. Johnson 361

Effect of nitrogen rate and weed control treatments on quantitative and qualitative yield of sugar beet
Mohammad A. Dastorani and Mohammad Armin 362

The impact of two biological control agents, *Passalora ageratinae* and *Baeodromus eupatorii*, on *Ageratina adenophora* (crofton weed): synergism or antagonism?
Juliette Fontaine, Gavin C. Hunter and Louise Morin 363

Seed-set reduction of milk thistle (*Sonchus oleraceus* L.), wild turnip (*Rapistrum rugosum* L.) and flaxleaf fleabane (*Conyza bonariensis* L.) following a crop topping application of saflufenacil (Sharpen® WG Herbicide) in cereals
Ian Francis, Marco Montagna, Russell Ison and Adam Perfrement 364

Integrated weed management cotton (*Gossypium hirsutum* L.) in conventional and ultra narrow row spacing system
Ali R. Ghavi and Mohammad Armin 365

A useful *in vitro* bioassay for evaluation of weed suppression provided by cover crop residues
Saliya Gurusinghe, Sajid Latif, William Brown, Paul Weston and Leslie A Weston 366

Egeria (*Egeria densa* Planch.) presents an increasing threat to Australian waterways
Peter Harper and Nimal Chandrasena 367

Vacuum weeding: the removal of wild oats (*Avena ludoviciana*) and annual ryegrass (*Lolium rigidum*) seed from wheat stubble
Linda M. Heuke, Shona L. Robilliard, Asad Shabbir and Michael J. Walsh 368

Host testing and taxonomic clarification of a potential fungal biological control agent of sea spurge (*Euphorbia paralias*)
Gavin C. Hunter, John Lester, Isabel Zeil Rolfe, Mireille Jourdan, John K. Scott and Louise Morin 369

Evaluation of herbicide mixtures and manual weed control methods in maize (*Zea mays* L.) production
E.O. Imoloame 370

African boxthorn stakeholder survey: Impacts and management goals for biocontrol
Kylie B. Ireland, Michelle Rafter, Nagalingam Kumaran, S. Raghu and Louise Morin 371

Early invader Opuntioid cacti in Victoria
Bec James and Kate Blood 372

Using detection dogs as part of an integrated weed management strategy – A local government perspective
Melanie Kelly and Nicole Gill 373

Triticale competitive ability with annual ryegrass (*Lolium rigidum* Gaudin) as affected by variety, sowing rate and row spacing
Husam S.M. Khalaf, Brian M. Sindel, Paul E. Kristiansen, Robin S. Jessop and Craig Birchall 374

Effect of moisture stress on morphological and physiological attributes of *Amaranthus viridis* and *A. retroflexus*
Asad M. Khan, James P. Hereward, Jeff A. Werth, Gimme H. Walter and Bhagirath Singh Chauhan 378

Weed suppressive potential of selected pasture legumes against annual weeds in southeastern Australia
Sajid Latif, Paul A. Weston, John W. Piltz, Saliya Gurusinghe, William B. Brown, Jane C. Quinn and Leslie A Weston 379

Angled onion, *Allium triquetrum* L., a beautiful flower with a bitter taste
V. Lesieur, M. Jourdan, T. Thomann, A. Sheppard, J.F. Martin, M.S. Tixier, L. Morin, S. Raghu and B. Gooden 383

Diversity and extent of mutations endowing resistance to AHAS-inhibiting herbicides in *Sisymbrium orientale* in Australia
Weihua Long, Peter Boutsalis, Mahima Krishnan, Jenna M. Malone and Christopher Preston 384

Growth responses of two Australian biotypes of common sowthistle (*Sonchus oleraceus* L.) to varied soil moisture regimes
Gulshan Mahajan, Michael Walsh and Bhagirath S. Chauhan 385

Phalaris minor resistance is increasing and solutions urgently needed to farmers of Punjab and Haryana
Malwinder Singh Malhi and Pavitarpal Singh Pangli 386

Allelopathic activity of *Hosta alismifolia* against *Lactuca sativa* L.
Mohammad Shamim Hasan Mandal 387

The role of biochar pyrolysis temperature on hexazinone sorption-desorption in soil using the batch-equilibrium method
Kassio F. Mendes, Gustavo M. Chitolina, Mayara C. Arruda, Leonardo V. Junqueira, Cássio S. Almeida and Valdemar L. Tonisielo 388

Could *Phytophtora* species associated with declining populations of invasive European blackberry be used for biological control?
Louise Morin, John Lester, Patrick Gleeson, Treena I. Burgess, Giles E. St. J. Hardy and John K. Scott 389

Germination ecology of resistant and susceptible populations of barnyard grass (*Echinochloa colona* (L.) Link.) collected from Queensland
Navneet Kaur Mutti, Gulshan Mahajan and Bhagirath S. Chauhan 390

Field evaluation of selected canola competitive cultivars for suppression of natural weed populations
James M. Mwendwa, William B. Brown, Paul A. Weston, Shamsul K.M. Haque, Jeffrey D. Weidenhamer and Leslie A. Weston 391

An inventory of Queensland prioritised invasive plant species for management and research
Olusegun O. Osunkoya, Christine Perrett, Kerri Moore, Sam Nicol, Jens Froese, Kristy Gooding and Shane Campbell 392

Flupropanate resistant serrated tussock (*Nassella trichotoma*) identified at multiple locations in the Monaro, New South Wales
Jo Powells 397

The effect of weed interference with cotton in conventional and ultra narrow row spacing condition
Ahmad Rafeizade, Mohammad Armin and Matin Jamimoeini 398

Studies on tolerance of chickpea (*Cicer arietinum* L.) to some pre- and post-emergence herbicides
K. Rahamdad, I.A. Khan and S.S. Shah 399

Competitive wheat (*Triticum aestivum*) cultivars: A potential tool for managing herbicide resistant weeds
Abhimanyu Rinwa and Manpreet Singh 400

Efficacy of harvest weed seed control systems on wild oats
Shona L. Robilliard, Linda M. Heuke, Asad Shabbir and Michael J. Walsh 401

Interventional weed management methods in direct seeded rice
Pasoubady Saravanane 402

Bionomics and effectiveness of Aak fruit fly: A potential biocontrol candidate for calotrope in Australia
Asad Shabbir, Shahid Ali and Kunjithapatham Dhileepan 405

Managing barnyard grass through crop competition: The role of row spacing and crop density
Asad Shabbir and Michael J. Walsh 406

Successful biocontrol of *Echium* spp.
Paul R. Sullivan 407

Influence of resource conservation techniques on weed control in green manure maize-pulse cropping system
R. Thirumalaikumar, N.S. Venkataraman, K. Balakrishnan, R. Babu and A. Rathinasamy 408

Comparison of site-specific, whole field and seedbank physical weed control treatments
Michael J. Walsh, Salah Sukkarieh and Guy R. Coleman 409

Environmental weed advisory list update, Victoria
Matt White, Kate Blood and Bec James 410

Physiological and biochemical resistance cost of multiple herbicide resistance in *Echinochloa crus-galli*
Xia Yang, Yongfeng Li, Zichang Zhang and Tao Gu 411

Efficacy of pyrazosulfuron-ethyl to manage weeds of rainy season rice under non-puddled field condition and its effect on the subsequent wheat crop
Taslima Zahan, Abul Hashem and Md. Moshiur Rahman 412

Alterations in the 5 'untranslated region of the *EPSPS* gene influence *EPSPS* over-expression in glyphosate-resistant *Eleusine indica*
Chun Zhang, Li Feng and Xing-shan Tian 416

Investigation of root exudation and phytotoxic potential of diverse *Sorghum* spp. through regulation of sorgoleone
Xiaocheng Zhu, Paul A. Weston, Ian D. Godwin, Sally Norton and Leslie A. Weston 417

AUTHORS' INDEX 418

KEYWORD INDEX 422

Herbicide resistance in Canada: Biggest threats, best management, and lessons learnt

Hugh J. Beckie
Agriculture and Agri-Food Canada, Saskatoon, Saskatchewan, Canada
Present address: School of Agriculture and Environment, University of Western Australia, Crawley, Western Australia 6009, Australia
(hugh.beckie@uwa.edu.au)

Summary I began my career in weed science dealing with the first case of herbicide resistance in western Canada in 1988: trifluralin-resistant green foxtail (*Setaria viridis* (L.P.) Beauv.). Thirty years later, there are 76 herbicide-resistant (HR) weed biotypes across Canada. The biggest threats to crop production are multiple-resistant wild oat (*Avena fatua* L.) and kochia (*Kochia scoparia* L. Schrad.) in western Canada and Canada fleabane or horseweed (*Conyza canadensis* (L.) Cronq.) in eastern Canada. Management of glyphosate plus ALS inhibitor-resistant Canada fleabane in soybean requires tank-mixing three herbicide modes of action, with significantly increased cost. Alternatively, Roundup Ready (RR) 2 Xtend™ (glyphosate + dicamba) soybean cultivars, introduced in 2017, were planted on 13% of crop area as a tool to control these populations. Glyphosate plus ALS inhibitor-resistant kochia in western Canada, similar to Canada fleabane, has spread very rapidly due to HR allele movement via seed and pollen, increasing reliance on auxinic herbicides. Many farmers cannot control wild oat post-emergence in wheat or barley due to resistance to both ACCase and ALS inhibitor herbicides, relying on pre-emergence tank-mixes. I outline best management practices recommended for HR weed management, such as crop diversity, crop competitiveness, and weed sanitation (including harvest weed seed control). Whereas near-monoculture RR soybean in eastern Canada has contributed to increasing occurrence of glyphosate-resistant weeds, genetically-engineered canola (Liberty Link™ (LL) or RR trait) in western Canada has positively impacted HR weed management over the past 20 years. I will share some lessons on HR weed management I have learnt as a weed scientist and as a farmer over the past 30 years.

Keywords Multiple resistance, integrated weed management.

BIGGEST THREATS

There are currently 76 herbicide-resistant (HR) weed biotypes across Canada, a rate of increase of about two new cases per year since 1988 (although the first case was reported in the 1950s). In eastern Canada, specifically the province of Ontario, Canada fleabane (*Conyza canadensis* (L.) Cronq.) resistant to both glyphosate and acetolactate synthase (ALS) inhibitors is the biggest threat to profitable crop production, where a typical three-year rotation is corn-soybean-winter wheat. Since the discovery of glyphosate-resistant Canada fleabane in eight fields in Essex County in 2010, it has now spread across all 30 counties (800 km) in southern Ontario, principally via seed movement by wind (Shields *et al.* 2006). Of those 30 counties, 23 have multiple-resistant (glyphosate + ALS inhibitor) populations. There are good control options in corn; in wheat, pyrasulfotole/bromoxynil provides best control. There are no good options in soybean (78% of crop area planted to Roundup Ready (RR) cultivars). The best treatment is a three-way mixture: glyphosate + metribuzin + saflufenacil, albeit inconsistent control at increased herbicide cost and yield loss (Pearce 2017). In 2017, RR2 Xtend™ (glyphosate + dicamba) soybean was introduced, which provides a new option to control multiple-resistant Canada fleabane. Adoption of these stacked (dual)-trait cultivars in the first year of introduction was 13% of seeded area.

In western Canada, the top five crops are canola (9.3 million ha), wheat (8.6), barley (2.2), lentil (1.8), and field pea (1.7), out of an annually-cropped area of 28 million ha. Of that area, 15 million ha (54%) are infested with HR weeds. The biggest threats to crop production are multiple-resistant wild oat (*Avena fatua* L.) and kochia (*Kochia scoparia* L. Schrad). Populations of wild oat resistant to both acetyl-CoA carboxylase (ACCase) and ALS inhibitors were found in 42% of randomly-surveyed fields in Manitoba in 2016 (up from 13% in 2008 and 8% in 2002); in Saskatchewan in 2014–2015, this biotype was found in 25% of surveyed fields (5% in 2009, 1% in 2003) (Beckie *et al.* 2017a,b). In 2017, five-way resistance in wild oat was reported: ACCase inhibitor, ALS inhibitor, triallate, sulfentrazone, and pyroxasulfone (Mangin *et al.* 2016), despite no history of use of the latter two herbicides. There are no post-emergence herbicide options for multiple-resistant (ACCase + ALS inhibitor) wild oat in wheat or barley; pre-emergence options include triallate or trifluralin + triallate (suppression by trifluralin or pyroxasulfone). The only options in field pea are pre-emergence trifluralin or triallate; in lentil, the only option is pre-emergence-applied trifluralin. Kochia, resistant to both glyphosate and ALS inhibitors in western Canada, similar to Canada fleabane, has spread very rapidly due to HR allele movement via seed and pollen (Beckie *et al.* 2016), increasing reliance on auxinic herbicides (especially fluroxypyr and dicamba). For example, three counties in southern Alberta had kochia populations resistant to both glyphosate and ALS inhibitors in 2012, or 5% of ca. 300 populations (Hall *et al.* 2014). By 2017, this biotype had spread via tumbleweed across all 16 southern Alberta counties, comprising over 50% of the ca. 300 populations sampled.

BEST MANAGEMENT

Although proactive and reactive HR weed management practices have both been recommended to growers in different agroecoregions globally, there was a need to identify and priorise those having the most impact in mitigating or managing herbicide selection pressure in the Northern Great Plains of North America (Table 1, Beckie and Harker 2017). Our perspective on this issue is based on collaborative research, extension activities and dialogue with growers, or farming experience (cereal, oilseed and pulse crop production) during the past 30 years. The foundation of HR weed management is crop diversity and competitive crops and practices that promote competitiveness. Although the most common crop rotation is not diverse (canola-wheat), canola allows different in-crop herbicide modes-of-action (glufosinate, glyphosate) to manage

Table 1. Top 10 herbicide resistance weed management practices in western Canada (as reviewed in Beckie and Harker 2017).

10	Maintaining a database
9	Strategic tillage
8	Field- and site-specific weed management
7	Weed sanitation
6	In-crop wheat-selective herbicide rotation
5	Herbicide site-of-action rotation
4	Herbicide mixtures or sequences
3	Pre- and post-herbicide scouting
2	Competitive crops and environment
1	Crop diversity

ACCase or ALS inhibitor resistance. The majority of canola area is planted to Liberty Link (LL) cultivars, with the remainder mainly RR. Hybrid canola is similar to barley in weed suppression ability. Growers who routinely adhere to good weed sanitation practices are less likely to have HR weeds. Harvest weed seed control falls under this category; on-farm research using the Harrington Seed Destructor started in 2017 in Alberta to evaluate its efficacy on key economic weed species (Tidemann *et al.* 2017).

Growers in western Canada are using a number of recommended practices, based on responses of 600 Saskatchewan growers in 2015; in particular scouting has rapidly increased in adoption since the last grower management questionnaire in 2003 (Beckie *et al.* 2017a). Other top 10 weed management practices used by growers, such as crop rotation, tank-mixing herbicides, and herbicide mode-of-action (group) rotation, are among the top 10 recommended practices. For those growers managing HR weeds, four practices are preferentially used compared with growers without HR weeds (Figure 1): crop rotation, herbicide site-of-action rotation, tank-mixing herbicides, and pre-emergence herbicide treatment. Overall, those with HR weeds rely more on herbicides at all application windows, and have greater adoption of scouting before in-crop herbicide treatment, tank-mixing herbicides, herbicide site-of-action rotation, growing weed-competitive crops, and increasing crop seeding rates vs. those without resistance (G-statistic).

LESSONS LEARNT

A dozen lessons or perhaps perspectives I have gained over the last 30 years as a farmer and weed scientist are the following: (1) the more you look, the more you find (surveillance); (2) lack of weed management diversity (or as Prof. Steve Powles says, 'herbicide-only syndrome'); (3) growers still expect a new mode-of-action herbicide to save them; (4) growers reactively, not proactively, manage HR weeds; (5) integrated weed management is only practised after multiple-HR weeds evolve; (6) short-term profitability and/or cash flow drive crop management decisions and growers greatly discount future potential rewards relative to an immediate gratuity; (7) stewardship is not as important for rented or leased land; (8) grass HR weed management is generally more difficult than broadleaf HR weed management; (9) gene flow, the second part of the HR weed equation, is usually neglected or underappreciated; (10) genetically engineered canola (LL or RR) has been a net benefit in proactively or reactively managing HR weeds in western Canada; (11) there is no silver bullet in managing HR weeds: synergistic 'stacked' cultural weed management practices have

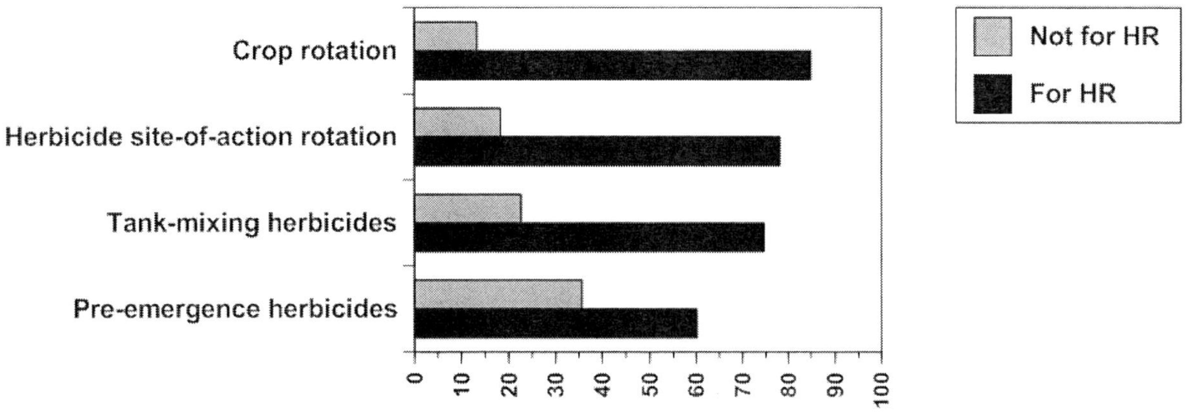

Figure 1. Four practices preferentially used by growers for HR weed management vs. those growers who do not have HR weeds (2015 Saskatchewan survey; n = 250 respondents).

been proven effective; and (12) HR weeds can be managed successfully and profitably if growers are flexible and adaptable.

REFERENCES

Beckie, H.J., Blackshaw, R.E., Hall, L.M. and Johnson, E.N. (2016). Pollen- and seed-mediated gene flow in kochia (*Kochia scoparia*). *Weed Science* 64, 624-33.

Beckie, H.J. and Harker, K.N. (2017). Our top 10 herbicide-resistant weed management practices. *Pest Management Science* 73, 1045-52.

Beckie, H.J., Shirriff, S.W. and Leeson, J.Y. (2017a). Saskatchewan weed survey of herbicide-resistant weeds in 2014-2015. Weed Survey Series Publication 17-1, Agriculture and Agri-Food Canada, Saskatoon, SK. 37 pp.

Beckie, H.J., Shirriff, S.W. and Leeson, J.Y. (2017b). Manitoba weed survey of herbicide-resistant weeds in 2016. Weed Survey Series Publication 17-3, Agriculture and Agri-Food Canada, Saskatoon, SK. 39 pp.

Hall, L.M., Beckie, H.J., Low, R., Shirriff, S.W., Blackshaw, R.E., Kimmel, N. and Neeser, C. (2014). Survey of glyphosate-resistant kochia (*Kochia scoparia* L. Schrad.) in Alberta. *Canadian Journal of Plant Science* 94, 127-30.

Mangin, A.R., Hall, L.M. and Beckie, H.J. (2016). Triallate-resistant wild oat (*Avena fatua* L.): unexpected resistance to pyroxasulfone and sulfentrazone. *Canadian Journal of Plant Science* 97, 20-5.

Pearce, R. (2017). Tough news on fleabane. *Country Guide*. https://www.country-guide.ca/2017/03/13/ (accessed 20 March 2018).

Shields, E.J., Dauer, J.T., Van Gessel, M.J. and Neumann, G. (2006). Horseweed (*Conyza canadensis*) seed collected in the planetary boundary layer. *Weed Science* 54, 1063-67.

Tidemann, B.D., Hall, L.M., Harker, K.N. and Beckie, H.J. (2017) Factors affecting weed seed devitalization with the Harrington Seed Destructor. *Weed Science* 65, 650-8.

National priority list of exotic environmental pests and diseases: an update

Sandra Parsons, Jessica Evans, Bo Raphael and Marwan El Hassan
Australian Bureau of Agricultural and Resource Economics and Sciences, GPO Box 858, Canberra, Australian Capital Territory 2601, Australia
(environmentalpestlist@agriculture.gov.au)

Summary In 2017, the independent review into the capacity of the national biosecurity system and its underpinning intergovernmental agreement delivered its final report, *Priorities for Australia's biosecurity system*. The report recommended the development of a national priority list for environmental pests and diseases, to be developed in partnership with system participants (recommendation 11). The Department of Agriculture and Water Resources and the Australian Bureau of Agricultural and Resource Economics and Sciences (ABARES) are leading the development of a national priority list of exotic environmental pests and diseases. The priority list will provide an understanding of those exotic species that pose the greatest risk to the Australian environment and social amenity and provide for enhanced awareness and preparedness of environmental biosecurity risks. The list will include within its scope consideration of exotic environmental pests and diseases across all taxa, including weeds. ABARES will be collaborating with weed experts in governments, research and environmental stakeholder groups to develop the priority list. The list will be developed over several stages, including two expert workshops held in early and mid-2018, and the project is due for completion in 2019. Here, we will present the current status of the project.

Keywords Environmental weeds, priority, exotic, prevention.

Next generation testing of herbicide resistance: Rapid detection, multiple herbicides, a single assay

Aaron L. Preston[1] and James E. Pratley[2,3]
[1] NSW Department of Primary Industries, Pine Gully Road, Wagga Wagga, New South Wales 2650, Australia
[2] Graham Centre for Agricultural Innovation, (NSW Department of Primary Industries and Charles Sturt University), Wagga Wagga, NSW 2678, Australia
[3] School of Agricultural and Wine Sciences, Faculty of Science, Charles Sturt University, Locked Bag 588, Wagga Wagga, New South Wales 2678, Australia
(aaron_preston@outlook.com)

Summary Annual ryegrass is the major weed in Australian farming systems costing farmers \$96 million annually, with control primarily by herbicides. This weed has evolved resistance to many herbicides making control extremely difficult with herbicide-based management programs. Using the genotyping technology DArTseq, a genetic assay has been developed that detects resistance to multiple herbicides in a single test within weeks of field observation. Over 350 individual ryegrass plants were screened with one or more herbicides (diclofop-methyl, chlorsulfuron and glyphosate) using conventional glasshouse based resistance testing procedures. DNA from these samples was then analysed with DArTseq, and herbicide resistance predicted based on the presence/absence of genetic markers. DArTseq analysis predicted herbicide resistance to each of these herbicides with 50–59% accuracy relative to glasshouse testing. This assay surpasses other resistance testing methods which require individual tests, a 9-month timeframe for results, and offers a new tool to combat resistance.

Keywords *Lolium rigidum*, annual ryegrass.

INTRODUCTION

Weeds cost the Australian grains industry an estimated \$3.3 billion each year. Annual ryegrass (*Lolium rigidum* Gaudin) is the most costly weed to manage, largely due to the additional costs of controlling resistant populations (Llewellyn *et al.* 2016). These costs can be reduced with the identification of herbicide resistance prior to herbicide application, ensuring that only effective herbicides are used. Currently, commercial detection of herbicide resistance is conducted by either whole plant or germination assays (Boutsalis and Broster 2006). Whole plant testing procedures used for commercial herbicide resistance testing are limited in their ability to provide timely results to identify herbicide resistance in season. Other methods, such as germination, seedling, and plant tissue assays have attempted to reduce the time required for testing, and despite a more rapid delivery of results, these methods are still constrained by seed dormancy or the inability to provide broad testing of herbicides due to the specificity of their assays. In addition to these methods, many technologies have been implemented in identifying different resistance mutations in annual ryegrass using PCR, dCAPS and sequencing (Tan *et al.* 2007, Yu *et al.* 2008, Bostamam *et al.* 2012, Gaines *et al.* 2014, Duhoux *et al.* 2015). These assays are also rapid, precise and can provide detailed information on the mechanism of resistance and are not constrained by seed dormancy. The highly specific nature of these tests has prevented them being implemented in broad detection of herbicide resistance as they are currently limited to screening for well documented mutations, usually in individual assays. Of these, resistance to only one or two modes of action have been detected simultaneously. If a method could be developed using genetic screening, the prospects for the simultaneous testing of multiple forms of herbicide resistance are increased. This could be up-scaled to commercial herbicide resistance testing and it would provide more timely, detailed, and actionable results than current testing methods. As the incidence of herbicide resistance is increasing (Boutsalis *et al.* 2012, Owen *et al.* 2015, Shergill *et al.* 2015) and no new herbicide modes of action commercialised in the last thirty years besides Tirexor™ (expected release date 2020), there is a demonstrated need for better herbicide resistance detection methods to facilitate better resistance management (Duke and Dayan 2015, BASF 2018).

A genetic screening technology yet to be explored for resistance phenotyping is DArTseq. As DArTseq can detect and identify multiple traits simultaneously, it is an ideal technology for resistance phenotyping, as it has the potential to screen for multiple forms of resistance once they are properly characterised. Annual ryegrass is an ideal weed for study as it is genetically diverse, has multiple forms of herbicide resistance and is a widespread weed of economic significance. If DArTseq can be used successfully to identify multiple forms of herbicide resistance in annual ryegrass in a quick, cheap and accurate assay, the applicability to other less diverse weeds will be demonstrated and a new paradigm in resistance testing will be established. The aim of this study was to identify genetic markers linked to herbicide resistances in populations of annual ryegrass, and to assess the ability of DArTseq to discriminate between herbicide resistant and susceptible phenotypes.

MATERIALS AND METHODS

Seventeen populations were selected with known resistance to diclofop-methyl, glyphosate and chlorsulfuron were selected for study. Seeds from each population were placed within 90 mm plastic Petri dishes lined with filter paper (Advantech No. 2) and covered with 4 mL of deionised water. The Petri dishes were sealed with laboratory film and stored in a growth chamber (12 h 15/25°C day/night). After 6 days germination was complete and seedlings planted into 40 cell plastic containers (8 × 58 × 5 tray, 93 mL cell volume) filled with a 1:3 mix of sand and peat moss and transferred to the glasshouse. Glasshouse conditions were maintained at a day/night temperature cycle of 25/10°C. To enable both herbicide resistance testing and DNA extraction of individual plants, seedlings were cloned via tiller propagation. This resulted in an untreated plant (for DNA extraction) and multiple separate tillers suitable for herbicide application at different rates and herbicides. To encourage tillering, plants were repeatedly cut to a height of approximately 40 mm. Plants were watered and fertilised with commercial fertiliser as required until multiple tillers formed (approximately 9 weeks). Whole plants were then excised from the soil with a scalpel. Tillers were separated by incision between the tillers at the crown. The separated tillers (henceforth clones) were trimmed to 40 mm and 10 mm leaf and root length respectively, measured from crown. Clones were then replanted into new 40 cell plastic containers filled with a 1:3 mix of sand and peat moss. Clones are watered and fertilised as required. As the clones are genetically identical, these can be tested for resistance or used for DNA extraction with confidence that each will have the same resistance mechanism. Multiple clones were taken from each individual plant to allow for dose response determinations to each herbicide. One clone from each seedling was retained for DNA analysis, four were used for glyphosate testing, and two for chlorsulfuron testing. Where an abundance of tillers occurred, one or two additional clones were taken to test for resistance to an additional herbicide, diclofop-methyl. Dose response evaluations were conducted with four concentrations of glyphosate representing 0.5×, 1×, 2× and 4× recommended label rate. Where there were additional tillers for cloning, similar dose response evaluations were conducted for resistance to chlorsulfuron and diclofop-methyl at 1× and 2× recommended herbicide label rates of each herbicide.

To determine herbicide resistance status, seedlings and clones were sprayed with the selected herbicide at the 2–3 leaf stage or one week after propagation. Herbicide treatments, at the recommended label rates and adjuvants (Table 1), were applied using a cabinet sprayer at a 50 cm height with two TeeJet (XR11001) flat fan nozzles at an operating speed of 5 km h^{-1}. Output from the sprayer was calibrated at 85 L ha^{-1} at a pressure of 250 kPa. After treatment, the plants were allowed to dry before being transported to a temperature-controlled glasshouse and commencing 24 hours after treatment, watered as required. To ensure accurate phenotyping, plants were assessed 30 days after treatment to confirm their resistance status. Surviving plants were deemed resistant to the applied herbicide.

DNA extraction was completed by collecting fresh, young, healthy leaf tissue (3–4 leaf stage) from samples and flash freezing them with liquid nitrogen. Samples were then ground into a fine powder with a mortar and pestle before undergoing DNA extraction using the DNeasy Plant Mini Kit (Qiagen). Extracted DNA was tested for quality with gel electrophoresis for quantity and RNA presence with a Nanodrop 1000 (Thermo Scientific, DE, USA). Extracted DNA was submitted to DArT Pty Ltd for DNA complexity reduction and development of SNP markers. A subset of these markers was created by removing non-informative monomorphic markers and selecting high quality markers with call rate >80% and reproducibility >95%.

Table 1. Herbicide and adjuvant application rates for glyphosate, chlorsulfuron and diclofop-methyl.

Herbicide	Application rate (g ai L^{-1})				Adjuvant and rate (% v/v)
Glyphosate*	180	360	720	1440	Wetter TX (0.2%)
Chlorsulfuron	–	20	40	–	BS1000 (0.1%)
Diclofop-methyl	–	375	750	–	BS1000 (0.25%)

* Glyphosate application rate is expressed in g ae L^{-1}.

Frequency of specific markers was compared and correlated with herbicide resistance phenotypes. Markers that did not have a discriminating presence (>30% difference in occurrence between phenotype groups) were discarded. Genome wide association analysis on DArTseq SNP data was performed using ridge regressed best linear unbiased prediction (rrBLUP) using R Studio (Endelman 2011) package rrBLUP. The suitability of rrBLUP phenotype prediction analysis with DArTseq markers have been established by dos Santos *et al.* (2016), Roorkiwal *et al.* (2016) and de Azevedo Peixoto *et al.* (2017). The predictive ability of rrBLUP was validated using a training population to teach marker linkage to phenotype, followed by comparing the prediction of these against the remaining samples (hereafter the validation dataset) and comparing the accuracy of this phenotyping to that of the whole plant analysis. Samples used in the training population are selected at random and removed from the validation dataset to prevent prediction bias. Sampling, training and testing of prediction accuracy was run 1000 times and the average taken. Five training population sizes were trialled to determine the effect on predictive accuracy.

RESULTS

In total, 345 individual plants were tested for glyphosate resistance. Of these, 188 individuals had enough clones for chlorsulfuron testing, and 130 had enough for diclofop testing (Table 2). DNA from plant samples was analysed with DArTseq, producing over 69,000 single nucleotide polymorphism (SNP) markers. Marker association analysis was conducted to link markers to phenotype using rrBLUP. rrBLUP training results increased with accuracy with increasing training population size. Starting training population size was 50 samples, with each additional assessment of training population increasing this number by an additional 50 samples, with the maximum number of samples used being 250. Predictive accuracy from these varying population sizes ranged from 41–50% for glyphosate, 48–50% for diclofop-methyl and 50–59% for chlorsulfuron resistance respectively. Only glyphosate resistance could be tested with a training population of 150+ due to a lack of samples phenotyped with the other herbicides.

DISCUSSION

These results establish the use of DArTseq for herbicide resistance testing for annual ryegrass and proves its ability to detect resistance to multiple herbicides in one assay. This was achieved with greater accuracy than what was achieved with the previous exploratory study using microarray DArT. The predictive power of this assay could be increased with analysis of an inbred F_2 population coupled with greater sample numbers (Peixoto *et al.* 2016). Additionally, specific knowledge of the resistance mutations present in each of the selected samples may have enhanced the accuracy of the resistance phenotype predictions, and allowed for possible sub-categorisation of the different populations. Inference could then be made about the ability of this assay to predict cross-resistance to untested herbicides. The maximum accuracy of 59% for herbicide resistance detection achieved in this study means that a DArTseq assay will predict resistance level with 59% accuracy with the current marker set. Both false negatives and false positives are likely until the many diverse forms of herbicide resistance and the genetic background of *L. rigidum* can be analysed to eliminate spurious marker associations. Resistance prediction may not need to occur with 100% accuracy for a test to be useful, or superior to existing testing methods. Future research could examine the accuracy threshold required of a DArTseq assay to provide meaningful diagnostic information to assist in on-farm agronomic choices.

This proof of concept study establishes DArTseq as a tool for broad-spectrum evaluation of herbicide resistance, having been demonstrated across three

Table 2. Herbicide resistance results for individual annual ryegrass samples at differing herbicide dosage. R indicates the number of resistant samples, S the number of susceptible.

Herbicide	Glyphosate*				Diclofop-methyl		Chlorsulfuron	
Application rate (g ai L^{-1})	180	360	720	1040	375	750	20	40
Phenotype								
R	321	296	321	18	321	13	321	321
S	5	49	153	116	49	18	73	10
Total	326	345	338	134	130	31	188	12

* Glyphosate application rate is expressed in g ae L^{-1}.

herbicide groups at varying doses. This analysis occurred in a three-week time frame, but could be completed in as little as a week in a single assay. This rapid diagnosis allows for early in-season detection of herbicide resistance, remedying the slowness of conventional glasshouse screening for herbicide resistance. Feasibly, early emerging annual ryegrass could be sampled and analysed with DArTseq and its resistance to a range of herbicides and dosages known within a week. In the Australian farming context, weeds emerging with the first March rain could be analysed, and resistance known before sowing in early April. This would facilitate application of an effective herbicide. Alternatively, ryegrass could be sampled and analysed whenever it is found in field, at any growth stage, as long as viable DNA can be extracted from its tissue. Further study is required before this assay will be appropriate for commercial use, to improve the accuracy of phenotyping and identification of key herbicide resistance markers. Once phenotyping accuracy has been enhanced, additional analysis will be required to validate the method across wider Australian ryegrass populations. These results expand this use of genetic assays for the detection of herbicide resistance and contribute to the overall goal of the development of the next generation of resistance screening assays for better weed management and herbicide stewardship.

ACKNOWLEDGMENTS

We wish to thank GRDC for scholarship top-up funding and Dr John Broster for his advice and assistance.

REFERENCES

BASF. (2018). BASF submits regulatory dossiers for two new herbicide active ingredients. https://www.basf.com/en/company/news-and-media/news-releases/2018/06/p-18-242.html (accessed 10 July 2018).

Bostamam, Y., Malone, J.M., Dolman, F.C., Boutsalis, P. and Preston, C. (2012). Rigid ryegrass (*Lolium rigidum*) populations containing a target site mutation in EPSPS and reduced glyphosate translocation are more resistant to glyphosate. *Weed Science* 60, 474-9.

Boutsalis, P. and Broster, J.C. (2006). Herbicide resistance testing of *Lolium rigidum* by commercial institutions. Proceedings of the 15th Australasian Weeds Conference, eds. C. Preston, J.H. Watts and N.D. Crossman, pp. 488-90. (Weed Management Society of South Australia, Adelaide).

Boutsalis, P., Gill, G.S. and Preston, C. (2012). Incidence of herbicide resistance in rigid ryegrass (*Lolium rigidum*) across southeastern Australia. *Weed Technology* 26, 391-8.

de Azevedo Peixoto, L., Laviola, B.G., Alves, A.A., Rosado, T.B. and Bhering, L.L. (2017). Breeding *Jatropha curcas* by genomic selection: A pilot assessment of the accuracy of predictive models. *PloS One* 12, e0173368

dos Santos, J.P.R., Pires, L.P.M., de Castro Vasconcellos, R.C., Pereira, G.S., Von Pinho, R.G. and Balestre, M. (2016). Genomic selection to resistance to *Stenocarpella maydis* in maize lines using DArTseq markers. *BMC Genetics* 17, 86.

Duhoux, A., Carrère, A., Gouzy, J., Bonin, L. and Délye, C. (2015). RNA-Seq analysis of rye-grass transcriptomic response to an herbicide inhibiting acetolactate-synthase identifies transcripts linked to non-target-site-based resistance. *Plant Molecular Biology* 87, 473-87.

Duke, S.O. and Dayan, F.E. (2015). Discovery of new herbicides modes of action with natural phytotoxins. *American Chemical Society Symposium Series* 1204, 79-92.

Endelman, J.B. (2011). Ridge regression and other kernels for genomic selection with R package rrBLUP. *The Plant Genome* 4, 250-5.

Gaines, T.A., Lorentz, L., Figge, A., Herrmann, J., Maiwald, F., Ott, M.C., Han, H., Busi, R., Yu, Q. and Powles, S.B. (2014). RNA-Seq transcriptome analysis to identify genes involved in metabolism-based diclofop resistance in *Lolium rigidum*. *The Plant Journal* 78, 865-76.

Llewellyn, R.S., Ronning, D., Ouzman, J., Walker, S., Mayfield, A. and Clarke, M. (2016). Impact of weeds on Australian grain production: the cost of weeds to Australian grain growers and the adoption of weed management and tillage practices. GRDC and CSIRO.

Owen, M.J., Goggin, D.E. and Powles, S.B. (2015). Intensive cropping systems select for greater seed dormancy and increased herbicide resistance levels in *Lolium rigidum* (annual ryegrass). *Pest Management Science* 71, 966-71.

Peixoto, L.A., Bhering, L.L. and Cruz, C.D. (2016). Determination of the optimal number of markers and individuals in a training population necessary for maximum prediction accuracy in F2 populations by using genomic selection models. *Genetics and Molecular Research* 15, doi:10.4238/gmr15048874

Roorkiwal, M., Rathore, A., Das, R.R., Singh, M.K., Jain, A., Srinivasan, S., Gaur, P.M., Chellapilla, B., Tripathi, S. and Li, Y. (2016). Genome-enabled prediction models for yield related traits in chickpea. *Frontiers in Plant Science* https://doi.org/10.3389/fpls.2016.01666

Shergill, L.S., Fleet, B., Preston, C. and Gill, G. (2015). Incidence of herbicide resistance, seedling emergence and seed persistence of smooth barley (*Hordeum glaucum*) in South Australia. *Weed Technology* 29, 782-92.

Tan, M.K., Preston, C. and Wang, G.X. (2007). Molecular basis of multiple resistance to ACCase-inhibiting and ALS-inhibiting herbicides in *Lolium rigidum*. *Weed Research* 47, 534-41.

Yu, Q., Han, H. and Powles, S.B. (2008). Mutations of the ALS gene endowing resistance to ALS-inhibiting herbicides in *Lolium rigidum* populations. *Pest Management Science* 64, 1229-36.

Multiple herbicide resistance in winter grass (*Poa annua*) in Australia

Peter Boutsalis[#], Jenna Malone and Christopher Preston
The University of Adelaide and [#]Plant Science Consulting, PMB 1, Glen Osmond,
South Australia 5064, Australia
(peter.boutsalis@adelaide.edu.au)

Summary Winter grass (*Poa annua*) is a genetically diverse winter annual species. A native of Europe it is widespread across most continents, predominantly in temperate countries infesting orchards, nurseries, seed crops, horticultural crops, turf and golf courses. Winter grass colonises of disturbed areas including annual crops in higher rainfall areas. It has adapted to heavy grazing and close mowing. Plants can mature and set seed at 6 mm height.

Winter grass is the most challenging weed species in the USA, Japan and Australia in turf and golf courses. It is more prevalent between early autumn and late spring in Australian golf courses. The most common form of weed control in turf and golf courses are herbicides because alternative weed control methods are limited. Heavy herbicide pressure in golf courses (predominately in the fairways) has resulted in the confirmation of resistance to several mode of action herbicides in South Australia, Victoria and New South Wales. Most resistance has occurred to post-emergent herbicides selective in couch, the dominant turf species of fairways. Resistance has been confirmed to Group B, C, D, M and Z herbicides with some biotypes resistant to multiple mode of action herbicides. Management of resistant biotypes includes strategically using diverse mode of action herbicides.

Keywords Winter grass, *Poa annua*, multiple herbicide resistance, golf, turf.

Eradication or mitigation? How to manage patches of glyphosate resistant *Echinochloa colona* L.

Jeff A. Werth[1], David F. Thornby[2], Michelle D. Keenan[1], James P. Hereward[3] and Bhagirath S. Chauhan[4]
[1] Queensland Department of Agriculture and Fisheries, PO Box 2282, Toowoomba, Queensland 4350, Australia
[2] Innokas Intellectual Services, 7 Purlingbrook Street, Upper Coomera, Queensland 4209, Australia
[3] The University of Queensland, Goddard Building, St Lucia, Brisbane, Queensland 4072, Australia
[4] Queensland Alliance for Agriculture and Food Innovation, PO Box 1218, Gatton, Queensland 4343, Australia
(jeff.werth@daf.qld.gov.au)

Summary Glyphosate-resistant *Echinochloa colona* L. (Link) is a common weed in non-irrigated cotton systems in Australia. *E. colona* is a small-seeded species that is not wind-blown and has a relatively short seed bank life. These characteristics make it a potential candidate to attempt to eradicate resistant populations when they are detected. A four year systems experiment established in 2012 was developed to determine the practicality of eradicating glyphosate resistant populations using different tactics under field conditions. Treatments ranged from glyphosate only to two additional tactics in crop and fallow with other treatments containing further eradication tactics. The glyphosate only treatment had consistently higher emergence and plants remaining at the end of each season throughout the experiment. In this experiment, the 'Best Management Practice' (BMP) treatment which contained two non-glyphosate tactics in crop and fallow were found effective at managing glyphosate resistant *E. colona* populations. Treatments that included additional eradication tactics were significantly better at times; however, this was not always the case. The final seed bank counts at the end of the experiment were 5648 seeds m^{-2} for the glyphosate only treatment, compared to 124 seeds m^{-2} for the BMP treatments, and ranged from 25–272 seeds m^{-2} for the BMP + Eradication treatments. It is recommended that the BMP approach of incorporating two additional non-glyphosate tactics in crop and fallow should be used as a baseline combined with monitoring, so that additional eradication tactics can be included as required.

Keywords Glyphosate resistance, *Echinochloa colona*, eradication.

Control of thiocarbamate resistant annual ryegrass (*Lolium rigidum*) in lentils (*Lens culinaris*) in southern Australia

David J. Brunton, Peter Boutsalis, Gurjeet Gill and Christopher Preston
The University of Adelaide, Waite Campus, PMB 1, Glen Osmond, South Australia 5064, Australia.
david.brunton@adelaide.edu.au

Summary Annual ryegrass (*Lolium rigidum*) is the most significant weed of cropping systems in southern Australia. Widespread resistance to trifluralin and post-emergent herbicides has increased reliance on thiocarbamate (group J) herbicides prosulfocarb and triallate for ryegrass control. Recent random weed surveys have confirmed resistance to thiocarbamate herbicides in South Australia, New South Wales and Victoria. A field study was undertaken in a lentil crop at Urania, South Australia in 2017 to investigate the performance of alternative pre-emergent herbicides in a population of ryegrass that had evolved resistance to prosulfocarb and triallate. Annual ryegrass density and seed production was reduced by herbicide treatments. Trifluralin did not reduce ryegrass seed set compared to the untreated. Prosulfocarb at 2400 g ha^{-1} reduced seed set by 52%, which was similar to the level of control achieved with pyroxasulfone and triallate 56% and 63% respectively. In contrast pyroxasulfone + triallate reduced seed set by 95%. Propyzamide provided the highest level of weed control, which was similar to pyroxasulfone + triallate. These results indicate thiocarbamate resistant ryegrass can be effectively managed with alternative pre-emergent herbicides.

Keywords Pre-emergent, herbicide resistance, prosulfocarb, triallate, management.

INTRODUCTION

Annual ryegrass (*Lolium rigidum* Gaud.) is a problematic weed species present in cropping fields across southern Australia with the capacity to germinate multiple times throughout autumn and winter and produce a large amount of seed. The development of wide-spread herbicide resistance in annual ryegrass has made management more difficult. Annual ryegrass control in field crops over the past 40 years has largely been through the use of herbicides. Resistance to various herbicides has increased significantly over this period including cross- and multiple resistance to diverse mode of action herbicides (Owen *et al.* 2007).

Annual ryegrass populations resistant to post-emergent herbicides has resulted in a greater reliance on pre-emergent herbicides. The pre-emergent herbicide trifluralin has been widely adopted across many parts of southern Australia with resistance detected in a large proportion of randomly collected populations (Boutsalis *et al.* 2012). The registration of prosulfocarb in 2008 provided growers with an alternative mode of action (group J thiocarbamate family), which enabled them to successfully manage trifluralin resistant annual ryegrass and was quickly adopted in medium to high rainfall environments across southern Australia. A recent survey of resistance has shown that 66% of ryegrass populations in the Mid north and Yorke Peninsula (South Australia) are now resistant to trifluralin. Similarly 78% of annual ryegrass populations in the South East are resistant to trifluralin (Boutsalis *et al.* 2012). Repeated use of group J herbicides (prosulfocarb, triallate) on trifluralin resistant annual ryegrass has increased selection pressure for group J resistance which has been reflected in some early reports of resistance. Triallate and prosulfocarb resistance was detected in a ryegrass populations on the Eyre Peninsula, South Australia in 2014. Surveys show triallate resistance present in 3% and 11% of ryegrass populations collected in Western and Southern Victoria respectively (Gill *et al.* 2016).

Evolution of resistance to group J herbicides prosulfocarb and triallate has not been widely studied in annual ryegrass. Resistance to triallate has been documented in wild oat (*Avena fatua L.*) in Canada (Beckie *et al.* 2012). Busi and Powles (2013) reported development of cross-resistance to group J herbicides triallate and prosulfocarb in a recurrently selected population of annual ryegrass. However, field evolved cross-resistance to group J herbicides has not been reported so far. This paper discusses the management options for controlling thiocarbamate resistant ryegrass.

MATERIALS AND METHODS

A replicated field trial was established in a lentil crop at Urania, Yorke Peninsula, South Australia to investigate the management of group J resistant annual ryegrass. Site selection was based on background resistance to the herbicides trifluralin, triallate, prosulfocarb and prosulfocarb + S-metolachlor and annual ryegrass density of 100–150 plants m^2. Plot size was 2 m × 10 m with four replicates in a randomised complete block design.

Lentils were sown using farmer no-till seeding equipment. A range of treatments at the recommended label rates including trifluralin, triallate, prosulfocarb, prosulfocarb + S-metolachlor, propyzamide and pyroxasulfone were applied. Herbicides were applied using a Hardi CO_2 pressurised hand operated plot sprayer equipped with T-jet 110020 nozzles spaced 50 cm apart. The sprayer was calibrated to apply 100 L ha^{-1} at 2 bar pressure at a walking speed of 3 m s^{-1}. Incorporated by seeding (IBS) treatments were applied immediately prior to seeding. Early post-emergent treatments (EPE) were applied to 2-leaf annual ryegrass.

Ryegrass and crop establishment counts were taken 28 days after IBS treatments. Ryegrass counts were taken a further 28 days after EPE treatments. Final spike counts were taken 150 days after herbicide treatments. Crop establishment counts were taken using a 50 cm rod and ryegrass counts using an 80 cm × 40 cm quadrat. Annual ryegrass spikes were also collected to calculate total seed production. Plots were not harvested for grain yield due to below-average rainfall conditions. Weather data was collected from the local Bureau of Meteorology weather station.

RESULTS

Weed density and seed production was reduced by herbicide treatments ($P<0.05$, Table 1). Herbicide treatments trifluralin, triallate, prosulfocarb, prosulfocarb + S-metolachlor and pyroxasulfone applied alone failed to provide acceptable levels of weed control (<50%) 60 days after treatment (DAT). Higher levels of weed control was achieved with propyzamide and pyroxasulfone + triallate providing 93% and 89% control respectively (Table 1). Pyroxasulfone + prosulfocarb + S-metolachlor (EPE) provided a good level of control (82%). Mixtures of products were considerably more effective compared to products alone. Prosulfocarb + S-metolachlor (EPE) resulted in poor control (59%). Prosulfocarb at 2400 g ha^{-1}, pyroxasulfone, triallate and prosulfocarb + S-metolachlor achieved poor levels of seed set reduction (<65%) 180 DAT. In contrast propyzamide and pyroxasulfone + triallate provided the highest seed set reduction (95%), which was closely

Table 1. Effect of herbicide treatments on annual ryegrass plant density and seed production in lentils grown at Urania, South Australia in 2017.

Treatment[b]	Rate (g ai ha^{-1})	Weed density[c] (plants m^{-2})	Control % (60 DAT)	Seed production[c] (seed m^{-2})	Control % (180 DAT)
		July 2017		November 2017	
Untreated	–	323 g		16,174 de	
Triallate PPI	1,500	164 cde	49	5,906 bc	63
Trifluralin PPI	1,200	343 g	0	17,878 e	0
Prosulfocarb + S-metolachlor PPI	2,000 + 300	162 cde	50	5,764 bc	64
Prosulfocarb + S-metolachlor EPE	2,000 + 300	133 cd	59	3,656 ab	77
Pyroxasulfone PPI	100	197 def	39	7,095 bc	56
Propyzamide PPI	750	24 a	93	815 a	95
Prosulfocarb PPI	2,000	280 fg	13	10,778 cde	33
Prosulfocarb PPI	2,400	265 efg	18	7,697 bcd	52
Trifluralin + triallate PPI	1,200 + 1,500	158 cd	51	6,936 bc	57
Pyroxasulfone + triallate PPI	100 + 1,500	34 a	89	788 a	95
Prosulfocarb + S-metolachlor + triallate PPI	2,000 + 300 + 1,500	93 bc	71	3,121 ab	81
Prosulfocarb + triallate PPI	2,000 + 1,500	177 cde	45	6,875 bc	58
Pyroxasulfone PPI fb prosulfocarb + S-metolachlor EPE	100 + 2,000 + 300	58 ab	82	1,310 a	92

[a] Means within the same column followed by the same letter are not significantly different according to LSD at P=0.05.
[b] Abbreviations: fb, followed by; PPI, pre-plant incorporated; EPE early post-emergent application of herbicides; DAT days after treatment.
[c] Annual ryegrass plant density m^{-2} and seed production m^{-2} data were square root transformed before mean comparisons. Data presented are the non-transformed means.

followed by pyroxasulfone followed by prosulfocarb + S-metolachlor (EPE) treatment (92%).

DISCUSSION

Development of resistance to the thiocarbamate herbicides significantly reduces the available pre-emergent herbicide options to manage these populations. Continued reliance on trifluralin in southern Australia has led to widespread resistance (Boutsalis *et al.* 2012). Management of trifluralin resistant populations has seen increased reliance on thiocarbamate herbicides (prosulfocarb) and now resistance has been detected to these herbicides. These developments highlight the risks associated with overreliance on any single mode of action herbicides (Gill *et al.* 2016). The use of newer mode of action herbicides such as pyroxasulfone in wheat and propyzamide in canola has allowed growers to manage resistant populations while rotating between other mode of action herbicides (Boutsalis *et al.* 2014). Superior weed control with pyroxasulfone and mixtures with thiocarbamates highlight the benefit of applying herbicides with different modes of action and residual activity.

Low rainfall conditions during the period of May-June in 2017 reduced the performance of pre-emergent herbicides investigated. Rainfall greatly influences herbicide behaviour (Preston 2014) and this was observed with herbicides that require rainfall post-application. In addition, the weed population at the site displayed resistance to multiple pre-emergent herbicides as previously determined through dose-response experiments. Herbicides most affected were trifluralin, triallate, prosulfocarb and prosulfocarb + S-metolachlor. Herbicides, such as pyroxasulfone and propyzamide, were also affected by the lack of early season rainfall; however, these herbicides have been successfully used to control multiple resistant ryegrass in cereal, pulse and oilseed crops respectively (Boutsalis *et al.* 2014).

Further research is required to develop effective management strategies to control thiocarbamate resistant ryegrass populations. Ineffective post-emergent options in cereals challenge growers to rely heavily on pre-emergent herbicides. Of greatest concern is management in cereals, particularly wheat where currently control of multiple resistant populations is being achieved with pyroxasulfone. Current management of multiple resistance ryegrass in pulse and oilseed crops is achieved using propyzamide. The level of crop competition from cereals in combination with a robust herbicide strategy would greatly reduce ryegrass, which currently remains a major issue in less competitive crops such as pulses. This research has shown that there are options available to growers to manage multiple resistant annual ryegrass. However, alternative management strategies such as harvest weed seed control and non-herbicide tactics would need to be introduced to avoid overreliance on remaining herbicides (Borger *et al.* 2016).

ACKNOWLEDGMENTS

This work was funded by the Grains Research and Development Corporation.

REFERENCES

Beckie, H.J., Warwick, S.I. and Sauder, C.A. (2012). Basis for herbicide resistance in Canadian populations of wild oat (*Avena fatua*). *Weed Science* 60, 10-8.

Borger, C., Riethmuller, G. and D'Antuono, M. (2016). Eleven years of integrated weed management: Long-term impacts of row spacing and harvest weed seed destruction on *Lolium rigidum* control. *Weed Research* 56, 359-6.

Boutsalis, P., Gill, G.S. and Preston, C. (2012). Incidence of herbicide resistance in rigid ryegrass (*Lolium rigidum*) across southeastern Australia. *Weed Technology* 26, 391-8.

Boutsalis, P., Gill, G.S. and Preston, C. (2014). Control of rigid ryegrass in Australian wheat production with pyroxasulfone. *Weed Technology* 28, 332-9.

Busi, R. and Powles, S.B. (2013). Cross-resistance to prosulfocarb and triallate in pyroxasulfone-resistant *Lolium rigidum*. *Pest Managment Science* 69, 1379-4.

Gill, G.S., Boutsalis, P., Kleemann, S. and Preston, C. (2016). Herbicide resistance challenges and management strategies southern GRDC Grower and Advisor Update Adelaide 2016. School of Land, Food & Wine, The University of Adelaide.

Owen, M.J., Walsh, M.J., Llewellyn, R.S. and Powles, S.B. (2007). Widespread occurrence of multiple herbicide resistance in Western Australian annual ryegrass (*Lolium rigidum*) populations. *Australian Journal of Agricultural Research* 58, 711-8.

Preston, C. (2014). Understanding pre-emergent cereal herbicides. Grains Research and Development Corporation Update 2014. School of Land, Food and Wine, The University of Adelaide.

Glyphosate resistant common sowthistle (*Sonchus oleraceus* L.)

Jenna M. Malone[1], Tony Cook[2], Mahima Krishnan[1] and Christopher Preston[1]
[1] School of Agriculture, Food & Wine, University of Adelaide, PMB1, Glen Osmond, South Australia 5064, Australia
[2] Department of Primary Industries, Tamworth Agricultural Institute, 4 Marsden Park Road, Calala, New South Wales 2340, Australia
(christopher.preston@adelaide.edu.au)

Summary A consequence of the widespread use of glyphosate for weed control has been the evolution of glyphosate-resistant weeds, with resistance to glyphosate having evolved in 40 weed species worldwide. The first case of glyphosate resistant common sowthistle (*Sonchus oleraceus* L.) was reported in 2014 when an application of glyphosate failed to control field populations of sowthistle in northern New South Wales. Sowthistle is an annual broad-leaf weed of both crops and pastures in Australia, particularly prevalent in grain and cotton crops in northern New South Wales and Queensland.

Environmental factors have an impact on glyphosate activity and may affect the expression of glyphosate resistance in plants. Temperature is one such factor, with some studies finding reduced glyphosate efficacy at higher temperatures, particularly in the case of glyphosate resistance. Preliminary results suggest this may be the case in glyphosate resistant sowthistle, with resistant biotypes more easily controlled at lower temperatures. Susceptible and intermediate resistant biotypes had an LD_{50} of 135 and 132 g ai ha^{-1} respectively at 20°C, and two resistant biotypes had an LD_{50} of 389 and 439 g ai ha^{-1} respectively. However at a higher temperature of 30°C, the LD_{50} of both the intermediate resistant and two resistant biotypes increased 2–3 fold, while the susceptible remained around the same.

This study investigates glyphosate resistance in sowthistle with the aim of determining the level of resistance in glyphosate resistant populations, the mechanism of resistance, and whether temperature has an influence on the expression of glyphosate resistance.

Keywords *Sonchus oleraceus*, glyphosate, resistance mechanism, temperature.

Extent of herbicide resistant common sowthistle (*Sonchus oleraceus*) in southern Australia

Alicia Blyth Merriam, Peter Boutsalis, Jenna Malone, Gurjeet Gill and Christopher Preston
School of Agriculture, Food and Wine, University of Adelaide Waite Campus, PMB 1 Glen Osmond, South Australia 5064, Australia
(alicia.merriam@adelaide.edu.au)

Summary Herbicide resistant weeds are a major problem in crop production. Random weed resistance surveys are a tool to determine the extent of this problem. In Australia, common sowthistle (*Sonchus oleraceus* L.) has been identified as resistant to ALS inhibiting herbicides, synthetic auxin and glyphosate. Resistance surveys in the southern region show that resistance to ALS inhibitors is widespread. In the Mid-north, Mallee and Yorke Peninsula of South Australia and the Northeast, Wimmera and Mallee regions of Victoria, over 80% of randomly collected samples have exhibited resistance to sulfonylurea herbicide chlorsulfuron. Common sowthistle has also been screened with the synthetic auxin herbicide 2,4-D with a low frequency of resistance detected. No resistance to EPSPS inhibitor glyphosate or the PSII inhibitor atrazine has been detected in the southern region.

Of the samples screened to both sulfonylurea and imidazolinone ALS inhibitors, 57% were resistant to both chemical families. These results indicate a high incidence of resistance to both sulfonylureas and imidazolinones in common sowthistle. A significant proportion of lentils grown in the southern region are ALS inhibitor tolerant varieties and growers rely heavily on these herbicides to control broadleaf weeds in this crop. Losing control of common sowthistle using this chemical group will be a major weed management challenge.

Keywords ALS inhibitor resistance, cross-resistance, herbicide resistance monitoring surveys.

INTRODUCTION

Weeds negatively impact global crop production by reducing yields, affecting grain quality or serving as alternate hosts to insect pests or disease vectors. Many weed species have evolved resistance to herbicides, making them an even bigger threat to sustainable crop production (Powles and Yu 2010). In southern Australia, common sowthistle (*Sonchus oleraceus* L.) control is an emerging issue, particularly in weakly competitive broadleaf crops such as lentils. The shift to reduced tillage practice in Australia has accelerated resistance to several herbicide modes of action commonly used to control it. ALS inhibitor resistance in common sowthistle was first discovered in 1991 in a population from Goondiwindi in Queensland (Boutsalis and Powles 1995) but has since been discovered in South Australia (SA) and Victoria. More recently, glyphosate and synthetic auxin resistance have been detected in New South Wales and southern Australia, respectively (Heap 2018).

Random weed resistance surveys are an important tool for monitoring the distribution of herbicide resistant weeds over time. In southern Australia, resistance surveys have been conducted regularly since 2005, covering each of the major cropping regions in South Australia and Victoria on a five year rotation. Common sowthistle samples have been collected and screened through this program since 2010, and the results have indicated widespread ALS inhibitor resistance and emerging, low-frequency synthetic auxin resistance.

MATERIALS AND METHODS

Field surveys This study is based on common sowthistle data from monitoring surveys conducted from 2010–2016. Surveys were timed to coincide with crop and weed maturity (but before harvest) for the given region that year. Cropping paddocks were chosen at random and surveyed as described in Boutsalis *et al.* (2012). Mature common sowthistle seed was collected and pooled from plants within a paddock to form a representative sample of the population. Seed was collected in labelled paper bags, air dried, and stored at ambient temperature until screening.

Herbicide screening Weed samples were grown and screened at the University of Adelaide Waite Campus in Urrbrae, SA the autumn/winter following their collection (approximately six months after maturity). A small number of seeds of each sample were sown in pots on the surface of a coco-peat potting mix (Boutsalis *et al.* 2012) for germination and then transplanted at the first-leaf stage. In earlier surveys (2010–2013), as many seedlings as were available for each sample were screened (average = 15 ± 11 plants). In subsequent years the protocols were standardised to 10 plants per sample in 0.9 L pots at a density of 5 plants per pot (435 plants m^{-2}). After transplanting, plants were left outdoors and watered as needed. Herbicide application was conducted at approximately the 4–5 leaf stage using a laboratory moving boom cabinet sprayer equipped with TeeJet flat fan (015-110) nozzles and an output of 118 L ha^{-1}. Herbicide screening rates (Table 1) are based on previously established discriminating doses between susceptible (S) and resistant (R) plants and, in most cases, are close to the field rate of the herbicide.

Data analysis A sample was considered to be from a resistant population if at least 20% of plants survived the screening rate. Resistance levels in different areas of southern Australia were compared by grouping data points into nine regions and comparing the proportion of resistant samples from each region. Confidence intervals were calculated using a one-proportion z-test. For regions with n<30, an exact binomial test was used. Proportions of samples for which confidence intervals did not overlap were considered to be significantly different. Temporal comparisons between cropping regions were done by two-proportion z-test. A Chi-squared test was used to test for independence of sulfonylurea and imidazolinone resistance. All statistical analyses were performed using R statistics package (R Core Team 2018). Survey data were mapped and analysed for spatial autocorrelation (Moran's I) using ArcMap 10.5.1.

RESULTS

None of the samples tested were resistant to glyphosate or atrazine (Table 1). In 2014, screening with 2,4-D identified four resistant samples (2%). One of these was from the Eyre Peninsula (34.3680, 135.3638), while the other three were from southwest Victoria near Shelford (8.0153, 143.9238), Foxhow (38.0196, 143.4063), and Nerrin Nerrin (37.7787, 142.9747). Most of these exhibited low-level resistance (20% survival), except the sample from Nerrin Nerrin (50% survival). The Eyre Peninsula and the Nerrin Nerrin sample both had high levels (>90% survival) of resistance to sulfonylureas.

Analysis of the survey results showed that ALS inhibitor resistance to both the sulfonylureas (SU) and the imidazolinones (IMI) is common in southern Australia (Table 1). Available SU resistance data for most major cropping regions shows that resistance is very common (>65% of samples) in all regions (Figure 1A). In most cases the proportions from different regions were not significantly different from each other. Available IMI resistance data from 2015 and 2016 for surveys done in Victoria (Figure 1B) suggests that IMI resistance in Victoria could be just as common. Among samples screened with an SU and an IMI, more than half (57%) were resistant to both chemical families and few were susceptible to both (13%) (Figure 2). A Chi-square test of independence between SU resistance and IMI resistance indicated that resistance to the two chemical families was not independent (P = 0.02553).

The major cropping regions of Victoria have now been surveyed twice at 4–5 year intervals, allowing the analysis of resistance shifts over time (Table 2). Most areas returned a higher percentage of resistant samples in the more recent survey compared to the earlier ones; however, none of these differences were statistically significant. Interestingly, the only statistically significant difference was a decrease in the proportion of resistant samples from 2011–2016 in Northeast Victoria.

The spatial relationships between the data were examined by testing for spatial autocorrelation (Moran's I) of SU resistance. The percentage of plants surviving chlorsulfuron application in each sample was used as the input and returned a Moran's Index of 2.029035 and a P-value of 0.001161, indicating that clustering had not occurred due to random chance.

Table 1. History of herbicide screening, rate, sample size and total percentage of resistant samples identified from all years of screening for each herbicide.

Herbicide	Survey years	Rate (g ai ha^{-1})	n	Resistance (% samples)
Chlorsulfuron	2010–2016	15	355	78
2,4-D	2013–2016	455	169	2
Glyphosate	2013–2016	300	169	–
Atrazine	2014, 2015	900	53	–
Imazamox + imazapyr	2015, 2016	24.75 + 11.25	84	68

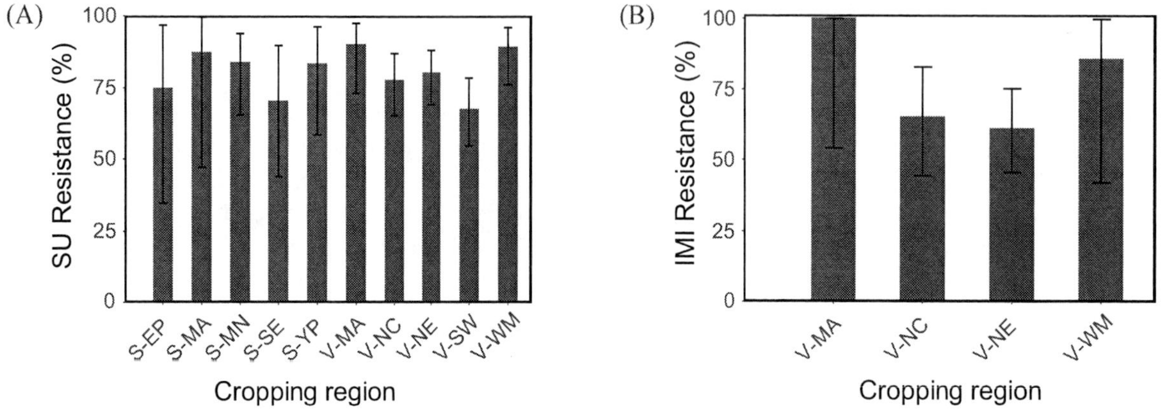

Figure 1. Percentage of populations resistant to (A) sulfonylurea (SU) and (B) imidazolinone (IMI) herbicides are shown by cropping region in South Australia (S–) and Victoria (V–). Cropping regions are the Eyre Peninsula (S-EP, SU n = 8), the SA Mallee (S-MA, SU n = 8), the Mid-north (S-MN, SU n = 31), the Southeast (S-SE, SU n = 17), the Yorke Peninsula (S-YP, SU n = 18), the Victorian Mallee (V-MA, SU n = 31, IMI n = 6), North-central (V-NC, SU n = 63, IMI n = 26), Northeast (V-NE, SU n = 76, IMI n = 44), and Southwest (V-SW, SU n = 65) Victoria, and the Wimmera (V-WM, SU n = 47, IMI n = 7). Errors bars represent 95% confidence intervals.

Table 2. Comparison between Victorian cropping regions with repeat survey data. P-values are the result of two-proportion z-tests.

Survey	n	Resistance (% samples)	P
Southwest			
2010	47	66	0.8517
2014	18	72	
North-central			
2010 + 2011	33	79	1
2015 + 2016	30	77	
Mallee			
2010	19	84	0.4095
2015	12	100	
Wimmera			
2010	34	85	0.3504
2015	13	100	
Northeast			
2011	30	97	0.009141
2016	46	70	

DISCUSSION

Herbicide resistance monitoring surveys are a valuable monitoring tool for growers, advisors and researchers. The resistance surveys in southern Australia have shown that common sowthistle has widespread resistance to ALS inhibitor herbicides. It is also clear that cross-resistance between the sulfonylurea and imidazolinone herbicides occurs widely. If current weed management practice does not effectively adapt to this fact, there will be serious consequences for the sustainability of lentil production in Australia.

When analysed by cropping region, the current dataset results in relatively low sample sizes for some areas which make it difficult to identify significant differences between regions. As more data is added (from the SA Mallee and Southeast survey in late 2017, and the Mid north and Yorke Peninsula in late 2018), analytical power of this dataset will continue to increase. However, the magnitude of the problem in general is still apparent, especially for the sulfonylureas. In the same program of surveys in southern Australia, sulfonylurea herbicides were also the most frequent case of resistance for rigid ryegrass (*Lolium rigidum* Gaudin) (Boutsalis *et al.* 2012). The frequency of resistance in this dataset was relatively high when compared to surveys of other weeds from other cropping areas around the world (Beckie *et al.* 2008).

The fact that SU and IMI resistance failed a test of independence is an indicator of cross-resistance between these two chemical families. It is well known that ALS inhibitor resistance can be conferred by a target-site mutation in the ALS gene, and at least five such mutations have been established in resistant weeds (Tranel and Wright 2002). Different mutations may confer a different profile of resistance across chemical families of ALS inhibiting herbicides and the pattern of resistance seen in the survey (Figure 2) could be the result of these differing mutations.

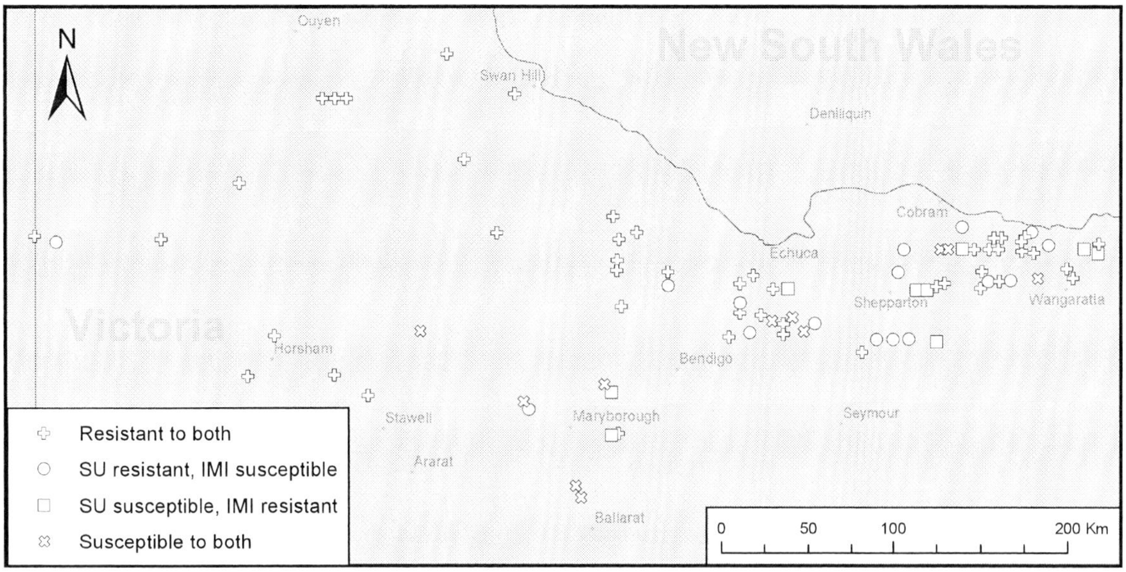

Figure 2. Map showing the distribution of ALS inhibitor resistance in regions of Victoria surveyed in 2015 and 2016. Resistance to both the sulfonylurea (SU) and imidazolinone (IMI) chemical families was present in 48 samples (✣), 16 samples were SU resistant and IMI susceptible (○), 9 samples were SU susceptible and IMI resistant (□), and 11 samples were susceptible to both chemical families (✕).

While the frequency of resistance did increase from the 2010–2011 round of surveys to the 2015–2016 round, most of these differences were not significant (Table 2). Again, larger sample sizes as surveys continue will help to elucidate these differences. During the later (2015–2016) surveys, there was increased interest in screening roadside populations of common sowthistle and this could have influenced the proportion of resistant samples.

The presence of spatial correlation in SU resistance is an indication of gene flow. Common sowthistle is a wind-dispersed weed capable of travelling long distances (Hutchinson *et al.* 1984), but the extent to which this phenomenon contributes to resistance gene flow is not well understood. Data from this survey indicate that it could be a relevant factor and present the possibility of using ALS mutations as a marker to study gene flow. The results of the survey indicate that further investigations into the biology and ecology of common sowthistle will be crucial to the development of effective management strategies that will help to maintain the sustainability of lentil production systems.

ACKNOWLEDGMENTS

Herbicide resistance monitoring surveys are supported by the Grains Research and Development Corporation (GRDC) (Project UCS00020). This research is supported by an Australian Government Research Training Program Scholarship and a GRDC Research Scholarship.

REFERENCES

Beckie, H.J., Leeson, J.Y., Thomas, A.G., Brenzil, C.A., Hall, L.M., Holzgang, G., Lozinski, C. and Shirriff, S. (2008). Weed resistance monitoring in the Canadian Prairies. *Weed Technology* 22, 530-43.

Boutsalis, P., Gill, G.S. and Preston, C. (2012). Incidence of herbicide resistance in rigid ryegrass (*Lolium rigidum*) across Southeastern Australia. *Weed Technology* 26, 391-98.

Boutsalis, P. and Powles, S.B. (1995). Resistance of dicot weeds to acetolactate synthase (ALS)-inhibiting herbicides in Australia. *Weed Research* 35, 149-55.

Heap, I. (2018). The International Survey of Herbicide Resistant Weeds. http://www.weedscience.org/ (accessed January 2018).

Hutchinson, I., Colosi, J. and Lewin, R.A. (1984). The biology of Canadian weeds. 63. *Sonchus asper* (L.) Hill and *Sonchus oleraceus* L. *Canadian Journal of Plant Science* 64, 731-44.

Powles, S.B. and Yu, Q. (2010). Evolution in action: Plants resistant to herbicides. *Annual Review of Plant Biology* 61, 317-47.

R Core Team (2018). R: A language and environment for statistical computing. (R Foundation for Statistical Computing, Vienna, Austria).

Tranel, P.J. and Wright, T.R. (2002). Resistance of weeds to ALS-inhibiting herbicides: what have we learned? *Weed Science* 50, 700-12.

Genome sequencing of glyphosate resistant fleabane (*Conyza bonariensis*) and sowthistle (*Sonchus oleraceus*)

James P. Hereward[1], Jeff A. Werth[2], David F. Thornby[3], Michelle Keenan[4], Bhagirath Singh Chauhan[5] and Gimme H. Walter[1]
[1] School of Biological Sciences, The University of Queensland, Brisbane, Queensland 4068, Australia
[2] Leslie Research Centre, Queensland Department of Agriculture and Fisheries, Toowoomba, Queensland 4350, Australia
[3] Innokas Intellectual Services, Coomera, Queensland 4209, Australia
[4] The Centre for Plant Science. Queensland Alliance for Agriculture and Food Innovation (QAAFI), The University of Queensland, Gatton, Queensland 4343, Australia
[5] Queensland Alliance for Agriculture and Food Innovation, PO Box 1218, Gatton, Queensland 4343, Australia
(j.hereward@uq.edu.au)

Summary Genome sequencing is the process of digitising the complete set of genetic instructions for an organism. The cost of sequencing has been dropping steadily, but until very recently it has still been expensive and difficult to assemble (piece together) the genome of a species that has not been previously characterised. Some weeds present additional challenges because they have large genomes and polyploidy. New technologies promise to reduce this challenge to one that can be accomplished within a single project at a reasonable cost. This talk will discuss the progress made to date in genome sequencing projects initiated for flaxleaf fleabane (*Conyza bonariensis*) and sowthistle (*Sonchus oleraceus*). Both of these species have evolved glyphosate resistance in Australia. A genome sequence is like a map for any genetic studies, and a well-resolved assembly will enable the component genomes in these polyploid species to be disentangled.

Keywords Genome sequencing, assembly, sowthistle, fleabane.

Mechanisms of resistance to penoxsulam and quinclorac in *Echinochloa crus-galli*

Yongfeng Li[1], Xia Yang[1], Mingchao Dong[1], Zichang Zhang[1], Tao Gu[1], Jingjing Cao[1] and Qin Yu[2]
[1] Institute of Plant Protection, Jiangsu Academy of Agricultural Sciences, Nanjing 210014, China
[2] Australian Herbicide Resistance Initiative (AHRI), School of Agriculture and Environment, The University of Western Australia, 35 Stirling Highway, Crawley, Western Australia 6009, Australia
(liyongfeng_2010@hotmail.com)

Summary Barnyard grass (*Echinochloa crus-galli*) is one of the major economically important weeds in the world rice production. ALS- and quinclorac-resistant barnyard grass populations are widespread in rice fields in China. To unravel the multiple-herbicide resistance mechanisms in *E. crus-galli*, the population with resistance (R) to penoxsulam and quinclorac was confirmed and characterised. Dose-response experiments showed that the R population has evolved a high level resistance to penoxsulam (RI>95) and quinclorac (RI = 158), relative to the susceptible (S) population. ALS gene sequencing revealed the known Trp574Leu mutation in the ALS2 gene in the R population is responsible for penoxsulam resistance. The full coding sequences of 1-aminocyclopropane-1-carboxylate oxidase (ACO) gene involved in ethylene biosynthesis were cloned from the R and S plants. Gene point mutations resulting in five amino acid substitutions (F98→Y, G101→D, Q206→R, G270→V, A272→G) were identified in the ACO-R versus ACO-S sequences, and three substitutions (F98→Y, G101→D, Q206→R) were located in the putatively conserved domains. The *E. crus-galli* ACO-R and ACO-S genes were expressed in *E. coli*, the ACO purified, and in vitro ethylene production measured. Results showed that ethylene production from ACO-R was 2-fold lower than that from ACO-S. The ACO gene expression of the R and S populations showed no significant difference. Therefore, reduced ethylene production from the R biotype, likely associated with ACO gene mutations, might be involved in quinclorac resistance in barnyard grass. In conclusion, the ALS Trp574Leu mutation and ACO gene mutations might endow multiple resistance in *E. crus-galli*.

Keywords *Echinochloa crus-galli*, penoxsulam, quinclorac, ethylene biosynthesis, 1-aminocyclopropane-1-carboxylate oxidase (ACO).

Understanding the mechanisms of 2,4-D resistance in *Sonchus oleraceus* (common sowthistle)

Mahima Krishnan, Tijana Petrovic, Alicia Merriam, Geetha Velappan and Christopher Preston
The University of Adelaide, Adelaide, South Australia 5005, Australia
(Mahima.krishnan@adelaide.edu.au)

Summary Common sowthistle is an increasingly problematic weed in fallows and legume crops. This species has evolved resistance to Group B herbicides and glyphosate across Australia leading to an increase in the reliance on Group I herbicides for control. Failure of 2,4-D to control common sowthistle was reported at several sites in South Australia in 2015. Dose response experiments showed these populations were resistant not only to 2,4-D, but across all Group I herbicides tested. Studies have been undertaken to determine the mode of inheritance to 2,4-D and the mechanism of 2,4-D resistance in these populations.

Keywords Common sowthistle, Group I herbicides, 2,4-D, inheritance.

BAS684H a new tool for selective control of multi-resistant ryegrass

Roberto Busi and Stephen Powles
Australian Herbicide Resistance Initiative, School of Agriculture and Environment,
The University of Western Australia, Crawley, Western Australia 6009, Australia
(roberto.busi@uwa.edu.au)

Summary The response of herbicide susceptible and multiple resistant annual ryegrass populations and wheat (cv Magenta) to the herbicide BAS684H was examined. BAS684H was applied pre-emergence under carefully controlled conditions at different dosages. Plant survival (emergence followed by active growth) and plant growth (aboveground biomass) was assessed 28 days after herbicide treatments. A concomitant study to assess cross-resistance profile to the key pre-emergence herbicides prosulfocarb + metolachlor (Boxer Gold), pyroxasulfone (Sakura) and trifluralin was conducted. BAS684H was found to be safe to wheat. Good control (>85% kill rate and 89% biomass reduction) of herbicide-susceptible and multi-resistant ryegrass was achieved at 400 g BAS684H ha^{-1}. We have observed some variability in herbicide efficacy and control of ryegrass, whereas crop selectivity was maintained at high doses in wheat. In addition to assess the risk of resistance to the new herbicide BAS684H we have also assessed the response of 200 ryegrass populations to BAS684H. Survivors in key populations with known profile of herbicide resistance were recurrently selected. After two generations we have information on the potential of ryegrass to evolve resistance to BAS684H. We conclude that BAS684H may have a significant commercial opportunity in the Australian grain-belt.

Target-site and metabolic resistance mechanisms to dinitroaniline herbicides in a *Lolium rigidum* population

Jinyi Chen, Heping Han, Danica Goggin, Qin Yu and Stephen Powles
Australian Herbicide Resistance Initiative (AHRI), School of Agriculture and Environment,
University of Western Australia, 35 Stirling Highway, Crawley, Western Australia 6009, Australia
(jinyi.chen@research.uwa.edu.au)

Summary The dinitroaniline herbicides (particularly trifluralin) have been used in Australia as a pre-emergence herbicide to effectively control weeds that have evolved resistance to common post-emergence herbicides. One *Lolium rigidum* population (hereafter as R) collected from Western Australian grain belt was found to be resistant to various dinitroaniline herbicides as compared to a standard susceptible (S) population SVLR1. Individual plants surviving the field rate trifluralin (960 g ha^{-1}) were sequenced and most of the survivors (35 out of 39) were identified to possess the Val-202-Phe α-tubulin mutation.

Rice genetic transformation was used to test if the Val-202-Phe mutation confers resistance. Rice calli transformed with the mutant 202-Phe α-tubulin gene were found to be more tolerant to trifluralin and other dinitroaniline herbicides than calli transformed with the wild type α-tubulin gene from *L. rigidum*.

In addition, enhanced trifluralin metabolism was also detected in the R population. ^{14}C-trifluralin was introduced to S and R plants via rooting media, total ^{14}C radioactivity was extracted and partitioned against 80% methanol (polar phase) and 100% hexane (nonpolar phase), percentage of radioactivity distribution in the two phases was calculated and compared. Results showed that more of the total radioactivity from the R plants was distributed in the methanol layer than S plants, indicating enhanced metabolism of the parent trifluralin in R plants. HPLC work confirmed a higher percentage of polar metabolites in the methanol phase in R versus S plants.

Our work demonstrated that both target-site and non-target-site metabolic resistance are involved in the trifluralin-resistant *L. rigidum* population investigated. This is for the first time we demonstrate metabolic resistance to trifluralin in weedy plants. We consider that metabolic resistance to dinitroaniline herbicides is common in *L. rigidum*.

Keywords α-tubulin mutation, trifluralin resistance, dinitroaniline, *Lolium rigidum*, trifluralin metabolism.

A novel *psbA* mutation (Phe-274-Val) confers resistance to PSII herbicides in wild radish (*Raphanus raphanistrum*)

Huan Lu, Qin Yu, Heping Han, Mechelle J. Owen and Stephen B. Powles
Australian Herbicide Resistance Initiative, School of Agriculture and Environment,
University of Western Australia, Western Australia 6009, Australia
(huan.lu@research.uwa.edu.au)

Summary Wild radish (*Raphanus raphanistrum*) is a globally important weed of crops. Two atrazine-resistant wild radish populations (R1 and R2), collected from the Western Australia grain belt, were investigated for resistance to PSII herbicides. Sequencing of the full-length *psbA* gene revealed a novel Phe-274-Val mutation in population R1, in addition to the well-known Ser-264-Gly mutation in population R2. Herbicide dose-response studies confirmed that the population with the novel Phe-274-Val mutation exhibited modest level of resistance to atrazine, metribuzin and diuron, but were bromoxynil susceptible. The population with the Ser-264-Gly mutation exhibited high level resistance to atrazine, but super sensitivity to bromoxynil. Structural modelling of the mutant D1 proteins predicts the Ser-264-Gly mutation endows atrazine resistance by abolishing H-bonds, but confers bromoxynil super sensitivity by enhancing H-bonding. The Phe-274-Val substitution confers resistance to atrazine and diuron by indirectly affecting H-bond formation between the Ser-264 residue and the herbicides. The results demonstrate that the Phe-274-Val mutation is likely responsible for resistance to PSII-inhibiting triazine and urea herbicides. To our knowledge, this is the first evidence of the *psbA* Phe-274-Val mutation in higher plants conferring resistance to PSII herbicides.

Keywords PSII herbicide, wild radish, *psbA* gene mutation, D1 protein.

Metabolic herbicide resistance gene discovery in wild oat (*Avena fatua*)

Qiong Peng[1,2], Qin Yu[1], Heping Han[1], Lianyang Bai[2] and Stephen Powles[1]
[1]Australian Herbicide Resistance Initiative (AHRI), School of Agriculture and Environment, The University of Western Australia, 35 Stirling Highway, Crawley, Western Australia 6009, Australia
[2] Hunan Agricultural Biotechnology Research Institute, Hunan Academy of Agricultural Sciences, Changsha 410125, China
(183749137@qq.com)

Summary Weedy plant species that have evolved resistance to herbicides due to enhanced metabolic capacity to detoxify herbicides are a major issue threatening herbicide sustainability and global crop production. Metabolism-based herbicide resistance (referred to as metabolic resistance) in weeds has not been well characterised at the genetic level. Diclofop-methyl resistant (R) wild oat lines only possessing metabolic resistance were isolated and used for RNA-Seq transcriptome analysis for metabolic resistance gene discovery. Forty contig genes that showed constitutive expression differences between the R and S samples were selected for validation, and these include 22 CYP P450 genes, 9 Glycosyl transferase genes and 9 GST genes related to herbicide metabolism. After validation with RNA-seq samples, the candidate genes were reduced to 31 including 17 CYP P450, 8 Glycosyl transferase family genes, and 6 GST genes. The 31 contigs were then evaluated using additional samples from the R and S lines, and from a single wild oat line segregating R at 4000 or 6000 g ha^{-1} and S at 1000 g ha^{-1} of diclofop-methyl. Next, a physiological validation experiment was conducted, in which 2,4-D pre-treatment was used to induce gene expression providing diclofop protection in S individuals due to increased metabolism. This way only two candidate genes (CYP450) were significantly induced in the S samples. Therefore these two CYP450 genes (CYP72A15 and CYP71C4) are very likely associated with metabolic resistance to diclofop-methyl in the R wild oat line and prioritised for further characterisation using rice genetic transformation.

Keywords RNA-sequencing, wild oat, diclofop, metabolic resistance, CYP450.

Abutilon theophrasti – its biology and management in New Zealand

Trevor K. James[1] and Heidi M. Pene[2]
[1] AgResearch, Private Bag 3123, Hamilton 3240, New Zealand
[2] PestPlants, PO Box 704, Cambridge 3450, New Zealand
(trevor.james@agresearch.co.nz)

Summary *Abutilon theophrasti* Medikus is a declared 'Unwanted Organism' and has been subject to a control programme on 41 properties in Waikato, New Zealand since it's discovery in 2011. In 2015 *A. theophrasti* seed was introduced into New Zealand as a contaminant in fodder beet (*Beta vulgaris* L.) seed and later found to have grown on at least 250 farms throughout the country. In response, the Ministry for Primary Industries objective is 'containment with the long-term goal of eradication'. To assist with achieving this, a research programme commenced in 2016 to better understand critical aspects of the biology of *A. theophrasti* and to develop improved management programmes that also aid surveillance.

Field trials demonstrated excellent post-emergence control of *A, theophrasti* in maize with topramezone + atrazine + crop oil, in barley with MCPA + carfentrazone, and in pasture with flumetsulam + bentazone (small plants only) or 2,4-D. Results from the laboratory showed *A. theophrasti* seed emerging from depths up to 120 mm and germination commencing at 6°C. However, results from the field demonstrated germination typically occurred at higher temperatures, peaking in January in open ground and February in pasture.

Keywords Indian mallow, velvetleaf, control.

INTRODUCTION

Abutilon theophrasti Medikus, known variously as Indian mallow, velvetleaf or butter print, is a weed in many countries in North America, Europe, Asia and Africa (Warwick and Black 1988, Sattin *et al.* 1992). It was introduced to North America in the 18th century from China and has now become the main broadleaf weed in maize (*Zea mays* L.) and soyabean (also known as soybean (*Glycine max* (L.) Merr.)) crops there.

Abutilon theophrasti is a member of the mallow family (Malvaceae) and is easily distinguished by its tall, erect habit, its large, alternate, heart-shaped leaves which are soft and velvety to the touch and have a musky odour. It has small yellow to yellow-orange flowers which only open for a few hours during the hottest part of the day. It has distinctive black seed pods containing about 40 large, hard black seeds that show great persistence in the soil seed bank (Warwick and Black 1988, Cardina and Sparrow 1997). It is a summer-growing annual plant which has the ability to germinate throughout the warmer months. Growing to 3 m tall, its biological and ecological characteristics make it particularly competitive and persistent in cultivated fields. It has been reported to cause severe crop losses particularly in fields of maize, sorghum, soya bean and cotton (Spencer 1984). As *A. theophrasti* can germinate throughout the warm season, it is very difficult to eradicate and once established, it is almost impossible to avoid seed production (Warwick and Black 1988, Sattin *et al.* 1992).

The first reports of it in New Zealand are herbarium records from 1929 and 1948 where it was introduced for evaluation as a crop, the latter sourced from United States of America (Anon 2018). Since then it has been accidently imported with soyabean seed and as a contaminant of other grains. It is presently unknown if any of these persisted to the present day.

In 2011, a paddock size infestation of *A. theophrasti* was found in Waikato (James and Cooper 2012) and resulted in the Ministry for Primary Industries (MPI) declaring it an 'Unwanted Organism'. This initiated an incursion response and *A. theophrasti* has since been tracked to more than 40 other locations in the upper North Island. In 2015, *A. theophrasti* seed was introduced as a contaminant in six lines of fodder beet (*Beta vulgaris* L.) seed from Italy and was subsequently found in about 250 paddocks spread across New Zealand. Rapid response searches and effective tracing quickly delimited both the Waikato and the new fodder beet incursions. Tailored farm management plans were written for all affected properties and their implementation along with continuing surveillance appears to have contained the weed. The Ministry for Primary Industries current objective is 'containment with the long-term goal of eradication'. To assist with achieving this, a research programme commenced in 2016 to better understand critical aspects of its biology and to develop improved management programmes that also aid surveillance. Results from this programme are reported here.

MATERIALS AND METHODS

All the field trials reported here were carried out at the AgResearch trial site near Matamata in Waikato, where there was a natural soil seed bank of *A. theophrasti* from an uncontrolled incursion.

Soil seed bank depletion At the trial site, an area of land 30 × 10 m was treated with glyphosate and then cultivated on 31 October 2016, 22 December 2016, 18 January 2017 and 1 March 2017 using a power harrow (3–4 passes). The following season the trial area was extended into the pasture by 10 m (now 40 × 10 m), treated with glyphosate and the total area cultivated on 17 November 2017, 9 December 2017, 9 January 2018 and 27 February 2018. About 4 weeks after each cultivation all *A. theophrasti* plants in the old (2016/2017) or both old and new (2017/2018) areas were separately counted and removed.

Control of *A. theophrasti* in pasture On 15 February 2017, a randomised, four replicate trial with 2 × 6 m plots and eight treatments (Table 1) was pegged out and sprayed. In two of the replicates *A. theophrasti* plants were 20–250 mm high with 3–7 leaves while the other two replicates had *A. theophrasti* plants only 20–130 mm high with 3–6 leaves. All treatments were applied in 200 L ha^{-1} water with a CO_2 powered backpack precision sprayer fitted with a 2 m boom and four TeeJet AI-11002 nozzles. At the time the ambient temperature was 22–23°C and there had been light rain (2–3 mm) the previous day. Plant counts per plot were made immediately prior to spraying and again on 7 March 2017 at which time all the *A. theophrasti* plants were removed as some had commenced flowering.

Table 1. Control of *A. theophrasti* in pasture.

Herbicide	Rate g ai ha^{-1}	Plants plot^{-1} Pre-spray	Plants plot^{-1} Final
Untreated	–	23	42
Flumetsulam + bentazone	51 + 1440 g	40	11
Triclopyr + picloram + aminopyralid	150 + 50 + 4 g	41	80
Flumetsulam	52 g	32	3
2,4-D	2280 g	33	0
MCPA+MCPB	100 + 1500 g	20	20
Triclopyr + aminopyralid	400 + 60 g	41	52
Aminopyralid	60 g	45	71
LSD (P<0.05)		22.4	15.2

Control of *A. theophrasti* in maize and barley On 21 November 2017, a randomised, four replicate trial with 2 × 6 m (barley (*Hordeum vulgare* L.)) and 3 × 6 m (maize) plots and five treatments (Table 2) was pegged out, planted and pre-emergence treatments applied (sunny, 23°C). On 13 December 2017 early post-emergence treatments were applied (sunny, 24°C) to the plots previously treated with the pre-emergence treatments. On 8 January 2018 the late post-emergence treatments were applied (sunny, 25°C) to the plots that had not received any previous herbicide treatments. All treatments were applied as above with either a 2 m or 3 m boom (TeeJet AI-11003 nozzles) to match plot width. Counts of *A. theophrasti* were made on 13 December 2017, 8 January 2018 and 15 February 2018 using four random throws of a 0.1 m^2 quadrat. At this time all *A. theophrasti* plants were removed to prevent seeding.

Depth of emergence Burial depth from which *A. theophrasti* seed can emerge was investigated using. 60 (15 depths × four replicates) PVC tubes each 100 mm in diameter and from 20 to 200 mm long. These were filled with Horotiu silt loam soil on 30 August 2016 and 10 *A. theophrasti* seeds placed on top. Further tubes were added above the seed level to give a total height of 200 mm, and filled with the same soil. This resulted in columns where the seeds were on the surface (0 mm) and where they were buried at 10 mm increments to 100 mm deep and 20 mm increments to 180 mm deep. Thus when assembled, each column was the same height and contained the same amount of soil. There were four replicate columns for each depth. The columns were maintained with both top- and sub-irrigation to keep the soil moist. The tubes with buried seeds were maintained in a controlled glasshouse with a night time minimum of 15°C and daytime maximum of 25°C. They were checked for emergence daily for 42 days, although emergence ceased after 17 days. Once the emergence date had been recorded, the newly-emerged plants were severed 5 mm above the ground so that further emergence of other seeds was not inhibited.

Critical temperatures for emergence A LinTek GR10D thermogradient table was used to evaluate germination temperatures for *A, theophrasti*. The table was set to run a one-way gradient, using a temperature range of 5°C to 35°C. Eight replicates of 25 seed were placed on separate blotters at each 1°C increment across the table. Observations were made daily for 3 weeks and germinated seeds (2 mm radicle) counted and removed at this time.

Table 2. *A. theophrasti* populations after pre- and post-emergence herbicides applied on 23.11.2017 and 8.1.2018 respectively.

Pre-emergence Crop/Herbicide	Rate (g ai ha^{-1})	Plant counts (# m^{-2}) 13.12.2017	Post-emergence Crop/Herbicide	Rate (g ai ha^{-1})	Plant counts (# m^{-2}) 8.1.2018
No crop, no herbicide	–	84	No crop, no herbicide	–	107
Maize, no herbicide	–	103	Maize, no herbicide	–	113
Maize + acetochlor + saflufenacil	2520 + 105 g	89	Maize, topramezone + atrazine	67 + 500 g	0
Barley, no herbicide	–	98	Barley, no herbicide	–	102
Barley + flufenacet + diflufenican	140 + 70 g	84	Barley, MCPA + carfentrazone	1170 + 17 g	0
LSD (P<0.05)		27.1			36.8

Table 3. Number of *Abutilon* seedlings at about 4 weeks after each cultivation.

Cultivation date	*A. theophrasti* seedlings	
	Original area Plants 300 m^{-2}	New area Plants 100 m^{-2}
31.10.16	119	–
22.12.16	497	–
18.1.17	149	–
1.3.17	15	–
17.11.17	354	2443
9.12.17	569	3522
9.1.18	29	296
27.2.18	117	526

RESULTS

Soil seed bank depletion The summer of 2016/2017 was 2–3°C cooler than average while that of 2017/2018 was 2–3°C warmer and wetter than average and too wet to cultivate the site until mid-November. The warmer summer of 2017/2018 is reflected in the emergence of *A. theophrasti* after cultivation with more seedlings emerging after the four cultivations in 2017/2018 compared to 2016/2017, even though no new seed was introduced to the site (Table 3).

Peak emergence occurred after the December cultivation in both years. In the new area cultivated for the first time after a year in pasture, large numbers of *A. theophrasti* still germinated in late February.

Control of *A. theophrasti* in pasture The results (Table 1) show two treatments, flumetsulam ± bentazone, controlling small plants but failing to control all the *A. theophrasti* in the two replicate plots containing larger plants (data not presented). In comparison, 2,4-D killed all the plants in all four replicate plots. Triclopyr, picloram, aminopyralid and MCPB+MCPA failed to achieve any control of *A. theophrasti.*

Control of *A. theophrasti* in maize and barley There was no rain within the two weeks of the application of the pre-emergence treatments, and consequently, they failed to control the *A. theophrasti* as it emerged (Table 2). In contrast both the early post-emergence treatments gave 100% control of the *A. theophrasti*. By the time the late post-emergence treatments were applied the *A. theophrasti* plants were 200–300 mm tall and although the treatments caused severe damage and defoliation, most of the *A. theophrasti* plants soon regrew (data not presented).

Depth of emergence All of the *A. theophrasti* seed placed at 20 mm depth emerged (Table 4). However, 80% or greater emerged from all depths up to 90 mm and 70% emerged from 100 mm depth. For the greater depths, only 8% emerged from 120 mm, and none emerged from 140, 160 or 180 mm depth. The cost of emergence from the greater depths was a delay in emergence time. Seeds buried between 10 and 30 mm emerged after 5 days while those planted at greater depths took longer to emerge (Table 4). For the seed on the surface and the seed at greater planting depths, emergence was more variable. The seed placed at 90 and 100 mm depth emerged over a 7-day period, 8–14 days after planting.

Critical temperatures for emergence On the thermogradient table, *A. theophrasti* had 17% germination at 5°C with germination commencing after 13 days (Table 5).

Table 4. Depth from which *Abutilon theophrasti* seedlings emerged (Number emerged is the sum of four replications (n = 40) seeds) at 42 days after planting.

Depth (mm)	Number emerged[1]	Time to emergence (d)	
		First	Last
0	37 ± 1.0	5	9
10	37 ± 1.0	5	5
20	40 ± 0	5	5
30	35 ± 0.5	5	6
40	37 ± 0.5	6	6
50	35 ± 0.5	6	8
60	38 ± 0.6	6	8
70	37 ± 1.0	6	12
80	32 ± 0.8	8	8
90	33 ± 1.0	8	14
100	28 ± 2.4	8	14
120	3 ± 1.5	12	12
140	0	–	–
160	0	–	–
180	0	–	–

[1] Cumulative total ± standard deviation.

Table 5. Germination of *Abutilon theophrasti* at different temperatures (Number emerged is the sum of eight replications (n = 200) seeds) at 21 days after planting.

Temp. (°C)	Number emerged[1]	Time to first emergence (d)
5	34 ± 3.3	13
6	63 ± 5.7	7
7	88 ± 7.9	6
8	116 ± 7.3	5
9	117 ± 6.8	4
10	132 ± 6.7	4
11	104 ± 8.0	4
12	109 ± 7.3	4
13	124 ± 6.9	4
14	112 ± 5.9	4
15	135 ± 4.8	4
16	148 ± 4.9	4
17	156 ± 3.5	4
18	168 ± 3.9	4
19	154 ± 5.2	4
20	183 ± 2.5	4
21–35	176–191	4

[1] Cumulative total ± standard deviation.

Total germination steadily increased up to about 20°C. Between 20 and 35°C the total germination remained constant and occurred at high numbers (176–191/200 seed). Time to germination was slower at 5°C, as the temperature increased, the time taken to germination decreased until 9°C. From 9°C until 35°C the time to germination at all temperatures was 4 days.

DISCUSSION

Although some mature *A. theophrasti* plants were found in the South Island after the fodder beet incursion of 2015/2016, no subsequent plants were found in 2016/2017 but 35 plants have been found at two sites in Canterbury in 2017/2018 (S. Beaumont, pers. comm.). This low number of subsequent finds could be due to the infested fields being returned to pasture and the cooler soil temperatures under pasture not being conducive to germination of *A. theophrasti*. In the North Island small numbers of *A. theophrasti* have been found in fields after the fodder beet incursion even though those fields were returned to pasture. This is supported by the observations of delayed emergence of this weed in pasture at the Waikato, North Island trial site. This field evidence also runs contrary to the results from the thermogradient table where *A. theophrasti* germinated at the low temperature of 6°C suggesting that other critical factors are involved. In controlled environment studies evaluating germination of *A. theophrasti* in soil, Mester and Buhler (1991) found that no seed germinated at 5°C, 30% at 10°C but peak germination was at 20°C. Horowitz and Taylorson (1984) also found 24–30°C to be optimal for germination of *A. theophrasti* and further that germination commenced to decline above 35°C, the limit of our experiment. This also fits with Sadeghloo *et al.* (2013) who found the three cardinal temperatures; base, optimum and ceiling, for *A. theophrasti* to be 5, 35 and 48°C respectively.

This project has demonstrated good control of *A. theophrasti* in pasture, maize and barley although maize is not a favoured crop as searching for escapes is difficult. Barley (for silage) may be a better alternative as any escaped *A. theophrasti* plants would be easier to find. Pasture is also an alternative for managing this weed but as results have demonstrated this crop reduces or prevents germination and the seed is likely to remain viable in the soil seed bank for a long time (Lueschen and Andersen 1980). Constantly reworking the soil should ultimately reduce the soil seed bank if fresh inputs of seed are denied but over the two-year period of this study there was little evidence of this.

Although we have demonstrated improved tools and identified important ecological data for *A. theophrasti* it will still remain a constant and long-term battle to eliminate this pernicious weed from New Zealand. To further assist farmers and authorities in this effort, over the next 3 years the research programme will:

- study the growth and development of *A. theophrasti* in different regions to determine Growing Degree Days to reach maturity and identify habitat limitation;
- investigate the genetics of various populations to determine whether there have been several incursions or simply recurring incidents;
- determine the interaction of *A. theophrasti* with other forage crops and develop management regimes to give farmers more options and allow for easier locating of isolated *A. theophrasti* escapes;
- extend the soil seed bank studies to other soil types and investigate means to more rapidly deplete the seed reservoir; and
- establish the criteria which would need to be met in order to declare a site free of *A. theophrasti*.

ACKNOWLEDGMENTS

This project was funded by Ministries for Primary Industries, Foundation for Arable Research, DairyNZ, Waikato Regional Council and Environment Canterbury. Thanks to Claire Dowsett, Ashley Wade and Jessie Davys for assistance with the project.

REFERENCES

Anon. (2018). Ngā Tipu o Aotearoa – New Zealand Plants. Manaali Whenua – Landcare Research database. https://www.landcareresearch.co.nz/resources/data/nzplants (accessed 18 April 2018).

Cardina, J. and Sparrow, D.H. (1997). Temporal changes in velvetleaf (*Abutilon theophrasti*) seed dormancy. *Weed Science* 45, 61-6.

Horowitz, M. and Taylorson, R.B. (1984). Hardseededness and germinability of velvetleaf (*Abutilon theophrasti*) as affected by temperature and moisture. *Weed Science* 32, 111-5.

James, T.K. and Cooper J.M. (2012). Control of the recently-introduced weed butterprint (*A. theophrasti*) in maize. *New Zealand Plant Protection* 65, 64-8.

Lueschen, W.E. and Andersen, R.N. (1980). Longevity of velvetleaf (*Abutilon theophrasti*) seeds in soil under agricultural practices. *Weed Science* 28, 341-6.

Mester, T.C. and Buhler, D.D. (1991). Effects of soil temperature, seed depth, and cyanazine on giant foxtail (*Setaria faberi*) and velvetleaf (*Abutilon theophrasti*) seedling development. *Weed Science* 39, 204-9.

Sadeghloo, A., Asghari, J. and Ghaderi-Far, F. (2013). Seed germination and seedling emergence of velvetleaf (*Abutilon theophrasti*) and barnyardgrass (*Echinochloa crus-galli*). *Planta daninha* 31, 259-66.

Sattin, M., Zanin, G. and Berti, A. (1992). Case history for weed competition/population ecology: Velvetleaf (*Abutilon theophrasti*) in corn (*Zea mays*). *Weed Technology* 6, 213-9.

Spencer, N.R. (1984). Velvetleaf, *Abutilon theophrasti* (Malvaceae), history and economic impact in the United States. *Economic Botany* 38, 407-16.

Warwick, S.I. and Black, L.D. (1988). The biology of Canadian weeds. 90. *Abutilon theophrasti*. *Canadian Journal of Plant Science* 68, 1069-85.

Germination ecology of two Australian populations of African turnip weed (*Sisymbrium thellungii* L.)

Gulshan Mahajan[1], Amar Matloob[2], Michael Walsh[3] and Bhagirath S. Chauhan[1]
[1] Queensland Alliance for Agriculture and Food Innovation (QAAFI), The University of Queensland, Gatton, Queensland 4343, Australia
[2] Muhammad Nawaz Shareef University of Agriculture, Multan, 11 Pakistan
[3] Sydney Institute of Agriculture, School of Life and Environmental Sciences, The University of Sydney, Narrabri, New South Wales 2390, Australia
(b.chauhan@uq.edu.au)

Summary African turnip weed is an annual winter weed of the northern grain region of Australia. No information is available on the seed germination ecology of this weed species. Experiments were conducted to determine the effect of various environmental factors on seed germination and seedling emergence of two populations (cropped and fenceline) of African turnip weed. The response of both populations to various environmental factors was similar. In both populations, germination was greater in the light/dark regime than in darkness at the optimum temperature condition (20/10°C). An osmotic potential of −0.1 MPa inhibited germination of both populations by more than 50% and complete inhibition occurred at −0.8 MPa. The seeds of both populations were germinated in sodium chloride concentrations ranging from 50–100 mM and beyond that concentration germination was completely inhibited. Seeds placed on the soil surface had maximum germination and emergence did not occur even with increase in seed burial depth of 1.0 cm. The addition of 6 t ha^{-1} of wheat residue on the soil surface reduced the emergence of both populations by more than 60%. The results could help in developing strategies for integrated management of African turnip weed.

Keywords Weed biology, seed germination, burial depth, osmotic potential, integrated weed management.

Response of common sowthistle, *Sonchus oleraceus* L. to simulated herbivory

Nagalingam Kumaran[1], Michelle A. Rafter[1], Kylie B. Ireland[2], Gavin Hunter[2], Louise Morin[2] and S. Raghu[1]
[1] Commonwealth Scientific and Industrial Research Organisation (CSIRO), Ecosciences Precinct, 41 Boggo Road, Dutton Park, Queensland 4102, Australia
[2] Commonwealth Scientific and Industrial Research Organisation (CSIRO), Black Mountain Science & Innovation Park, Clunies Ross Street, Acton, Australian Capital Territory 2601, Australia
(kumaran.nagalingam@csiro.au)

Summary Common sowthistle, *Sonchus oleraceus* L. is a major weed in grain growing regions of Australia. This weed is prevalent in conservation tillage systems, and its control is exclusively dependent on herbicides. However, *S. oleraceus* has developed resistance to herbicides chlorsulfuron and glyphosate, which makes it a difficult weed to control. Given its ability to develop resistance to herbicides and inefficiency of other tools, biocontrol is being developed as an additional tool for the management of *S. oleraceus*. Simulated herbivory studies on *S. oleraceus* were conducted under glasshouse conditions to understand the weed's ecology and inform the selection and prioritisation of potential biocontrol agents. We subjected *S. oleraceus* to 25, 50 and 75% leaf removal treatments (*sensu* defoliation) at early rosette stage and recorded effects on plant growth and physiology. Defoliation exerted significant negative effect on growth: plant height was reduced and fewer leaves were produced than non-defoliated plants, especially at 50 and 75% leaf removal. Regarding physiology, negative effect on transpiration and stomatal conductance was observed. While there was no effect of defoliation on CO_2 assimilation per unit area of leaves, a reduced number of leaves recorded in defoliated plants suggests there is likely a reduction in overall CO_2 assimilation per plant. These results suggest biocontrol agents that can reduce leaf numbers and leaf areas are likely to have a negative impact on *S. oleraceus*.

Keywords Biocontrol, CO_2 assimilation, respiration, defoliation, leaf area, photosynthesis.

Allelopathic potential of pasture legume species and selected winter crops on barnyard grass and rice

Jhoana Opena[1], Hanwen Wu[1,2], Sergio Moroni[1], Deirdre Lemerle[1] and James Pratley[1]
[1] Graham Centre for Agricultural Innovation (Charles Sturt University and NSW Department of Primary Industries), School of Agricultural and Wine Sciences, Locked Bag 588, Wagga Wagga, New South Wales 2678, Australia
[2] NSW Department of Primary Industries, Wagga Wagga 2650, New South Wales, Australia
(jopena@csu.edu.au)

Summary Allelopathic potential of pasture legumes and selected winter crops on barnyard grass and rice were investigated. Three annual pasture legume species (antas subclover, shaftal persian, and viper balansa clovers), canola and barley were used for the evaluation of potential allelopathic effects of residues on the germination and early growth of barnyard grass and rice varieties used for drill sown rice such as YRK5, YRM70, Opus and Sherpa. Results of the study will be presented during the conference. This information on the use of pasture legume species and/or winter crops in rotation with rice may provide a non-chemical alternative method to manage barnyard grass in drill-sown rice.

Keywords Allelopathy, barnyard grass, pasture legumes.

Inter-specific variations in seed germination biology and seedling emergence of bladder ketmia (*Hibiscus* spp.)

Md Asaduzzaman and Eric Koetz
NSW Department of Primary Industries, Wagga Wagga, New South Wales 2650, Australia
(asad.asaduzzaman@dpi.nsw.gov.au)

Summary Bladder ketmia (*Hibiscus* spp.) is a widespread and troublesome weed found throughout the Australian cotton industry. The species is closely related to cotton plants in terms of phenological and physiological traits. There are two different species of bladder ketmia, wide leaf (*Hibiscus trionum* var. *vesicarius*) and narrow leaf (*Hibiscus trionum* var. *trionum*). Understanding the difference in seed dormancy and biology between the bladder ketmia varieties can influence the choice of control options in some situations. This study was undertaken to quantify the inter-specific differences in germination biology of *Hibiscus* spp. We also postulated that glyphosate's action may differ between two ketmia species. Experimental results suggest 20 and 25 minutes sulphuric acid scarification is required to release 60% seeds of wide and narrow leaf ketmia respectively. The model also suggests wide leaf ketmia was twice as salt tolerant as the narrow leaf and at 100 mM NaCl, growth medium, about 70% seeds of wide leaf ketmia are still able to germinate. Narrow leaf ketmia is more sensitive to water stress and its germination was completely inhibited at −2 Mpa water potential. Glyphosate spray (1.5 L ha^{-1}) at pod development stage demonstrates, narrow leaf ketmia is 3-fold more tolerant to glyphosate than wide leaf ketmia and 1.6 L ha^{-1} glyphosate was required to kill 50% plants from the population of narrow leaf ketmia. We conclude that seed dormancy and biology variation exists between ketmia species. This result may fits well with growers' experiences of controlling both varieties of bladder ketmia in Roundup Ready crops.

Keywords Germination, tolerant, glyphosate, cotton.

Predicting the mass emergence of *Vulpia bromoides*

Claire A. Dowsett[1], Christopher E. Buddenhagen, Richard J. Chynoweth[2] and Trevor K. James[1]
[1] AgResearch, Private Bag 3123, Hamilton New Zealand 3240
[2] Foundation for Arable Research, PO Box 23133, Hornby Christchurch New Zealand 8441
(Claire.dowsett@agresearch.co.nz)

Summary *Vulpia bromoides* (L.) Gray (hair grass) is a serious grass weed of New Zealand cereal crops, contaminating seed and reducing yield. Optimal management of this weed may be achieved when the environmental factors that affect its emergence are well understood. Emergence may then be predicted and manipulated through land management techniques, e.g. through the promotion of germination (irrigation) or the prevention of it (soil inversion), thus enabling optimisation of weed control. The hydrothermal time for germination of *V. bromoides* was determined in order to better understand when germination and emergence occurs and how environmental conditions including soil temperature (T) and water potential (Ψ) may influence this. The germination profile of *V. bromoides* was determined on a thermogradient table. Seeds were also incubated under constant factorial combinations of temperature and water potential in order to better understand the water potential required for the germination of this species. The rate of germination of *V. bromoides* was slowest from 5 to 7°C and most rapid from 13 to 32°C. At temperatures of 5 to 15°C *V. bromoides* experienced reduced germination rates at water potentials between –0.75 to –1.00 MPa. At temperatures of 20 and 25°C *V. bromoides* germinated rapidly across all of the water potentials assessed. These results will inform the development of optimised weed management on farm.

Keywords Hair grass, hydrothermal time, germination, base temperature.

INTRODUCTION

Grass weeds such as *Vulpia bromoides* (L.) Gray (hair grass) are a continuing problem in New Zealand cereal crops, especially in the eastern South Island. Reduced tillage systems favour winter annual grasses which, as with *V. bromoides*, have short lived seeds with little innate dormancy (germinate readily). Reduced tillage systems especially favour winter annual grass weeds because seed remains in the top layer of soil (not buried with soil inversion) which favours germination (Melander *et al.* 2013). Reduced tillage though is common practice due to benefits of soil quality and, at least until recently, herbicides successfully controlled most weeds. Cases of herbicide tolerance by *V. bromoides* and herbicide resistance by *Lolium* spp. (Gunnarsson *et al.* 2017) as well as market demands for reduced herbicide use are fuelling investigations into alternative methods of weed management (Melander *et al.* 2013).

Seed germination and seedling emergence is arguably the most important phenological event that influences the success of a weed (Benech-Arnold *et al.* 2000, Forcella *et al.* 2000). As such, weed management strategies often focus on the seeds and seedlings of problem weeds, as this is when they are most easily controlled (Bradford 2002). Soil temperature, soil water, soil air and light quality are the main environmental factors affecting seedling emergence, factors which may be manipulated indirectly through land management techniques which either promote germination (irrigation) or prevent it (soil inversion) (Forcella *et al.* 2000). The ability to predict weed emergence could facilitate the implementation of effective weed management by optimising the timing of weed control (Dastgheib and Poole 2010, García *et al.* 2013).

Seeds germinate across a temperature range that can be defined in terms of three cardinal temperatures, the base (T_b), the optimum (T_o) and the maximum (T_m) (Monks *et al.* 2009). The cardinal temperature range represents species-specific responses to environmental conditions and can be used to predict the time of germination and emergence of weeds (Moot *et al.* 2000a, Alvarado and Bradford 2002).

Water availability is similarly an important factor controlling germination. Changes in soil water content affect the soil water potential and are directly involved in water transport to germinating seeds such that base water potential (ψ_b) is also required to predict germination and emergence of weeds (Bloomberg and Watt 2008, Bloomberg *et al.* 2009, Guillemin *et al.* 2013).

To determine the germination emergence period of *V. bromoides* in New Zealand cereal cropping two separate experiments were performed in the laboratory. The first was to determine the temperature profile and the second was to determine the water potential required for germination of this species.

MATERIALS AND METHODS

Preparation of seed Seed was supplied by the Foundation for Arable Research in 2015. The initial seed quality was assessed following the standards set in the International rules for seed testing (ISTA 2015): this included an assessment of purity, germination and seedling health.

The initial seed germination was assessed by germinating four replicates of 50 seeds on steel blue seed germination blotters (Anchor Paper Company, St. Paul, Minnesota) at alternating temperatures of 20°C for 16 hours with no light and 30°C for eight hours with light (ISTA 2015). A seedling was scored as normal when it showed a well-developed healthy primary root, a coleoptile free of splits at the base or less than one-third the coleoptile length if split from the top, and an intact primary leaf emerging through the coleoptile (ISTA 2009).

Cardinal temperature range A Grant Temperature Gradient Plate (Grant Instruments Limited, Cambridge, United Kingdom) was used to determine the cardinal temperature range for germination. The temperature gradient plate was set to run a one-way gradient, using a temperature range of 5°C to 35°C. A single layer of K-22 Kimpak® (Anchor Paper Company, St. Paul, Minnesota) that had been soaked in water was placed on the plate surface and two layers of steel blue seed germination blotters, also soaked in water, placed on top. The minimum and maximum temperatures on the plate were verified by measuring the temperature on the germination blotters at each end of the plate with a calibrated thermometer. The plate was located in a laboratory and received natural light.

The plate was split into two blocks. Each block was divided into three sub-blocks and one species randomly assigned to each sub-block. Fifty seeds were placed at even increments up the plate with each increment representing a 1°C change in temperature. Seed germination was assessed twice on the plate to provide data from four blocks in total. Germination on the plate was scored as radicle emergence. Germinated seeds were left on the plate until normal development of the emerging seedling could be seen.

Once germination had slowed on the plate (after 30–40 days) the plate was set to run at 21°C at one end and 24°C at the other, and remaining un-germinated seed allowed to germinate for at least a further 30 days. Any seed that had not germinated on the plate after 30 days was then determined as not-viable.

Hydrothermal time to germination Seed were incubated for up to 42 days in constant controlled factorial combinations of temperature; 5, 10, 15, 20, and 25°C and water potentials 0, −0.10, −0.30, −0.50, −0.75 and −1.00 (±0.05 MPa). Solutions with different water potentials were created using polyethylene glycol (PEG) as described by Hardegree and Emmerich (1990). The water potential of the solutions was checked using a VAPRO vapour pressure osmometer at the start of the experiment and again at the end to ensure that the osmotic potential of the solutions did not change significantly due to evaporation during the experiment.

Lidded containers (264 × 238 × 85 mm) were filled with 65 mL of each of the six different solutions. Four pre-soaked Whatman Grade 181 90 mm seed grade filter papers, were placed into the containers on top of cut glass squares (80 × 80 × 3 mm), so that they were touching but not free floating in the solution. Fifty seeds were counted onto the filter paper within these containers. Containers were carefully sealed and placed into the incubators held at the different temperatures.

Seed germination was counted, initially every few hours as germination commenced and peaked, and then as germination slowed these were checked once a day and finally once every two days. Seed were considered to have germinated when the radicle protruded more than 2 mm from the seed coat and were discarded immediately after counting. The containers were left in the incubators for up to six weeks or when all seed within them had germinated.

Statistical analysis All germination data was transformed (arcsine square root transformation) prior to analysis. A test for normal distribution (Kolmogorov-Smirnov) was applied to all data after transformation, where $P \geq 0.05$ was considered normal distribution. For data not normalised by the transformation, a non-parametric ANOVA was used to analyse the data, followed by the Bonferroni (Dunn) t-test for multiple comparisons. For germination data that was normally distributed after the arcsine square root transformation when a significant treatment effect was observed, treatment means were compared using the Tukey test.

RESULTS

Initial seed quality The *V. bromoides* seed was contaminated with an *Agrostis* sp., the seed was cleaned and the percentage of normal germination was recorded as 98%. The remaining 2% was made up of either dead or underdeveloped seed.

Cardinal temperature range Germination was high, reaching 96–98% between 8–28°C. The seed was able to germinate over the entire temperature range assessed, indicating that the base temperature of *V. bromoides* is below 5°C. There was no significant difference in germination percentage of *Vulpia* seed between 5°C and 32°C but germination began to decline after 32°C dropping from 80% at 32°C to 60% at 33°C and down to 7% at 35°C. The rate of germination of *V. bromoides* was slowest from 5–12°C and above 32 °C. Faster rates were seen between 13–32°C (Figure 1).

Hydrothermal time to germination The lowest water potential (−1.00 MPa) was not able to prevent germination. Germination was significantly slower when exposed to the more severe reductions in water potentials (−0.75 and −1.00 MPa) at the lower temperatures of 5 and 10°C, where the time to 50% germination was unable to be measured as 50% germination was not reached (Figure 2). The study also found that at the higher temperatures of 20 and 25°C, the water potentials tested here were not severe enough to restrict germination. This indicates that the optimal germination temperature range is between 15–25°C (Figure 2).

DISCUSSION

The base temperature of *V. bromoides* was not determined from this study. The base temperature of the closely related *V. myuros* (L.) C.C. Gmel. was determined by Monks *et al.* (2009) as 3.4°C although Scherner *et al.* (2017) determined 0.6°C. Other studies have also determined that it is not unusual for winter annual grass species to germinate from as low as 0°C (Moot *et al.* 2000b). It is reasonable to suggest that the base temperature of *V. bromoides* would be similar.

Ball (2008) also found that germination of *V. myuros* declined at temperatures above 35°C. Monks *et al.* (2009) found a lower maximum temperature (25°C) at which germination percentage began to decline. In our study, it was not until the temperature increased above 31°C that abnormal seedling development of *V. bromoides*, predominantly stunted or no seedling roots, increased. Abnormal seedlings when observed above 31°C ranged from 2–20% (average 12%). Monks *et al.* (2009) also found a similar decline in germination rate at 30°C. In this study, germination was 60% at

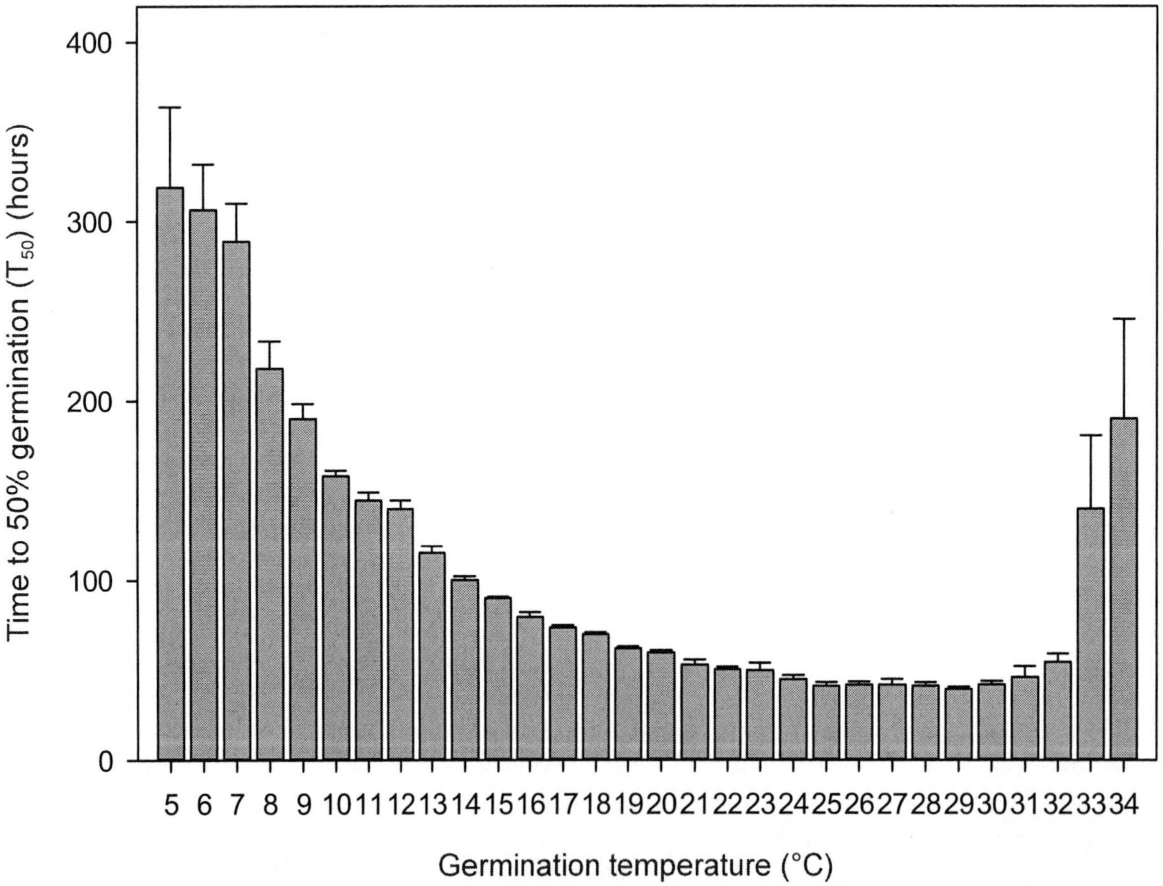

Figure 1. Time to 50% germination of *Vulpia bromoides* seed at a range of temperatures from 5°C to 34°C. Bars are the standard error of the mean.

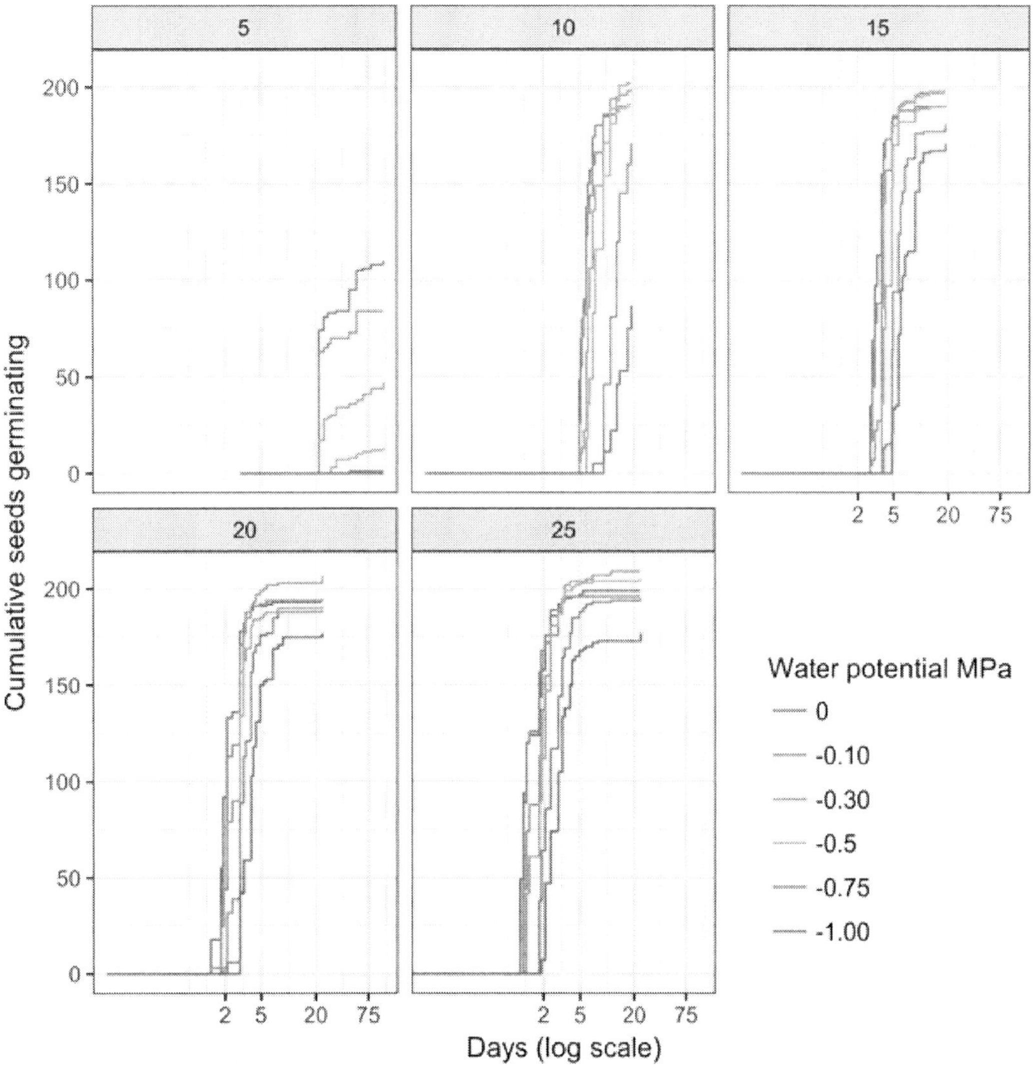

Figure 2. Cumulative numbers of *Vulpia bromoides* seed germinating over time (200 were planted) under different temperatures 5, 10, 15, 20 and 25°C (panels) and under exposure to normal or reduced water potentials (legend).

33°C and 33% at 34°C declining to only 7% at 35°C, suggesting that the seed was under stress at these higher temperatures.

Germination of *V. bromoides* occurred at the fastest rate between 13 and 32°C. The hydrothermal time to germination study found that reductions of water potential were not limiting to percentage germination inside this range either. Though this was not always the case; often high or low potentials are very limiting at all temperatures.

Cordeau *et al.* (2018) found that *V. myuros* is capable of tolerating fairly extreme hydric stress. Similarly, *V. bromoides* is able to tolerate reductions in water potential well, with germination possible across all reductions of water potential examined in this study.

Weed seed banks are the primary source of annual weed infestation in most crop production systems and can be depleted through germination, predation, or death. In New Zealand *V. bromoides* will shed seed in mid-summer. Seeds typically remain dormant for up to three months but germinate readily following rainfall during March through to May (Dillon and Forcella 1984, FAR 2008). In Canterbury the average temperature and rainfall during this time is: March – 16°C and 45 mm; April – 13°C and 46 mm; and May – 10°C and 64 mm (NIWA 2018).

Our data show that germination rates are slower in cool and dry conditions (Figure 2). The amount of seed germinated over 100 days at 5°C was less than half that observed at 10°C. Unfortunately this is not enough to prevent hair grass establishment in winter crops. The soil seed bank is known to be short lived and only a small proportion, 1–7% may survive for three years (FAR 2008). Using the stale seed-bed approach, with good irrigation in early autumn, while temperatures are high and before the winter crop is sown, will be a quickly effective method to deplete the seed bank, given any remaining seed should be short lived. Our study showed that close to 100% *V. bromoides* will germinate between 20–25°C (Figure 1) especially under low hydric stress (Figure 2). Normal 2–3 month delays in seed ripening could reduce the effectiveness of this approach, and give farmers problems later in the season. Therefore monitoring of seed-shed (and preventing seeding) is important for the success of this management strategy.

Other authors suggest that ongoing issues with winter annual grass weeds are best managed by making (temporary) changes from reduced to conventional tillage systems, and by rotating crop species (Melander *et al.* 2013). Growing broad-leaf crops during summer could give farmers access to better chemical control options, and provide better crop to weed competition scenarios.

ACKNOWLEDGMENTS

This study was funded by The Foundation for Arable Research.

REFERENCES

Alvarado, V. and Bradford, K. (2002). A hydrothermal time model explains the cardinal temperatures for seed germination. *Plant, Cell and Environment* 25, 1061-9.

Ball, D.A., Frost, S.M., Fandrich, L. and Tarasoff, C. (2008). Biological attributes of rattail fescue (*Vulpia myuros*). *Weed Science* 56, 26-31.

Benech-Arnold, R.L., Sánchez, R.A., Forcella, F., Kruk, B.C. and Ghersa, C.M. (2000). Environmental control of dormancy in weed seed banks in soil. *Field Crops Research* 67, 105-22.

Bloomberg, M., Sedcole, J.R., Mason, E.G. and Buchan, G. (2009). Hydrothermal time germination models for radiata pine (*Pinus radiata* D. Don). *Seed Science Research* 19, 171-82.

Bloomberg, M. and Watt, M.S. (2008). Base water potential for germination of radiata pine and buddleia seeds. *Agronomy Society of New Zealand Special Publication No.* 13, 113-5.

Bradford, K.J. (2002). Applications of hydrothermal time to quantifying and modeling seed germination and dormancy. *Weed Science* 50, 248-60.

Cordeau, S., Wayman, S., Reibel, C., Strbik, F., Chauvel, B. and Guillemin, J.-P. (2018). Effects of drought on weed emergence and growth vary with the seed burial depth and presence of a cover crop. *Weed Biology and Management* 18, 12-25.

Dastgheib, F. and Poole, N. (2010). Seed biology of brome grass weeds (*Bromus diandrus* and *B. hordeaceus*) and effects of land management. *New Zealand Plant Protection* 63, 78-83.

Dillon, S.P. and Forcella, F. (1984). Germination, emergence, vegetative growth and flowering of two silvergrasses, *Vulpia bromoides* (L.) SF Gray and *V. myuros* (L.) CC Gmel. *Australian Journal of Botany* 32, 165-75.

FAR, Foundation for Arable Research (2008). Hairgarss (*Vulpia* spp.). What do we know? A review of the literature. https://www.far.org.nz/assets/files/uploads/X78_Vulpia.pdf (accessed 9 April 2018).

Forcella, F., Benech Arnold, R.L., Sanchez, R. and Ghersa, C.M. (2000). Modeling seedling emergence. *Field Crops Research* 67, 123-39.

García, A.L., Recasens, J., Forcella, F., Torra, J. and Royo-Esnal, A. (2013). Hydrothermal emergence model for ripgut brome (*Bromus diandrus*). *Weed Science* 61, 146-53.

Guillemin, J.-P., Gardarin, A., Granger, S., Reibel, C., Munier-Jolain, N. and Colbach, N. (2013). Assessing potential germination period of weeds with base temperatures and base water potentials. *Weed Research* 53, 76-87.

Gunnarsson, M., James, T., Chynoweth, R. and Rolston, M. (2017). An evaluation of the resistance of annual and perennial ryegrass to herbicides. *New Zealand Plant Protection* 70, 165-70.

Hardegree, S.P. and Emmerich, W.E. (1990). Effect of polyethylene glycol exclusion on the water potential of solution-saturated filter paper. *Plant Physiology* 92, 462-6.

ISTA. (2009) ISTA Handbook on Seeding Evalution, 3rd Edition, CH-8303. (ISTA Secretariat, Bassersdorf Switzerland)

ISTA. (2015). International Rules for Seed Testing. CH-8303. (ISTA Secretariat, Bassersdorf Switzerland).

Melander, B., Munier-Jolain, N., Charles, R., Wirth, J., Schwarz, J., Van Der Weide, R., Bonin, L., Jensen, P.K. and Kudsk, P. (2013). European perspectives on the adoption of nonchemical weed management in reduced-tillage systems for arable crops. *Weed Technology* 27, 231-40.

Monks, D.P., Sadat Asilan, K. and Moot, D.J. (2009). Cardinal temperatures and thermal time requirements for germination of annual and perennial temperate pasture species. *Agronomy New Zealand* 39, 95-110.

Moot, D., Scott, W., Roy, A. and Nicholls, A. (2000a). Base temperature and thermal time requirements for germination and emergence of temperate pasture species. *New Zealand Journal of Agricultural Research* 43, 15-25.

Moot, D.J., Scott, W.R., Roy, A.M. and Nicholls, A. C. (2000b). Base temperature and thermal time requirements for germination and emergence of temperate pasture species. *New Zealand Journal of Agricultural Research* 43, 15-25.

NIWA, National Institute of Water and Atmospheric Research (2018). Mean Monthly Rainfall (mm) 1981–2010.

Rasmussen, I.A. (2004). The effect of sowing date, stale seedbed, row width and mechanical weed control on weeds and yields of organic winter wheat. *Weed Research* 44, 12-20.

Scherner, A., Melander, B., Jensen, P.K., Kudsk, P. and Avila, L.A. (2017). Germination of winter annual grass weeds under a range of temperatures and water potentials. *Weed Science* 65, 468-78.

Fact or Fiction: Do glyphosate resistant cropping systems reduce crop or soil health?

Martin M. Williams II[1], Stephen O. Duke[2], Agnes M. Rimando[2], Jude E. Maul[3], Krishna N. Reddy[4] and James V. Cizdziel[5]
[1] USDA-ARS, Global Change and Photosynthesis Research, 1102 S. Goodwin Avenue, Urbana, IL, USA
[2] USDA-ARS, Natural Products Utilization Research, PO Box 8048, University, MS, USA
[3] USDA-ARS, Sustainable Agricultural Systems Laboratory, 10300 Baltimore Avenue, Beltsville, MD, USA
[4] USDA-ARS, Crop Production Systems Research, PO Box 350, Stoneville, MS, USA
[5] Department of Chemistry and Biochemistry, University of Mississippi, University, MS, USA
(martin.williams@ars.usda.gov)

Summary Over the past quarter century, glyphosate and transgenic glyphosate resistant (GR) cultivars have transformed weed management in major agronomic crops throughout much of the Americas. Bold, unsubstantiated claims have been made that 1) a glyphosate toxicity in GR crops enhances their vulnerability to fungal diseases following glyphosate application, 2) glyphosate causes deficiencies in cationic minerals in GR crops, and 3) glyphosate use on GR crops has a deleterious 'legacy' effect on the soil microbial community. These claims were tested at locations in the USDA-ARS network of research stations in the Eastern, Southern, and Midwestern regions of the United States. Incidence of Goss's wilt, a fungal disease of corn, was very low until the widespread adoption of GR corn. However, field studies in susceptible sweet corn found no relationships between the GR transgene and/or glyphosate application to Goss's wilt incidence. In soybean and corn, neither glyphosate nor the GR transgene affected the mineral content of leaves or seed or crop yield. In order to test the 'legacy' hypothesis, bulk and rhizosphere soils in corn and soybean roots were collected from plots with and without a 15 year history of glyphosate use. Although the 'legacy' effect of glyphosate application on the root-associated microbiome is difficult to disentangle from confounding differences in cropping systems among locations, we observed insignificant changes in root-associated potential fungal pathogens. Claims that glyphosate and the GR transgene are detrimental to crop and soil health appear to be categorical myths.

Keywords Amino acid, corn, *cp4 epsp*, glyphosate, glyphosate-resistant crops, mineral, rhizosphere, soybean.

Allelopathic interference of annual ryegrass (*Lolium rigidum*) on lucerne nodulation

H.M. Zubair[1,2], J.E. Pratley[1,2], G.A. Sandral[2,3], A.W. Humphries[4], S.G. Nielsen[5] and A. Price[2,3]

[1] School of Agricultural and Wine Sciences, Faculty of Science, Charles Sturt University, Wagga Wagga, New South Wales 2650, Australia
[2] Graham Centre for Agricultural Innovation (an alliance between Charles Sturt University and New South Wales Department of Primary Industries), Wagga Wagga, New South Wales 2650, Australia
[3] Department of Primary Industries, Wagga Wagga Agricultural Institute, New South Wales 2650, Australia
[4] South Australian Research and Development Institute, Waite Campus, Adelaide, South Australia 5001, Australia
[5] Quantitative Consulting Unit, Research Office, Charles Sturt University, Wagga Wagga, New South Wales 2650, Australia
(jpratley@csu.edu.au)

Summary Laboratory and glasshouse experiments were undertaken to evaluate the allelopathic effects of annual ryegrass (*Lolium rigidum* Gaudin, ARG) root exudates against growth and nodulation of lucerne genotypes. This paper reports the outcomes of the nodulation investigation. The impact of ARG on lucerne nodulation was evaluated using SARDI Five and Titan 9 lucerne cultivars (in both laboratory and glasshouse) at 6 rhizobium (*Sinorhizobium* spp.) concentrations (0, 0.6, 1.5×10, 1.5×10^2, 1.5×10^3, 1.5×10^4 colony forming units mL^{-1}). At the highest rhizobium concentration, a fivefold reduction in nodule number was observed in the laboratory when lucerne genotypes were grown with ARG. Increasing annual ryegrass density also reduced nodulation under glasshouse conditions. The variability identified within and between lucerne populations suggests that there is scope to reduce allelopathic interference of ARG on seedling lucerne growth and improve the compatibility of these species in pasture mixtures at establishment.

Keywords Allelopathy, annual ryegrass, inhibition, lucerne, nodulation.

INTRODUCTION

Plant communities evolve as a result of the interactions of individual plants with their neighbours. This will be influenced by characteristics such as relative competitiveness and chemical interference. The chemical interference, known as allelopathy, refers to a mechanism whereby a plant releases, into the surrounding environment, bioactive metabolites that negatively affect the germination, establishment and growth of other plants growing in the vicinity (Anaya 1999, Asaduzzaman *et al.* 2014b). This interaction occurs in both directions with the outcome depending on relative chemical potencies and plant tolerances (Moore *et al.* 2015). The effect is usually on root growth and, in legumes, it might be anticipated that nodulation might be compromised.

Leaf leachates and root exudates of some weeds have been shown to affect nodulation adversely in white clover (*Trifolium repens* L.), red kidney beans (*Phaseolus vulgaris* L.) and Korean lespedeza (*Lespedeza stipulacea* Maxim.) (Rice 1968, 1971, Weston and Putnam 1985). Annual ryegrass is the most common weed in temperate crop production in Australia and is a regular companion plant in temperate pastures. It is well reported to have allelopathic impacts on crops (e.g. Wu *et al.* 2000, Asadazzaman 2014a). However, there are no recent studies that have assessed the allelopathic potential from root exudates of annual ryegrass (ARG, *Lolium rigidum* Gaudin) on lucerne as a companion plant in a pasture sward.

The aim of this study was to evaluate the allelopathic capability of ARG against lucerne genotypes to investigate whether such inhibition might extend to root nodulation in the legume.

MATERIALS AND METHODS

Laboratory experiment A modification of Gibson's tube culture technique (Gibson 1963) was used for growing the lucerne plants. Agar (25 mL of 0.5% nutrient free water agar) was added to each glass tube (150×25 mm). The tubes were plugged with foam bungs. After autoclaving, the tubes were sloped so that the agar extended to the top of the tube. Pre-germinated seeds of ARG at the density of four seeds per tube were aseptically sown on the middle position of the agar surface in one row. The tubes were then kept in a growth incubator. After seven days, pre-germinated seeds of lucerne, cultivars SARDI Five and Titan 9, were introduced at a single seed per tube. Six different rhizobium concentrations (0, 0.6, 1.5 × 10, 1.5 × 10^2, 1.5 × 10^3, 1.5 × 10^4 colony forming units mL^{-1}) at 1 mL per tube were added to the agar surface of the tube which was then placed back in the growth chamber for seven days. 'The plate count method' was used for rhizobium enumeration (Vincent 1970). After 7 days of growing in concert, the ARG seedlings were removed and 2 mL of nutrient solution were included per glass tube and incubated for another four weeks in the growth incubator. Half strength Jensen's (1942) nutrient solution was used. Nodule numbers were counted after four weeks. The control treatment comprised growth of lucerne seedlings without the presence of ARG. The tubes were arranged in a randomised complete block design along with eight replications under controlled condition.

Glasshouse experiment Two lucerne genotypes, cultivars SARDI Five and Titan 9, were used as test species in this experiment based on previous findings. Pre-germinated seeds of annual ryegrass were sown in cylindrical PVC pots (90 mm diameter × 210 mm height). Each pot was filled with 1.6 kg (oven dry basis) brickie sand. The sand was pasteurised with steam. All pots were maintained at a water content of approximately 80% of field capacity by daily weighing and watering with deionised water. ARG densities used were 0, 5, 10 and 20 seedlings and there were four circles of these densities per pot. After the growth of ARG seedlings for a week, single pre-germinated seed of each of two lucerne genotypes were sown in the middle of the each circle. Pots were inoculated with six different rhizobium concentrations (0, 0.6, 1.5 × 10, 1.5 × 10^2, 1.5 × 10^3, 1.5 × 10^4 colony forming units mL^{-1}) at 1 mL per lucerne plant. After 7 days of growing in concert, the ARG seedlings were uprooted and N-free nutrient solution was applied. The growth of lucerne seedlings without ARG was the control. Plants were grown under natural lighting in the glasshouse from April to June in Wagga Wagga, New South Wales. Temperatures were regulated between 25°C (max) and 15°C (min). Three replicates were used and pots were arranged in a randomised block design. Lucerne plants were harvested four weeks after removal of ARG. Shoots were separated from roots. Roots were washed from the sand and the root data were assessed by scanning and image analysis using WinRhizo software (Regent Instruments, Montreal). Nodules were counted for accuracy using scan images. In addition, images of fully elongated root hairs were taken using a Nikon SMZ 25 Stereo microscope fitted with a Nikon DS-Fi2 camera. Ten images were captured per plant and root hair lengths of 10 root hairs per image were measured using ImageJ software (Rasband 1997–2014). Root and shoot dry mass were determined after drying at 70°C.

Data were subjected to analysis of variance using Genstat v17 (VSN International, Hemel Hempstead, United Kingdom). The model assumptions used were: i) the residuals are normally distributed; ii) the residuals are independent; iii) the residuals have a mean of zero; iv) the residuals have constant variance; v) the treatment groups are normally distributed; and vi) homogeneity of variances of the data within factor levels. These model assumptions were checked and met in every case. Where significant treatment differences were found, a Tukey's pairwise multiple comparison test (Tukey 1949) was used.

RESULTS

Laboratory experiment Root nodule numbers of SARDI Five and Titan 9 lucerne cultivars were reduced (P<0.01) by the root exudates of ARG relative to the control (Figure 1). Nodule numbers of Sardi 5 were reduced sixfold and Titan 9 eightfold relative to the controls due to the presence of ARG.

Glasshouse experiment The main effects of lucerne genotype, rhizobium concentration and their interactions on root hair length were not significant. However annual ryegrass density had a significant (P<0.05)

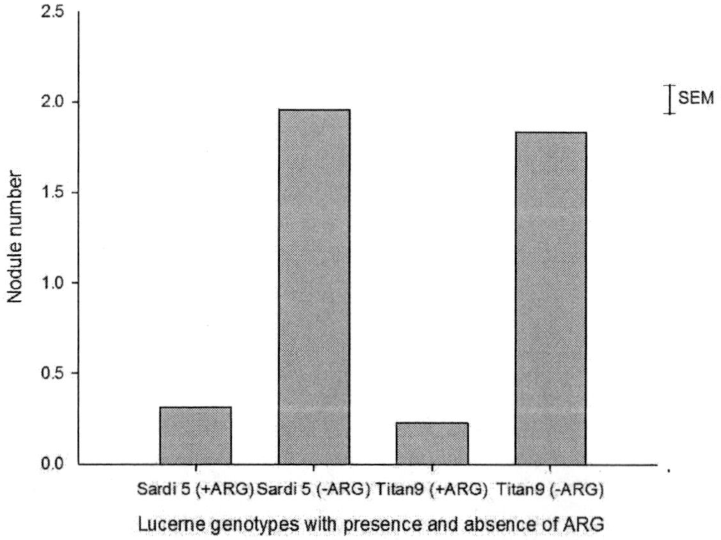

Figure 1. Impact of annual ryegrass presence of nodulation number on roots of lucerne grown in tube culture.

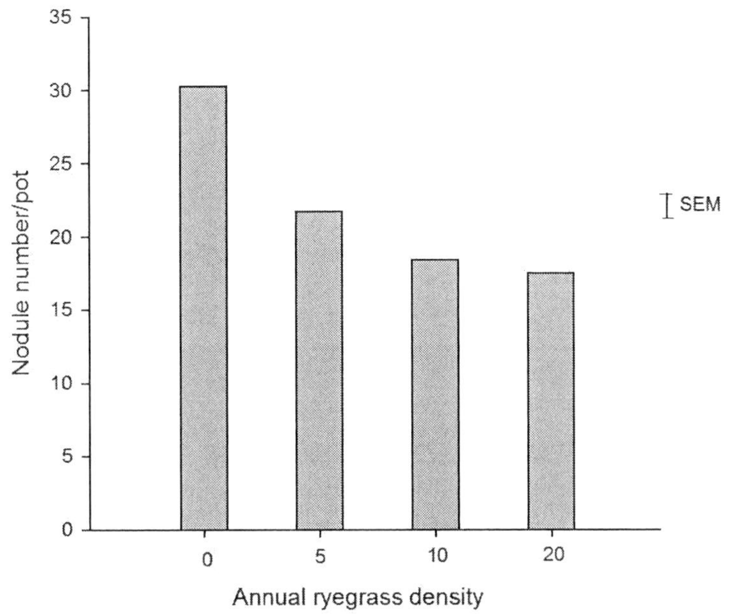

Figure 2. The effect of annual ryegrass seedling densities on lucerne nodule numbers per pot.

effect on lucerne root hair length (data not shown). The lowest root hair lengths recorded were 436 μm and 433 μm at the densities of 10 and 20 seedlings per beaker respectively compared with control (475 μm). ARG density significantly ($P<0.01$) reduced lucerne nodule number (Figure 2) compared with lucerne growing alone. At the highest ARG density, nodule density was approximately halved at 17 nodules per pot relative to 30 nodules per pot for the control.

DISCUSSION

Nodule development involves interaction between rhizobia and legume root hairs (Schultze and Kondorosi 1998, Limpens and Bisseling 2003). The development of a root nodule is under the control of nodulation (nod) genes (Eckardt 2006). Legume roots secrete flavonoids which perform as signals to activate the expression of nod genes resulting in the production of rhizobial nod factors that cause root cortical cells to breakdown and develop into nodules (Schultze and Kondorosi 1998, Spaink 2000). Both laboratory and glasshouse experiments have shown that root exudates from ARG affect the development of nodules in lucerne genotypes. The reduction in root hair length was around the order of 10% whereas nodule number reduction was nearly 50% due to the association with ARG allelochemicals, suggesting that root structures have been compromised. This finding has implications for productivity in the field where effective nodulation is a fundamental to legume-based pastures and pulse crops. Related work (H. Zubair, unpubl. data) has shown that there is variable tolerance in lucerne genotypes in their susceptibility to root growth inhibition due to ARG allelopathy. It may also be a factor in nodulation susceptibility as well, although genotype screening has not been done. This work also raises the question of species and cultivar compatibility in mixed pasture swards and the need to select companions which minimise the incompatibility.

ACKNOWLEDGMENTS

The authors are grateful to Charles Sturt University for the award of Charles Sturt University Postgraduate Research Scholarship. We wish to acknowledge PGG Wrightson Seeds, Seed Distributors seed companies and SARDI for the provision of lucerne variety seeds.

REFERENCES

Anaya, A.L. (1999). Allelopathy as a tool in the management of biotic resources in agroecosystems. *Critical Reviews in Plant Sciences* 18, 697-739.

Asaduzzaman, Md., An, M., Pratley, J.E., Luckett, D.J. and Lemerle, D. (2014a). Canola (*Brassica napus*) germplasm shows variable allelopathic effects against annual ryegrass (*Lolium rigidum*). *Plant and Soil* 380, 47-56.

Asaduzzaman, Md., Luckett, D.J, Cowley, R.B., An M., Pratley, J.E. and Lemerle, D. (2014b). Canola cultivar performance in weed-infested field plots confirms allelopathy ranking from in vitro testing. *Biocontrol Science and Technology* 24, 1394-411.

Eckardt, N.A. (2006). The role of flavonoids in root nodule development and auxin transport in *Medicago truncatula*. *Plant Cell* 18, 1539-40.

Gibson, A.H. (1963). Physical environment and symbiotic nitrogen fixation.1. Effect of root temperatures on recently nodulated *Trifolium subterraneum* L. plants. *Australian Journal of Biological Sciences* 16, 28-42.

Jensen, H.L. (1942). Nitrogen fixation in leguminous plants. I. General characters of root-nodule bacteria isolated from species of *Medicago* and *Trifolium* in Australia. *Proceedings of the Linnean Society of New South Wales* 66, 98-108.

Limpens, E. and Bisseling, T. (2003). Signalling in symbiosis. *Current Opinion in Plant Biology* 6, 343-50.

Moore, J.R., Asaduzzaman, Md. and Pratley, J.E. (2015). Dual direction allelopathy: the case of canola, wheat and annual ryegrass. Proceedings of 17th Australian Agronomy Conference, Hobart, eds T. Acuña, C. Moeller, D. Parsons and M. Harrison. http://agronomyaustraliaproceedings.org/images/sampledata/2015_Conference/pdf/agronomy2015final00328.pdf (accessed 1 August 2018).

Rasband, W.S. (1997–2014). ImageJ, U.S. National Institutes of Health, Bethesda, Maryland, USA, http://imagej.nih.gov/ij/. (accessed 1 August 2018).

Rice, E.L. (1968). Inhibition of nodulation of inoculated legumes by pioneer plant species from abandoned fields. *Bulletin of the Torrey Botanical Club* 95, 346-58.

Rice, E.L. (1971). Inhibition of nodulation of inoculated legumes by leaf leachates from pioneer plant species from abandoned fields. *American Journal of Botany* 58, 368-71.

Schultze M. and Kondorosi A. (1998). Regulation of symbiotic root nodule development. *Annual Review of Genetics* 32, 33-57.

Spaink, H.P. (2000). Root nodulation and infection factors produced by rhizobial bacteria. *Annual Review of Microbiology,* 54, 257-88.

Tukey, J. (1949). Comparing individual means in the analysis of variance. *Biometrics* 5, 99-114.

Vincent, J.M. (1970). 'A manual for the practical study of the root-nodule bacteria.' (Blackwell Scientific Publication, Oxford and Edinburgh).

Weston, L.A. and Putnam, A.R. (1985). Inhibition of growth, nodulation, and nitrogen fixation of legumes by quackgrass. *Crop Science* 25, 561-5.

Wu, H., Pratley, J., Lemerle, D. and Haig, T. (2000). Evaluation of seedling allelopathy in 453 wheat (*Triticum aestivum*) accessions against annual ryegrass (*Lolium rigidum*) by the equal-compartment-agar method. *Australian Journal of Agricultural Research* 51, 937-44.

Seed germination ecology of *Sesbania cannibana* (Retz.) Poir.

Nadeem Iqbal[1], Sudheesh Manalil[2], Bhagirath S. Chauhan[2] and Steve Adkins[1]
[1] School of Agriculture and Food Sciences (SAFS), The University of Queensland, Gatton 4343, Queensland, Australia
[2] Queensland Alliance for Agriculture and Food Innovation (QAAFI), The University of Queensland, Gatton 4343, Queensland, Australia
(nadeem.iqbal@uq.net.au)

Summary Sesbania pea (*S. cannibana* (Retz.) Poir.), also known as yellow pea bush, is present in abundance along water channels and other wet areas in central and northern coastal regions of Australia. In cotton (*Gossypium hirsutum* L.), this weed is highly competitive and dramatically reduces cotton yield. It also creates a problem at harvest time due to the development of dense populations and its robust stature. Up till now, little information has been available concerning its seed production and germination ecology. Therefore, experiments were conducted in the laboratory and greenhouse to evaluate the impact of different environmental factors on its seed germination. Two populations of sesbania pea were collected from Saint George and the Dalby regions of south-east Queensland based on their differential rainfall patterns. Saint George is the low rainfall area while Dalby is the medium rainfall area. To overcome physical dormancy in the freshly harvested seed a hot water treatment (100°C for a period of 5 minutes) was used to promote seed germination. Highest germination was achieved under a temperature range of 30/20°C, and 35/25°C (alternating day/night temperature) for both populations. Germination was not affected by pH in the range of 7 to 10, however, below 6 there was poor germination. Sesbania pea showed salt tolerance of up to 50 mM NaCl but beyond that, germination was severely affected. In the greenhouse, both populations showed maximum germination at 1 cm burial depth and it decreased gradually with increasing burial depth up till 8 cm, and no seedling emerged more than 16 cm.

Keywords Seed germination, temperature, light, burial depth, salt stress.

Developing a framework to prioritize plant biosecurity threats to New Zealand's dairy industry

David Hodges[1], Trevor James[2] and Petra Muellner[3]
[1] DairyNZ Ltd, Private Bag 3221, Hamilton 3240, New Zealand.
[2] AgResearch Ltd, Ruakura Agricultural Centre, Hamilton 3214, New Zealand.
[3] Epi-interactive Ltd, 127 Park Road, Miramar 6022, Wellington, New Zealand

Summary Recent incursions of exotic organisms into New Zealand continues to demonstrate the need for robust identification and risk assessment of potentially hazardous exotic organisms to help prioritise efforts to prevent and prepare for incursions. DairyNZ Ltd has been working with experts within New Zealand Ministry for Primary Industries, AgResearch Ltd., and Epi-interactive Ltd to develop a risk evaluation framework.

DBRiEF (the Dairy Biosecurity Risk Evaluation Framework) has been developed to rank the risk of pastoral weeds, insect pests, and animal disease. This presentation will focus on the areas of DBRiEF relevant to weeds, describe the processes incorporated into the framework, progress made to date at ranking hazardous exotic weeds, and the challenges faced and overcome by the DBRiEF project team.

Keywords Risk assessment, ranking, prioritisation, biosecurity, weeds.

Exploring and extending integrated weed management opportunities in the Australian vegetable industry

Michael Coleman, Christine Fyfe, Sita Tiwari Pokhrel, Graham Marshall, Brian Sindel and Paul Kristiansen
School of Environmental and Rural Science, Faculty of Science, Agriculture, Business and Law,
University of New England, Armidale, New South Wales 2351, Australia
(mcolema8@une.edu.au; cfyfe3@une.edu.au)

Summary Weeds are a persistent problem for vegetable producers because of the favourable growing conditions, regular soil disturbance and the lack of registered herbicides available to selectively control broadleaf weeds. The potential for weed growth within and between rows of vegetable crops is therefore high. Since 2011, we have been conducting research for Horticulture Innovation Australia to identify the key issues faced by vegetable producers in all the main growing regions of Australia with respect to weed impact and management, including priority weed species and their impact, and innovative approaches that warrant further exploration. This initial work identified a series of research priorities: understanding the ecology of high priority weeds within the industry, including their germination and early growth, and response to management strategies; quantifying weed seed banks on vegetable farms and linking them to farming practices; evaluating the weed management effectiveness of supplementary cultural methods (with a focus on biofumigant and green manure cover crops, and various hand weeding implements); and conducting economic analyses of weed management costs and benefits. These research priorities are the subject of an ongoing project that will address some gaps in the industry's understanding of weed impact and management, and contribute to the first comprehensive integrated weed management manual for the industry. We discuss these research priorities in more detail, including our progress regarding a national baseline seed bank assessment, establishment of cover crop trial sites, and planned extension activities.

Keywords Horticulture, extension, seedbanks, hand weeding, cover crops.

INTRODUCTION

Weeds are a persistent problem for vegetable producers because of the favourable growing conditions, regular soil disturbance and the lack of registered herbicides available to selectively control broadleaf weeds. Given these factors, weeds can have a significant impact on crop profitability, yield, quality and management practices in the Australian vegetable industry, and crop losses can be high if effective weed control is not implemented (Blaesing 2013, Charles 2013, Coleman *et al.* 2015, Henderson and Bishop 2000). There have been substantial advances in sustainable weed management in cotton and broadacre grains production in Australia in the last 20 years (e.g. Charles 2013, McGillion and Storrie 2006). However, despite some limited earlier studies in the vegetable industry (e.g. Henderson 2000), little recent attention has been paid to developing such comprehensive weed control strategies in vegetables.

A recent scoping study (Kristiansen *et al.* 2014) identified the most common weeds in vegetable crops, as well as current industry weed control strategies and research priorities. Based on these priorities, the current project seeks to address the following research objectives:

- understanding the ecology of high priority weeds within the industry, their germination and early growth, and response to management strategies;
- quantifying weed seed banks on vegetable farms and linking them to farming practices;
- evaluating the weed management effectiveness of supplementary cultural methods, including biofumigant and green manure cover crops, and various hand weeding implements; and
- conducting an economic analysis of weed management costs and benefits.

This project commenced in October 2016, and is due for completion in July 2020. Activities relating to each of the project objectives are underway, and in this paper we discuss our progress with regard to each project objective, planned research and extension activities, and some preliminary results from our baseline study of the weed seed bank on Australian vegetable farms.

MATERIALS AND METHODS

Priority weed species Based on extensive nationwide consultation with vegetable producers and industry experts (on-ground agronomy, research and extension), Kristiansen *et al.* (2014) sought to determine the most important weed species on farms and across diverse vegetable growing regions. Farmer interviews and focus group discussions were conducted in the Lockyer Valley (Queensland, Qld),

Richmond (New South Wales, NSW), Werribee South (Victoria, Vic), Hobart and Devonport (Tasmania, Tas), Adelaide Hills and Virginia (South Australia, SA), Gingin (Western Australia, WA), and Darwin and Katherine (Northern Territory, NT). The interviews and discussions collected information about the major weeds encountered.

From this work, based on the frequency of reporting, as well as the perceived seriousness of their impact on vegetable production, Kristiansen *et al.* (2014) identified a list of high priority weeds for the Australian vegetable industry. These have been taken to be a particular target of research and extension activity over the course of the current project.

Weed seed bank assessment An assessment is underway of the baseline weed seed bank on sample farms in key growing regions across Australia, including the Lockyer Valley (Qld), Sydney Basin (NSW), Cranbourne and East Gippsland (Vic), Hobart and Devonport (Tas), Adelaide Hills and Virginia (SA), Gingin and Myalup (WA), and Darwin (NT).

A review of the global literature was completed to identify suitable methods of weed seed bank assessment in vegetable production. The review focused on paddock sampling approaches, coring depth, subdivision of cores to assess vertical distribution of seed, and sample representativeness (e.g. Buhler *et al.* 2001, Kashe *et al.* 2010, Wortman *et al.* 2010). The sampling method selected following this review involved taking 10 × 50 mm soil cores to a depth of 20 cm, systematically using a single diagonal transect spanning the length and width of the selected paddock. The cores were stratified at depths of 0–5 cm, 5–10 cm and 10–20 cm to assess vertical distribution of the weed seed bank. Five farm sites were sampled in each vegetable growing region (six in Western Australia). For each farm, soil samples from each depth increment were combined to give one bulked sample per depth. Above-ground weed biomass was also recorded within 10 × 0.5 m^2 quadrats using the same systematic paddock sampling approach as for the soil cores. In addition to biomass, data on number of weeds by species were recorded to provide a picture of the above-ground weed load within different crops.

The weed seed enumeration approach (e.g. Hartley and Rahman 1995) was selected to quantify the weed seed bank, involving soil seed germination and counting rather than more complex seed separation techniques (Auškalnienė and Auškalnis 2009, Widderick *et al.* 2010) because of the number of farms sampled and likely weeds present. For each farm, bulked soil samples from each depth increment were mixed, and three replicate samples of approximately 400 g were spread in plastic trays to a depth of 2 cm over a 2 cm layer of vermiculite, with a gauze layer in between. Trays were kept in a glasshouse with a controlled 15°C to 25°C diel temperature cycle. As seedlings emerged over a period of 2–4 months, they were removed and counted by species for each tray. Several dormancy-breaking activities (soil stirring, drying and refrigeration; e.g. Sjursen 2001, Swanton *et al.* 2000) were applied consistently to all trays to maximise the germination of all viable seeds.

Paddock management history data over the two years preceding soil sampling were collected to allow the weed load, species present and depth distribution to potentially be linked to particular management practices or vegetable crops. The management data will be utilised in detailed analysis once the seed bank assessment is complete.

Cultural weed management evaluation Cultural methods of weed control in vegetable production which will be assessed include green manure/biofumigant cover crops, crop rotation, irrigation, crop orientation and hand weeding.

Cover crops have been shown to have some benefit in reducing the weed burden in vegetable production systems, either through suppression and competition for resources (e.g. Bond and Grundy 2001), or through their biofumigant effects which may inhibit seed germination and weed growth (e.g. Al-Khatib *et al.* 1997, Alcántara *et al.* 2011). Replicated field trials have commenced to evaluate cover crop effects on weed seed banks and weed growth, with trials expected to take in two growing seasons of either winter or summer cover cropping. Each site is monitored fortnightly to measure above-ground weed species and biomass, the weed seed bank, and cover crop biomass, to establish the relative performance of the cover crop varieties in suppressing weeds.

A review of the literature was completed to evaluate the suitability of cover crop varieties considered for field trials. This included an assessment of their growing season, climatic and soil-type suitability, use of cover crops in Australia, and their pros and cons from cover crop management and weed control perspectives.

A preliminary literature review of available hand weeding implements has also been completed. Field trials to determine the efficiency and user-comfort of different hand weeding implements and approaches were in the early stages of planning at the time of writing.

Economic analysis A review of the literature suggested that the 'partial budgeting' approach is the most suitable economic evaluation technique to determine the economic impact of weeds and innovative weed control practices in a cross-section of Australian vegetable production enterprises (e.g. Cho *et al.* 2012, Roth and Hyde 2002). An interview schedule to collect these data from farmers was prepared and piloted. In it we ask farmer interviewees to focus on a single vegetable-growing 'enterprise' within their farm (for example, a particular crop grown in a particular way). The data collected will focus on: the impact of weeds and weed control on the profitability of the enterprise; the profitability of recently implemented/trialled or otherwise innovative weed control practices; and whether the decision to implement innovative weed control practices and their subsequent profitability depended on the weed control efforts of neighbours. At the time of writing the economic analysis had not yet commenced. The results will be published and appear in industry extension materials.

RESULTS AND DISCUSSION

Priority weed species Of all weeds of significance to the vegetable industry identified by Kristiansen *et al.* (2014), approximately 50% were short-lived (annual and/or biennial) broadleaf species. While grasses were commonly found on vegetable farms, they were less likely to be considered a problem from the point of view of vegetable production due to the availability of selective herbicide options. All identifiable weed species will be accounted for in our research, however the following priority weed species have been identified from this earlier research as a focus of our research and extension efforts to improve understanding of their ecology, impact and management in the Australian vegetable industry.

- Fat hen (*Chenopodium album* L.)
- Dwarf nettle (*Urtica urens* L.)
- Mallow (*Malva parviflora* L.)
- Nutgrass (*Cyperus rotundus* L.)
- Pigweed (*Portulaca oleracea* L.)
- Amaranth (*Amaranthus* spp., with a focus on *A. retroflexus* L. and *A. viridis* L.)
- Wild radish (*Raphanus raphanistrum* L.)
- Blackberry nightshade (*Solanum nigrum* L.)
- Milk thistle (*Sonchus oleraceus* L.)
- Chickweed (*Stellaria media* (L.) Vill.)

For each of these weed species, we have completed a detailed review of the literature focusing on: identification of each weed; their ecological characteristics (including form/s of reproduction, required conditions for germination, seed viability and production, preferred soil types, competitive advantages and distribution throughout Australia); their specific impacts in vegetable production systems; and the effectiveness of various management methods. Methods used by vegetable growers include tillage, herbicides, hand weeding, and other cultural practices that may be relevant in some situations such as cover cropping, higher plant densities, stale and false seedbeds, mowing, thermal control, and mulching.

With the exception of the sedge *C. rotundus*, each of the priority weeds for the Australian vegetable industry is a heavy-seeding annual or perennial broadleaf species that can thrive in vegetable cropping systems if not managed diligently. Common impacts of these weeds were reported by farmers to include their capacity to compete with crop plants for resources due to early emergence and rapid growth, interference with harvesting activities and contamination of produce depending on crop (particularly relevant in leafy vegetables and brassicas), and acting as a host for crop viruses, pest insects and diseases.

Weed seed bank assessment At the time of writing, weed seed bank assessment had been completed for soil samples collected from all key vegetable growing regions visited. Preliminary exploration of the data suggests that burial depth in the soil has a relationship with the quantity of viable seeds observed on vegetable farms. In some cases (e.g. *Eleusine indica* (L.) Gaertn., *Lolium* spp.) the vast majority of viable seed was found closer to the soil surface (0–5 cm), while in others (e.g. *Polygonum aviculare* L., *Phalaris* spp.) the largest portion of viable seed was found deeper in the soil profile at 10–20 cm. The data also show inconsistent findings between regions, for example *C. album* appeared more likely to be present as viable seed nearer the soil surface in SA, but more likely at depth in WA. These findings are expected to reflect different paddock management histories on the farms sampled, though they may also be a strategy of the survivability of viable seeds of various weed species at different depths in the soil. With highly disturbed soils in vegetable production systems, long-lived seeds may equally be found at these various depths. Detailed analysis of the data will be completed at a later stage of the project to explore the relationship between viability at different depths, the characteristics of the weeds and their seed, and paddock management history.

Cultural weed management evaluation Four replicated cover crop trial sites were established in 2017/2018 to evaluate the effectiveness of various cover crop species in suppressing weed germination and growth. Three of these sites, Gatton Qld, Peats Ridge NSW and Forthside Tas, are being managed by a partner HIA-funded project ('VG16068 Optimising

Cover Cropping for the Australian Vegetable Industry'). The fourth in Myalup WA, is being managed by our project team. The cover crop varieties under consideration at each site are as follows:

- Gatton, Qld, summer cover crop trial: sorghum (*Sorghum bicolor* (L.) Moench), Caliente (*Brassica juncea* (L.) Czern.), Nemat (*Eruca sativa* Mill.), BQ Mulch (a mix of *Brassica nigra* (L.) Kohl ex Koch, *B. abyssinica* Brummitt and *B. carinata* A.Braun), tillage radish (*Raphanus sativus* L.), bare fallow control.
- Peats Ridge, NSW, summer cover crop trial: millet (*Echinochloa esculenta* (A.Braun) H.Scholz), sunn hemp (*Crotalaria juncea* L.), buckwheat (*Fagopyrum esculentum* Moench), ryegrass (*Lolium* spp.), millet/cowpea/ buckwheat mix, bare fallow control.
- Forthside, Tas, winter cover crop trial: Caliente (*B. juncea*), ryegrass (*Lolium* spp.), bare fallow control.
- Myalup, WA, winter cover crop trial: cereal rye (*Secale cereal* L.), Italian ryegrass (*Lolium multiflorum* Lam.), Caliente (*B. juncea*), BQ Mulch, field pea (*Pisum sativum* L.).

Industry extension The research completed throughout the course of this project will result in the first Integrated Weed Management (IWM) Manual for the Australian vegetable industry, due for publication in 2020. Reviews of literature and industry consultation have contributed to development of a draft structure for the IWM Manual. The Manual will provide background on the weed management problem within vegetable crops, details on the effectiveness of the various weed management methods available to vegetable growers and how to implement these effectively in different crops, and ways to bring these techniques together and implement IWM on the farm.

Additionally, weed management guides for approximately 10 priority weed species of vegetable production will be produced and included with the IWM Manual. These coincide with the priority species listed above, though there is potential to add other commonly encountered species such as potato weed (*Galinsoga parviflora* Cav.). At the time of writing, draft weed management guides for *C. album* and *U. urens* have been produced and provided to the industry for comment.

ACKNOWLEDGMENTS

The authors gratefully acknowledge Horticulture Innovation Australia for funding this research under project VG15070 – A strategic approach to weed management for the Australian vegetable industry. We are very grateful to all farmers, farm advisors and industry extension and support staff who have contributed to our research, particularly those farmers who are participating in the on-farm trials, and who have invited us on to their farm to collect soil and plant samples.

REFERENCES

Al-Khatib, K., Libbey, C. and Boydston, R. (1997). Weed suppression with *Brassica* green manure crops in green pea. *Weed Science* 45, 439-45.

Alcántara, C., Pujadas, A. and Saavedra, M. (2011). Management of *Sinapis alba* subsp. *mairei* winter cover crop residues for summer weed control in southern Spain. *Crop Protection* 30, 1239-44.

Auškalnienė, O. and Auškalnis, A. (2009). The influence of tillage system on diversities of soil weed seed bank. *Agronomy Research* 7, 156-61.

Blaesing, D. (2013). Plant Health Desk Top Study. VG12048. Horticulture Australia, Sydney.

Bond, W. and Grundy, A.C. (2001). Non-chemical weed management in organic farming systems. *Weed Research* 41, 383-405.

Buhler, D.D., Kohler, K.A. and Thompson, R.L. (2001). Weed seed bank dynamics during a five-year crop rotation. *Weed Technology* 15, 170-6.

Charles, G. (2013). Integrated Weed Management (IWM) Guidelines for Australian Cotton Production. In 'WEEDpak: a guide for integrated management of weeds in cotton.' (Cotton CRC, Narrabri).

Cho, A.H., Hodges, A.W. and Chase, C.A. (2012). Partial budget analysis of summer fallows for organic nutrient and weed management in Florida. *HortTechnology* 22, 258-62.

Coleman, M.J., Sindel, B.M., Kristiansen, P. and Henderson, C.W.L. (2015). Survey of weed impact, management, and research priorities in Australian cucurbit production. *Plant Protection Quarterly* 30, 12-20.

Hartley, M.J. and Rahman, A. (1995). Optimising conditions for weed seed emergence from soil trays in a glasshouse. *Proceedings of the New Zealand Plant Protection Conference* 48, 181-5.

Henderson, C. (2000). 'Integrated Weed Management, Cover Crop, Mulch and Rotation Management in Vegetable Production Systems.' (Horticultural Research and Development Corporation, Sydney)

Henderson, C.W.L. and Bishop, A.C. (2000). Vegetable weed management systems. *In* 'Australian weed management systems.' ed. B Sindel, pp. 355-72. (R.G. & F.J. Richardson, Melbourne).

Kashe, K., Sindel, B., Kristiansen, P. and Jessop, R. (2010). Effect of tillage on weed seed bank and weed flora in maize (*Zea mays*). Proceedings of the 17th Australasian Weeds Conference, ed. S.M. Zydenbos, pp. 67-70. (New Zealand Plant Protection Society, Christchurch).

Kristiansen, P., Coleman, M., Fyfe, C. and Sindel, B. (2014). Weed Management for the Vegetable Industry – Scoping Study. Horticulture Innovation Australia, Sydney.

McGillion, T. and Storrie, A. (2006). 'Integrated Weed Management in Australian cropping systems.' (CRC for Weed Management, Adelaide).

Roth, S. and Hyde, J. (2002). Partial budgeting for agricultural businesses. http://agmarketing.extension.psu.edu/Business/PDFs/PartlBudgetAgBus.pdf (accessed 1 August 2018).

Sjursen, H. (2001) .Change of the weed seed bank during the first complete six-course crop rotation after conversion from conventional to organic farming. *Biological Agriculture & Horticulture* 19, 71-90.

Swanton, C.J., Shrestha, A., Knezevic, S.Z., Roy, R.C. and Ball-Coelho, B.R. (2000). Influence of tillage type on vertical weed seedbank distribution in a sandy soil. *Canadian Journal of Plant Science* 80, 455-7.

Widderick, M.J., Walker, S.R., Sindel, B.M. and Bell, K.L. (2010). Germination, emergence, and persistence of *Sonchus oleraceus*, a major crop weed in subtropical Australia. *Weed Biology and Management* 10, 102-12.

Wortman, S.E., Lindquist, J.L., Haar, M.J. and Francis, C.A. (2010). Increased weed diversity, density and above-ground biomass in long-term organic crop rotations. *Renewable Agriculture and Food Systems* 25, 281-95.

Developing new options for weed control in Australian cotton systems

James Neilsen, Meredith Conaty and Kate Lang
Monsanto Australia, Level 12, 600 St Kilda Road, Melbourne, Victoria 3004, Australia

Summary Glyphosate plays a critical role for Australian cotton growers through its use in the Roundup Ready® Flex crop system. A 2017 survey of 180 Australian cotton growers and crop consultants showed that while almost 90% of cotton growers rated their weed control with glyphosate as good or excellent, 74% of cotton growers have observed glyphosate resistant weeds on their farms, with the resistance appearing both in crop and on other parts of the farm. In addition, 2 of the top 5 weeds growers are looking to control are not listed on the Roundup Ready® Herbicide with PLANTSHIELD® label, despite growers and consultants believing they should achieve control with glyphosate. Increasing rates of glyphosate resistance and the prevalence of difficult-to-control weeds means that growers must diversify their weed control system and ensure they are rotating chemistries. The Xtend® Crop System Monsanto's new weed control tool that will help growers manage glyphosate resistant and difficult-to-control weeds in their cotton crop and will play a role in on-farm weed management through the ability to use multiple in-crop modes of action. This will provide better management of their current weed situation as well as protecting glyphosate into the future.

Keywords Cotton, dicamba, XtendFlex, glyphosate resistance.

An integrated approach to managing Noogoora burr in maize crops

Shane R. Hona[1], Trevor K. James[2] and Andrew Blayney[3]
[1] Bay of Plenty Regional Council, PO Box 364, Whakatane 3158, New Zealand
[2] AgResearch, Private Bag 3123, Waikato Mail Centre, Hamilton 3240, New Zealand
[3] Boffa Miskell Ltd, PO Box 1094, Hamilton 3240, New Zealand
(shane.hona@boprc.govt.nz)

Summary This paper summarises the integrated approach taken to control Noogoora burr in a large maize crop. The ecology of both the weed and the crop were studied in detail to determine which changes to the cropping regime could be made to reduce the number of Noogoora burr plants, with the long term goal of eradication.

Keywords Herbicide trial, ecology, life cycle.

INTRODUCTION

Noogoora burr (*Xanthium strumarium* L.) is a pest plant of relatively limited distribution in New Zealand, but is a significant pest in Australia where it has impacts on cropping and wool production areas (Government of South Australia n.d.). It is classified as an eradication pest in the Bay of Plenty Region of New Zealand, where the goal is immediate control leading to eradication (Bay of Plenty Regional Council 2011).

The largest Noogoora burr site in the Bay of Plenty Region an area of arable land approximately 100 hectares used for maize cropping. In the 2012–2013 growing season, 33,450 Noogoora burr plants were found and hand-pulled by manual surveillance teams surveying the whole site, row by row. This was both a time-consuming and expensive exercise, and has not halted the spread of this pest plant. A change of approach was required if the number of Noogoora burr plants at this site were to be reduced and achieve the stated goal of eradication.

This paper outlines the integrated approach taken to better manage Noogoora burr at this site, and summarises unpublished research trials completed by AgResearch for the Bay of Plenty Regional Council. To support the efficacy of the new maize crop regime at this site, additional data was collected and analysed by the Bay of Plenty Regional Council.

The ecology and lifecycle of both Noogoora burr and the maize regime at this site were studied to determine whether changes could be made to aid the control of this pest plant. The aim was to achieve better pest management by concentrating on the system as a whole, rather than the pest plant in isolation. To do this, knowledge gaps would have to be filled with a mix of observational and empirical research.

It was observed that the long-rotation maize-for-grain regime at this site allowed two generations of Noogoora burr to grow and set seed each year. In addition, it was also noted that the greatest number of Noogoora burr plants were emerging in the small furrow caused by the side cutting of urea fertiliser into the soil (possibly due to breaking the residual pre-emergence herbicide layer in the soil). A pre- and post-emergence spray regime was in use at this site but it was not effectively controlling Noogoora burr.

Deep cultivation can be an effective mechanical method of weed control for many species (Pennsylvania State University 2004) and could possibly be utilised at this site. However the maximum depth of Noogoora burr emergence was not known.

To aid control of Noogoora burr at this site, the maize regime was switched from a long-rotation maize-for-grain regime to a short-rotation maize-for-silage regime. This had the benefit of harvesting before the second generation of Noogoora burr could germinate, and allowed for the sowing of a ryegrass or a summer oats crop for the remainder of the growing season. Any Noogoora burr that germinated in this second crop could then be effectively controlled by the application of a broadleaf-specific herbicide. In addition, the fertiliser application method was changed from side cutting to surface broadcasting, to prevent the cutting of furrows that appeared to be aiding Noogoora burr germination. Rigorous machinery hygiene practices were adopted, to prevent movement of seed from the site via contaminated soil. These practices included keeping the seed planting machine on site longer to allow full cleaning, and harvesting the maize with a site-based tractor and trailer, followed by a transfer of maize to trucks in a designated load-out area adjacent to the maize crop.

MATERIALS AND METHODS

Herbicide trial A field trial was conducted to evaluate several post-emergence herbicide options (Tables 1 and 2) for their efficacy on Noogoora burr. Plot sizes were 3 m wide by 12 m long, containing four rows of maize, set up in a randomised block design with four replicates. The treatments included an untreated experimental control, current selected post-emergence herbicides, as well as a new herbicide from BASF and an experimental product from Bayer. Two post-emergence treatments of herbicide were applied to the plots at week two and week five, respectively. Counts of Noogoora burr plants in each plot were conducted pre-treatment, and three times post-treatment. Visual assessments of overall weed control were also undertaken twice.

Depth of emergence trial The aim of this trial was to determine whether burying Noogoora burr seeds to a depth where they could not germinate would be a viable mechanical control method at this site. Fresh Noogoora burr seeds (burrs) were collected from this site, and scarified to aid germination. Ten scarified burrs were placed in columns at various depths between 0–100 mm to evaluate depth of emergence. Each depth was replicated four times. The columns were regularly checked for Noogoora burr emergence, and once recorded the germinated plants were cut off 10 mm above ground so that further emergence was not inhibited.

Noogoora burr density pre- and post-regime-change Point data was collected for two growing seasons pre regime-change for the locations of all Noogoora burr plants found and removed from the site using handheld Global Positioning System (GPS) units.

To monitor the efficacy of the new regime at this site, eight transects were set up in January 2015, through field areas where the highest density of Noogoora burr was found in previous years. Noogoora burr plants were counted several times during the growing season between 2015 and 2016 in one row either side of each transect. As the length of each transect was dictated by paddock size, transects varied in length from 200–276 m. In order to standardise the data from each transect, the number of plants per 100 m was calculated.

Table 2. Herbicide products/active ingredients used (see Table 1).

Product	Active ingredient	Source
Adengo™	isoxaflutole/ thiencarbazone-methyl/ cyprosulfamide (experimental product)	Bayer
Arietta™	topramezone	BASF
Astound Ultra™	nicosulfuron	Orion
Banvel™	dicamba	Orion
Callisto™	mesotrione	Orion
Gesaprim™	atrazine	Orion
Roustabout™	acetochlor	Nufarm

From Rahman *et al.* (2014), reprinted with permission.

Table 1. Post-emergence treatments for a herbicide trial to control Noogoora burr.

Tr. #	Post-Em 1	Rate (mL ha^{-1})	Post-Em 2	Rate (mL ha^{-1})
1	Control	–	Control	–
2	Banvel™	1500	–	–
3	Astound Ultra™	1500	Astound Ultra™	1500
4	Astound Ultra™ + Banvel™	1500 + 1500	–	–
5	Callisto™ + Gesaprim™	200 + 1000	Callisto™ + Gesaprim™	200 + 1000
6	Callisto™ + Gesaprim™ + Banvel™	200 + 1000 + 1500	–	–
7	Arietta™ + Gesaprim™ + Hasten™	200 + 1000 + 0.5%	Arietta™ + Gesaprim™ + Hasten™	200 + 1000 + 0.5%
8	Arietta™ + Gesaprim™ + Hasten™ + Banvel™	200 + 1000 + 0.5% + 1500	–	–
9	Adengo™	440	Adengo™	440
10	Adengo™ + Banvel™	440 + 1500	–	–
11	Gesaprim™	3000	Gesaprim™ + Synoil™	3000 + 1%
12	Gesaprim™ + Banvel™	3000 + 1500	–	–
13	Atsound Ultra™ + Callisto™	1500 + 200	Atsound Ultra™ + Callisto™	1500 + 200
14	Atsound Ultra™ + Callisto™ + Banvel™	1500 + 200 + 1500	–	–

From Rahman *et al.* (2014), reprinted with permission.

To compare the pre and post regime-change data, a geographical information system analysis (using ArcGIS™) of the point data was collected in the two growing seasons pre regime-change period. The transect lines were overlaid on this point data and a buffer of 2 m was applied to simulate the area searched during the transect surveys, post-regime-change. All points that fell within the 2 m buffer of each transect were counted, then the data was standardised to plants per 100 m. (Note that due to the eradication status of Noogoora burr at this site, an untreated area could not be left to use as an experimental control to compare the new management regime with the old regime).

RESULTS AND DISCUSSION

Herbicide trial Eight of the treatments controlled >80% of Noogoora burr plants after the first post-emergence treatment. Treatment 3 (Astound Ultra™) was the least effective. Banvel™ was also added to each product, as well as the product being used without, as it was expected to be especially effective on Noogoora burr, however control only improved with its addition to Astound™ (Table 3).

The second herbicide treatment was applied to the treatments that did not receive Banvel™ in the first application. Contrary to the results observed with the first application, the second application of Astound alone (Treatment 3), or in combination with Callisto (Treatment 13), was quite effective. Noogoora burr was also effectively controlled by the second application of the other treatments (5, 7, 9 and 11).

The other main weeds present at this site were assessed 10 days and 34 days after the second herbicide application (Table 4). Five of the treatments provided

Table 4. Weed Control scores (0 = no control; 100 = complete control) for weeds other than Noogoora burr 10 and 34 days after the second application: Post-Em 2 (Scores are the average of 4 replicates).

Treatment	16/12/2013	09/01/2014
1	0	0
2	71	50
3	90	96
4	83	75
5	99	99
6	93	86
7	100	99
8	91	85
9	87	93
10	88	87
11	71	48
12	72	64
13	98	99
14	87	86
LSD (P<0.05)	6.6	10.1

From Rahman *et al.* (2014), reprinted with permission.

Table 3. The number (sum of 4 replicates) of Noogoora burr plants per plot after treatments with various herbicides (Table 1).

Treatment	14/11/2013	04/12/2013	% reduction	16/12/2013	09/01/2014	% reduction[1]
1	93	101	–	109	120	–
2	9	1	89	1	0	100
3	40	37	9	27	3	92
4	6	1	83	6	4	33
5	13	2	85	0	0	100
6	17	1	94	8	1	94
7	8	0	100	0	0	100
8	7	4	43	2	5	29
9	27	4	85	0	0	100
10	236	25[2]	89	36	12	95
11	119	8[2]	93	3	0	100
12	4	2	50	0	1	75
13	26	9	65	3	1	96
14	1	3	0	0	0	100

[1] % reduction from the original population after Post-Em 2 application.
[2] Mostly seedlings appearing in the urea furrow.
From Rahman *et al.* (2014), reprinted with permission.

excellent efficacy (scores of 93–99), while five provided medium efficacy (scores of 75–87); and three provided low/poor efficacy.

Based on these results, both Banvel™ and Gesaprim™ provided excellent control of Noogoora burr alone and in combination with other products. Callisto™ (with Gesaprim™) and Adengo™ provided good control of Noogoora burr alone as well as in combination with Banvel™. These treatments also provided excellent control of all other weeds.

Depth of emergence trial Seedlings emerged from all depths except for 100 mm deep (Figure 1). More, rapid seedling emergence occurred from shallower depths. At depths of 30 mm or less, Noogoora burr plants started emerging after 8 days, and after 14 days from 90 mm depth.

Approximately half of the seedlings emerged in the first three weeks of this study. Emergence did however continue, for a further 90 days duration the study (Figure 2).

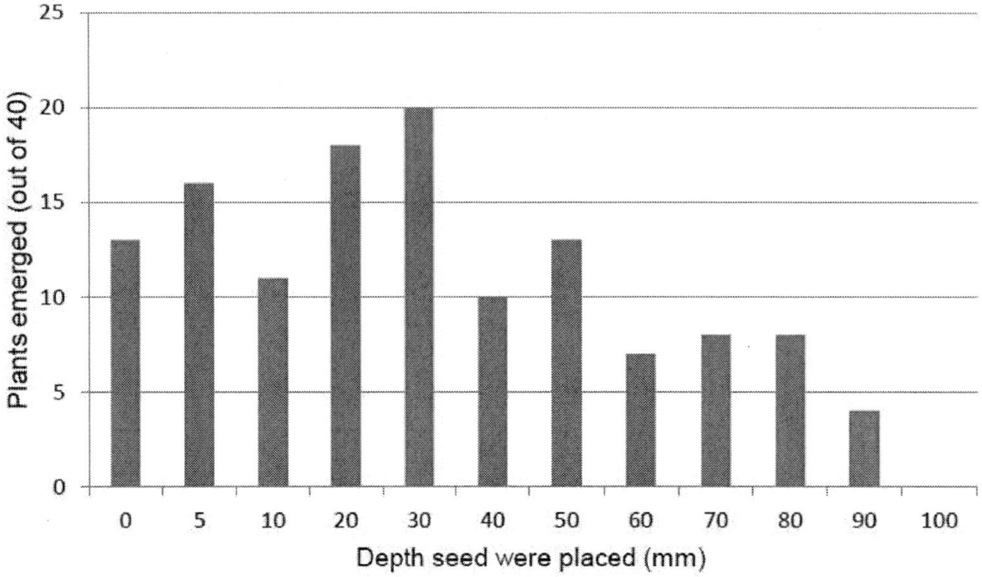

Figure 1. Total emergence of Noogoora burr plants (out of 40 burrs), buried at various depths, after 3 months. From James and Dowsett (2015) and reprinted with permission.

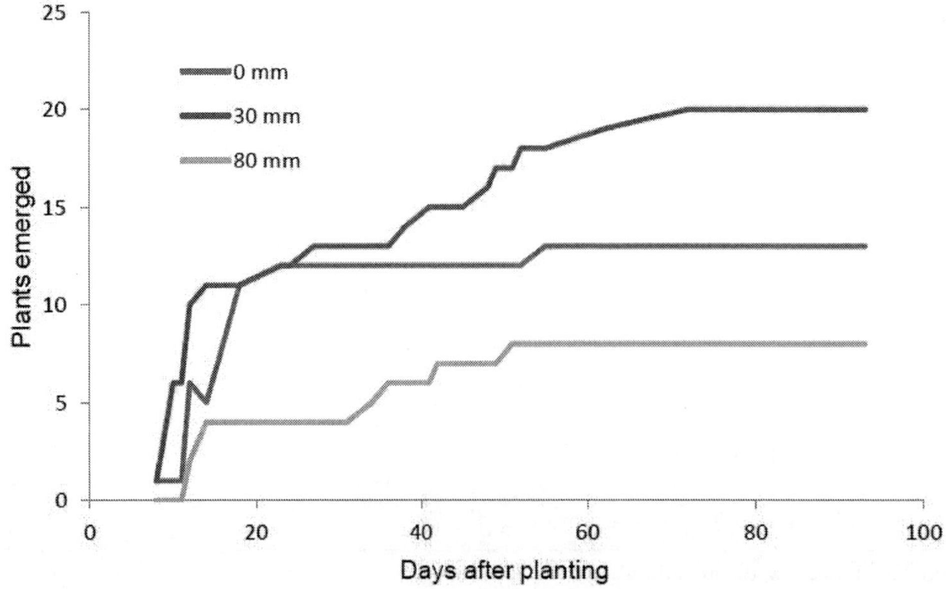

Figure 2. Sequential emergence of Noogoora burr for three selected depths (0, 30 and 80 mm) from a total of 40 burrs. From James and Dowsett (2015) and reprinted with permission.

Cultivation would probably not be an effective control method for Noogoora burr, as seedlings readily emerged from a depth of up to 90 mm.

Noogoora burr density pre- and post- regime-change There was a marked decrease in Noogoora burr plants found in the transects post regime-change when compared with the pre-regime-change data (Table 5). As well as the average decrease across all transects, Noogoora burr plants were only ever found in one or two of the eight transects after the move to the new regime compared to the pre regime-change, where Noogoora burr plants were found in 15 of the 16 transects. This data suggests that the new maize-for-silage regime is helping reduce Noogoora burr numbers at the site.

Table 5. Noogoora burr plants found 100 m^{-1}.

Transect #	2012–2013	2013–2014	16/01/2015	17/04/2015	20/01/2016	04/04/2016	20/04/2016
1	12.1	22.2	0	0	6.1	10.1	3.3
2	0	1.0	0	0.5	0	0	0
3	11.0	6.0	0	0	0	0	0
4	0	0.8	4.3	0	0	0	0
5	12.8	6.8	0	0	0	0	0
6	4.4	5.1	0	0	0	0	0
7	2.8	5.9	0	0	0	0	0
8	1.1	8.3	0	0	0	0.4	0
Average of all transects	5.4	7.1	0.6	0.1	0.8	1.3	0.4

REFERENCES

Bay of Plenty Regional Council. (2011). Keeping pests out – Regional pest management plan for the Bay of Plenty 2011–2016. Bay of Plenty Regional Council.

Government of South Australia. (n.d.). Noogoora burrs (*Xanthium strumarium*). Declared Plant Policy under the Natural Resources Management Act 2004. http://www.pir.sa.gov.au/biosecurity/weeds_and_pest_animals/weeds_in_sa/plant_policies/pest_weed_policies/declared_plants_2/Noogoora_burrs_policy.pdf (accessed 10 April 2018)

James, T. and Dowsett, C. (2015). Depth of emergence for Noogoora bur seed. Report for Bay of Plenty Regional Council. AgResearch Publication No 2415. Unpublished report.

Pennsylvania State University. (2004). Weed Management in organic cropping systems. Penn State Extension. Agronomy Facts 64. https://extension.psu.edu/weed-management-in-organic-cropping-systems (accessed 10 April 2018).

Rahman, A., James, T. and Trolove, M.K. (2014). Best options for management of Noogoora bur Chemical control options. Report for Foundation of Arable Research; Environment Bay of Plenty and Waikato Regional Council. Client Report No. RE400/2014/510. Unpublished report.

From little things big things grow – A review of behaviour change initiatives for weeds management in NSW

Birgitte Verbeek[1], Elissa van Oosterhout[2] and Wendy Gibney[3]
Invasive Plants and Animals Unit, New South Wales Department of Primary Industries,
[1] 4 Marsden Park Road, Calala New South Wales 2340 Australia
[2] Trenayr Road, Junction Hill, New South Wales 2460, Australia
[3] 1243 Bruxner Highway, Wollongbar, New South Wales 2477, Australia
(birgitte.verbeek@dpi.nsw.gov.au)

Summary This paper reviews progress towards implementing effective behaviour change programs for weeds management in New South Wales (NSW). Supported programs include the application of the community-based social marketing (CBSM) framework to weeds issues and behaviourally effective delivery of information.

Keywords Behaviour change, community-based social marketing, CBSM, weeds, training.

INTRODUCTION

For over two decades, community-based social marketing (CBSM) has been used extensively by government and non-government programs to improve community participation with environmental issues. There are many examples and strategies that are shared worldwide through the Fostering Sustainable Behaviour website (www.cbsm.com).

In 2014 the Invasive Plants and Animals Unit (IPA) in the NSW Department of Primary Industries (DPI) began facilitating the application of CBSM to improve weeds management in NSW (Verbeek *et al.* 2014).

Despite its global adoption there was initial scepticism that CBSM could accommodate the diversity and complexity surrounding weeds issues. The lack of known applications of CBSM in weeds management contexts was also an initial barrier.

Four years on, and a number of weeds-related programs have used CBSM successfully and more programs are being developed.

In addition to facilitating the application of CBSM, training on behaviourally effective communications is enhancing the delivery of weeds management programs.

Most weeds professionals have agricultural or environmental science backgrounds and the use of technical language and scientific styles occurs almost by default. This style is not the most effective for delivering key messages, or for increasing desired behaviours within our target audiences. It is widely accepted that behaviourally effective communication requires a sophisticated approach informed by the behavioural sciences (Hine *et al.* 2014, NSW DPI 2016).

As practitioners we have been guided on these scientific approaches through collaborations with the Invasive Animals Cooperative Research Centre (IA CRC) Community Engagement – Program Four. As a result, engagement efforts in weeds management are evolving through the use of a wider range of behaviour change techniques to improve efficacy of programs.

CBSM – CAPACITY BUILDING 2014–2018

Community-based social marketing is a structured framework made up of five steps: selecting behaviours; identifying barriers and benefits to the behaviours; developing strategies to encourage the behaviours; piloting the strategies; and broad-scale implementation of successful strategies with evaluation.

A number of IPA staff and regional weeds professionals have been trained in CBSM with founder Dr Doug McKenzie-Mohr (Driver *et al.* 2015, Verbeek *et al.* 2014).

Since the initial introduction of weeds professionals to CBSM a number of support initiatives have been deployed to help interested staff become more behaviourally effective in their programs. These included:

- organisation of CBSM training workshops with founder Dr. McKenzie-Mohr for weeds professionals in NSW-DPI and other partner organisations;
- development of a database to collate a comprehensive set of potential weed management behaviours with a schema of 'sectors' and 'categories' such as site constraints. The database can be accessed by weeds professionals to inform and assist them in the application of CBSM. The intent of the database is also to grow institutional expertise and share results of CBSM projects over time;
- employment of a dedicated staff member to facilitate the application of CBSM by weeds professionals and their organisations;
- collaboration through a Google forum and the DPI Weeds Extranet; and

- sponsoring and assisting weeds professionals to undertake the CBSM Master Class coaching with Doug McKenzie-Mohr, applying the framework in full to practical weeds issues.

COMPLETED MASTER CLASSES

Tropical soda apple (*Solanum viarum* Dunal) control by landholders in the Clarence Valley Since its identification in 2010, government authorities have been working with landholders to control tropical soda apple (TSA) on private land in an attempt to eradicate it. TSA is regulated by a Biosecurity Control order (under the *Biosecurity Act 2015*). While some good progress had been made, CBSM was trialled to increase the levels of control undertaken by landholders. The main target audience was landholders with the weed in the Tallawudjah Creek area south of Grafton (van Oosterhout *et al.* 2017).

Six end-state, non-divisible control behaviours (in behaviour chains) were identified. The target audience surveyed and most frequently described barriers to the behaviours were *preference for other methods* and *not convinced of the need*. The most frequent benefits were *stops spread/regrowth* and *see no barriers*.

Informed with this insight, proven social marketing techniques were incorporated into the following strategy that addressed the barriers, and promoted the benefits:

- coaching and call-backs by weeds officers (incorporating goal setting, prompts and feedback loops);
- control cards (incorporating communicating, framing and convenience); and
- signs and maps (incorporating goal-setting, norming and diffusion).

The strategy was piloted and then implemented in May/June 2017. It was highly successful resulting in a doubling of landholders doing as advised and a decrease by two thirds of landholders doing nothing.

Lantana (*Lantana camara* L.) control by landholders in Eurobodalla Lantana presents a serious threat to the biodiversity of the southern part of the Eurobodalla Shire Council area, having recently expanded its range between Narooma and Wallaga Lake, invading riparian zones and other high value forest environments.

CBSM was sought as a way to increase and improve control by landholders (Martin *et al.*2017).

The target audience was interviewed, and collectively across all the behaviours, the most frequently described barriers were *access*, *difficult terrain* and *time poor*, and the most frequently described benefits were *easy/quick, kills it* and *property management*.

These insights resulted in the decision to concentrate on one desirable end-state behaviour – spraying lantana with a gas-powered splatter gun. The strategy developed to address the barriers and maximise the benefits included:

- the Council purchasing and providing splatter guns and herbicides for the initial knockdown of lantana as well as providing instruction on the use of the splatter guns;
- the Council engaging contractors to make access tracks to improve access to difficult to reach lantana infestations on private property;
- weeds officers making follow up visits to review progress by landholders;
- prompts were sent via SMS, email or phone messages advising landholders to splatter-gun their lantana whenever conditions were good; and
- landholders endorsing signs and hanging them on property boundaries publically displaying their commitment to lantana control.

After piloting and evaluating this strategy the results revealed a 53% increase in lantana control. The strategy is now being delivered across all of the Council's lantana control area.

OTHER CBSM PROJECTS

In addition to the two successful Master Class outcomes, other projects have commenced including:

- green cestrum (*Cestrum parqui* L'Hér.) in rural and urban areas in Maitland City Council area;
- African lovegrass (*Eragrostis curvula* (Schrad.) Nees) in Snowy Monaro Regional Council area;
- Chilean needle grass (*Nassella neesiana* (Trin. & Rupr.) Barkworth) in Queanbeyan Palerang Regional Council area;
- preventing spread of hawkweeds (*Hieracium* species) in Kosciusko National Park with National Parks and Wildlife Services;
- gorse (*Ulex europaeus* L.) control in Queanbeyan Palerang Regional Council area;
- alligator weed (*Alternanthera philoxeroides* (Mart.) Griseb.) hygiene and containment in the Hunter Region;
- tropical soda apple control in the New England Weeds Authority area; and
- tropical soda apple control in the Kempsey Shire Council area.

Two state-wide projects informed by the experience gained in applying CBSM to date are now underway: *Barriers and drivers of control of widespread weeds on small acreages* and *Increasing community capacity in using biocontrol for weeds*.

The first project will apply the CBSM framework to 22 widespread weeds across the 11 Local Land

Service (LLS) regions in NSW with a focus on peri-urban areas. The project will begin with a phone survey of 2200 peri-urban landholders as Step 1 of the CBSM framework, and will then move through Steps 2–5 over the next 18 months, drilling down to address 2 specific widespread weed species in each region. From this first step, insights into the motivations of peri-urban audiences in relation to general weed control behaviours including: mowing/slashing; herbicides; pasture improvement; burning; biocontrol; physical removal; and checking and containment, will be gained.

The second project is being delivered in partnership with Landcare groups across NSW. Step 1 in this project incorporates consultation meetings with regional Landcare facilitators, Landcare coordinators, LLS regional weed coordinators and weeds officers to determine the desired behaviours. Step 2 is using online surveys distributed via the Landcare networks to determine the barriers and benefits for using biocontrol both in general terms and specifically for each agent and weed. Four surveys are underway with 305 respondents so far who are members of over 90 Landcare, Coastcare or farming groups surveyed. The collated data will be presented to regional groups to collaboratively develop strategies to address the barriers and benefits for the behaviours relating to biocontrol (Step 3). Desired behaviours include collecting, releasing, monitoring and raising biocontrol agents.

BEHAVIOURALLY EFFECTIVE COMMUNICATION

The delivery and framing of messages for target audiences can have a big impact on how they are received.

In the world of advertising to sell products or services, this is everyday business. Millions of dollars are spent on pitches designed to appeal to various characteristics of the target audiences.

Weeds professionals are not in the business of selling products, but we are in the business of selling ideas and most of the same principles apply and have been developed in the field of social marketing (B. Howie pers. comm.).

The application and facilitation of this approach in the weeds space has manifested in a new training short course, and efforts to improve weeds information, as follows:

Training As part of their lead role in the IA CRC Vocational Education Training *Training packages on Strategic Pest Management* Project, the IPA unit has collaborated with Bruce Howie to develop a two day accredited course 'Community Engagement – Moving people towards action' (BSBPMG418 Apply project stakeholder engagement techniques, training.gov.au). The course was piloted and refined in consultation with invasive species professionals in NSW, Victoria, Western Australia and Queensland, and is supported by three key publications *Behaviourally effective communications for invasive animals management: A practical guide* (Hine *et al.* 2015); the online *Invasive Action Tool* (Pestsmart 2018); and *How to use persuasion skills to enhance community engagement* (Howie 2017).

The course provides structured hands-on training to trigger practitioners to think differently about how they design communication material to gain audience attention as well as developing a planned approach to community engagement. In NSW alone this course has been delivered to over 90 invasive species professionals with excellent feedback.

An example of how this short course has influenced the approach taken to deliver weeds messages includes the design of Control Cards used in the tropical soda apple CBSM strategy.

Positive motivational messaging and imagery was used as the most prominent content on the Control Cards appealing to the values of the landholders in the Clarence Valley. Text minimisation and plain language were used to describe the desired behaviours, decrease barriers and promote benefits. A total of 75% of landholders found the control cards made a difference to their control efforts, and 80% of landholders described them as *great*, *helpful* and *informative*.

Plain language and improvements to digital delivery of weeds information A recent audit of the behavioural effectiveness of weeds content (NSW DPI 2016) made recommendation to improve aspects of digitally delivered information. In collaboration with Dan Kaufman, a custom-built content toolkit was developed to guide the writing style for the information in NSW WeedWise, and other DPI weeds publications (NSW DPI 2017). It guides the structure of the information, giving priority to how the reader is directly affected by the information, writing headings in the first person, and clearly stating the main reason why the reader would want, or need to know about a weed. It calls for a style that helps readers (rather than talks down to them); thinks from the reader's perspective; avoids unnecessary information and repetition; uses bullet points wherever possible; and above all uses plain language, avoiding confusing terms and industry and government jargon.

DISCUSSION

The evolution of how we deliver weed management programs in NSW has occurred over decades. Increased knowledge about behaviour change informed by the social and behavioural sciences and the field of social marketing is further enabling weeds professionals to improve their current approaches. The IPA unit is providing strong leadership in this area and is addressing the recommendations made by the Natural Resources Commission that the NSW government should ensure that regional plans *'promote behavioural change and adoption of integrated land management practices'* (NRC 2014).

The uptake of behaviour change is becoming more evident at a strategic and planning level in NSW. The Regional Strategic Weed Management Plans include promoting behavioural change as a strategy in Goal 1. Behaviour change including CBSM is incorporated into the NSW Weeds Action Program and Invasive Species Plans. Further to this the performance audit on LLS communications made the following recommendation *'Design communication programs from the outset to drive behavioural change'* (NRC 2016).

Interest and engagement from the industry is strong, indicated by the voluntary access to the NSW Weeds Extranet *Behaviour change* pages and the associated links provided on this page. Since March 2016, 843 weeds professionals have visited the online content. Also, 102 weeds professionals have engaged with the interactive Prezi presentation *'Roadmap through behaviour change initiatives for weeds in NSW'* (NSW DPI 2018) over the last six months. With increasing experience in the application of behaviour change techniques there is no going back. We can no longer sit comfortably with the assumption that the provision of information and education alone leads our target audiences to undertake our desired weed control and management activities.

REFERENCES

Driver, A., Verbeek, B., van Oosterhout, E. and Maguire, A. (2015). 2 steps back five steps forward. Proceedings of the 18th NSW Weeds Conference. (NSW Weeds Society, Cooma).

Hine, D.W., Please, P., McLeod, L. and Driver, A. (2015). Behaviourally effective communications for invasive animals management: A practical guide, https://www.pestsmart.org.au/behaviourally-effective-communications-for-invasive-animals-management/ (accessed 24 April 2018).

Howie, B. (2017). How to use persuasion skills to enhance community engagement. (C-Qual Agritelligence, Cabarita).

Martin, P., van Oosterhout, E. and Verbeek, B. (2017). CHAIN, CHAIN, CHANGE – PART TWO Applying a behaviour change framework to lantana control in NSW – insights from the Far South Coast. Proceedings of the 19th NSW Biennial Weeds Conference. (Armidale).

NRC, Natural Resources Commission (2014). Weeds – Time to get serious. Review of weed management in NSW. http://www.nrc.nsw.gov.au/publications (accessed 24 April 2018).

NRC, Natural Resources Commission (2016). Performance Audit Local Land Services Communications. http://www.nrc.nsw.gov.au/audit-and-assurance (accessed 3 March 2017).

NSW DPI, New South Wales Department of Primary Industries (2016). Biosecurity CBSM – Weeds Initiative Audit Report https://extranet.dpi.nsw.gov.au/weeds/cbsm/Audit-Report-Behaviour-change-metrics.pdf (accessed 24 April 2018).

NSW DPI, New South Wales Department of Primary Industries (2017). *NSW WeedWise Content Toolkit*. Internal NSW DPI report. Unpublished.

NSW DPI, New South Wales Department of Primary Industries (2018). Roadmap through behaviour change initiatives for weeds in NSW. https://prezi.com/view/w5NZnaISEBF0oNReKIp1/ (accessed 24 April 2018).

Pestsmart (2018). Invasives Action Tool. https://community.pestsmart.org.au/ (accessed 24 April 2018).

van Oosterhout, E., Luxton, R., Gibney, W. and Verbeek, B. (2017). CHAIN, CHAIN, CHANGE – PART ONE Applying a behaviour change framework to tropical soda apple control in NSW – insights from the North Coast. Proceedings of the 19th NSW Biennial Weeds Conference. (Armidale).

Verbeek, B., van Oosterhout, E. and Grantley, J. (2014). Social science methods to improve community participation in weed management programs in New South Wales. Proceedings of the 19th Australasian Weeds Conference, ed. M. Baker, pp. 342-6. (Tasmanian Weed Society, Hobart).

Building capacity for weed management in the Lao PDR

Deirdre Lemerle[1], John Burley[2], Sarah Hain[3], Brendon Bangma[3], Nicholas Pain[3], Sengphet Phantavong[4], Kylie Ireland[5], Madaline Healey[6] and Lester Burgess[7]
[1] Graham Centre for Agricultural Innovation, Charles Sturt University, Wagga Wagga, New South Wales 2678, Australia
[2] John Burley Consulting, Balwyn North, Victoria 3104, Australia
[3] Agriculture Section, Provincial Agriculture and Forestry Office, Savannakhet, Lao PDR
[4] Agriculture Section, Provincial Agriculture and Forestry Office, Pakse, Lao PDR
[5] CSIRO Health and Biosecurity, Black Mountain Science and Innovation Park, Clunies Ross Street, Acton, Australian Capital Territory 2601, Australia
[6] Forest Industries Research Centre, University of the Sunshine Coast, Queensland 4556, Australia
[7] Sydney Institute of Agriculture, University of Sydney, New South Wales 2006, Australia
(dlemerle@csu.edu.au)

Summary Weed control in crops, mainly hand-weeding and hoeing, has traditionally been managed by women on small-holder farms. This is unsustainable due to labour scarcity, and this limits the adoption of productive systems in the Lao Peoples Democratic Republic (PDR). Herbicides are available but knowledge on careful and effective chemical use is virtually non-existent. In addition, policies on the effective use of herbicides are needed to minimise environmental and health risks. For these reasons there is a need to build capacity in weed management and integrate non-chemical practices with new herbicidal options as they become available.

The Crawford Fund has made a significant contribution to capacity-building in crop pest and disease management, and the aim is to extend this to weed management. Australian Government volunteers have worked in this program at specific locations with farmers and local government advisers and with support from key mentors. This paper recommends the key priorities for increasing capacity in weed management practices and policies. Furthermore, feed-back from volunteers provides guidance on ways to improve the professional and personal outcomes of the volunteering experience.

Keywords Integrated weed management, small-holder farmers, volunteer support, government regulation.

INTRODUCTION

Laos is one of the least developed countries in East Asia. It is land-locked between Vietnam, Thailand, China, Myanmar and Cambodia with a population of around seven million people. Subsistence agriculture is practised by 80% of the population and accounts for 50% of the country's Gross Domestic Product. Foreign aid provides a significant support to agricultural development and the economy. Although poverty has been reduced in recent years, malnutrition in children is still a concern. Rice is the main crop, with about 80% of the arable land area used for growing rice by small-holder farmers.

Approximately 77% of Lao farm households are self-sufficient in rice for domestic consumption. Vegetables are also important for domestic consumption as well as for cash crops.

Weeds reduce crop yield and quality, degrade pastures, host pests and diseases, and decrease biodiversity of natural ecosystems, all of which can have negative economic impacts. These costs have not been well quantified in the Lao Peoples Democratic Republic (PDR), and so it is difficult to engender support for improved practices and policies for weed management.

For some years, the Crawford Fund (www.crawfordfund.org) in collaboration with Australian Government-funded volunteers in the Australian Volunteers for International Development (AVID) program, has supported capacity-building for crop pest and disease management in a range of rice, fruit and vegetable crops (e.g. Ireland 2014, Ireland *et al.* 2016). A scoping study was undertaken in November 2016, mainly in vegetables, and to a lesser extent rice, to assess the need to extend the capacity building to include weeds (Lemerle 2017). In addition, previous experience with AVID volunteers was assessed as a basis for optimising the recruitment of volunteers to work in weed management in the future.

NEED FOR BUILDING CAPACITY IN PRACTICES

The scoping study involved visits to representative arable areas in the southern region in Savannakhet and Champasak Provinces. Activities included: a) field tours using weed identification field guides (Martin and Chanthy 2007, Caton *et al.* 2010) and expert assistance (R. Martin, pers. comm.) to identify the most common species; and b) discussions with 'leading' farmers and provincial and district staff about priority weed species, current management practices, and needs for the future.

The weed flora in vegetable crops (e.g. cabbage, watermelon, long-bean, eggplant, tomato and chilli) is generally very diverse. Many species are common in both lowland and highland areas, mainly in the Asteraceae, Amaranthaceae, Fabaceae and Poaceae families. Symptoms of diseases, viruses and pests were observed on some weed species, an indication of the importance of weeds as foci of infection and spread. The most common and abundant species observed in dry-season rice crops (transplanted, broadcast and direct seeded) were barnyard grasses (*Echinochola* spp.), various sedges (*Cyperus* spp. and *Fimbristylis* spp.), and a broad-leaf water primrose (*Ludwigia octovalvis* (Jacq.) P.H.Raven).

Farmers' and government staff's knowledge of weed identification and management is variable. Many plant species are considered both weeds and useful (for wild-harvest food, medicinal uses, and grazing). The economic cost of weeds is often not recognised leading to low levels of management, especially in vegetable crops. The main weed control management practices followed by farmers are traditional non-chemical practices such as hand-weeding, hoeing, cultivation, cutting, rotations, burning and grazing. However, many farmers want mechanisation and herbicides for weed control to save labour and speed up operations. Herbicide use by small-holder farmers is low due to lack of availability and registration of products in Laos. Some products have been imported from Thailand, Vietnam and China through registered traders, but there is no labelling in the local language so farmers cannot read application instructions. The demand for chemical control is growing with declining labour availability. Glyphosate is available for non-selective and pre-planting weed control in the larger regional centres, but there has been no research or extension on potential applications of selective herbicides.

The priority needs for weed management in crops as identified by provincial and district advisory staff and farmers is rice followed by vegetables. The development of drilled and dry-seeding in rice which is intended to save labour and water, will exacerbate weed problems. This will lead to even greater future need for capacity in integrated weed management (IWM) (Matloob *et al.* 2015, Kumar *et al.* 2017). Components of IWM for rice (Kumar *et al.* 2017) and other crops include:

- promoting 'fatal germination' of weeds using stale seedbed techniques;
- using clean seeds, such as certified seeds free from weed seed contamination;
- managing fallows to prevent weed seed production;
- using soil mulch and crop residue mulch for weed suppression;
- improving crop competitiveness against weeds by choice of cultivar, adjusting water (in rice) and nutrient management, and manipulating seed rate and crop geometry;
- rotating crops and establishment methods;
- utilising grazing by animals, and 'cutting and carrying' weeds for animal food;
- using mechanical tools and need-based hand weeding to remove escaped weeds and prevent seed production; and
- using appropriate herbicides and their combinations.

In summary, the key activities and resources needed in Laos for sustainable weed management include: research and development for cost-effective IWM strategies and practices; correct weed identification and herbarium facilities; training and extension materials for small-holder farmers and advisers in the Lao language; quantification of the costs of weeds to farmers and the broader impact of weeds to Lao economy; and effective policy and regulation for herbicide distribution and usage.

POLICY CONTEXT

The Lao Department of Agriculture has been proactive in providing an effective policy framework for plant protection through promulgation of the new Law on Plant Protection and Quarantine. This law is primarily focused on providing a contemporary and comparable world-class legislative basis for policy formulation and action to protect crops from pests and diseases, along with mechanisms to prevent the introduction of new pests and diseases into the country. In addition, the scope of policy and action under this law could be readily expanded to cover other areas of action to combat the economic and environmental impacts of weeds. However, there is still a considerable support needed to provide a suitable regulatory and policy environment for this law to achieve its desired outcomes.

The need for enhanced crop protection has to match the increasing recognition to produce crops free from unacceptable levels of chemical residues. There has recently been a concerted effort to manage agricultural chemical use responsibly and safely. Programs such as the Good Agricultural Practices

(GAP), developed under the framework of the Lao-Thai-German Trilateral Cooperation Programme, and the ASEAN Sustainable Agrifood Systems, are increasing awareness of, and commitment to responsible chemical use. Lao PDR expects to increase the number of GAP farms from 300 in 2016 to 100,000 in 2020. This indicates the priority given by the Government to reach food security and safety. This increased level of landholder adherence to GAP will clearly need considerable financial and policy commitment from the Government.

VOLUNTEERS FOR BUILDING CAPACITY

The clustering of AVID volunteers on assignment for 12 months with appropriate mentors has previously focused on plant pests and diseases. This will be extended to include weeds. However, there is variation in the benefits of, and risks to, deploying international volunteers (e.g. retirees, students, people on vacation) for capacity building in Laos. Volunteers with different skills at specific locations, can achieve synergies, on-going continuity, and economically sustainable support. In addition, they can value-add to Australian (or other country) aid from other agencies such as the Australian Centre of International Agricultural Research (ACIAR), universities, and state governments.

The professional and personal experiences gained, as identified by volunteers, include:

- improved knowledge of local small-holder farming systems, constraints to production at a subsistence level, and the options available to improve their livelihoods;
- exposure to a broader range of pathogens, pests, crops, and farming techniques than would normally be experienced;
- building one's skills and knowledge quickly to work independently, perform under challenging conditions, learn problem solving techniques with minimal resources, teaching, project management and negotiation, and improving communication skills to interact with people from different cultures, languages and educational backgrounds;
- having a mentor providing support and encouragement, helping in identifying key priorities and pests, and encouraging career development;
- working in an environment with an established infrastructure, networks and knowledge built up over the duration of the Crawford program in Laos;
- enjoyment of working with a diverse range of highly motivated colleagues, and developing personal and professional relationships and networks, in a culturally and socially different country;
- an opportunity for 'time out' to broaden personal and professional horizons; and
- knowing, from feedback, that the volunteer work is making a difference in the livelihood for farmers in Laos.

Ways to make the volunteering experience and assignment more effective and enjoyable also need to be identified. Success is underpinned by providing volunteers with sufficient support through strong links and effective communication between the partner organisations in Australia and Laos. The strong commitment of volunteer, mentor and host organisation in Laos is essential, as are clearly defined roles and responsibilities, with structured goals and activities, for all involved. High expectations of the outputs from volunteers, who are constrained by time and inadequate provision of required resources, can cause difficulties to achieve the desired outputs. On-going monitoring and flexibility is required as assignments might change and each party should be kept up-to-date. Volunteers need to be adequately prepared prior to placement; this could include language training, streamlining administrative processes (including visas), and pre-departure connections established with peers working on similar projects. Finally, consideration and planning are needed to help volunteers return to their workplace in Australia following completion of an assignment.

CONCLUSIONS

We have a good opportunity in Laos to gradually, but significantly improve the quality and productivity of farming systems by increasing farmers' capacity to manage weeds. This includes optimising the introduction of herbicides where appropriate, and their integration with non-chemical options for IWM. This will reduce the risks of developing significant problems of herbicide resistance, and chemical residues in food, the environment (soil, water, air, biodiverse) as well as human and animal health issues, that are prevalent in other Asian countries. There is also an opportunity to utilise the beneficial properties of weeds when developing management strategies. Capacity building with volunteers and mentors grouped at specific locations has proven effective for supporting farmers and advisers. This can value-add to research and development projects (e.g. ACIAR), universities, and other public and private sector organisations. The overall aim is to increase farm profitability, reduce poverty and poor human nutrition, and enhance healthy and sustainable food production systems.

ACKNOWLEDGMENTS

We thank the following for their support: the Crawford Fund; the Australian Centre for International Agricultural Research; staff at Scope Global and AVID (now Australian Volunteers International); Gerry Duckitt (mentor); Judith Paquin (Canadian Volunteer program); Dr Phoudalay (National Agriculture and Forests Research Institute); Mr Somlit (Provincial Agriculture and Forestry Office, Champasack); Mr Khamseng (Provincial Agriculture and Forestry Office, Savannakhet); and the many kind and generous farmers who open their fields, homes, minds and hearts to us.

REFERENCES

Caton, B.P., Mortimer M., Hill, J.E., Johnson, D.E. (2010). 'A practical field guide to weeds of rice in Asia.' Second Edition. (International Rice Research Institute, Los Baños (Philippines). 118 pp.

Ireland, K.B. (2014). 'Building capacity to combat crop diseases in Laos to improve agriculture'. (Scope Global and Department of Foreign Affairs and Trade (DFAT). Blog post for Australian Volunteers for International Development (AVID). http://www.australianaidvolunteers.gov.au/meet-the-volunteers/case-studies/building-capacity-to-combat-crop-diseases-in-laos-to-improve-agriculture.aspx (accessed 6 August 2018).

Ireland, K.B., Weir, B.S., Cother, E.J., Phantavong, S., Phitsanoukane, P., Vongvichid, K., Vongphachanh, P.P., Songvilay, P., Chittarath, K., Sayapatha, S., Walsh, J., Turner, S., Park, D., Tesoriero, L.A., Vilavong, S., Duckitt, G.S. and Burgess, L.W. (2016). First report of *Ralstonia pseudosolanacearum* in the Lao PDR. *Australasian Plant Disease Notes* 11, 36.

Kumar, V., Opena, J., Valencia, K., Le Thi, H., Nguyen, H.S., Donayre, D.K., Janiya, J. and Johnson, D.E. (2017). Rice weed management in southeast Asia. *In* 'Weed management in rice in the Asian-Pacific region', eds A.N. Rao and H. Matsumoto, pp. 282-307. (Asian-Pacific Weed Science Society, The Weed Science Society of Japan, and The Indian Society of Weed Science, India).

Lemerle, D. (2017). Opportunities for capacity building in weed management – Lao PDR. Proceedings of the 26th Asian-Pacific Weed Science pp. 158. (Asian-Pacific Weed Science Society, Kyoto, Japan).

Martin, R. and Chanthy, P. (2007). 'Weeds of upland crops in Cambodia' (New South Wales Department of Primary Industries, Orange).

Matloob, A., Khaliq, A. and Chauhan, B.S. (2015). Weeds of direct-seeded rice in Asia: Problems and opportunities. *Advances in Agronomy* 130, 291-336.

Floating Landcare: Bringing a bit of WOW to weedy hotspots on the Hawkesbury Estuary and NSW Central Coast waterways

Rebecca Mooy[1] and Tegan Burton[2]
[1] Greater Sydney Local Land Services, Level 4, 2–6 Station Street, Penrith, New South Wales 2750, Australia
[2] NSW National Parks and Wildlife Service, Max Allen Drive, Lindfield, New South Wales 2070, Australia
(Rebecca.Mooy@lls.nsw.gov.au)

Summary Floating Landcare, a program formally known as Workers on Water (WOW) was a concept realised in 2012 with funding from the NSW Environmental Trust. Now in its seventh year, it is a successful volunteer program working steadily to restore weedy hotspots at 20 locations within national parks and on other public lands along the Hawkesbury estuary, Pittwater and waterways on the Central Coast. Floating Landcare is a multi-partner program coordinated by Greater Sydney Local Land Services with the support of National Parks and Northern Beaches, Hornsby Shire and Central Coast Councils. In focussing the volunteers' effort on weedy hotspots at locations largely surrounded by good condition vegetation, and carrying out litter clean ups along estuarine beaches, the program has a strong positive impact. Vegetation condition has improved and there is a reduction of weed volume and spread. Endangered Ecological Communities, including Swamp Oak Floodplain Forest, Coastal Saltmarsh, and Pittwater Spotted Gum Forest, have responded well to the programs' strategic approach.

Floating Landcare is attractive to highly skilled and experienced bushcare volunteers through to 'first timers'. It provides an opportunity for those passionate about managing weeds to do so with their own two hands, and is also an adventure, as volunteers travel by boat through amazing river scenery to discover special spots, largely restricted in access to the general public. It gives existing volunteers a chance to work on new sites, learn new weeding techniques, be introduced to new native plants, and enjoy the company of other like-minded volunteers from across Greater Sydney.

Keywords Floating Landcare, weeds, boats, partnerships, adventure volunteers, Hawkesbury estuary.

Using the Atlas of Living Australia to support weed biological control

Greg Lefoe[1], Raelene Kwong[1] and Kerinne Harvey[2]
[1] Agriculture Victoria Research, AgriBio, 5 Ring Road, Bundoora, Victoria 3083, Australia
[2] NSW Department of Primary Industries, 1447 Forest Road, Orange, New South Wales 2800, Australia
(rae.kwong@ecodev.vic.gov.au)

Summary Farmers, natural resource managers, community volunteers and others contribute to weed biological control efforts by identifying infestations suitable for biological control, releasing biological control agents, monitoring biological control agent release sites, and recording field observations of agents as they spread across the landscape. However, significant barriers can prevent data being captured, shared and used. Maintaining and accessing long-term monitoring data is particularly difficult. The Atlas of Living Australia (ALA) is a national on-line biodiversity database that has potential to address these barriers. The ALA contains more than 70 million user-submitted species occurrence records from field observations, collections and surveys. Data such as location coordinates, photographs, and species and host information, can be uploaded and later accessed in spreadsheets or species distribution maps. This resource has tremendous potential for biological control, as it can be used to:

1. record biological control agent release and establishment data;
2. capture observations of biological control agent spread;
3. ensure biological control agent distribution data is readily accessible, and
4. provide a one-stop shop for biological control extension material.

A custom-made app for iPhone and Android devices further facilitates the collection of data from the field. The Atlas of Living Australia, combined with smartphone technology and a custom-made app, has potential to transform the assessment of biological control agent establishment and spread, and facilitate adoption of biological control by end-users.

Practical best practice management of opuntioid cacti – from desk to dirt

Shauna Potter[1], Matthew R. Sheehan[1], Kate Blood[2] and Kay Bailey[3]
[1] Wild Matters, 11 Templeton Street, Castlemaine, Victoria 3450, Australia
[2] Department of Environment, Land, Water and Planning, PO Box 7, Beaufort, Victoria 3373, Australia
[3] Department of Primary Industries and Regional Development, Locked Bag 4, Bentley DC, Western Australia 6983, Australia
(shauna@wildmatters.com.au)

Summary Opuntioid cacti, comprising of 27 species from three genera in Australia, are often described as cryptic plants. Their distribution ranges from widespread to isolated. Growth form varies from low growing shrubs to 8-metre-tall trees. For some species seed germination and longevity remain knowledge gaps. Some species are kept in check by one of Australia's most successful biological control agents (*Cactoblastis cactorum*), whilst others require specific, labour intensive chemical or physical control. Field identification is challenging, and common and scientific names are often misapplied. What is certain is that, largely, opuntioid cacti remain difficult to control and can out persist the resources and will of land managers.

Approximately five years after their listing as Weeds of National Significance, a best practice control manual for opuntioid cacti has been developed. The manual goes a long way in demystifying this group of 27 species by providing a comprehensive synthesis of known information on the weeds' biology, ecology, distribution and impacts. Details on planning and case studies from across Australia provide useful ideas to consider before control work begins.

Additionally, the manual offers practical, easy to follow field advice through identification guides; descriptions of control methods; registered herbicides; and tips for safe working conditions. A decision support tool helps weed managers chose and implement best practice control options based on:

- cacti growth form;
- weed density;
- site sensitivity; and
- site accessibility.

Combined, the use of these tools and other resources, provides the best chance of successfully managing opuntioid cacti in Australia.

Keywords Opuntioid cacti, Weeds of National Significance, best practice management, decision support tools.

Introducing the Plant Sure Ornamental Plant Decision Support Tool

Michelle Leishman, Nola Hancock, Victoria Graham, Vanessa Adams, Tim Maher and Anthony Manea
Macquarie University, North Ryde, New South Wales 2109, Australia
(michelle.leishman@mq.edu.au)

Summary Horticulture accounts for a substantial proportion of the plant species introduced into Australia. It is therefore no surprise that horticulture is the source of the majority of exotics that have become invasive. Whilst strict importation controls currently regulate the arrival of new exotic plant species, the sale of exotics already present are less regulated and there is the potential that some of these will become garden 'escapes' and invasive. Thus, a cost-effective way to restrict the spread of potentially invasive plants is to intercept them at the nursery level before they 'jump the garden fence'. The Plant Sure Environmentally Safe Ornamental Plant Scheme project is a collaborative undertaking that aims to limit the supply or dissuade the use of non-native plants with high potential for invasiveness. The Plant Sure Ornamental Decision Support Tool (OPDST) was developed as part of the Scheme to provide a voluntary screening tool for the green life industry that discourages the sale of plants with a high risk of becoming invasive. The OPDST comprises a set of 24 questions and separates plant species into invasion risk categories. Species that are assessed as low invasive risk are accepted into the Scheme and species assessed as high invasive risk are rejected. In a recent trial, 58 species were evaluated to test the accuracy of the OPDST. The OPDST achieved 100% accuracy in accepting non-invasive species and rejecting invasive species, where the 4 species assessed as 'needs further evaluation' were not accepted. This presentation introduces the OPDST and discusses its role within the green life industry.

Keywords Horticulture, screening tool, invasive.

Refining the environmental weed risk assessment for non-indigenous plants which may have agricultural potential

Christine J. Munday[1], Geoff Moore[2], Rachel Whitsed[3], Margaret Byrne[4] and Clinton Revell[2]
[1] Centre for Plant Genetics and Breeding, University of Western Australia, 35 Stirling Highway, Crawley, Western Australia 6009, Australia (formerly at Department of Biodiversity, Conservation and Attractions, Dick Perry Avenue, Kensington, Western Australia 6151 Australia)
[2] Department of Primary Industries and Regional Development, Locked Bag 4, Bentley DC, Western Australia 6983, Australia
[3] Charles Sturt University, Elizabeth Mitchell Drive, Thurgoona, New South Wales 2640, Australia
[4] Department of Biodiversity, Conservation and Attractions, Dick Perry Avenue, Kensington, Western Australia 6151 Australia
(geoff.moore@dpird.wa.gov.au)

Summary Traditionally, in the rangelands of Western Australia (WA), the pastoral industry has relied on grazing of native vegetation. However, in recent years there has been considerable interest in mosaic agriculture based on irrigated forages. There is also a potential role for improved dryland pastures in medium to high rainfall areas of the Kimberley.

Under these conditions there is increased interest in the introduction of non-indigenous species which may improve the productivity and sustainability of agricultural systems. However, there is also clear evidence that a number of introduced non-indigenous plant species have become established in non-target areas and some have become environmental weeds.

An Environmental Risk Framework and weed risk assessment (WRA) protocol to assess species with agricultural potential in southern Australia was developed for the Future Farm Industries Co-operative Research Centre (FFI CRC) (Stone *et al.* 2012). This protocol has been adapted for use in the rangelands of northern Western Australia. The refined methodology introduces the concept of 'Inherent weed risk'. The potential distribution or land capability is considered as a separate, but integral step in the process to provide a flexible system which can assess the risk at the regional scale and also take account of the operational context.

This environmental WRA system is designed to improve the access, transparency, rigour and consistency of information available to policy makers and those considering the use of non-indigenous species and their introduction to novel environments in the WA rangelands.

Keywords Non-indigenous plants, rangelands, environmental weeds.

INTRODUCTION

Western Australia's (WA) rangelands experience a variety of extreme climatic conditions and contain multiple unique ecosystems with high environmental, economic, social and cultural values. It is an area with a rich endemic flora and fauna recognised with two World Heritage listed sites and many National Parks. However, the WA rangelands are also the source of great mining wealth and contain large areas under pastoral production, together with associated infrastructure (roads, rail, and ports). In some regions there has been a long history (>100 years) of grazing by domestic and feral animals. As a result, the condition of the native vegetation varies widely from pristine to highly denuded and disturbed areas due to overgrazing and soil erosion.

Traditionally agriculture in the WA rangelands has predominantly relied on the grazing of stock on native vegetation, with the development of irrigation precincts around Carnarvon and on the Ord River near Kununurra. However, in recent years there has been considerable interest in mosaic agriculture based on irrigated forages using groundwater or surface water resources in the west Kimberley and from mine dewatering in the Pilbara. There is also a potential role for improved dryland pastures in medium to high rainfall areas of the Kimberley.

The introduction of non-indigenous species may improve the productivity and sustainability of agricultural systems, including pastoral enterprises, and they have been used to successfully rehabilitate unstable, eroded and disturbed sites in the northern Australian rangelands (Payne *et al.* 2004). However, there is clear evidence throughout Australia that some non-indigenous species introduced for forage have become established in non-target areas and some have become environmental weeds (Lonsdale 1994, Cook and Dias 2006). For example, Virtue *et al.* (2004) reported that 23% of species introduced into Australia as potential forage species were weeds of natural ecosystems. Introduced plants can change hydrological conditions, increase the frequency and intensity of fire regimes, provide cover and nutrition for feral animals, making control more difficult, and reduce the diversity of the native flora and fauna.

Many of the characteristics that make a species useful as a pasture species may also increase the risk that it will become a weed capable of invading natural environments. A weed risk assessment (WRA) can provide information for species selection and management to help achieve a balance between the risk to the natural environment and potential agricultural productivity.

POST-BORDER WEED RISK ASSESSMENT

Post-border WRA in Australia has predominantly focused on prioritising species identified as weeds for management/control). The Future Farm Industries Co-operative Research Centre (FFI CRC) environmental WRA system was designed to assess the weed risk of forage and agro-forestry species to natural ecosystems across southern Australia (Stone *et al.* 2008, 2012) using the principles of the National Post-Border Weed Risk Management Protocol (Anon. 2006) and the South Australian Weed Risk Management Guide (Virtue 2004).

Both current commercial and prospective species were assessed and species assessed as having a 'very high' weed risk were not evaluated or promoted by the FFI CRC. The key characteristics of the FFI CRC Environmental WRA protocol were that:

- three criteria were used to assess weed risk: invasiveness, impact and potential distribution;
- the WRA were undertaken by pasture agronomists and weed scientists who had a good understanding of the species;
- the WRA were fully documented including references and sources for unpublished data;
- the WRA were critically reviewed by independent subject experts prior to being published on-line;
- the potential distribution maps were based on the combination of three factors: areas of native vegetation with a strong climate match (CLIMATE match top 20%) and suitable soils; and
- the WRA was part of an environmental weed risk strategy which included a 'not recommended' category. (Stone *et al.* 2008, 2012).

Interest in the introduction of non-indigenous into the WA rangelands suggested the need for a WRA for the region which could be based on the established and peer reviewed FFI CRC Environmental WRA system, but with some changes to reflect the different environment and objectives.

The WA Rangelands WRA protocol retains the three sections outlined in the FFI CRC Environmental WRA protocol, together with the key characteristics as outlined above. The main change in the WA Rangelands WRA protocol is the relationship with, and methodology for the potential distribution component.

The information explored for the Invasiveness and Impacts sections is essentially the same, however some questions have been rephrased as they were previously aligned with the FFI CRC forage improvement program and examples provided are from a rangelands context. The five weed risk levels in the FFI CRC system have been reduced to 4 inherent weed risk categories reflecting four recommendations or management guidelines to minimise weed risk to the natural environment.

The key refinement is the introduction of the term, 'Inherent weed risk' which is determined from the Invasiveness score multiplied by the Impacts score (Equation 1). In essence this reflects the relevant inherent characteristics of a plant species and applies irrespective of where it is grown or whether it is well suited, or poorly suited to a particular environment.

$$\text{Invasiveness score } (0\text{–}10) \times \text{Impacts score } (0\text{–}10) = \text{Inherent weed risk } (0\text{–}100) \qquad (\textit{Equation 1})$$

The Inherent weed risk scores fall along a continuum, but these scores have been divided into categories, so that appropriate management guidelines can be developed. The four categories of Inherent weed risk and the class categories were computed from the calibrated scores used in the FFI CRC Environmental WRA system (Stone *et al.* 2012).

Inherent weed risk category	Weed risk score*
1	>38.4*
2	23–38.4
3	13.7–22.9
4	≤13.6

*Weed risk values computed from Stone *et al.* (2012).

Example: A species has a weighted Invasiveness score of 4.2 out of 10 and an Impacts score of 5.2 out of 10, which when multiplied gives an inherent weed risk score of 21.8. This equates to Inherent weed risk Category 3 and applies irrespective of whether it is broadly suited, a niche species, or poorly suited to a particular environment.

The Land capability – potential distribution assessment is the final step – how well suited is the species to the assessment area? The assessment area could be a region, sub-region or 'site specific'. It is important to consider both the proposed site and surrounding land systems plus the broader region. In addition, the proximity to high value environmental assets should also be considered.

POTENTIAL DISTRIBUTION

Stone and Byrne (2011) compared five post-border WRA systems and reported that there was considerable variation in the determination of species distribution and that it significantly impacts on the final outcome of the WRA. Likewise, within the FFI CRC Environmental WRA, the potential distribution maps were often the main point of debate within the reviewing Technical Committee.

The Department of Primary Industries and Regional Development (DPIRD) in WA have been producing potential distribution maps for species based on climate for the rangelands of Western Australia. This has involved an overall Australian climatic profile developed using CLIMATCH (Anon. 2008) based on each species' overseas distribution and a final climate outcome score determined for the Kimberly and Pilbara regions. The CLIMATCH analysis is undertaken using data from a global distribution map from the Global Biodiversity Information Facility (GBIF 2014).

Similar methods have been used elsewhere (Brown *et al.* 2006), however some maps appear to: (a) either over- or under-estimate the potential distribution for species for which the soil and climate requirements are well understood, or which are naturalised in the Kimberley and Pilbara; or (b) the potential distribution for Australia has a weak correlation with the naturalised current distribution from the Atlas of Living Australia (ALA 2018) for species with a long history under cultivation.

These issues suggest that there is scope to consider new ways to determine potential distribution. A series of new potential distribution maps were produced through Charles Sturt University for 30 species considered to have potential for dryland or irrigated agriculture in the Kimberley and Pilbara regions.

The methodology to produce the potential distribution maps based on soils (no fertiliser) and climate was adapted from that used for the Tropical Forages Database potential distribution maps (Cook *et al.* 2005). The maps were produced using a GIS system to combine four climate layers together with soils information (Table 1). For each species the climate information was classified using the FAO UN Ecocrop database for annual rainfall, growing season temperature, elevation and climate zone (Anon 2013).

The soils information was taken from the DPIRD Soils database which for the Kimberley and Pilbara uses the proportion of WA Soil Super-groups (Schoknecht and Pathan 2013) within each map unit which are based on land system mapping @1:250,000. For each species the land capability class was assessed for the WA Soil Super-groups (no fertiliser) using a standard five class capability system adapted from FAO UN (1976) and van Gool *et al.* (2005) for a rangelands context.

The five layers of digital information were classified (e.g. optimal, absolute, or neither for both annual rainfall and growing season temperature) and then combined using consistent rules to determine the final percentage in each capability class. The maps display the percentage of an area which is suitable, or suitable to marginally suitable.

A matrix table of the Inherent weed risk and the land capability–potential distribution determines the four categories of management guidelines. Species assessed as Category 4 have a negligible to low risk of becoming environmental weeds and as a consequence it is not essential to determine their potential distribution.

Table 1. A summary of the input data sets for the potential distribution maps for the Pilbara-Kimberley regions.

Input data	Source
Annual rainfall	Bureau of Meteorology (BoM) 1961–1990 averages, resolution 0.025 degrees (approx. 2.7 km)
Average minimum growing season temperature	BoM 1961–1990 minimum average minimum temperature Dec-Mar, resolution 0.025 degrees (approx. 2.7 km)
Elevation	Shuttle Radar Topographic Mission (SRTM) digital surface model resolution 0.000833 degrees, resampled to 0.025 degrees
Climate zone	Trewartha climate zones, calculated from monthly rainfall and temperature long term averages (1961–1990)
Soils	Land System map units with % of Soil Super-groups for Kimberley and Pilbara (shapefile)
Species requirements	FAO Ecocrop data sheets plus Tropical species X WA soil groups (land capability)

DISCUSSION

The WA Rangelands WRA protocol has been specifically developed as a post-border assessment of the risk of non-indigenous species which may have agricultural value in the WA rangelands becoming environmental weeds.

With the FFI CRC Environmental WRA system if there are no areas with a good climate match, suitable soils and native vegetation, the potential distribution is attributed a score of 0.5 (Stone *et al.* 2012). The context for the FFI CRC was dryland or rainfed perennial pastures and agro-forestry species. In low rainfall rangelands (e.g. Pilbara) most pastures and fodders require irrigation to persist, so the potential distribution score would be 0.5. In this scenario any species could potentially be grown under irrigation as the maximum potential score would be 50 (i.e. 10 × 10 × 0.5) and in practice likely to be less than 32 which equates to a low weed risk (Stone *et al.* 2012).

In the WA Rangelands most of the agricultural development is based on irrigation, but there is the risk of propagule movement over long distances through stock or vehicle movement (Randall 2014). The introduction of 'Inherent weed risk' overcomes this issue as it identifies higher risk species. For example, species with an inherent weed risk score in category 1 (i.e. >38.4) require special management conditions even when grown in areas where they would not persist without irrigation.

This is illustrated by two hypothetical species – species A has an FFI CRC weed risk score (Invasiveness score x Impacts score x Potential distribution) of 4 × 4 × 10 = 160 (high), while species B also has a FFI CRC weed risk score of 160 (8 × 8 × 2.5 = 160). However, the inherent weed risk scores are 16 (Category 3) and 64 (Category 1), respectively, for species A and B, which better reflects the risk to the environment.

To date, 10 species have been assessed using the refined Rangelands WRA protocol and reviewed by independent assessors with technical knowledge of the species. Examples of the Inherent weed risk score results versus FFI CRC weed risk scores are presented (Table 2).

This environmental WRA system is designed to add to the transparency, scientific rigour and consistency of information available to those considering the selection or use of non-indigenous species and their introduction to novel environments in the WA rangelands. For example, pastoralists could use this information when considering the selection of non-indigenous species. In addition a high proportion of the WA rangelands are currently under pastoral lease, and under this tenure pastoralists are required to apply for a diversification permit from the Pastoral Lands Board (PLB) to grow non-indigenous plants. The PLB through the Department of Lands then seeks advice from State Government agencies including the Department of Primary Industries and Regional Development (DPIRD) and the Department of Biodiversity, Conservation and Attractions (DBCA). These WRAs provide information to aid the decision making process. The new methodology is under final stages of development and is yet to be adopted across all agencies.

ACKNOWLEDGMENTS

Financial support for this project was provided by the Northern Beef Development Project funded by the WA State Government Royalties for Regions program and the Meat and Livestock Australia Donor Company. We thank B. Cook and G. Keighery (DBCA) for the technical reviews of the WRAs.

Table 2. A comparison of Inherent weed risk versus the Invasiveness × Impacts × Potential Distribution WRA scores for key species in the Pilbara and Kimberley of Western Australia.

Species	Common name	Inherent weed risk category	Pilbara Inv × Imp × PD	Kimberley Inv × Imp × PD
Cenchrus setiger Vahl	birdwood grass	2	3	2
Cenchrus ciliaris L.	buffel grass	1	2	1
Leucaena leucocephala (Lam.) de Wit	leucaena	1	3	2
Megathyrsus maximus (Jacq.) B.K.Simon & S.W.L.Jacobs	panic grass	3	4	4
Cenchrus amercianus (L.) Morrone	pearl millet	4	4	4
Chloris gayana Kunth	Rhodes grass	3	4	4
Stylosanthes hamata L. (Taub)	Caribbean stylo	3	4	2

REFERENCES

ALA (2018). Atlas of Living Australia. http://www.ala.org.au/australias-species/ (accessed 14 August 2018).

Anon. (2006). HB 294:2006. National post-border weed risk management protocol. (CRC Australian Weed Management, Adelaide and Standards Australia, Sydney and Standards New Zealand, Wellington).

Anon. (2008). CLIMATCH. Department of Agriculture, Fisheries and Forestry, Bureau of Rural Sciences. http://data.daff.gov.au:8080/Climatch/climatch.jsp (accessed 14 August 2018).

Anon. (2013). Ecocrop database. FAO Rome, Italy http://ecocrop.fao.org/ecocrop/srv/en/home (accessed 14 August 2018).

Brown, L., Barry, S., Cunningham, D. and Bomford, M. (2006). Current practice in applying CLIMATE for weed risk assessment in Australia. Proceedings of the 15th Australian Weeds Conference, eds C. Preston, J.H. Watts and N.D. Crossman, pp. 703-6. (Weed Society of South Australia.

Cook, B.G., Pengelly, B.C., Brown, S.D., Donnelly, J.L., Eagles, D.A., Franco, M.A., Hanson, J., Mullen, B.F., Partridge, I.J., Peters, M. and Schultze-Kraft, R. (2005). Tropical Forages: An interactive selection tool, CSIRO, DPI and F (Queensland), CIAT and ILRI, Brisbane, Australia.) http://www.tropicalforages.info/ (accessed 14 August 2018).

Cook, G.D. and Dias, L. (2005). Turner Review No. 12 It was no accident: deliberate plant introductions by Australian Government Agencies during the 20th century. Australian *Journal of Botany* 54, 601-25.

FAO UN, Food and Agriculture Organisation of the United Nations (1976). A framework for land evaluation. Soils Bulletin 32. (Food and Agriculture Organisation of the United Nations, Rome).

GBIF (2014). Global Biodiversity Information Facility. http://www.gbif.org/ (accessed 14 August 2018).

Lonsdale, M.W. (1994). Inviting trouble – introduced pasture species in northern Australia. *Australian Journal of Ecology* 19, 345-54.

Payne, A.L., Watson, I.W., and Novelly, P.E. (2004). Spectacular recovery in the Ord River catchment. Miscellaneous Publication No. 17/2004. Department of Agriculture and Food, Western Australia Perth.

Randall, R. (2014). Weed threats to the Kimberley Region, Pathways and the risk of incursion. Department of Agriculture and Food, Western Australia.

Schoknecht, N.R., and Pathan, S. (2013). Soil groups of Western Australia: A simple guide to the main soils of Western Australia (4th Edition). Resource Management Technical Report 380. Department of Agriculture and Food, Western Australia, Perth.

Stone, L.M. and Byrne, M. (2011). Comparing the outputs of five weed risk assessment models implemented in Australia: are there consistencies across models? *Plant Protection Quarterly* 26, 29-35.

Stone, L.M., Byrne, M., and Virtue, J.G. (2008). An environmental weed risk assessment model for Australian forage improvement programs. *Australian Journal of Experimental Agriculture*, 48, 568-74.

Stone, L., Munday, C. and Bettink, K. (2012). Future Farm Industries CRC Weed Risk Assessment.

Thorpe, J.R. and Lynch, R. (2000). The determination of Weeds of National Significance. National Weeds Strategy Executive Committee, Launceston.

van Gool, D., Tille, P.J. and Moore, G.A. (2005). Land evaluation standards for land resource mapping: Assessing land qualities and determining land capability in south-western Australia. Report 298. Department of Agriculture and Food, Western Australia, Perth.

Virtue, J. (2004). SA Weed Risk Management Guide. Animal and Plant Control Commission, South Australia.

Virtue, J.G., Bennett, S.J. and Randall, R.P. (2004), Plant introductions in Australia: how can we resolve 'weedy' conflicts of interest? Proceedings of the 14th Australian Weeds Conference, eds B.M. Sindel and S.B. Johnson pp. 42-8. (Weed Society of New South Wales, Wagga Wagga).

WRASP: A spatial weed risk assessment tool reveals critical variations in weed risks

Darren J. Kriticos[1], Josef Beautrais[2] and Mike B. Dodd[2]
[1] CSIRO Entomology, GPO Box 1700, Canberra, Australian Capital Territory 2601, Australia
[2] AgResearch, Grasslands Research Centre, Private Bag 11008, Palmerston North 4442, New Zealand
(darren.kriticos@csiro.au)

Summary Due to the overwhelming number and diversity of introduced invasive plants, coupled with limited weed management budgets, government agencies are typically required to employ systems to prioritise weeds for management attention. To help tackle this problem, an analytical protocol and spreadsheet tool was previously developed for post-border weed risk management (PBWRM). The popular PBWRM tool utilises a simple framework that ignores any spatial variation in risk factors within the geographical area of risk concern. However, invasive plants vary spatially in risk factors such as invasiveness, potential impacts, and feasibility of control. The PBWRM system requires the analyst to integrate each of these spatially-variable factors into a single risk score. This process is inherently subjective, difficult to implement and conceals the variations in weed risks. To address these concerns our trans-Tasman research team took the PBRWM logic and spatialised it, allowing weed managers to assess weed risks and management classes across geographical space. We illustrate this new spatial system using a case study of *Senecio glastifolius* L.f. in New Zealand, comparing the results of a spatial and an aspatial analysis of the risks it poses, and the consequent logical management options. The spatial portrayal of risks revealed locations of both higher and lower risk and suitability for management attention that were concealed within the aspatial weed risk scores of the current PBWRM system. The overall national level risk assessed using the spatial tool was also appreciably higher than that indicated by the aspatial scoring system. The spatial tool, WRASP, takes its name from Weed Risk Assessment SPatial. In New Zealand, WRASP forms part of an integrated set of tools for strategic weed management. This decision-support toolkit is described in Bourdôt *et al.* (2018).

Keywords CLIMEX, invasive plants, prioritisation tool, strategic weed management, Weed Risk Assessment, Weed Risk Management.

INTRODUCTION

Post-border weed risk management poses a set of costly and challenging problems to both public and private land managers. Taxonomically, exotic plants now dominate many national flora. For example, some 26,000 of the 46,000 known vascular plants present in Australia are exotic (Randall 2007). Of these, a little over 10% (~2700) are naturalised, and of these, approximately 30% (798) are considered a significant threat to the environment or agriculture. In New Zealand, the situation is similar, with the number of naturalised exotic plant species being similar to the number of indigenous species (~2500), and these being approximately 10% of the total number of introduced species (Howell 2008). Clearly, in any given jurisdiction there is a need to assess the threat posed by weeds as a step toward ensuring that appropriate management strategies are prepared and implemented for species posing the most significant threats.

Since the late 1990s, there has been a considerable interest in pre- and post-border risk assessment tools for invasive plants (Pheloung *et al.* 1999, Groves *et al.* 2001, Gordon *et al.* 2008). These tools provide a systematic method for considering the weed risk factors in both relative and absolute terms. The patterns of weed characteristics that are associated with weediness and invasiveness (Reichard 2001) form the basis for a spreadsheet-based point-scoring system to assess the likely weed risks plants might pose under different circumstances.

The Australian and New Zealand National Post-Border Weed Risk Management Protocol HB 294:2006 was developed on the back of a pre-border weed risk assessment system (Pheloung *et al.* 1999). It was developed into a standard protocol for use in Australia and New Zealand (PBWRM, Anon. 2006). Since that time the Food and Agriculture Organization have adopted it, and it has become an important tool in the management of weeds, being applied in Australia, New Zealand, Latin America, South America and North Africa (FAO 2011, Auld 2012). The logic of the system is clear, and the responses can be reviewed and critiqued; attributes that have doubtless contributed to it becoming popular throughout Australia, New Zealand and elsewhere. This system was updated recently to reflect developments in: '...risk management practice and in indicating the reliability of predictions; the management of contentious plants; and the translation of

WRM results into policy and management responses.' (Auld *et al.* 2012).

One of the inherent difficulties with using this type of aspatial scoring system is the need to account for the heterogeneous nature of the weed threats and the production or natural resource assets at risk.

This introduces a significant source of instability into the risk assessment method, in terms of the operator-specific or subjective nature of the factors taken into consideration when making their broad estimate. In the worst case they may simply apply a score transposed from a previous assessment in another region without considering the context-specific factors. The risk assessment results are highly sensitive to how the individual risk analyst transforms the heterogeneous risk factors into a single response (e.g. averaging or taking the extreme case) (see discussion and references in Barry and Lin 2010). A second consequence of the point based nature of the PBWRM tool is that each jurisdiction is required to complete the assessment *de novo*, resulting in wasted effort and inconsistent assessments.

Several authors have recognised that weed risks are inherently spatial, and some effort has been expended on developing spatial decision support systems for identifying weed management actions for invasive species (Crossman 2004, Crossman and Bass 2008, Januchowski-Hartley *et al.* 2011, Skurka Darin *et al.* 2011). The analytical frameworks of Januchowski-Hartley *et al.* (2011) and Skurka Darin *et al.* (2011) are each designed to optimise the allocation of resources to tactical weed management, rather than a broader assessment of relative risks. As observed in Auld (2012), they also require detailed information that may not be readily available for newly invading species, thereby limiting their utility for prioritising the management of weed communities that include invasive plants that are newly-arrived through to those that are well-established.

Here we describe the development of a spatialised version of the PBWRM tool to provide an analytical framework for identifying strategic weed management priorities. We compare the results of applying the spatial and aspatial versions of the PBWRM tool to assess the risks posed by *Senecio glastifolius* L.f. and the management options for it in New Zealand.

MATERIALS AND METHODS

Case study plant *Senecio glastifolius* (Asteraceae: Pink ragwort, Holly-leaved ragwort) is native to the Cape region of South Africa (Wells *et al.* 1986). It is an annual or short-lived perennial herb that grows to a height of 1–1.5 m and is presently increasing its distribution in both New Zealand (Figure 1) and south-western Western Australia (Hussey *et al.* 1997, Beautrais 2013). The foliage is palatable to livestock, and hence the plant typically invades ungrazed waste areas, roadside batters and coastal dunes. Its impact is regarded as limited in terms of alteration of natural biota in those ecosystems where it establishes (Williams *et al.* 1999). It has been the subject of localised but considerably labour-intensive management campaigns, commonly using manual removal.

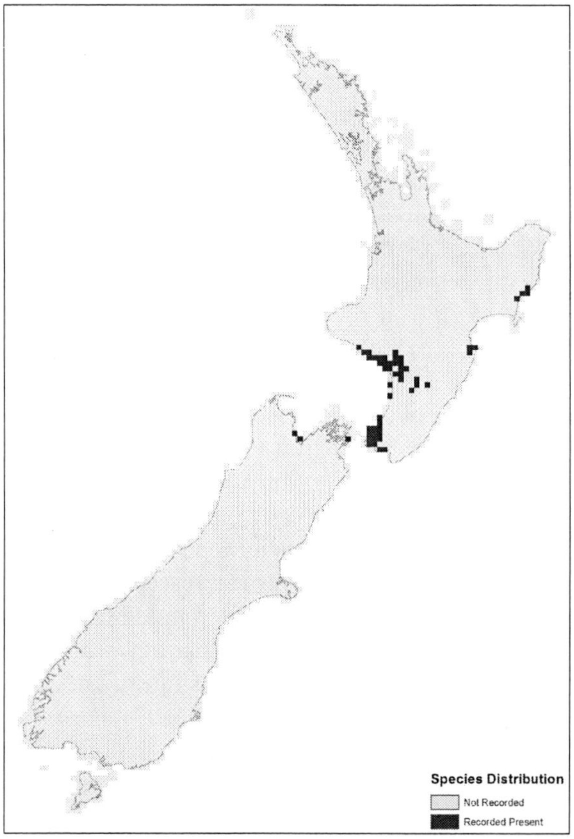

Figure 1. Current known distribution of *Senecio glastifolius* in New Zealand (source: herbarium records, Regional Council databases, Department of Conservation databases, and the observations of the authors).

Post-Border Weed Risk Management Protocol The Post-Border Weed Risk Management (PBWRM) protocol was published as an Australian and New Zealand Standard (Anon. 2006). It is based on 43 questions covering aspects of the history, biogeography, ecology, biology and impacts of a given species. The scoring system requires either ratings (0, 1 or 2) or yes/no answers to questions and is constructed such that equal weight is given to most questions. This aspatial system is implemented as an Excel spreadsheet. The PBWRM system was used to assess the risks and management options for *S. glastifolius* in the context of New Zealand.

Spatial weed risk management system (WRASP)
The logic of the PBWRM system was translated directly into ArcGIS (10.3) using model-builder. Where questions had answers that concerned the biology of the species and were spatially uniform, their non-spatial form was retained as in the current spreadsheet model (e.g. questions concerning dispersal mechanisms). Where a question could be answered spatially, the system prompts for a spatial dataset in raster form. The relevant spatial dataset for responding to each question is classified using exactly the same scoring schema as that used in the aspatial system.

RESULTS

The aspatial PBWRM system resulted in a risk score of 6, indicating that *S. glastifolius* posed a negligible risk to New Zealand as a whole. Combined with a feasibility of containment score of 27, the weed risk category was low and the corresponding recommended action was *limited action*. The risk score for *S. glastifolius* was substantially lowered because its potential distribution in New Zealand is limited.

In contrast, the spatial WRASP system indicated an average risk score of approximately 20. More importantly, it provided maps indicating that whilst most of New Zealand was under negligible threat, some areas were under low, medium and even high threat (Figure 2a, 3). The risk maps indicate that the greatest threats lie in the southern half of the North Island apart from the central highlands, and in the peri-coastal arc across the northern quarter of the South Island. The corresponding management actions range from *monitor populations* through most of the country through to *destroy infestations*, *protect sites* and *contain spread* (Figure 2b).

DISCUSSION

WRASP revealed significant sub-national spatial variation in weed risk and the technically prudent management strategy for the case study weed. Given New Zealand's legislative mandate for Regional Councils to manage pests, the WRASP system provides an economical and effective means for Regional Councils to identify risks and technically appropriate management strategies, perhaps as an input into a further economic screening process (Bourdôt *et al.* 2015). The single spatial analysis conducted here for *S. glastifolius* can be overlain by Regional Council (NZ) or Natural Resource Management Area boundaries (Australia) to Figure 3 highlight the specific risks to each management unit and the range of management options that

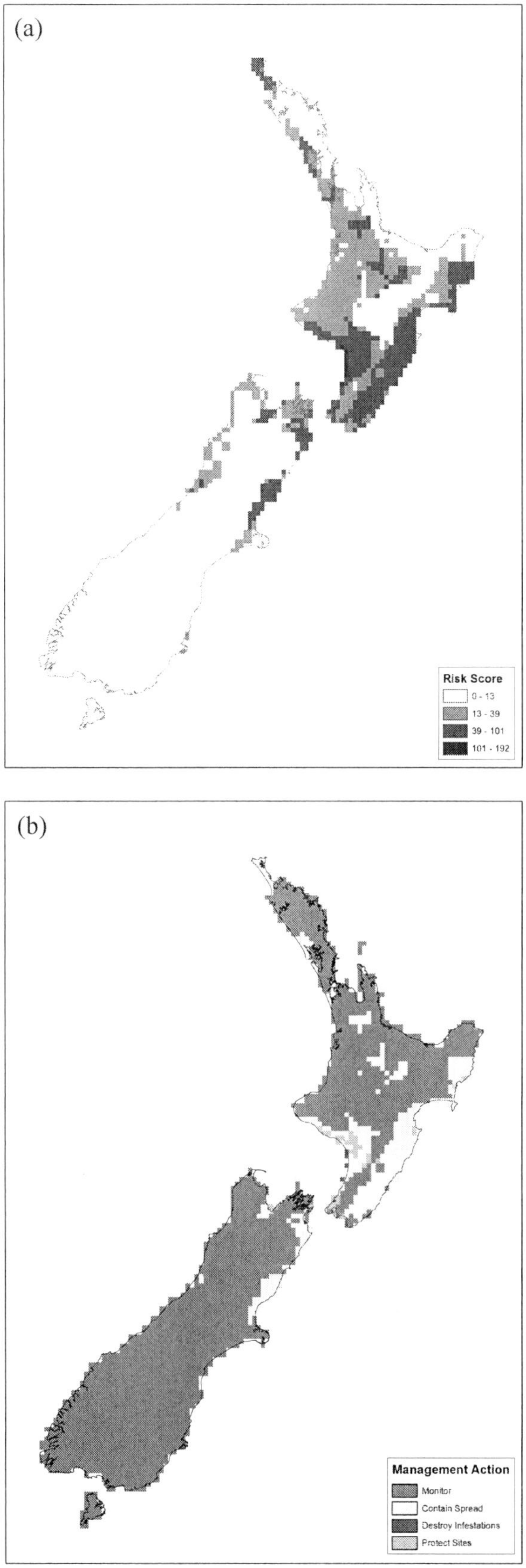

Figure 2. Spatialised weed risk (a) and management options (b) *Senecio glastifolius* in New Zealand developed using WRASP.

should be considered. At a glance, each biosecurity manager could immediately appreciate the threat patterns they are responsible for managing. Further, such a picture can highlight opportunities for transboundary coordinated efforts for weed management. For example, efforts to contain the spread of a weed in one region may stop or slow the spread to another region where it could generate significant impacts.

The fact that the single analysis can provide answers for all of the councils in New Zealand suggests that there is an economy of scales to be gained by undertaking the analyses in a centralised, or at least coordinated manner. The heterogeneity of the resulting risks also highlights the importance of considering how the risk factors vary across the country, and the folly of applying the results of the aspatial WRM system throughout the country.

The WRASP system is inherently more 'truthful' than the PBWRM; allowing the analyst to express their understanding of the various risk components using spatially-explicit evidence at a scale that is relevant for planning and executing strategic weed management. The degree of additional detail revealed in a WRASP analysis c.f. a PBWRM analysis is likely to vary depending upon the availability of information regarding the species being analysed, and the ability of the analyst to reveal the information spatially. WRASP reveals the outcome of combining the risk components in a spatially-explicit form. Conversely, the PBWRM requires the analyst to reduce the spatial complexity of the risk factors subjectively, increasing the potential for uncontrolled biases to be introduced into the analysis. At the extreme simplest limit, the WRASP model gives the same result as the PBWRM when no information is spatialised. In our case study we give examples of techniques for generating spatial datasets to answer questions in WRASP using common tools such as CLIMEX (Kriticos *et al.* 2015) to estimate the potential for spread of the weed.

WRASP is presently being made available to Regional Councils in New Zealand as part of a comprehensive Decision Support System (Bourdôt *et al.* 2018). When we have a number of completed spatial risk assessments we will turn our attention to a means of prioritising weeds at different levels of scale. Australia can, and should benefit from the development of WRASP.

ACKNOWLEDGMENTS

We thank the New Zealand Foundation for Research, Science and Technology for funding this research (Undermining Weeds, C10X0811).

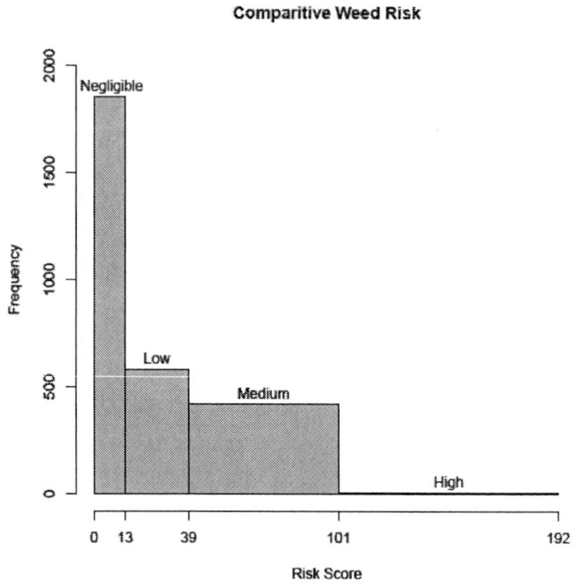

Figure 3. Histogram of risk scores for *Senecio glastifolius* in New Zealand developed using WRASP.

REFERENCES

Anon. (2006). HB 294:2006. National post-border weed risk management protocol. (CRC for Australian Weed Management, Adelaide and Standards Australia, Sydney and Standards New Zealand, Wellington).

Auld, B. (2012). An overview of pre-border weed risk assessment and post-border weed risk management Protocols. *Plant Protection Quarterly* 27, 105-11.

Auld, B., Johnson, S., Ainsworth, N., Brunel, S., Csurhes, S., Downey, P., Ferdinands, K., Hannan-Jones, M., Hennecke, B., Keese, P., Kriticos, D., Low, D., Munday, C., Panetta, F.D., Robold, A., Setterfield, S., Steele, J., Stewart, K. and Virtue, J. (2012). Further development of the National Weed Risk Management Protocol. Proceedings of the 18th Australasian Weeds Conference, ed. V. Eldershaw, pp. 317-9. (Weed Society of Victoria, Melbourne)

Barry, S. and Lin, X. (2010). Point of Truth calibration: Putting science into scoring systems. Australian Centre of Excellence for Risk Analysis (ACERA).

Beautrais, J.R. (2013). The geographic distribution of *Senecio glastifolius* in New Zealand: past, current and climatic potential, Victoria University of Wellington, Wellington, New Zealand.

Bourdôt, G., Basse, B., Kriticos, D. and Dodd, M. (2015). Cost-benefit analysis blueprint for regional weed management: *Nassella neesiana* (Chilean needle grass) as a case study. *New Zealand Journal of Agricultural Research* 58, 325-38.

Bourdôt, G.W., Kriticos, D.J. and Dodd, M.B. (2018). A national weed management decision-support system. Proceedings of the 21st Australasian Weeds Conference, 9–13 September, Sydney, New South Wales, eds S. Johnson, L. Weston, H. Wu and B. Auld, pp. 82-5. (Weed Society of New South Wales).

Crossman, N.D. (2004). The role of spatial analyses in post-border weed risk management. Proceedings of the 14th Australian Weeds Conference, eds B.M. Sindel and S.B. Johnson, pp. 450-3. (Weed Society of New South Wales, Wagga Wagga)

Crossman, N.D. and Bass, D.A. (2008). Application of common predictive habitat techniques for post-border weed risk management. *Diversity and Distributions* 14, 213-24.

FAO (2011) 'Procedures for post-border weed risk management', 2nd edition. (Plant Production and Protection Division, Food and Agriculture Organization of the United Nations, Rome).

Gordon, D.R., Onderdonk, D.A., Fox, A.M. and Stocker, R.K. (2008). Consistent accuracy of the Australian weed risk assessment system across varied geographies. *Diversity and Distributions* 14, 234-42.

Groves, R.H., Panetta, F.D. and Virtue, J.G. (2001) 'Weed Risk Assessment' (CSIRO Publishing, Melbourne, Australia).

Howell, C. (2008) 'Consolidated list of environmental weeds in New Zealand' (Science and Technical Publication, Department of Conservation, Wellington).

Hussey, B.M.J., Keighery, G.J., Cousens, R.D., Dodd, J. and Lloyd, S.G. (1997) 'Western weeds. A Guide to the weeds of Western Australia' (Plant Protection Society of Western Australia, Perth).

Januchowski-Hartley, S.R., Visconti, P. and Pressey, R.L. (2011). A systematic approach for prioritizing multiple management actions for invasive species. *Biological Invasions* 13, 1241-53.

Kriticos, D.J., Maywald, G.F., Yonow, T., Zurcher, E.J., Herrmann, N.I. and Sutherst, R.W. (2015) 'CLIMEX Version 4: Exploring the Effects of Climate on Plants, Animals and Diseases' (CSIRO, Canberra).

Pheloung, P.C., Williams, P.A. and Halloy, S.R. (1999). A weed risk assessment model for use as a biosecurity tool evaluating plant introductions. *Journal of Environmental Management* 57, 239-51.

Randall, R.P. (2007) 'The introduced flora of Australia and its weed status' (CRC for Australian Weed Management Adelaide).

Reichard, S.H. (2001). The search for patterns that enable prediction of invasion. *In* 'Weed Risk Assessment', eds R.H. Groves, F.D. Panetta and J.G. Virtue, pp. 10-20. (CSIRO Publishing, Melbourne).

Skurka Darin, G.M., Schoenig, S., Barney, J.N., Panetta, F.D. and Di Tomaso, J.M. (2011). WHIPPET: A novel tool for prioritizing invasive plant populations for regional eradication. *Journal of Environmental Management* 92, 131-9.

Wells, M.J., Balsinhas, A.A., Joffe, H., Engelbrecht, V.M., Harding, G. and Stirton, C.H. (1986). A catalogue of problem plants in southern Africa. *Memoirs of the Botanical Survey of Southern Africa* 53, 1-658.

Williams, P.A., Ogle, C.C., Timmins, S.M., La Cock, G. and Reid, V. (1999). Biology and ecology of *Senecio glastifolius* and its spread and impacts in New Zealand. Department of Conservation.

A national weed management decision-support system

Graeme W. Bourdôt[1], Darren J. Kriticos[2] and Michael B. Dodd[3]
[1] AgResearch, Lincoln, Private Bag 4749, Christchurch 8140, New Zealand
[2] CSIRO Entomology, GPO Box 1700, Canberra, Australian Capital Territory, 2601 Australia
[3] AgResearch Grasslands, Private Bag 11008, Palmerston North 4442, New Zealand
(graeme.bourdot@agresearch.co.nz)

Summary In New Zealand, Regional Councils have the primary governmental responsibility for coordinating the management of weeds and other pests. Similarly, in Australia, it is the States that have this responsibility. In both countries, the national coordination and harmonisation of weed control efforts has been a perennial challenge, leading to initiatives such as the Weeds of National Significance in Australia and the National Policy Direction for Pest Management in New Zealand. Further, in both countries, the number of weeds that affect natural and productive ecosystems far exceeds the resources available to manage them such that prioritisation is necessary. In New Zealand, a favourable cost benefit analysis is required for a weed to be considered for inclusion in a Regional Pest Management Plan. The ability of New Zealand's Regional Councils to undertake such analyses has been hampered by a lack of crucial biophysical data regarding the target weed, including its current distribution, potential future distribution and its rate of spread in the absence of control. To redress this, and to facilitate the required nationally consistent approach, scientists from AgResearch and CSIRO have collaborated to develop a platform of five tools: a national weed occurrence database (NWDD) that automatically harvests regional weed occurrence data; a database of CLIMEX models and their projections for New Zealand (CLIMENZ); a weed spread model (MDiG); a spatial weed risk assessment model (WRASP); and a cost benefit model (CBA for regional pest management). In this paper, we illustrate this decision-support system and comment on its adoption by Regional Councils in New Zealand and its contribution to targeting resources to the most deserving weed problems. We then consider how easily it might be extended for use in Australia.

Keywords Biosecurity, weed, cost benefit analysis, CLIMEX, weed risk assessment, spread, spatial occurrence.

INTRODUCTION

New Zealand's *Biosecurity Act 1993* provides for the eradication or effective management of harmful organisms that are present in New Zealand (NZ) through 'Pest Management' (national and regional) and 'Pathway Management' Plans (Ministry for Primary Industries 2016). These plans must be consistent with the National Policy Direction for Pest Management 2015 (NPD) to ensure that they provide for the best use of available resources for New Zealand's best interests, and are aligned with one another when necessary (Ministry for Primary Industries 2015). To this end, the NPD firstly requires, and provides direction for, an analysis of the benefits and costs of a proposed pest management plan. The analyses must quantify the impacts of the pest and the benefits of alternative management options. They must also state assumptions and account for uncertainties. Secondly, the NPD requires that a weed (or other pest) be managed under one of four defined program types (exclusion, eradication, progressive containment, sustained control) (Ministry for Primary Industries 2015).

The ability of New Zealand's Regional Councils to comply with the requirements of the NPD in their Regional Pest Management Plans (RPMP), in particular, regarding the analysis of benefits and costs for the proposed management of a weed species, has been hampered by a lack of availability of crucial biophysical data. This includes the species' current distribution, potential future distribution and its probable rate of spread and impact in the absence of the proposed regional management program. To redress this, and to facilitate the nationally consistent approach and standardised classification of weed management programs specified in the NPD, scientists from AgResearch and CSIRO have collaborated to develop a platform of five tools. Two are currently available. Three require further development for which funding is being sought. In this paper we describe these tools, present usage statistics for one that is currently available (a cost benefit analysis web application), and consider how the latter, and the platform as a whole, might be applied in both New Zealand and Australia.

THE DECISION SUPPORT SYSTEM

Overview The weed management decision support system that we are developing is structured around informing the cost benefit analysis (CBA) required for a weed (or other pest) proposed for inclusion in a Regional Pest Management Plan in New Zealand (Figure 1). The system recognises that the CBA must account for the expected benefits and costs for managing the species of interest under the proposed Program Type (Exclusion, Eradication, Progressive Containment, Sustained Control).

The component WRASP, a spatially explicit weed risk analysis tool, determines which of the four Program Type(s) best suits the species. The costs of the program are determined externally, based on the type and frequency of interventions required to achieve the outcome for the program. The benefits are informed jointly by the three component system tools NWDD, CLIMENZ and MDiG (Modular Dispersal in GRASS). Respectively, they estimate the species' current distribution (ha), potential distribution (ha) and the rate (spread rate) at which the potential distribution is expected to be realised in the absence of the management program. The CBA returns the Net Present Value (NPV) and Internal Rate of Return (IRR) supporting the decision to proceed (or not) with the proposed management program for the species of interest.

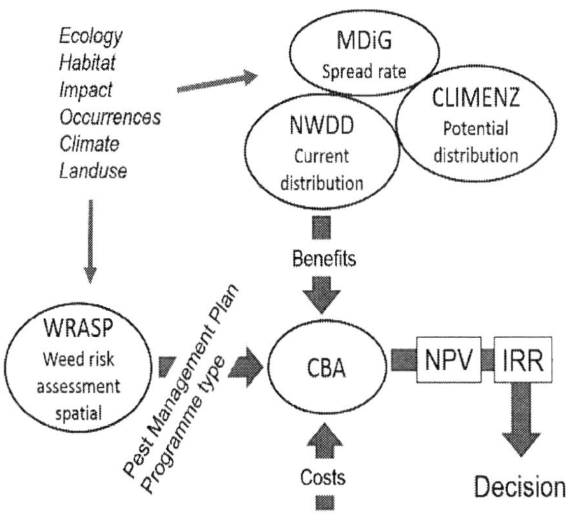

Figure 1. Regional weed management decision support system under development in New Zealand.

NWDD There is significant interest in New Zealand in establishing a National Weed Distribution Database (NWDD) that would automatically harvest weed occurrence data from the disparate databases held by regional councils, Department of Conservation and other organisations (Basse *et al.* 2008). In addition to the many benefits perceived in a recent study (viewing changes in weed distributions, understanding threats, understanding weed spread, improved decision making, long-term archival service for weed distribution data, identifying national priorities for coordinated action) (Cooper *et al.* 2010), a NWDD could provide the data necessary to quantify the current infested area of a weed in a region as required by the CBA tool (Figure 2). A prototype NWDD has been developed, based on the work of Clayson Howell in the NZ Department of Conservation (DoC), which combined NZ data on 180 weeds from several sources – herbaria records, Regional Council databases, DoC databases and personal interviews with biosecurity staff in all Regional Councils. This data set was completed in 2008 and has subsequently been lodged in an online resource under development.

CLIMENZ Climenz, a web-based tool approximates the CLIMEX algorithm (Kriticos *et al.* 2015, Sutherst and Maywald 1985). It allows authorised users to generate high resolution maps for the projected climate suitability of New Zealand for particular taxa under current and future climates (Kean *et al.* 2015). The projections are made from published CLIMEX models of which 118 were available for plants in 2015; models for other weed species can be added when available. The output from CLIMENZ includes land areas (ha) for the species that are climatically Optimal, Suitable, Marginal and Unsuitable. CLIMENZ is available at (http://b3.net.nz/climenz) and these estimates inform the potential infested area parameter in the CBA tool (Figure 2).

MDiG MDiG is a dynamic, stochastic dispersal model that simulates temporally discrete dispersal processes within an open-source GIS system (Neteler and Mitasova 2004). A raster map of the current distribution of a given pest with an appropriate cell size is updated annually through two dispersal processes – local spread (based on knowledge of population boundary movement) and long-distance spread (by sampling from a Poisson probability distribution based on historic distribution data, such as might be obtained from GBIF, http://data.gbif.org/datasets). Newly establishing infestations are restricted to a climatic suitability envelope, obtained from CLIMEX, and a geographic envelope (a GIS layer defining unsuitable habitats such as water bodies for terrestrial species). The model produces a large set of realised distributions from many stochastic replications and thus the output is an occupancy probability map over time. The impact of RPMP actions can be simulated by modifying key parameters such as the success of new infestations, spread rates etc. The model has been applied to two organisms to date, Argentine ant (Pitt *et al.* 2009) and the shrub weed *Buddleja davidii* Franch. (Pitt *et al.* 2011).

WRASP This tool is a spatialised version of the widely-used aspatial Post Border Weed Risk Management tool, PBWRM (Anon. 2006). It allows the risks posed by a weed to be characterised using spatial variables, and indicates the most appropriate type of management program (exclusion, eradication etc.) as a spatial variable. This new spatial system for weed risk assessment has been illustrated using a case study in New Zealand, comparing the results of a spatial and an aspatial analysis of the risks it poses and the logical management program options (Kriticos *et al.* 2018). In the case study, the spatial view of risks revealed locations of higher and lower risk and suitability for management attention that were hidden by blanket, aspatial weed risk scores of the PBWRM system (Kriticos *et al.* 2018). Development of a web-based version of WRASP will help complete the decision support system by adding a tool enabling the user to objectively determine the most suitable management program type for the weed in question.

CBA This tool employs a standard discounted cash flow analysis to calculate the NPV and IRR for the proposed weed management program. It is accessible free of charge at the AgPest website (https://www.agresearch.co.nz/cba/cba.php) and its underpinning logic is described elsewhere (Bourdôt *et al.* 2015). The user begins by setting the species, region, program type to be simulated (exclusion, eradication, progressive containment, sustained control), and the simulation period (up to 100 years). The expected invasion trajectory of the weed in the region in the absence of the program is then set by the input of the three parameters of a logistic spread model (Current area infested (ha), Potential area infested (ha) and Spread rate (ha y^{-1})). These three parameters are informed, respectively by NWDD, CLIMENZ and MDiG. The weed-free earnings expected from the land at risk of invasion, the expected reduction in these earnings due to the weed, the discount rate to be applied, and the probability of success of the program, are then entered. These parameters are then used by the model, along with the defined invasion trajectory, to calculate the present value of the loss prevented in each year of the program. The costs incurred by the proposed program in each year are then entered and the model returns the NPV as the difference between the present value of the losses prevented ($Cost_0$) and the present value of the costs incurred for the management program ($Cost_1$) (Figure 2). The IRR is also calculated, a useful metric for prioritising species for inclusion in a RPMP. All parameters in the CBA tool, apart from the current area infested (which is assumed to be known with certainty), are adjustable using sliders for sensitivity testing, enabling the uncertainties to be addressed as required by the NPD. The results of each analysis are presented on-screen in graphs and in a downloadable report file.

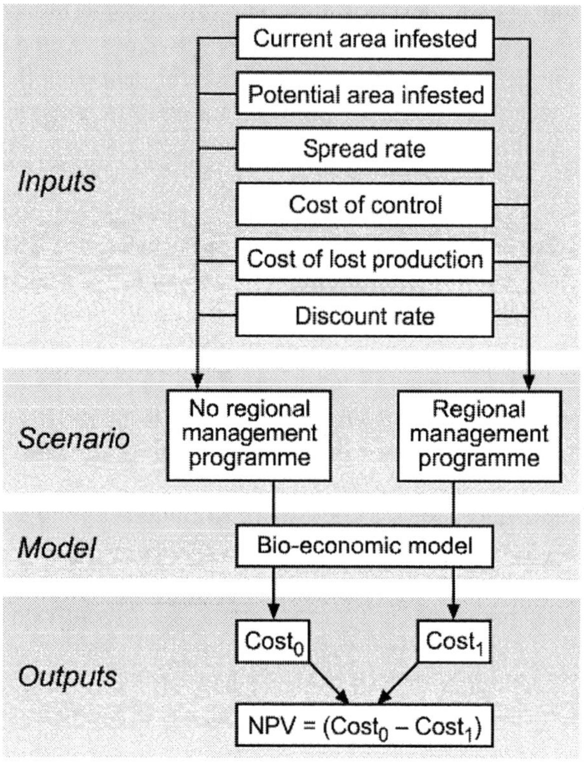

Figure 2. The logic underpinning the CBA model in the system (Bourdôt *et al.* 2015).

ADOPTION

The CBA tool is available to users without restriction whilst CLIMENZ is licensed for use by the New Zealand Ministry for Primary Industries and AgResearch. Whilst the other three supporting tools are yet to be completed, the CBA tool is being used by regional councils throughout New Zealand (Table 1). Usage is increasing with a 30% increase in users and a 220% increase in the number of sessions between 2016 and 2017 following launching in 2015 (Table 1).

Table 1. Usage of the regional weed management CBA web application in New Zealand since launching in 2015. The application is available at https://www.agresearch.co.nz/cba/cba.php

	2016	2017
Users	80	104
Sessions	127	280
Sites	18	18

ADAPTING THE WEED MANAGEMENT DECISION-SUPPORT SYSTEM FOR AUSTRALIA

Conceptually, the Regional Councils in New Zealand and the Australian States are similar in many respects. Of particular interest here is the fact that most post-border biosecurity responsibilities fall on these governance bodies. The system described here could be adapted to the Australian context relatively easily. The NWDD could, for example, easily be extended off the Atlas of Living Australia platform (www.ala.org). Indeed, the concept for the NWDD was drawn from a prototype system developed in Australia for the National Land and Water Audit (Auricht Projects). The CLIMENZ database provides an estimate of the climatic niche for each weed species using CLIMEX models. Because CLIMEX models can be run using globally available climate data, this information could easily be generated for Australia at a suitable spatial resolution to underpin strategic weed management. MDiG is an open-source generic spread modelling platform that can be applied to any jurisdiction. WRASP is presently coded as a model within ARCGIS, and can be run using whatever geographical data the analyst wishes to use. The CBA tool is inherently independent of jurisdiction, and the results are likely to be universally relevant to decisions involving resource allocation to weed management in situations where there is market failure.

CONCLUSION

Adapting these tools to the Australian context offers potentially significant savings in analytical effort, as well as improvements in the quality of information to support decisions regarding strategic weed management. This system offers the prospect of a robust, defensible, cost-effective system for strategic weed management.

ACKNOWLEDGMENTS

We thank the New Zealand Foundation for Research, Science and Technology for funding this research (Undermining Weeds, C10X0811).

REFERENCES

Anon. (2006). HB 294:2006. National post-border weed risk management protocol. (CRC for Australian Weed Management, Adelaide and Standards Australia, Sydney and Standards New Zealand, Wellington).

Basse, B., Bourdôt, G., Brown, J. and Lamoureaux, S. (2008). New Zealand national weeds distribution database: a feasibility study – Report for Environment Southland, Envirolink medium advice 499-ESRC212. University of Canterbury, Christchurch.

Bourdôt, G.W., Basse, B., Kriticos, D.J. and Dodd, M. (2015). A cost-benefit analysis blueprint for regional weed management: *Nassella neesiana* (Chilean needle grass) as a case study. *New Zealand Journal of Agricultural Research* 58, 1-14.

Cooper, J., Carver, J. and Browne, M. (2010). National Weeds Distribution Database scoping study. Landcare Research, Lincoln.

Kean, J.M., Brockerhoff, E.G., Fowler, S.V. and Gerard, P. (2015). Effects of climate change on current and potential biosecurity pests and diseases in New Zealand. Report for Ministry for Primary Industries.

Kriticos, D.J., Beautrais, J.R. and Dodd, M.B. (2018). WRASP: A spatial strategic weed risk analysis tool reveals important sub-national variations in weed risks. *Weed Research* (in press).

Kriticos, D.J., Maywald, G.F., Yonow, T., Zurcher, E.J., Herrmann, N.I. and Sutherst, R.W. (2015) 'CLIMEX Version 4: Exploring the effects of climate on plants, animals and diseases' (CSIRO, Canberra).

Ministry for Primary Industries (2015). National Policy Direction for Pest Management 2015, ed. J. Mateparae, pp. 10. Ministry for Primary Industries, New Zealand, Wellington.

Ministry for Primary Industries (2016). *Biosecurity Act 1993*. http://www.legislation.govt.nz/act/public/1993/0095/latest/DLM314623.html (accessed 25 July 2018).

Neteler, M. and Mitasova, H. (2004) 'Open Source GIS: A Grass GIS approach' (Kluwer Academic Publishers, Springer, Boston).

Pitt, J.P.W., Kriticos, D.J. and Dodd, M.B. (2011). Temporal limits to simulating the future spread pattern of invasive species: *Buddleja davidii* in Europe and New Zealand. *Ecological Modelling* 222, 1880-7.

Pitt, J.P.W., Worner, S.P. and Suarez, A. (2009). Predicting Argentine ant spread over the heterogenous landscape using a spatially-explicit model *Ecological Applications* 19, 1176-86.

Sutherst, R.W. and Maywald, G.F. (1985). A computerised system for matching climates in ecology. *Agriculture, Ecosystems & Environment* 13, 281-99.

Futures in risk assessment: Challenges in preventing and mitigating weed threats

Stephen B. Johnson[1], Belinda Mitterdorfer[2], Dorjee[3,4], Tony Buckmaster[3,5] and Paul O. Downey[3]
[1] Weed Research Unit, New South Wales Department of Primary Industries, Locked Bag 6006, Orange, New South Wales 2800, Australia
[2] Department of Agriculture and Water Resources, GPO Box 858, Canberra, Australian Capital Territory 2601, Australia
[3] Institute for Applied Ecology, University of Canberra, Australian Capital Territory 2601, Australia
[4] Ministry of Agriculture and Forests, Royal Government of Bhutan, PO Box 1095, Thimphu, Bhutan
[5] Centre for Invasive Species Solutions, University of Canberra, Australian Capital Territory 2601, Australia
(stephen.johnson@dpi.nsw.gov.au)

Summary Predicting the future is fraught with danger: this is no different in the weed management sphere. Yet as weed management professionals we try to in order to prevent an increased number and spread of weed species that negatively impact on the economic, environmental and societal concerns which are most precious to us. Pre-border weed risk assessment (WRA) and post-border weed risk management (WRM) systems seek to both predict and prioritise management responses for a less 'weedy' future. After initial Australian development, the use and further development of WRA systems has spread across the globe. Their use largely supports the exclusion (quarantine – pre-border) of non-native plant species from countries where they could become weeds. In comparison, post-border WRM prioritisation systems have not seen similar levels of development, testing and adoption, despite their importance in managing non-natives post-border (i.e. after they arrive).

Here we undertook an analysis of the future needs, strengths and shortfalls as well as opportunities associated with both the Australian WRA and WRM systems. Then we examined recent developments in WRA (pre-border) and WRM approaches from Australia and globally. From this we identified a series of key areas for future investment and research to improve both the WRA and WRM systems and their effectiveness in helping us reduce the future weed problem in Australia: for example given the rapid increase in environmental weeds in Australia, systems that predict future weeds already present in Australia are urgently needed (see Downey and Johnson, these proceedings).

Keywords Weed risk assessment, weed risk management, prediction, prioritisation.

Let's get strategic: Weed management for conservation in Western Australia

Kellie Passeretto
Ecosystem Health Branch, Parks and Wildlife Service, Department of Biodiversity, Conservation and Attractions, Locked Bag 104, Bentley Delivery Centre, Western Australia 6983, Australia
(kellie.passeretto@dbca.wa.gov.au)

Summary Ten years on from its implementation, the status of the Department of Biodiversity, Conservation and Attractions (DBCA) internal Weed Prioritisation Process was reviewed. This process was instigated to provide an integrated approach to weed management in Western Australia (WA) that was both species-led and asset-protection-based. These two elements of the process are used to inform regional priorities and strategic allocation of funds for the best return on investment and the most effective long-term outcomes for weed management and asset protection.

The outcomes of the species-led process are being actively utilised by staff and the process itself is being applied to determine priorities at different management levels across the department.

The asset-protection-based approach has been the most challenging part of the prioritisation process. While Department staff are still investigating a more detailed approach, we have implemented an interim process based on the ratings from the species-led prioritisation process and expert knowledge of the assets for which weeds are a key threatening process. The outcomes of this interim process are being applied across the state.

The new Weed Occurrence and Treatment App has provided the department with a coordinated, strategic approach to weed mapping and monitoring. It is assisting the department to map weed occurrences, plan weed management efforts and more effectively monitor management effectiveness.

Keywords Strategic weed management, management effectiveness, prioritisation, weed mapping application.

INTRODUCTION

With the continual decline in resources, it is imperative that we develop strategic approaches to weed management which allow us to determine our priorities, monitor our management effectiveness and apply effective adaptive management. In 2008, the Western Australia Department of Biodiversity Conservation and Attractions (DBCA, the Department) implemented an internal Weed Prioritisation Process designed to address the first of these issues by providing an integrated approach to weed management on DBCA-managed lands within Western Australia (WA). The aim of this process was to establish both a species led and an asset-protection-based approach to weed management.

The department recognised that implementing a coordinated, strategic approach to weed management involved more than just knowing what its priority weeds and locations/sites were. The new approach also involved establishment of an entirely new weed management framework for DBCA. This framework integrated a new policy direction that incorporated the prioritisation process for weed species and sites, and implementation of a new mapping system to standardise data collection and monitor management effectiveness of priority weeds on DBCA-managed lands.

Ten years on from its implementation, the Department reviewed the status of the prioritisation process and investigated how this process was assisting the department to prioritise its weed management efforts. This paper also touches on the department's weed management policy and the investigations into how its new Weed Occurrence and Treatment App is assisting the department to map weed occurrences and more effectively monitor management effectiveness.

POLICY DIRECTION

DBCA is the WA State Government agency responsible for managing WA's terrestrial, marine and estuarine parks and reserves, forests and islands to conserve biodiversity and manage key threatening processes. It has staff based around the state in nine administrative regions that deliver on-ground management activities and is supported by several specialist branches centralised in Perth and Bunbury.

In 2015, the Department updated its Weeds Management Policy (DPaW 2015) to focus on reducing the impacts of existing weed populations on key assets and values on department-managed lands (asset-protection), whilst also preventing new incursions and eradicating new (species-led) incursions, as cost-effectively as possible. These priorities are consistent with the Australian Weeds Strategy 2017 to 2027 (Invasive Plants and Animals Committee 2016).

The policy also has a focus on improving the performance of the department's weed management effort. It aims to achieve this through the monitoring, review and continuous improvement of the effectiveness and applicability of weed management techniques and strategies, and implementation of adaptive management as required.

SPECIES-LED PRIORITISATION PROCESS

The species-led prioritisation process is focused on weed species that are considered to be high impact, rapidly invasive, and still at a population size that is feasible to eradicate or contain to a manageable size. This approach rates weeds against their ecological impact, invasiveness, current and potential distribution, and feasibility of control. The resulting five ratings are then assessed using a series of matrices to provide an overall ranking for each weed species within a region. These matrices were based on the National Post-Border Weed Risk Management Protocol (Anon. 2006).

Outcomes This process enabled us to develop the following weed lists:

- species with a ranking (Very High, High, Medium, Low or Negligible);
- species ranked as 'further assessment required (FAR)' due to more than one unknown rating;
- species with no ranking (as they were not rated through the process for various reasons); and
- species that are 'alerts' for the region.

Several species with a ranking of very high and high have since been targeted for management. *Cylindropuntia kleiniae* DC. F.M.Knuth was targeted for eradication in the Goldfields, with all known plants removed and deep buried. Also in the Goldfields, a population of brown spine Hudson pear (*Cylindropuntia tunicata* (Lehm.) F.M.Knuth) was identified and targeted for management. The alert lists have also triggered a number of finds across department managed lands.

Since its implementation in 2008, the species led prioritisation process has continued to be utilised by DBCA regional staff to determine their weed control priorities. Although the intent was to conduct biennial regional species-led prioritisation review workshops, the timeframe between reviews has varied due to the availability of staff and other resources. Having said this, staff continue to update the spreadsheet when new species or infestations are identified.

Staff have also used the spreadsheets to determine priorities at a district and park level by adjusting the ratings as appropriate. This has provided staff with a great tool to prioritise their works programs at the various levels required by management.

ASSET-PROTECTION-BASED PROCESS

The aim of the asset-protection-based approach was to focus on identifying high value assets, the weeds that pose a threat to these assets, and the sites where control will have the greatest biodiversity (or other relevant) benefit.

This has been the most challenging part of the prioritisation process. Departmental staff have researched several different methodologies from around Australia, and the world, and are yet to determine the most appropriate process for the department. A key factor in applying an asset-protection-based process is having enough information on the condition of all our assets, as well as the threatening processes impacting those assets, so as to implement a detailed scoring process.

While there is incomplete information in many areas regarding the assets and/or the weed threats, we still need to manage the ever-increasing weed threat to department-managed lands. As such, we have used the rankings and ratings from the species-led prioritisation process, and expert knowledge of assets from each of the department's nine regions, to inform an interim asset-protection-based prioritisation process.

Based on this knowledge, staff have developed a list of priority weed species and the priority locations where they need to be managed in each region. This list includes the management goal (localised eradication, containment or density reduction), the priority for the region and/or district, the assets at risk and any relevant comments. Resources are then allocated to the highest priority weeds or sites on the list.

The comments field can be used to highlight secondary priorities by identifying those species and locations that should be targeted as a priority if additional funding opportunities became available.

The outcomes of this interim process are being used by staff across the state to prioritise their weed management programs.

WEED APP

Prior to 2017, the methods of collecting and recording weed management data and the type of data collected varied widely across DBCA, and largely relied on the interest and experience of staff in each area.

To address these issues and implement a coordinated, strategic approach to weed mapping and monitoring, the Weed Occurrence and Treatment Mapping App (Weed App) was launched in August 2017. This app was developed in consultation with the department's Geographic Information Services (GIS) Branch and regional staff.

The department developed a standard operating procedure (SOP) in 2011 which provided advice on techniques for detailed mapping of the distribution and cover of serious weeds in bushland and wetlands (DEC 2011). This SOP provided comprehensive advice and a list of standard attributes however, as it was developed while working in bushland reserves, of usually less than 1000 hectares in size, many staff found the methodology too onerous to apply to larger reserves.

To ensure the app would be useful across a State the size of Western Australia, and at a range of different scales, a more detailed list of attributes was developed based on those from the SOP and on previous discussions with staff around the state.

A consultation process was then undertaken to standardise this list for all weed management activities across the department and determine the most appropriate methodology for data collection. This consultation process also looked at the type of reporting required for weed management activities, which informed data capture needs.

To ensure staff were suitably trained to use the Weed App and allow for a final round of consultation before it was officially released in August 2017, a train-the-trainer workshop was held in Perth in June 2017 with at least one staff member from each of the department's nine regions being trained. These staff have since provided training to relevant staff in their work areas and an estimated 70% of work areas around the state are now using the Weed App to capture their weed management activities.

The Weed App, developed using Collector for ArcGIS (ESRI Australia 2018), provides several time-saving benefits allowing more efficient use of scarce resources. It allows staff to work off-line to collect weed occurrence and treatment information quickly and easily in the field. Once back in mobile or Wi-Fi range, staff can synchronise the map edits and the information is automatically uploaded to a corporate database. This information is instantly available for viewing by all named users on the system.

As staff accumulate information in the system, they will be able to report on weed occurrence and associated treatments in their region via predefined filters, also saving time. They will also be able to quickly report on various aspects of their weed management program including:

- weed species present in their region;
- weed distribution and/or density;
- what treatment methods were used;
- efficiency of treatment methods; and
- if weed management was effective in reducing distribution and/or density.

This information can then be used to report through a range of fora, such as the department's annual report, protected area management plans and conservation activity plans.

Into the future, it is planned that the system will be further developed to provide data analysis and reporting ability for weed management activities such as resources used (e.g. chemical type and quantity, control costs, person days), change over time in weed distribution and density, and aid in the development of regional weed management works programs.

The department is continuing to investigate opportunities to improve the Weed App and its accessibility to staff, with minor changes being made to the app throughout the first 12 months of operation. A full review will be scheduled now the Weed App has been in operation for a full 12 months.

LESSONS LEARNT

One of the idiosyncrasies of the species-led weed prioritisation approach is that high impact, rapidly invasive weeds that are already widespread across department-managed lands are ranked moderate or low. This is because these species are too widespread to control when viewed from a species-based approach across the landscape. These species would instead be dealt with through the asset protection-based approach. This has caused confusion and misrepresentation of the results both within and external to the agency, and highlighted the need to ensure that any material published regarding the Weed Prioritisation Process is very explicit when explaining how the rankings are to be interpreted and used to assist in determining weed management programs. As such, all public information was amended to only include the ecological impact and invasiveness ratings for each of the weed species rated for the region. The publicly available information also includes a list of priority alert species to keep an eye out for (WA, DBCA 2018).

Another key lesson learnt through this process is that asset-protection-based prioritisation is a big job. While implementation of a scoring system requires detailed information on both assets and key threatening processes, it is important not to let a lack of this information prevent the prioritisation process from occurring to inform development and implementation of works programs. Often staff have knowledge that is not formally recorded, and this can be utilised together with the information collected through interim processes to inform future prioritisation.

By gaining people's buy-in for the Weed App from the beginning, we now have a product that staff across the state are using and promoting to all involved in weed management in the department.

As more data is gathered on assets and weed threats via the Weed App, robust datasets will be provided that can be used to improve both the species-led and asset-protection-based prioritisation approaches.

WHERE TO FROM HERE

Species-led Prioritisation Process

DBCA will:

- continue with biennial reviews of regional weed lists; and
- investigate options of storing data and associated formulae for ease of maintenance of data into the future.

Asset-protection-based Process

DBCA will:

- continue to investigate processes to assist with the prioritisation of assets for which weeds are a key threatening process;
- work with regional staff to ensure all weed management programs include monitoring to assist in determining effectiveness of management on both the threat level and the condition of the asset; and
- continue to review current lists as more detailed information becomes available and use this information to implement adaptive management as required.

Weed App

DBCA will:

- continue to investigate opportunities to improve the Weed App and its accessibility to staff to ensure its continued use;
- ensure data collected using the Weed App is fed into the prioritisation processes and used to inform future management decisions;
- investigate reporting abilities within Collector for ArcGIS;
- undertake further research and development to meet remaining data analysis and reporting needs such as development of works programs and more detailed analysis of management effectiveness;
- explore alignment of DBCA weed mapping and monitoring information with that collected by external stakeholders, including potential for them to access the app or align attributes so data can more easily be shared; and
- investigate opportunities for publicly sharing DBCA collected weed data both within WA and nationally.

ACKNOWLEDGMENTS

I would like to thank all regional DBCA staff and external stakeholders for their participation in the prioritisation process and, more specifically, the regional workshops. A special thank you goes to Greg Keighery for providing his expertise throughout the prioritisation process and to Shane French for the time he has dedicated to work with me on both the development and implementation of the Collector Weed Occurrence and Treatment Mapping App.

REFERENCES

Anon. (2006). HB 294:2006. National post-border weed risk management protocol. (CRC Australian Weed Management, Adelaide and Standards Australia, Sydney and Standards New Zealand, Wellington).

DEC (2011). 'Techniques for mapping weed distribution and cover in bushland and wetlands.' SOP No: 22.1. (Department of Environment and Conservation, Perth).

DPaW (2015). Corporate Policy Statement No. 14. Weeds Management. Department of Parks and Wildlife, Perth.

ESRI Australia, ESRI Australia Pty Ltd (2018). Collector for ArcGIS._https://esriaustralia.com.au/collector-for-arcgis (accessed 13 August 2018).

Invasive Plants and Animals Committee (2016). 'Australian Weeds Strategy 2017 to 2027 (Australian Government Department of Agriculture and Water Resources, Canberra).

WA DBCA, Western Australia Department of Biodiversity, Conservation and Attractions (2018). How does Parks and Wildlife manage Weeds? https://www.dpaw.wa.gov.au/plants-and-animals/plants/weeds/156-how-do-we-manage-weeds (accessed 13 August 2018).

Buffel grass in South Australia: Progress and future directions

Ross Meffin[1], Ellen Ryan-Colton[2], John L. Read[3], Troy Bowman[4] and Michaela Heinson[5]
[1] Department of Primary Industries and Regions, GPO Box 1671, Adelaide South Australia 5001, Australia
[2] Research Institute for the Environment and Livelihoods, Charles Darwin University, Alice Springs, Northern Territory 0870, Australia
[3] School of Earth and Environmental Sciences, University of Adelaide, Adelaide, South Australia 5000, Australia
[4] Department of Primary Industries and Regions, GPO Box 469 Murray Bridge, South Australia 5253, Australia
[5] Department for Environment and Water, 205 Greenhill Road, Eastwood, South Australia 5063, Australia
(Ross.Meffin@sa.gov.au)

Summary Buffel grass is arguably the greatest invasive species threat to biodiversity across the Australian arid zone. It transforms ecosystems, forming dense monocultures, displacing native plants, altering habitat structure and diminishing resources that support arid zone fauna. Buffel grass invasion negatively impacts upon aboriginal culture and increases the frequency and intensity of bushfires.

In South Australia scattered populations are widely distributed across the north, with extensive infestations in the far north. The South Australian buffel grass strategic plan 2012–2017 provided a roadmap for buffel grass management and its successful implementation has seen some significant advances. Key challenges for the future are increased stakeholder engagement, improving distribution knowledge, maintaining funding and managing extensive, remote infestations.

Keywords Conflict species, stakeholder engagement, zoned management.

BACKGROUND

Buffel grass (*Cenchrus ciliaris* L.) is a perennial tussock grass native to Africa, India and Asia. Since its introduction into central and northern Australia for pasture improvement and dust control, buffel grass has spread widely across many landscapes. It is an ecosystem transformer (Bastin *et al.* 2008) capable of radically altering the species composition richness of Australian rangelands following invasion.

Buffel grass is considered one of Australia's worst environmental weeds (Humphries *et al.* 1991). It has been shown to increase the intensity and frequency of fires in native plant communities (Butler and Fairfax 2003). It is able to rapidly regenerate after fire and suppress regeneration of native species, generating positive fire invasion feedbacks (Miller *et al.* 2010). Numerous studies have shown that the cover of buffel grass is negatively associated with species richness (e.g. Clarke *et al.* 2005). High buffel grass densities are thought to lead to ongoing declines in species richness as the seed banks of native forbs and grasses are gradually depleted (Clarke *et al.* 2005).

Some specialist seed eating birds, such as finches and some parrots, do not include buffel seed in their diet and are relatively rare in buffel dominated landscapes (Franks *et al.* 2000). Endangered fauna species in northern South Australia threatened by buffel grass include black-footed rock-wallabies, great desert skinks, and mallee fowl as well as a suite of other birds such as chestnut quail thrush, dusk grass wren, and striated grass wren (Paltridge *et al.* 2009).

Buffel grass is recognised as a major threat to Country, negatively impacting Aboriginal cultural values and practices, including hunting and gathering through its impacts on key species as well as significant sites.

Buffel grass has a wide climatic tolerance and establishes on a range of soil types under various disturbance regimes, quickly forming self-sustaining populations (Franks 2002). The highly varied morphological and physiological characteristics of buffel grass across Australia contribute to this adaptability. Pasture introduction programs have brought in approximately 580 accessions, with many informal and formal releases in central, tropical and sub-tropical Australia (Hall 2000). Modelling suggests that the vast majority of South Australia is likely to be suitable for establishment of buffel grass (Marshall and Hobbs 2010), and more broadly 60% of mainland Australia may be suitable for buffel grass establishment (Lawson *et al.* 1994).

The difficulties and problems posed by buffel grass are compounded by its perceived benefits and widespread use as a pasture species. Buffel grass pasture is well regarded by many central and northern Australian cattle producers because of its palatability, responsiveness to limited rainfall, ability to colonise and its tolerance to drought, fire and heavy grazing (Fairfax and Fensham 2000). As with other 'conflict species', many of the same attributes contribute to the significance of buffel grass as both a pasture species and an environmental weed. These include ease of establishment, rapid growth rate, fast maturation, prolonged flowering periods, prolific seed production and high seed dispersal ability, coupled with relatively long seed dormancy (Franks 2002).

However, as well as evaluating the perceived value of buffel grass for livestock production with reference to its serious environmental and social impacts, there is a need to critically examine beliefs regarding its benefits. There is growing evidence that it may be nutritionally inadequate as the main dry feed for adult stock (Reseigh *et al.* 2014, NR SAAL 2017). Furthermore, it can lead to pasture degradation in the long term. While it responds to out of season rain when native species remain dormant (Hall 2000), it also displaces a large range of short-lived native grasses and forbs that are important in fattening cattle (Puckey and Albrecht 2004). Lack of diversity in pastures can limit the nutritive value available to stock during dry periods. Once buffel grass reaches high densities and displaces valuable native forage, reversing this process is extremely challenging. Buffel grass also has a high demand for available soil nitrogen and phosphorus. Over time, depletion of these nutrients leads to pasture 'run-down', with an associated decline in cattle live-weight gain (Puckey and Albrecht 2004). Buffel grass also contains oxalates and can cause acute oxalate poisoning in ruminants, most often in young and hungry sheep (Thomas 2004).

SOUTH AUSTRALIAN RESPONSE

To provide a coordinated response to buffel grass, in 2012, South Australia developed the State Buffel Grass Strategic Plan 2012–2017 (Biosecurity SA 2012), and formed the Buffel Grass Taskforce. The Taskforce comprises representatives with experience in weed policy and management from regional Natural Resource Management (NRM) authorities, Biosecurity SA, other state government agencies, conservation organisations, as well as ecologists and researchers. The Taskforce oversees and coordinates the plan, which aims to contain and effectively manage buffel grass to minimise its impact in South Australia.

To achieve this vision the plan took a zoned approach to guide state-wide planning and management (Figure 1). The use of a zoned approach recognises that while management of an invasive plant depends on local needs and priorities, these will be much more effective and efficient when they are part of a broader scale strategic approach (e.g. Grice *et al.* 2011). The zones were informed by a risk assessment in combination with expert knowledge and distribution data. The 2012 assessment determined a high weed risk in grazing rangelands and native vegetation, but also suggested a high feasibility of containment in the former and very high in the latter.

Based on this, the four goals of the plan were:

1. exclude the entry of buffel grass into South Australia and prevent its movement within the State;
2. manage impacts of buffel grass in Zone 1;
3. contain spread of buffel grass in Zone 2, and destroy infestations in Zone 3; and
4. build capacity to manage buffel grass.

In 2013, the Taskforce was successful in obtaining funding for the three-year Buffel Grass Control in Arid Rangelands Project. The Project included delivery of a number of strategic surveillance, control and capacity building activities, herbicide trials and development and dissemination of best practice buffel grass management tools. Specific outcomes include:

- strategic control of over 300 ha of buffel grass outlier infestations;
- herbicide research and a report identifying best practice options for control;
- five best practice management fact sheets (identification, hygiene, herbicides, integrated control, management planning) http://www.pir.sa.gov.au/biosecurity/weeds_and_pest_animals/weeds_in_sa/weed_id/plant_id_notes/buffel_grass
- over 4600 km of roadsides surveyed, documenting distribution and spread of buffel grass;
- a GIS database and control prioritisation tool; and
- community engagement activities involving over 270 stakeholders and media.

A major factor in the success of the Project was engagement of a dedicated Buffel Grass Coordinator, tasked with implementing the strategic plan and coordinating efforts of NRM regions, with support from the Taskforce.

Another success has been the declaration of buffel grass under South Australia's *Natural Resources Management Act 2004*. This prohibits sale and transport of buffel grass, and allows NRM regions to enforce control or destruction of infestations. At present no other Australian states or territories have regulated buffel grass under their legislation. This is likely to be in large part due to the different perceptions of buffel grass in those jurisdictions and overwhelmingly favourable views of it, particularly in summer rainfall production areas.

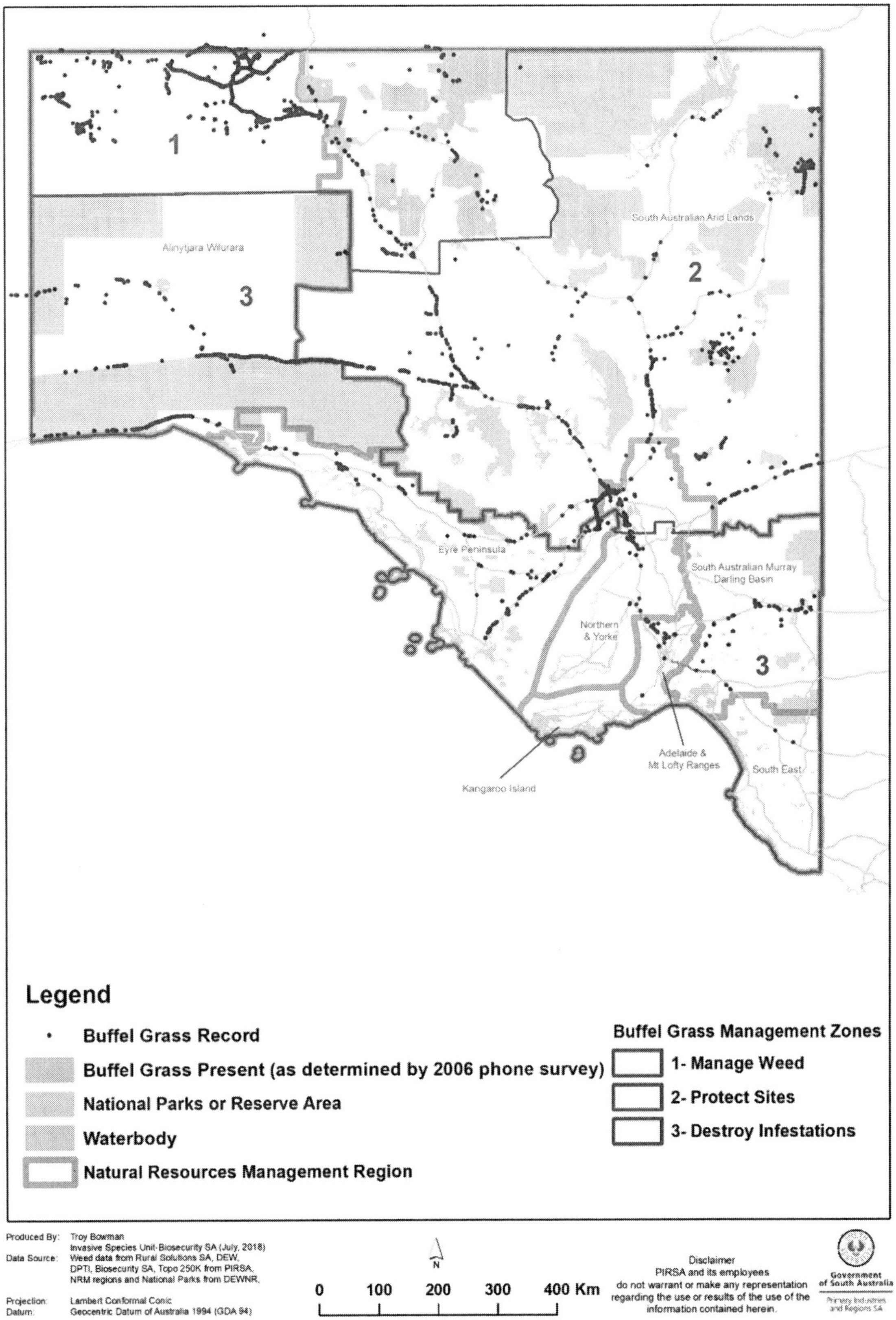

Figure 1. Buffel grass management zones and known distribution in South Australia in 2018.

CHALLENGES AND FUTURE DIRECTIONS

Despite these successes, much remains to be done and buffel grass distribution in South Australia continues to increase. As a result of this expansion and improved ecological and biological knowledge gained over the past six years, a revised risk assessment was undertaken in April 2018. This assessment upgraded the weed risk to very high, and reduced the feasibility of containment to low in native vegetation, and negligible in grazing rangelands.

This risk assessment formed part of a revised and updated strategic plan, designed to guide management through to 2023, sustaining and leveraging the outcomes of the 2012–2017 plan. At the time of writing the updated plan is yet to be finalised, but the focus will be to reduce the overall economic, environmental and social impacts of buffel grass, and to prevent spread of the weed to uninfested assets. In a state-wide context, with buffel grass primarily established in the north, this means protecting key environmental assets and preventing range expansion southwards into uninfested areas.

With best practice materials developed and a range of control options to suit most situations, future activities will be refocussed to fill gaps identified during the process of updating the plan. Key challenges include improved and expanded stakeholder engagement, maintaining funding and momentum for on-ground action and containment, better mapping and knowledge of distribution, and developing strategies and tools for managing buffel grass in remote areas with widespread, extensive infestations.

Further engagement of a range of stakeholders including the mining industry, road and rail corridor managers, and pastoralists is critical. Their engagement and active involvement will be essential to realising the aims of the strategic plan. This is likely to be particularly challenging with regards to pastoralists, but the information available on the deficiencies of buffel grass as a pasture species in arid areas will be a valuable tool in this respect.

Without a dedicated coordinator to drive statewide management there is a risk that the gains made to date may be difficult to maintain. A coordinated approach among NRM regions will be required. Furthermore, on-ground control works above and beyond those undertaken by NRM regions are likely to be necessary to prevent expansion southwards and into conservation assets such as the Flinders Ranges. As such securing a funding base for this will be a major focus of the Taskforce.

The remoteness of much of the country where buffel grass occurs also presents major challenges. Our knowledge of its current distribution is largely restricted to the sparse road network in northern South Australia. Better distribution data will be critical, both to accurately understand the current situation and risks and to plan management strategies. However, we do know there are large extensive infestations in the state's northwest. Managing these is extremely challenging. One possible option could be biological control, but without broad national support this is unlikely to be pursued.

ACKNOWLEDGMENTS

The authors wish to thank the members of the state Buffel Grass Taskforce, Tim Reynolds, Ben Shepherd, the Native Vegetation Council and NRM Boards across the state for their contributions to buffel grass management in South Australia.

REFERENCES

Bastin, G. and the ACRIS Management Committee (2008). 'Rangelands 2008 – Taking the Pulse', published on behalf of the ACRIS Management Committee by the National Land and Water Resources Audit, Canberra.

Biosecurity SA (2012). South Australia Buffel Grass Strategic Plan: A plan to reduce the weed threat of buffel grass in South Australia. Government of South Australia, Adelaide.

Butler, D.W. and Fairfax, R.J. (2003) Buffel grass and fire in a gidgee and brigalow woodland; a case study from central Queensland. *Ecological Management and Restoration* 4, 120-5.

Clarke, P.J., Latz, P.K. and Albrecht, D.E. (2005). Long-term changes in semi-arid vegetation: Invasion of an exotic perennial grass has larger effects than rainfall variability. *Journal of Vegetation Science* 16, 237-48.

Fairfax, R.J. and Fensham, R.J. (2000). The effects of exotic pasture development on floristic diversity in central Queensland, Australia. *Biodiversity Conservation* 94, 11-21.

Franks, A.J. (2002). The ecological consequences of buffel grass *Cenchrus ciliaris* establishment within remnant vegetation of Queensland. *Pacific Conservation Biology* 8, 99-107.

Franks, A.J., Butler, D. and Fairfax, R. (2000). A weed by any other name. *Wildlife Australia* 37, 24.

Grice, A.C., Clarkson, J.R. and Calvert, M. (2011). Geographic differentiation of management objectives for invasive species: a case study of *Hymenachne amplexicaulis* in Australia. *Environmental Science and Policy,* 14, 986-97.

Hall, T.J. (2000). History and development of buffel grass pasture lands in Queensland. *In* Buffel grass symposium. Proceedings of Buffel grass

Workshop, eds. J. Kyte, H. Lees, I. Loxton and B. Cook) pp. 2-12. (Department of Primary Industries, Brisbane).

Humphries, S.E., Groves, R.H. and Mitchell, D.S. (1991). Plant invasions of Australian ecosystems: a status review and management directions. (CSIRO, Canberra).

Lawson, B.E., Bryant, M.J. and Franks, A.J. (2004). Assessing the potential distribution of buffel grass (*Cenchrus ciliaris* L.) in Australia using a climate-soil model. *Plant Protection Quarterly* 19, 155-63.

Marshall, V. and Hobbs, T. (2010). The suitability for establishment of buffel grass in South Australia using BIOCLIM climatic modelling. Unpublished report.

Miller, G., Friedel, M., Adam, P. and Chewings, V. (2010) Ecological impacts of buffel grass (*Cenchrus ciliaris* L.) invasion in central Australia: does field evidence support a fire-invasion feedback? *The Rangeland Journal* 32, 353-65.

NR SAAL, Natural Resources South Australia Arid Lands (2017). Nutritional value of common pastoral grasses to livestock. Fact Sheet.

Paltridge, R., Latz, P., Pickburn, A. and Eldridge, S. (2009). Management Plan for rare and Threatened Flora in the Anangu Pitjantjatjara Yankunytjatjara Lands of South Australia. Department for Environment and Heritage, Adelaide.

Puckey, H. and Albrecht, D. (2004). Buffel grass (*Cenchrus ciliaris* L.) presenting the arid Northern Territory experience to our South Australian neighbours. *Plant Protection Quarterly* 19, 69-72.

Reseigh, J., Wurst, M. and Bennett, T. (2014) Upper North Farming Systems: Native Grass. Nutrition Fact Sheet.

Thomas, M. (2004). Perennial grasses – potential grazing issues. *Farmnote 29/2004*. Department of Agriculture, Perth, Western Australia.

An ounce of prevention: An accreditation scheme for ornamental plant industries

Hillary Cherry[1], Stephen B. Johnson[2] and Des Boorman[3]
[1] Pest and Weeds Team, National Parks and Wildlife Service, PO Box 1967, Hurstville, New South Wales 1481, Australia
[2] Weed Research Unit, New South Wales Department of Primary Industries, Locked Bag 6006, Orange, New South Wales 2800, Australia
[3] Nursery and Garden Industry of New South Wales and the Australian Capital Territory, PO Box 3013, Rouse Hill, New South Wales 2155, Australia
(hillary.cherry@environment.nsw.gov.au)

Summary Weeds are a significant threat to the environment. While the ornamental plant trade was historically a pathway for weed introductions, there is a strong desire within the industry, in government and the community to reduce the use of weedy ornamental plants and prevent future environmental weeds. The Plant Sure initiative aims to reduce the use of high risk (weedy) ornamental plants from production, supply and trade by engaging with government, industry and the community, to develop a voluntary accreditation scheme for green life industries. The scheme is underpinned by a decision support tool that allows plants to be categorised according to risk. Plants that pose a high risk of becoming weeds can be excluded from use, while the use of environmentally-friendly plants can be encouraged. Through education and training components, the scheme will seek to elicit long-term attitudinal and behavioural change in ornamental plant suppliers and consumers

Keywords Weeds, invasive plants, weed risk, environmentally-safe, assessment, green life industry, ornamental plants, decision support tool.

INTRODUCTION

Invasive plants (weeds) are a significant threat to biodiversity and the environment. The ornamental plant trade has been a predominant pathway of weed introductions: For example, over 65% of introduced plants that have naturalised in Australia are considered garden escapes (Groves *et al.* 2005). Approximately 5% of the 25,360 plants introduced into Australia for ornamental purposes have become environmental weeds, equating to some 1366 weed introductions (Virtue *et al.* 2004).

Accordingly, there is a strong desire in industry, government and the community to reduce the use of weedy ornamental plants and prevent future environmental weeds. There are many examples of proactive initiatives to reduce the use of weedy ornamental plants, including volunteer groups such as Sustainable Gardening Australia, and industry led initiatives such as 'Grow Me Instead' and the Invasive Plant Risk Assessment Tool, initiatives of the Nursery & Garden Industry Australia (NGIA) that aim to educate consumers about invasive plants (NGIA 2003).

To build on and progress these existing initiatives, in 2016, the New South Wales (NSW) Government through the NSW Environmental Trust provided support to develop and implement a voluntary scheme to promote the use of environmentally safe ornamental plants, as part of a 5 year project entitled 'Plant Sure'. The Plant Sure initiative aims to reduce the use of high risk (weedy) ornamental plants from production, supply and trade by engaging with government, industry and the community to develop a voluntary accreditation scheme for green life industries and other relevant ornamental plant users. The long-term vision of the project is to prevent future environmental impacts from weeds by reducing the availability of high risk (weedy) ornamental plants from supply and trade.

The project is initially being managed via a consortium led by the Nursery and Garden Industry of NSW and ACT (Australian Capital Territory) with representatives from the Australian Institute of Horticulture, the NSW Office of Environment and Heritage, and the NSW Department of Primary Industries. The consortium has a view to expanding more broadly as the project develops, to include a range of ornamental plant industry sectors. Consortium members are working in partnership with researchers, industry professionals, government and the community to co-design and develop the scheme.

While to be initially developed and trialled in NSW, the scheme will be designed to be 'National-ready' and 'sector ready', such that all relevant industries and jurisdictions can participate in future. The scheme will engage green life industries in promoting environmentally-safe plants and removing or avoiding use of plants that pose an environmental weed risk. The first phases of the project involve: a) identifying a suitable decision support process to assess the weed risk of ornamental plants; and b) investigating appropriate accreditation mechanisms and existing schemes to inform co-design of an appropriate scheme.

BUILDING ON EXISTING RISK-BASED MODELS

To determine the weed risk of plants, the scheme will be underpinned by a robust plant risk assessment and categorisation process. A rigorous risk assessment will provide confidence for industry and consumers that their plant choices are safe for the environment. The ornamental plant industry has a strong environmental stewardship ethic, and there are numerous examples of similar initiatives to reduce the use of invasive plants. For example, the Nursery and Garden Industry Australia currently support a risk-based approach using the Invasive Plant Risk Assessment Tool (NGIA 2013), and a robust plant risk assessment process was developed by Conser *et al.* (2015) that is used in Plant Right, a similar voluntary initiative in California, United States of America (PRC 2007).

The Plant Sure initiative is building on these and other best-practice plant and weed risk assessment tools to ensure a scientifically robust approach to assessing risk. In 2017, Macquarie University researchers completed a global review of existing plant and weed assessment tools and processes, and identified best practice models for the Plant Sure project. Based on the review outcomes, they have developed and tested a rigorous ornamental plant assessment and decision support tool that is being trialled for use in the scheme (Hancock *et al.* 2018). This includes a cultivar assessment process.

The risk assessment and decision support tool will allow plants to be categorised according to risk for use in the larger scheme. Plants that pose a high risk of becoming weeds can be excluded from use, while the use of plants that pose a low risk can be encouraged and promoted. The categorisation process will be designed to account for levels of uncertainty, and prompt re-assessment of species at appropriate intervals to allow incorporation of new information (e.g. if the species is found to be naturalising, or additional information is found on propagule longevity or dispersal mechanisms, etc.).

BUILDING A VOLUNTARY SCHEME

The project is working to co-design a scheme that allows a proactive approach to assessing new plant introductions, while helping industry to maintain diversity and consumer interest in environmentally-friendly ornamental plants. The scheme will highlight the environmental credentials of green life industries in avoiding the use of plants that pose an environmental weed risk, while promoting low-risk alternatives. The approach will support industry by showcasing their environmental stewardship and developing a strong brand to support a self-sustaining, independent scheme. It will include education and training components to elicit long-term attitudinal and behavioural change in ornamental plant suppliers and consumers, and aim to increase community knowledge of environmental weed issues.

The scheme will use a voluntary compliance approach; thus, it will need to create and promote a strong brand awareness to get the uptake required to achieve the desired result. While voluntary, it is being designed to complement existing government biosecurity regulation by providing the green life industry with mechanisms to demonstrate compliance with biosecurity obligations (for example, to comply with the General Biosecurity Duty under the NSW *Biosecurity Act 2015*). This can occur if green life entities are documented through the scheme as reducing or removing from sale plants that are assessed as having a weed risk. The project will work with government regulatory agency representatives to design a scheme that can be adapted to all Australian jurisdictions, and allow industry to pro-actively reduce weed-related biosecurity risks.

To develop a framework for designing the scheme, researchers at the University of Wollongong reviewed existing voluntary accreditation and certification programs, as well as other initiatives and standards for similar type projects (e.g. invasive species, sustainability, etc.) to determine what components should be included in the scheme and what type of scheme would be most suitable (Curtis *et al.* 2018).

Examples of two schemes that were reviewed include Plant Right California (PRC 2007) and the University of Florida, Institute of Food and Agriculture Sciences 'Assessment of Non-native Plants in Florida's Natural Areas' (UF/IFAS 2017). These two models are underpinned by robust plant risk assessment processes, and have successfully created linkages with industry and plant consumers to reduce the use of high-risk weedy ornamental plants. While these models have used different approaches, and have individual strengths and weaknesses, they demonstrate the need for broad cross-sectoral engagement, collaboration and co-design to ensure uptake. The Plant Right initiative, while created and managed through a non-government organisation, has been adopted by nursery industry professionals and the community (PRC 2007).

Based on the reviews of similar schemes, the report made several recommendations regarding set up, structure and development of the scheme. These included that scheme development consider the following components:

1. an educational and outreach program to educate green life industries on the problems associated with invasive (high risk) species, and mechanisms to promote the use of alternative plants;
2. a training program for green life organisations that enables them to effectively participate in the scheme and understand the key issues surrounding sale of invasive (high risk) plants;
3. a document (commitment) that a green life organisation signs that spells out its obligations in signing up as a scheme member;
4. sensitive recognition and acknowledgment of participating green life organisations, e.g. by allowing them to display the scheme's brand and logo on their promotional platforms;
5. a monitoring and evaluation system that enables the scheme to identify non-complying organisations, and to evaluate the success of the program;
6. a Charter or Standards document that spells out the approach of removing invasive (high risk) plants from sale and use, based on risk assessment;
7. a fee structure based on a business planning process that ensures that the scheme is self-supporting and inclusive of relevant businesses regardless of size, type, industry sector or location; and
8. a technical panel or panels encompassing independent expertise that undertakes the plant risk assessment process, identifies the plants that should not be sold, alternative plants, those that should be monitored, and those that have been successfully eliminated from sale.

In addition, the scheme must also: a) develop associated communication and engagement platforms, 'branding' and promotion campaigns; b) develop and implement behaviour change and public awareness campaigns to encourage consumer and industry uptake; and c) establish scheme governance, auditing and compliance protocols.

Model schemes and scheme components that could achieve project objectives are now being considered, and project partners will work together to co-design and trial a fit for purpose scheme that will achieve project objectives and be agreeable (and useful) for all stakeholders.

PARTNERSHIPS CRITICAL TO SUCCESS

The success of the Plant Sure project depends on extensive, cross-sector engagement and participation to enable a co-designed scheme that is appealing to industry, and respected by government and the community, while achieving the aim of reducing use of 'weedy' ornamental plants. This requires input from all affected sectors, and participation in scheme development to ensure the scheme is fit for each sector's purposes.

In early 2018, preliminary workshops were held with current partners to trial the plant assessment process and modify as necessary to develop a process that is suitable for a wide range of ornamental plant users. In addition, potential scheme components and 'model' schemes were discussed and options considered. All work to date has strongly recommended the need for more extensive, cross-industry and cross-sector stakeholder engagement and participation to increase the ability for robust co-design of the scheme for use by all sectors.

The project team is now focused on further engaging across multiple green life industry sectors (e.g. groups that recommend ornamental plants, such as landscape designers and other plant user groups, such as local government) to ensure scheme relevance for various sectors. We are also engaging with stakeholders from other States and Territories to build a scheme that is 'National ready' and adaptable for use by all jurisdictions and industries. With the appropriate stakeholders, we will co-design and develop the scheme. Once developed, we will pilot test the scheme, initially this is proposed to be within one industry sector (e.g. the nursery industry) in NSW, as a trial for broader implementation in future.

Current partners include representatives from the Australian Association of Bushland Regenerators, Local Government NSW, NSW Local Land Services, NSW Weed Officers Association, Royal Botanic Gardens and Centennial Parklands, Biosecurity SA (South Australia), ACT Parks and Conservation, and the Invasive Species Council. Additional partners are being sought from across green life industry sectors, the community and government (including other jurisdictions), to ensure wide consultation and participation.

Plant Sure project partners are keen to work with the community and all green-life industry groups across Australia that have an interest in the responsible use of ornamental plants, to ensure the plant assessment process and the scheme are relevant, user-friendly and available to everyone.

ACKNOWLEDGMENTS

Many thanks to the industry, non-government and community partners who are contributing their time to help co-design the scheme. The NSW Government supports this project through its Environmental Trust.

REFERENCES

Conser, C., Seebacher, L., Fujino, D.W., Reichard, S. and DiTomaso, J.M. (2015). The development of a Plant Risk Evaluation (PRE) tool for assessing the invasive potential of ornamental plants. *PLoS ONE* 10, e0121053.

Curtis, D., Gill, N., Atchison, J. and McKiernan, S. (2018) 'Plant Sure Environmentally Safe Ornamental Plant Scheme: Review of existing certification or accreditation schemes and recommendations for a model scheme'. University of Wollongong, Wollongong.

Groves, R.H., Boden, R. and Lonsdale, W.M. (2005). Jumping the garden fence: Invasive garden plants in Australia and their environmental and agricultural impacts. (WWF-Australia, Ultimo, NSW).

Leishman, M., Hancock, N., Graham, V., Adams, V., Maher, T. and Manea, A. (2018). Introducing the Plant Sure Ornamental Plant Decision Support Tool. Proceedings of the 21st Australasian Weeds Conference, 9–13 September, Sydney, New South Wales, eds S. Johnson, L. Weston, H. Wu and B. Auld, p. 234. (Weed Society of New South Wales).

NGIA, Nursery and Garden Industry Australia (2013). Invasive Plant Risk Assessment Tool (IPRAT). http://www.plantrisktool.com.au (accessed 1 May 2018)

PRC, Plant Right California (2007). http://www.plantright.org (accessed 1 May 2018)

UF/IFAS, University of Florida/Institute of Food and Agricultural Sciences (2017). Assessment of non-native plants in Florida's natural areas. https://assessment.ifas.ufl.edu/ (accessed 15 April 2018).

Virtue, J.G., Bennett, S.J. and Randall, R.P. (2004). Plant introductions in Australia: How can we resolve 'weedy' conflicts of interest? Proceedings of the 14th Australian Weeds Conference, eds. B.M. Sindel and S.B. Johnson, pp. 42-8. (Weed Society of New South Wales, Wagga Wagga).

A new approach to managing invasive pests at the local government level

Craig Magnussen
Southern Downs Regional Council, PO Box 26, Warwick, Queensland 4370, Australia
(craig.magnussen@sdrc.qld.gov.au)

Summary Southern Downs Regional Council (SDRC) has approved and is implementing an innovative invasive pest management policy through adoption of its Invasive Pests Control Scheme (IPCS).

Council expects all land owners to control invasive pests, so as to reduce their spread and the impacts they may impose to other lands and the owners of those lands. If those landholders who don't control pests are allowed to continue to do so, pests keep spreading and those doing the right thing will continue to have to address the impacts made by others' complacency.

Traditional invasive pest compliance processes have been resource intensive – often Council's enforcement costs will exceed the cost of control work – and, by and large, have been ineffective in reducing the impacts of invasive pests.

The IPCS introduces a levy on all rateable land within certain differential general rating categories – approximately 6000 properties. The levy is calculated according to unimproved land value, with a minimum set at $500. Council grants an upfront 100% rebate of the levy at rates issue. Owners of relevant land are required to submit invasive pest control plans and carry out documented control works, or their levy concession is revoked via issue of a supplementary rates notice.

2017/2018 is the first year of the IPCS but Council is very pleased with results achieved thus far. A 90% return rate of control plans by landowners, an unprecedented amount of invasive pest control work occurring across the region and stimulation of the local contract market and other related industries all point to success.

Keywords Invasive pests, compliance, levy.

Heritage or threat, conserve or destroy? The heritage plant management dilemma in protected landscapes

David C. Havell and Angela L. Scott
Department of Conservation, Private Bag 68908, Wellesley Street, Auckland 1141, New Zealand
(DHavell@doc.govt.nz)

Summary Many cultural heritage sites including archaeological site schedules and historic landmarks are protected in New Zealand by specific legislation administered by Heritage New Zealand and the Department of Conservation. Local authorities such as district councils protect cultural heritage and also maintain lists of protected trees, usually trees with significant horticultural and cultural merit. Under New Zealand law, cultural heritage includes trees, places and their setting, (the cultural heritage landscape). Protected natural areas in New Zealand such as national parks, conservation parks and nature reserves often contain cultural heritage sites sometimes dating back several hundred years. Associated with heritage sites are often a suite of exotic plants from multiple cultural periods which contribute to the heritage character of the site. Historic plants may also provide significant scientific archaeological evidence. However, some heritage plants are known to be invasive or left unmanaged may form extensive thickets. Consequently, there is some tension between teams over management of heritage plants. Biodiversity staff may be unaware or have little sympathy for cultural heritage. The best outcomes have been obtained where there is active engagement between both heritage and biodiversity teams, and where there has been a weed risk assessment and strategy, as clearly not all heritage plants threaten biodiversity values, and not all significant heritage plants are safe to leave in natural areas. Tactics to maintain the heritage character of sites when invasive heritage plants have been removed include extensive data gathering, interpretative signs or monuments, using less invasive cultivars, and transferring plants to heritage gardens.

Keywords Heritage plants, environmental pest plants.

The changing weedscape: Policy and management implications of unknown knowns

Paul O. Downey[1] and Stephen B. Johnson[2]
[1] Institute for Applied Ecology, University of Canberra, Australian Capital Territory 2601, Australia
[2] Department of Primary Industries, Locked Bag 6006, Orange 2800, New South Wales, Australia
(paul.downey@canberra.edu.au)

Summary Over the past few decades there has been uncertainty around the number of non-native weed species in Australia, despite numerous lists of species being produced. Additionally, the number of weeds that invade and impact upon different sectors is unclear. This data uncertainty has hampered management and policy decisions that reflect the Australian weedscape. We examined several weed datasets (e.g. Randall 2017), as well as records held in the Australian Virtual Herbarium to compile a list of weed species in Australia. From this we classified each species as either an environmental weed, an agricultural (primary production) weed, both, or as a weed of neither. We also tried to determine the introduction date, and used historical texts (e.g. Kloot 1987) to establish how these groups of weeds have changed over time in determining the changing weedscape of Australia. We compared these trends to policy and management decisions for weeds in Australia: this revealed significant mismatches, many of which were known (i.e. apparent to on-ground managers), but seemingly unknown at higher management and political levels. For example, whilst virtually all of the first weed species recorded in Australia were recognised as weeds of agriculture, such weeds only account for about 32% of today's weeds (with only 6% being exclusively weeds of primary production), whilst environmental weeds account for more than double that number today (66%; with >40% of all weeds exclusively environmental). The policy agenda remains firmly focused on weeds of primary production, despite the environmental weed tsunami already on our shores.

Keywords Weed lists, policy, management, agricultural weed, environmental weed, primary production, historical perspective, mismatches.

Lake Benmore islands wilding conifer control

Pete Caldwell
PO Box 110, Ground Floor, 4 Hazeldean Road, Christchurch 8140, New Zealand

Summary Wilding conifers are currently targeted through a national program to coordinate control on a wide scale, with the aim to reduce the extent of conifers and their adverse effects on biodiversity and agricultural values. Boffa Miskell Ltd is carrying out control of wilding conifers on islands within Lake Benmore (Canterbury, NZ) as part of their contract with Land Information New Zealand to deliver biosecurity services on unallocated crown land. Project management of this operation is complex, with the presence of water, steep terrain, endangered species and public opposition affecting the methodology and timeframes selected.

Pete will discuss the significance of wilding conifer control at this site and on a national scale, as well as the challenges and opportunities presented by this project.

Keywords Wilding conifers, islands, pine tree.

Managing WoNS for better drinking water

Dennis J. Gannaway[1] and Tim Odgers[2]
[1] Healthy Land and Water, PO Box 731, North Ipswich, Queensland 4305, Australia
[2] Seqwater, PO Box 328, Ipswich, Queensland 4305, Australia
(dennis.g@hlw.org.au)

Summary Cat's claw creeper (*Macfadyena unguis-cati* Gentry) and Madeira vine (*Anredera cordifolia* Steenis) are ubiquitous weeds found throughout South East Queensland (SEQ).

Previous efforts to manage them, including the release of biological control agents from both a government and community perspective have been predicated on the need to conserve native riparian flora and associated ecological communities. This narrower focus, given the lack of an agricultural or a broader natural resource management imperative has resulted in limited allocation of resources to tackle the growing problem.

A collaboration between the local NRM group, Healthy Land and Water (HLW) and the regional water utility, Seqwater will see a significant increase in riparian vine weed management focusing on the two Weeds of National Significance (WoNS) vines.

The objective is to reduce the density and distribution of the WoNS vines within prioritised catchments. The outcome sought is the preservation of deep rooted native flora which stabilise banks and reduce erosion, filter sediment and nutrients from overland flow, provide competition to other weed tree species and provide shade.

The motivation is the preservation of water quality in stream to avoid the increased cost of and pressure on purification systems delivering drinking water to SEQ. Biodiversity conservation is a bonus.

The change in reasoning from a pure conservation message to one of economic imperative creates interest within the community. This leads to increased participation and is useful in leveraging additional funding to expand efforts beyond current project boundaries to deliver long term sustainable weeding programs.

Keywords Community engagement, cat's claw creeper, Madeira vine, Weeds of National Significance, water quality, collaboration, biological control agents.

INTRODUCTION

Cat's claw creeper and Madeira vine are natives of the tropical Americas. These aggressive climbers were used as ornamental plants in older-style Queensland gardens. They have subsequently escaped and now grow extensively throughout South East Queensland's (SEQ) creeks and disturbed forested areas (DAF 2018).

Both are declared Weeds of National Significance (WoNS) (WoNS 2018). The vines are capable of completely smothering and eventually killing native riparian vegetation, even growing up and over ancient trees.

Both vine species have a vigorous root and tuber systems. Cat's claw creeper develops underground tubers capable of altering soil chemistry and its habit of growing along the ground acts as a physical barrier to native vegetation seedling recruitment. Madeira vine produces underground and aerial tubers that aid in rapid expansion of weed invasion when disturbed (WoNS 2018).

If left unmanaged, both vines will transform the current natural or marginally modified vegetation structure of riparian zones from creek banks covered by a diversity of medium to large deep-rooted trees and shrubs to potentially a monoculture of prostrate creeper and grass. The result will be a loss of biodiversity but more importantly from a water quality perspective the destabilisation of creek banks and loss of ecological services due to the death of the trees.

Loss of deep tree roots that bind the banks leads to mass erosion and bank slumps during moderate to high water flows resulting in increased sediment loads in stream.

Loss of the filtration mechanism provided by native vegetation means higher volumes of organic material and nutrients arriving at water treatment facilities. This poses a significant threat to drinking water quality increasing cost of purification and impacting on service delivery.

Investing in vine weed management to maintain natural vegetation stability in catchments areas above water treatment off take facilities is a cost-effective manner of reducing the immediate water purification cost, protecting facilities in high water flow events and reducing the need for costlier future expenditure to meet the challenge of decreasing source water quality.

Successful delivery of the projects requires close collaboration with the community. Access to weed infestations is usually through private property requiring landholder cooperation. Communities are empowered to rear and release biological control agents. This is an excellent engagement tool for the education of the public on weed issues and a viable alternative to physical removal in large inaccessible infestations.

The project investment provided by Seqwater is used by Healthy Land and Water (HLW) as leverage to acquire additional funding from other government and philanthropic sources. This money is in turn used for grants to equip and support community groups, effectively expanding the overall area managed.

DISCUSSION

The outcome sought in the delivery of this WoNS vine weed management project is the prevention of escalating costs to purify and deliver safe drinking water by preserving the native vegetation cover within riparian zones and benefiting from the ecological services they provide. Simply put, the investment in vine weed management now will alleviate the need for upgrading of equipment, costly restoration of degraded natural riparian systems or in the worst-case scenario the construction of engineering solutions for erosion control and bank stability.

The escalating costs of differing intervention approaches as determined by Seqwater are demonstrated (Table 1).

Table 1. Indicative costs of management.

Action	Indicative costs ($)[A]
Pre-emptive vine weed management (small scattered infestations)	$12,400 km^{-1}
Managing moderate infestations	$109,000 km^{-1}
Revegetation post riparian vegetation death	$280,000 km^{-1}
Stream bank erosion remediation (engineered solution)	$2,400,000 km^{-1}

[A] Tim Odgers, Seqwater 2018.

Delivery of the project through HLW, the local regional body has several advantages over self-management by Seqwater. HLW has a history of successful community engagement within the SEQ community collaborating with them to deliver many projects of this nature. They are trusted by the community and have a reputation of delivering quality work. Utilising this reputation overcomes resistance from the community to interventions or projects delivered by the water utility as they can be perceived to be an extension of government. HLW can negotiate and gain access to landholders across all land tenures.

Since 2016 Seqwater has invested $2 M in vine weed management on strategic reaches of creeks delivered through HLW. The areas targeted have been the Stanley River Catchment and the Upper and Mid Brisbane River catchments. The project has managed vine weeds to a secondary follow-up management stage over approximately 60 km of creek bank.

The future investment of $6 M will see an expansion of the project to include management sites in the Logan and Albert River catchments as well as continuation of works within the Brisbane and Stanley River catchments. Projections are that an additional 100 km of creek length will be managed by 2023. Reaching this target is very dependent on weather and ability to access infestations.

Management of riparian areas is undertaken by contractors chosen through a tendering process. They will undertake primary management within unmanaged areas before completing secondary and tertiary management in areas already under management. This approach will be repeated annually as they progress downstream within designated areas.

Physical control is a staged approach with initial control works focusing on the saving of significant trees and vegetation under direct threat of being killed. This is followed up by progressive retreatment of the re-emerging shoots from underground tubers and tendrils growing at ground level. Delivery over successive growing seasons progressively suppresses weed growth and ultimately eliminates the remainder of the weed from the targeted reach.

In areas that are unlikely to be managed within the scope of this project but pose a direct threat as future seed sources, releases of biological control agents are or will be undertaken to offer the weeds some resistance to future spread.

Under the current project approximately 2000 jewel beetles *Hylaeogena jureceki* Obenberger have been released at the Mid Brisbane management site targeting cat's claw creeper.

Over the next five years to 2023 the biocontrol program will be ramped up to see the release of potentially millions of jewel beetles across the region. The effectiveness of the effort will be monitored and findings used to inform future release initiatives. To achieve this large volume of beetle raising and release, the project will invest in community groups that currently or show interest in raising jewel beetles. This will allow for the expansion of their operations and through bulk purchases and the encouragement of other landholders and land managers to do the same. The objective is to develop a long term sustainable commercial market for the agents.

CONCLUSION

The advantage of pre-emptive vine weed management within riparian zones to preserve deep rooted vegetation and benefit from the ecological services they provide is clearly demonstrated (Table 1). Delivering the projects via the local regional body allows for greater community engagement and cooperation with the initiative and leveraging of other funds to develop collaborative projects with community ownership expanding the areas managed. The outcome of the project will strive for a sustainable future water supply curtailing unnecessary increasing purification costs, a more robust natural environment delivering ecological services and an engaged community.

REFERENCES

DAF, Department of Agriculture and Forestry (2018). www.daf.qld.gov.au/__data/assets/pdf_file/0003/63336/IPA-Cats-Claw-Creeper-PP139.pdf (accessed 10 April 2018).

WoNS, Weeds of National Significance (2018). https://weeds.ala.org.au/WoNS/catsclawcreeper/ (accessed 10 April 2018).

Containment of Western Cape bridal creeper in peri-urban Adelaide – an eleven-year review

Henry Rutherford[1], Michaela Heinson[1], Susan Ivory[1] and Phil Elson[2]
[1] Natural Resources Adelaide and Mt Lofty Ranges, 205 Greenhill Road, Eastwood, South Australia 5063, Australia
[2] Natural Resources South East, 11 Helen Street, Mount Gambier, South Australia 5290, Australia
(susan.ivory@sa.gov.au)

Summary The management program for Western Cape bridal creeper (*Asparagus asparagoides* (L.) Druce) in peri-urban Adelaide started in 2006 focusing on infestations in the Anstey Hill Recreation Park and surrounding suburbs in the Adelaide Hills. The management area covers 1400 ha including over 150 private properties. After a decade, the intent of the program shifted from eradication to containment when it became clear that the original goal of complete eradication by 2019 was not feasible due to the size of the infestation, the density of vegetation, steep terrain and a steady budget (not an increasing one). In this paper we discuss: adaptive management; improvements in data collection and analysis; actions that align with a containment goal; the use of zoning to prioritise control actions; and methods for engaging with private landholders. Considerations for future implementation are also described including the management of a designated buffer zone and improved methods for surveillance.

Keywords Western Cape bridal creeper, Adelaide, Weeds of National Significance, containment.

The research behind New Zealand's aquatic weed eradication programmes

Paul D. Champion and Deborah E. Hofstra
Freshwater and Estuaries Centre, National Institute of Water and Atmospheric Research Ltd,
PO Box 11-115, Hamilton 3251, New Zealand
(paul.champion@niwa.co.nz)

Summary Six aquatic plants were selected amongst 11 organisms included in the Ministry for Primary Industries national eradication programmes, known as National Interest Pest Responses (NIPR). The six weeds chosen were *Phragmites australis*, *Hydrilla verticillata*, *Zizania latifolia*, *Ceratophyllum demersum*, *Eichhornia crassipes* and *Salvinia molesta*. These species were selected based on several factors relating to the assessment of environmental, economic, cultural, recreational and health risks. This assessment included their current distribution, dispersal pathways, environmental and potential distribution and characterisation of likely impacts. Additionally, evaluation of current and future control tools was undertaken and the formulation of the best available tools needed for eradication was carried out, including an assessment of uncertainty. In some cases, the tools required for eradication were not approved for use in aquatic environments, and the research and advocacy steps needed to facilitate tool use, compliance and monitoring are outlined. Monitoring was also undertaken to ensure programme effectiveness and catalogue any unintended off-target impacts. All programmes have made significant progress since the inception of NIPR in 2008, with *C. demersum* eradicated from the target area.

Keywords Weed risk assessment, impact assessment, control tools, monitoring.

10 years after Victoria's major Mexican feather grass incursion: Assessing eradication methods and germination behaviour

Nicholas P. Moran[1], Alexei D. Rowles[2] and Angela B. Constantine[3]
[1] Agriculture Victoria, Department of Economic Development, Jobs, Transport and Resources, 475 Mickleham Road, Attwood, Victoria 3049, Australia
[2] Agriculture Victoria, Department of Economic Development, Jobs, Transport and Resources, PO Box 124, Benalla, Victoria 3672, Australia
[3] Agriculture Victoria, Department of Economic Development, Jobs, Transport and Resources, PO Box 5646, Cranbourne, Victoria 3977, Australia
(alexei.rowles@ecodev.vic.gov.au) (angela.constantine@ecodev.vic.gov.au)

Summary Mexican feather grass (*Nassella tenuissima* Trin. Barkworth) is a highly invasive weed, with a potential economic and environmental impact in Australia comparable to serrated tussock (*Nassella trichotoma* Nees Hack. ex Arechav.), a close relative and Weed of National Significance. In Victoria, *N. tenuissima* is declared as a State prohibited weed and is managed with the goal of eradication. In 2008, retailers sold approximately 4000 *N. tenuissima* plants throughout metropolitan Melbourne and elsewhere.

The decade-long response to this incursion provides an opportunity to assess both the management approach, and the species' reproductive behaviour in south-eastern Australia. Infestations are treated primarily by physical removal and then regularly monitored for re-emergence from seed germination. The management of over 400 infestations allows an assessment of the efficacy of this management approach towards eradication, with fewer than 1% of the infestations discovered from July 2008 to June 2010 having plants present in the 2016/2017 season. We also analysed incidences of seed germination to assess the risk of seedling emergence over time. For example, for infestations where seedlings have emerged, over 90% showed at least some germination occurring within the first two years of initial treatment, whereas over 5% showed germination occurring more than five years after initial treatment. This case demonstrates the importance of early intensive monitoring, as well as taking a long-term outlook in the design and execution of invasive plant eradication programs. It also highlights physical removal as a viable primary eradication method for *N. tenuissima* incursions of this type and scale.

Keywords *Nassella tenuissima,* biosecurity.

INTRODUCTION

A major pathway for species invasions is ornamental plants imported and cultivated for horticulture, such that the eradication of new invaders in their early stages can require management responses to be targeted towards private yards, parks and gardens. Estimates suggest most invasive plant species in countries like Australia (Virtue *et al.* 2004) and the United States of America (Reichard and White 2001) arrive via horticulture (e.g. nurseries, botanical gardens, seed trading). Gardens can therefore be a frontline in preventing the establishment of new invaders and their negative economic and environmental impacts, as is the case with Mexican feather grass (*N. tenuissima*) in south-eastern Australia.

Nassella tenuissima is native to the Americas, including Argentina, Chile, United States, and Mexico (Jacobs *et al.* 1998), and its appeal as an ornamental plant and broad climate tolerance have helped it become a successful invader. It is now an emerging invasive species in the Mediterranean region and South Africa (Brunel *et al.* 2010, Milton 2004). In Australia, the species' potential impact should it spread to agricultural regions is being recognised (McLaren *et al.* 1999). It shares a genus with two Weeds of National Significance, Chilean needle grass (*Nassella neesiana* (Trin. & Rupr.) Barkworth) and serrated tussock (*Nassella trichotoma*). Like *N. trichotoma*, it is also an unpalatable pastoral weed, and the two are challenging to distinguish in their vegetative state (Solarska *et al.* 2012). By some estimations, *N. tenuissima* has a greater potential invasive range than *N. trichotoma* (McLaren *et al.* 1999), highlighting the importance of managing this species for eradication.

In 2008, *N. tenuissima* was inadvertently sold by a number of retailers throughout Victoria and Australia. Investigations revealed that over 10,000 seeds may have been imported into Australia, with plants supplied to numerous retailers from January to May 2008. In Victoria, plants were traced using on-foot searches of over 150,000 front yards in suburban areas around these retails stores, voluntary product recalls, increased media coverage to promote self-reporting by the public, identification training for key stakeholders (e.g. Australia Post deliverers and utilities meter readers), and in 2009/2010, the use of credit card details to identify buyers. Over the 2008/2009 and 2009/2010 seasons, thousands of plants were recovered and almost 500 infestation sites, mostly in residential gardens on private property, were identified as requiring ongoing management. Most of these sites are still subject to monitoring. It is believed that over 2000 *N. tenuissima* plants sold during this period remain unaccounted for, although it is extremely rare for plants to be discovered outside of cultivation.

A standardised treatment method for *N. tenuissima* removal was developed and applied to infestation sites. Seed heads were first secured in a plastic bag to prevent accidental spread, and the tussock then chipped out, including the basal root ball (e.g. by mattock) to prevent regrowth. Entire plants were then bagged for removal. Seedlings were physically removed, or spot sprayed with herbicide (e.g. glyphosate). Seed germination suppressants (e.g. flupropoante 745 g L^{-1} as used in the Australian Capital Territory (ACT) response; (Connolly and Taylor 2016) were not advised due to risks identified with its widespread use in residential gardens. Sites are classified as 'active' or 'monitoring' at the end of each season, where active sites are those where target plants have been found in the current season, and monitoring sites had none found. Classifications determine the assessment schedule for the following season, with active sites assessed twice during the growth period covering both spring-summer and autumn germination periods. For monitoring sites, a single annual assessment is scheduled for the following season, and from 2012/2013, biennial assessments have been implemented for sites that have had a monitoring status for several years. Assessments only cease where there is proof of site extirpation (referred to here as eradication). This requires evidence that a seedbank does not exist (e.g. plants never seeded), or is unviable (e.g. based on a period of absence). We have currently set this period to be 10 years, although seed longevity is not known (Biosecurity Queensland 2016).

Here, we contribute to our ongoing review of this eradication program with analyses conducted to: (1) track the efficacy of treatment methods and adapt when necessary; and (2) determine patterns of seedling emergence to inform future management timelines and estimate the period of absence required to prove a seedbank is no longer at risk of germinating.

MATERIALS AND METHODS

Infestation sites included in this study are those recorded in Agriculture Victoria's State prohibited weed internal case management database ('BioWeb', based on the Microsoft SharePoint platform) with the date of first treatment/assessment between 1 July 2008 and 30 June 2010 (n = 488). Sites were excluded where infestations pre-date or were expressly not related to this incursion (n = 8), or where *N. tenuissima* was never confirmed to be present (n =). Sites at propagation or distribution centres for the original source businesses were also excluded (e.g. wholesalers, nurseries etc.; n = 4), as their treatment differed from other sites. The sites included in analysis (n = 474) are largely private residences where only a small number of plants were purchased (>75% had five or fewer plants).

Infestation characteristics and treatment efficacy At the initial and follow-up assessments of infestation sites, thorough searches of the property and surrounding areas were undertaken, with infestation data (e.g. number, age and location of seedlings) recorded. For this study, monitoring and treatment data for all *N. tenuissima* sites were extracted from BioWeb (as of 1 July 2017), and supplemented with operational records compiled during the initial response. Summary statistics were calculated based on this data at each site, describing overall characteristics of the infestations (e.g. proportion of sites that seeded, proportion of site with germinants, etc.).

Treatment efficacy was analysed qualitatively, by assessing the proportion of sites classified as 'active', 'monitoring' and 'eradicated' over time. This applied classifications determined at the end of each season from 2008/2009 to 2016/2017 based on each season's assessment data, e.g. active sites had *N. tenuissima* plants recorded in the current season, monitoring sites had not. Eradicated sites are largely those where seeding never occurred, or the seedbank is no longer considered a risk (e.g. where gardens have been built/concreted over).

Seed germination analysis Periodic assessment data allows the time between initial treatment and germination events to be analysed, as re-emergence of *N. tenuissima* plants after complete physical removal is assumed to occur solely by germination. Sites included in this analysis were restricted to those with a potentially viable seedbank, i.e. had seeded or were recorded as 'mature' (n = 369). To assess the likelihood that a site with a potential seedbank will begin to germinate after a period of monitoring, the time between each site's initial treatment and their first recorded re-emergence via germination was analysed. This seeks to inform future management of *N. tenuissima* sites with a potential seedbank, so that after a period with no germination detected, the risk of re-emergence may be considered low enough to confidently classify sites as 'eradicated'. However, any analysis should be viewed and applied cautiously due to uncontrolled factors, e.g. the chance that seedlings were removed by landholders without our knowledge, and also varying weather conditions from 2008/2009–2016/2017 which are an inherent confound in assessing germination risk over time.

Time-to-event (survival) analysis was used to account for the right censoring of data given that germination could still occur in the future, an issue analogous to laboratory-based seed germination experiments (Ritz *et al.* 2013). As germination events were not recorded at exact times, a non-parametric cohort life-table method was used to estimate the survival function ($S(t)$) based on time intervals (McNair *et al.* 2012). In this analysis, $S(t)$ is the probability that a site will not have begun to germinate by time t (or equally, that its first germination event will occur at time $\geq t$). One-year time intervals are used up to the third year as minimum annual monitoring was implemented, and two-year time intervals are used thereafter, as the sites found to be active in the fifth (n = 3) and seventh (n = 2) years had been subject to biennial monitoring prior to seedlings being found. Input data for cohort life-tables is the initial total number of sites (n = 369), the number of sites that first germinated in each time interval, the number of sites 'lost' in each time interval (i.e. sites where germination was no longer discoverable, e.g. had been extirpated/eradicated). $S(t)$ estimates with standard deviations were calculated with R Statistical Package (v3.1.2. R Foundation for Statistical Computing, Vienna, Austria) using the life table function of the 'KMsurv' package (Klein and Moeschberger 1997). To characterise the probability of an infestation first re-emerging at a site over time, a non-linear regression (one-phase exponential decay) of $S(t)$ was applied. Regression analysis and figures were produced using GraphPad Prism (v7.04. GraphPad Software, La Jolla California).

RESULTS

Infestation characteristics and treatment efficacy

Greater than three quarters of sites (77.85%) produced or were very likely to have produced seed at first inspection (Table 1). Germination was common, with over a third (37.97%) recording seedlings after first treatment. Secondary seeding was rare, with <5% of sites recording seeding plants after initial treatment, and the mean number of germination events recorded was less than two in sites where germination occurred (Table 1).

The progress of sites from active to monitoring then eradication shows the number of active infestations peaked in 2009/2010, coinciding with the influx of new cases stemming from tracing investigations in that year (Figure 1). Active cases have since rapidly and continuously declined, resulting in <1% of cases active at the end of 2016/2017. Meanwhile, the number of sites proven to be eradicated has steadily increased (Figure 1).

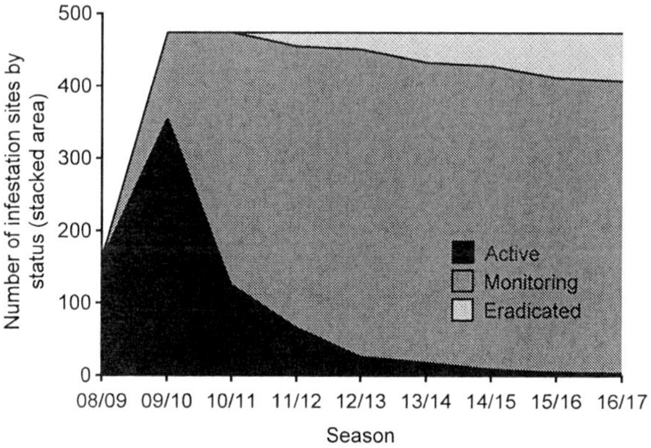

Figure 1. Infestation status across all sites as of the end of each season, 2008/2009–2016/2017.

Table 1. Site and germination summary statistics.

Number of sites	**474**
Seeding status of sites at first treatment	369 (77.85%) recorded as seeded or as 'mature'. 39 (8.23%) recorded as having not seeded. 66 (13.92%) status unknown.
Sites with seed germination after first treatment	180 (37.97%) sites recorded germinants found on follow-up inspections.
Sites with secondary seeding	15 (3.16%) sites recorded seed being set at follow-up inspections.
Average number of germination events recorded per site	0.861 (s.d. 1.469) at all sites. 1.958 (s.d. 1.031) at sites where post treatment germination occurred.

Seed germination analysis At infestations where germination has occurred, most (132 sites) had seedlings recorded within one year of initial treatment, however, two sites had their first germination event recorded 5–7 years after treatment (n = 2, Figure 2a). Furthermore, at 15 sites seedlings have been found more than five years after initial treatment (data not shown), suggesting seedbanks can remain viable for an extended period. Similarly, of the seeded/mature sites that did germinate, the majority did so within the first year after treatment, and more than 90% showed germination within two years (Figure 2b).

Non-linear regression ($R^2 = 0.9997$, Figure 3) was applied to the survivor function estimates (Table 2), giving a decay constant $K = 1.342$ (95% CI: 1.256, 1.437), and a lower limit of 0.5191 (95% CI: 0.5129 to 0.5252). In this study, the lower limit may be interpreted as the probability that a site will never germinate. Applying this K value to the program's current eradication period of 10 year continuous absence, corresponds to a <0.0001% chance that a site with a potential seedbank will first germinate more than 10 years after treatment. However, there is potential that germination has not yet occurred due to environmental factors, e.g. weather conditions or other local factors, that this kind of estimation must be considered with caution.

Table 2. Survivor function estimates for first recorded germination at seeded/mature infestations, based yearly time intervals since first treatment.

Time interval (years)	Number of sites recording first germination	*S(t)*, as of the start of the interval (standard deviation)
0–1	132	1.000 (0.000)
1–2	31	0.642 (0.025)
2–3	9	0.556 (0.026)
3–5	4	0.531 (0.026)
5–7	2	0.520 (0.026)
>7	0	0.514 (0.026)

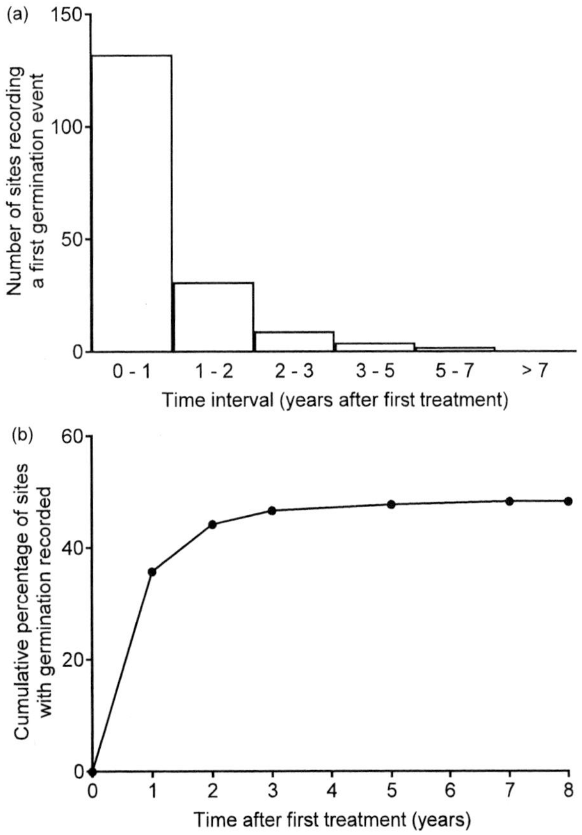

Figure 2. (a) Number of sites where germination first occurred within the time interval, taken from the time of initial treatment; and (b) the cumulative percentage of sites displaying germination since the first treatment (seeded/mature sites only).

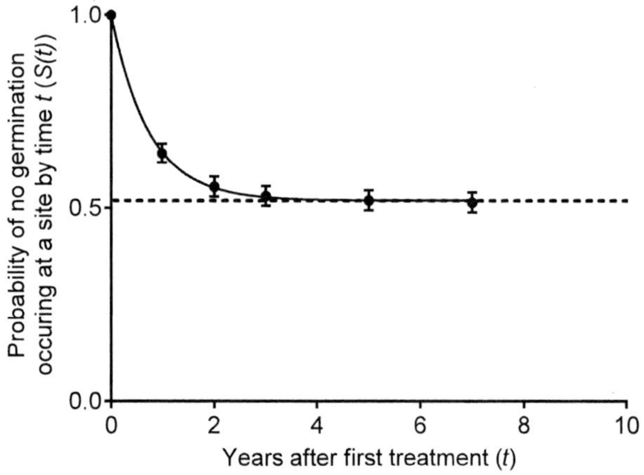

Figure 3. *S(t)* is the probability that a site will not have recorded germination by time *t*. The solid curve is a non-linear regression fit to *S(t)* estimates. The dotted line is its lower limit ($S(\infty) = 0.5191$).

DISCUSSION

The capacity of *N. tenuissima* to readily reproduce in Victoria is shown by most infestations having seeded when found, and seedlings later being found at around half of those sites. As the incursion was discovered fairly quickly and the data analysed here is that of sites found within two years of its discovery, much of the seed had a relatively short window to develop, and the seedbank only a short time to accumulate. While this study is not representative of infestations that have developed a larger seedbank over many years, it does show that a substantial proportion of these infestations rapidly developed viable seedbanks within a short period.

Nassella tenuissima does not appear to have an extensive history of invasiveness overseas, although it has proven difficult to eradicate after escaping from cultivation in California and New Zealand (Jacobs *et al.* 1998). It is also problematic in South Africa (Milton 2004). Predictions for its potential distribution in Australia range from 14.1 to over 100 million ha, including large areas of Victoria (McLaren *et al.* 1999, Biosecurity Queensland 2016). The capacity of *N. tenuissima* to reproduce in Victoria shown here highlights the real threat this species poses.

The physical removal treatment method used by Agriculture Victoria to destroy mature *N. tenuissima* plants has been effective, with few incidences of secondary germination and a clear trend towards eradication of sites. Nonetheless, many plants from this incursion remain unaccounted for, so effective management of this incursion also requires ongoing detection of undiscovered plants. The physical similarities to *N. trichotoma* are a barrier to this, and work improving identification (e.g. Solarska *et al.* 2012) and weed identification training (e.g. Agriculture Victoria's Weed Spotter program) are essential to preventing the establishment of *N. tenuissima* in Victoria.

A major issue for the future of the *N. tenuissima* program is the uncertainty surrounding seedbank longevity. With a lack of seed longevity data, this study can be valuable for informing management decisions despite the limitations of this dataset. Our results are consistent with the ACT experience (Conolly and Taylor 2016), showing infestations can remain active via germination for 7+ years and suggesting seed longevity is at least seven years.

Applying time-to-event (survival) analysis to the time between initial treatment and the first recorded germination event at sites, we could characterise the risk of infestations re-emerging following treatment. This showed the risk of re-emergence is particularly high in the first two years after removal, and that the risk of re-emergence after a prolonged absence (e.g. 10 years) is low. Were more specific data consistently collected over the course of the project (e.g. exact numbers of seedlings at each assessment and the presence/number of seedlings at initial inspections), predictions could be further refined. While our results are somewhat narrow in application, this shows how project data can be used to inform future planning in biosecurity programs.

ACKNOWLEDGMENTS

We acknowledge the work of Department staff in the incident response and the many years since, and specifically thank Annie Lamb for her contribution.

REFERENCES

Biosecurity Queensland (2016). Mexican feather grass (*Nassella tenuissima*) – Pest plant risk assessment, https://www.daf.qld.gov.au/__data/assets/pdf_file/0017/64160/IPA-Mexican-Feather-Grass-Risk-Assessment.pdf (accessed 28 March 2018).

Brunel, S., Schrader, G., Brundu, G., and Fried G. (2010). Emerging invasive alien plants for the Mediterranean Basin. *EPPO Bulletin* 40, 219-38.

Conolly, J. and Taylor, S. (2016). Mexican feather grass (*Nassella tenuissima*) control–the ACT experience. Proceedings of the 20th Australasian Weeds Conference, eds R. Randall, S. Lloyd and C. Borger, pp. 228-31. (Weeds Society of Western Australia, Perth).

Jacobs, S.W.L., Everett, J. and Torres, M.A. (1998). *Nassella tenuissima* (Gramineae) recorded from Australia, a potential new weed related to serrated tussock. *Telopea* 8, 41-6.

Klein, J.P. and Moeschberger, M.L. (1997). Survival analysis. Techniques for Censored and Truncated Data. (Springer, New York). https://cran.r-project.org/package=KMsurv (accessed 28 March 2018).

McLaren, D.A., Whattam, M., Blood, K., Stajsic, V. and Hore R. (1999). Mexican feather grass (*Nassella tenuissima*) a potential disaster for Australia. Proceedings of the 12th Australian Weeds Conference, eds A.C. Bishop, M. Boersma and C.D. Barnes, pp. 658–62. (Tasmanian Weed Society, Devonport).

McNair, J., Sunkara, A. and Frobish, D. (2012). How to analyse seed germination data using statistical time-to-event analysis: Non-parametric and semi-parametric methods. *Seed Science Research* 22, 77-95.

Milton, S.J. (2004). Grasses as invasive alien plants in South Africa. *South African Journal of Science* 100, 69-75.

Reichard, S.H. and White, P. (2001). Horticulture as a pathway of invasive plant introductions in the United States. *BioScience* 51, 103-13.

Ritz, C., Pipper, C.B. and Streibig, J.C. (2013). Analysis of germination data from agricultural experiments. *European Journal of Agronomy* 45, 1-6.

Solarska, S.E., McLaren, D.A., and Grech, C.J. (2012). Investigations into micro-morphological discrimination of vegetative Mexican feather grass, *Nassella tenuissima* from Serrated tussock, *Nassella trichotoma* and other stipoid species. Proceedings of the 18th Australasian Weeds Conference, ed. V. Eldershaw, pp. 378-81. (Weed Society of Victoria, Melbourne).

Virtue, J.G., Bennett, S.J. and Randal, R.P. (2004). Plant introductions in Australia: how can we resolve 'weedy' conflicts of interest? Proceedings of the 14th Australasian Weeds Conference, eds B.M. Sindel and S.B. Johnson, pp. 42-8. (Weed Society of New South Wales, Wagga Wagga).

Pushing 37 weed species towards extinction in Queensland

Steve Csurhes
Biosecurity Queensland, Queensland Department of Agriculture and Fisheries, EcoSciences Precinct, 41 Boggo Road, Dutton Park, Queensland 4102, Australia
(steve.csurhes@daf.qld.gov.au)

Summary After many years of concerted effort by state and local government staff, 21 species of high-risk emerging weeds appear to have been eradicated in Queensland, with another 16 tantalisingly close to eradication – a total of 37. The probability of eradication for an additional 12 species is less certain, while eradication efforts have been abandoned for another nine. High-risk targets include a range of species that are notorious problems overseas or interstate. Data have been collected and analysed to estimate likely duration and cost of eradication. The challenge is to maintain pressure on target populations and to prevent additional incursions.

Keywords Eradication, invasive plant.

INTRODUCTION

Early detection and eradication of potentially high-risk weed species is a worthy goal somewhat analogous to the early detection and removal of skin cancer. Once widely naturalised, pest populations become highly resilient and impossible to eradicate.

In the mid-1990s, a range of high-risk weed species known to have major negative impacts overseas or interstate were pre-emptively targeted for exclusion, prohibition on sale and early detection, while they were still absent from Queensland. Species were selected based on a review of the world's worst weeds by Csurhes (internal report, 1991). A key objective was to prevent entry and minimise propagule pressure, primarily by removing the commercial incentive to sell these species as garden ornamentals (this being the primary invasion pathway for at least 80% of species). High-risk targets include most non-native species of *Acacia/Vachellia/Acaciella*, including *V. karroo* (Hayne) Banfi & Glasso (Figure 1), all species of horsetails (*Equisetum* spp.), bitou bush (*Chrysanthemoides monilifera* subsp. *rotundata* (DC.) Norl.), non-native species of witchweed (*Striga* spp.) and

Figure 1. *Vachellia karroo* (Karroo thorn) detected on the Darling Downs, Queensland.

tropical soda apple (*Solanum viarum* Dunal). Weed risk assessments for these species are available at DAF (2018).

The primary aim of this paper is to outline progress towards early detection and eradication of high-risk weed species in Queensland.

METHODS

Data collection and analysis Since the mid 1990s suspect specimens of high-risk targets have been routinely sent to the Queensland Herbarium for positive identification. Whenever detection of a high-risk target was confirmed, eradication efforts commenced and field-staff began recording the species' location and status. Data on the species' full geographic extent was recorded, a process known as 'delimitation', often in parallel with on-ground eradication works. Over the years, as more species were detected, the list of eradication targets expanded. Data were periodically analysed, in 2009 and 2015, using a model called 'Weed-Search' developed by the University of New England (Cacho and Pheloung 2007), to assess the cost and duration of eradication (Panetta *et al.* 2011, Csurhes, internal report 2016). The model requires 24 input parameters, including the number of infested sites, estimated average plant density, search area, detectability, and a range of biological parameters such as seed longevity. Total search area is the primary factor that determines eradication cost. Modelling provides decision-makers with evidence-based information on likely cost and duration of eradication projects.

RESULTS

Eradication targets While pre-emptive restrictions on entry and sale at a state-level have prevented or delayed naturalisation of many high-risk taxa, 59 such targets have been detected in Queensland since the mid-1990s. Detections include isolated cultivated specimens as well as naturalised populations of varying size (area) and density. These 59 species can be placed into six broad categories: (1) cultivated specimens believed to have been eradicated (15 species); (2) naturalised populations believed to have been eradicated (6 species); (3) naturalised populations that appear to be declining and on-track for eradication (16 species); (4) naturalised populations that are proving resilient with a high level of uncertainty associated with eradication (12 species); (5) naturalised populations that were targeted for eradication for many years but subsequently transitioned to containment and ongoing management (9 species); and (6) unassessed species (1 species) (Table 1). Search areas have been estimated for each species, assuming that a 'search buffer' with a radius of at least 200 m needs to be routinely searched around each detection site.

DISCUSSION

Thirty seven weed species have either been eradicated or are 'on track' for eradication in Queensland. This paper acknowledges that 15 species were still restricted to gardens, at the time of detection, and argues that these are true examples of early detection. Eradication of naturalised populations is much more challenging and claiming success is fraught with uncertainty, due primarily to the difficulty associated with delimitation, ongoing incursions, detectability of the last few individuals in a population and the prolonged time required to exhaust soil-seed-banks. Species believed to have been eradicated are species that have not been seen for many years, despite follow-up surveillance. Species 'on-track' for eradication continue to be detected at low levels but with declining abundance. Eradication appears feasible for these species, particularly for those with small search areas, provided current surveillance and control efforts are maintained and provided additional 'outlier' populations are not detected. An additional 12 species continue to be targets for eradication, but carry a high level of uncertainty regarding feasibility of eradication. These species possess traits that make eradication particularly challenging, such as large search areas and multiple invasion sites, poor detectability, poor kill-rates, or inadequate delimitation following initial detection (and hence a low level of confidence regarding population size and extent). The policy objective for these species will be reviewed periodically.

Analysis of successful eradication programs from around the world provides evidence that, for eradication to be successful, the search area involved needs to be small, almost always less than 1000 ha and often only a few hectares (Waldendorp and Bomford 2004). Cost increases dramatically as search area increases. Based on an analysis of eradication programs undertaken by the California Department of Food and Agriculture, comprising 16 species and 50 infestations, Rejmánek and Pitcairn (2002) concluded that for infestations less than 1 ha in size, eradication was almost always possible; for infestations between 1 and 100 ha, approximately 30% were successfully eradicated; and for infestations between 100 and 1000 ha, only about 25% of eradication attempts were successful. Application of the Weed-Search model for eradication targets listed here produced similar results. For example, it predicted that eradication of *Vachellia robusta* (Burch.) Kyal. & Boatwr. listed as a category 1 species in Table 1, and with a total search area of 12 ha, is likely to cost $79,500 over 21 years. Eradication of *Vachellia karroo* (Karroo thorn), a category

Table 1. List of weed species that have been or are currently being pursued for eradication in Queensland (1 = cultivated specimens believed eradicated, 2 = naturalised population believed eradicated, 3 = naturalised population declining and 'on track' for eradication, 4 = eradication being pursued but success highly uncertain, 5 = species no longer being eradicated and 6 = unassessed species) (*species continues to be regularly detected in cultivation).

Species	Search area (ha)	Category
Austrocylindropuntia subulata (Eve's pin cactus)	12*	1
Cecropia pachystachya	12	1
Equisetum hyemale (horsetail)	12	1
Nassella tenuissima (Mexican feather grass)	46	1
Opuntia microdasys (bunny ears cactus)	n/a*	1
Opuntia robusta (wheel cactus)	12	1
Opuntia rufida (blind cactus)	24*	1
Opuntia santarita (Santa Rita prickly pear)	12	1
Prosopis laevigata (smooth mesquite)	12	1
Salix alba var. *alba* (white willow)	12	1
Senegalia nigrescens (knob thorn)	12	1
Senegalia chundra (syn. *Acacia catechu* var. *chundra*)	12	1
Vachellia gerrardii (grey-haired acacia)	12	1
Vachellia robusta (syn. *Acacia robusta* subsp. *clavigera*)	12	1
Vachellia xanthophloea (syn. *Acacia xanthophloea*) (yellow fever tree)	24	1
Chromolaena squalida	12	2
Miconia cionotricha	12	2
Opuntia elatior (red-flowered prickly pear)	12	2
Salix cinerea (grey willow)	12	2
Salix nigra (black willow)	38	2
Ulex europaeus (gorse)	25	2
Chrysanthemoides monilifera (bitou bush)	460	3
Cylindropuntia prolifera (jumping cholla)	>250	3
Helenium amarum (bitterweed)	150	3
Heterotheca grandiflora (telegraph weed)	>250	3
Limnocharis flava (limnocharis)	48	3
Miconia calvescens (miconia)	6,224	3
Miconia nervosa	190	3
Miconia racemosa	236	3
Mikania micrantha (mikania)	427	3
Mimosa pigra (giant sensitive tree)	1,000	3
Opuntia leucotricha (Aaron's beard cactus)	62	3
Opuntia puberula (puberula cactus)	25	3
Pithecellobium dulce (Madras thorn)	425	3
Solanum viarum (tropical soda apple)	867	3
Striga asiatica (red witchweed)	439	3
Vachellia karroo (syn. *Acacia karroo*) (Karroo thorn)	25	3
Acaciella angustissima (white-ball acacia)	375	4
Acaciella glauca (red wood)	192	4
Cecropia peltata complex (Mexican bean tree)	800	4
Cylindropuntia pallida (Hudson pear)	842	4
Cylindropuntia tunicata	>500	4
Gleditsia triacanthos (honey locust tree)	62,600	4
Gmelina elliptica (badhara bush)	871	4
Neptunia plena and *N. oleracea* (water mimosa)	>250*	4
Opuntia elata (Riverina pear)	600	4
Opuntia sulphurea (sulphur cactus)	2012	4
Senegalia insuavis (syn. *Acacia pennata* (cha-om)	125*	4
Senegalia rugata (syn. *Acacia concinna*) (soap pod)	138	4
Alternanthera philoxeroides (alligator weed)	>1,000	5
Asparagus asparagoides (bridal creeper)	>1,512*	5
Chromolaena odorata (Siam weed)	>1,000	5
Clidemia hirta (Koster's curse)	>1,000	5
Gymnocoronis spilanthoides (Senegal tea plant)	>1,000	5
Hedychium flavescens (white/yellow ginger)	>549*	5
Hygrophila costata (hygrophila)	>1,000	5
Nassella neesiana (Chilean needle grass)	>14,000	5
Thunbergia fragrans	>500*	5
Striga angustifolia (white witchweed)	not assessed	6

3 species with a search area of 25 ha, is estimated to cost $178,600 over 27 years. However, eradication of *Acaciella angustissima* (Mill.) Brit. & Rose (white-ball acacia), a category 4 species with a total search area of 338 ha (involving 27 separate locations) is estimated to cost $2,632,500 over 34 years. Hence, early detection while populations are very small is highly desirable.

Long timeframes are a common feature of eradication programs, primarily due to long-lived seed banks. This demands years of dedication and perseverance by field staff and partner agencies.

Surveillance for all targets will be on-going, as the risk of re-introduction and undetected incursions remains. A range of traditional and novel surveillance techniques are being applied, including the use of strategically positioned sentinel sites, the volunteer-based 'Weed Spotter Network Queensland' (coordinated by Melinda Laidlaw from the Queensland Herbarium), drones and, potentially, use of eDNA. By using novel techniques, the goal is to achieve detection much earlier than has tended to occur in the past, thereby improving the probability of eradication and reducing cost.

CONCLUSION

Eradication of high-risk invasive plant species from the state is a realistic objective, provided targets can be detected very early in the invasion process. Even so-called 'failed' eradication efforts have achieved significant reductions in population sizes and spread rates and it seems reasonable to assume that such reductions are likely to have generated benefits that exceed costs. Despite this, eradication targets need to be carefully selected to maximise the chances of a lasting outcome and to generate strong returns on investment.

ACKNOWLEDGMENTS

I acknowledge the combined efforts of local government staff, Biosecurity Queensland staff and many others to collect this information and to detect and eradicate high-risk weeds.

REFERENCES

Cacho, O. and Pheloung, P. (2007). WeedSearch: Weed Eradication Feasibility Analysis, Software Manual, CRC for Australian Weed Management. (University of New England, Armidale, New South Wales).

Csurhes, S.M. (internal report, 1991). 'Potential weed species that represent a threat to the Australian economy and environment'. Unpublished report prepared under contract to the Australian Government, Queensland Government, Brisbane.

Csurhes, S.M. (internal report 2016). A review of the cost and duration of eradication programs for declared high-risk emerging weed threats in Queensland. Internal report. Biosecurity Queensland, Queensland Government, Brisbane.

DAF (2018). Pest risk assessments, Biosecurity Queensland, Department of Agriculture and Fisheries, Queensland Government, Brisbane, https://www.daf.qld.gov.au/business-priorities/plants/weeds-pest-animals-ants/educational-resources-and-careers/publications/pest-risk-assessments (accessed 1 August 2018).

Panetta, F.D., Csurhes, S.M., Markula, A. and Hannan-Jones, M.A. (2011). Predicting the cost of eradication for 41 Class 1 declared weeds in Queensland. *Plant Protection Quarterly* 26, 42-6.

Rejmánek, M. and Pitcairn, M.J. (2002). When is eradication of exotic pest plants a realistic goal? *In* 'Turning the tide: The eradication of island invasives'. (eds C.R. Vietch and M.N. Clout), Occasional Paper of the IUCN Species Survival Commission No. 27, pp. 249-53. (IUCN – The World Conservation Union, Auckland, New Zealand).

Waldendorp, G. and Bomford, M. (2004). Weed eradication: Strategies, timeframes and costs, Bureau of Rural Sciences, Australian Government.

The effects of cyclones on a tropical weed eradication program

Simon J. Brooks[1] and Mick Jeffery[2]
[1] Tropical Weeds Research Centre, Biosecurity Queensland, Department of Agriculture and Fisheries, PO Box 187, Charters Towers, Queensland 4820, Australia
[2] Biosecurity Queensland, Department of Agriculture and Fisheries, PO Box 20, South Johnstone Queensland, 4859 Australia
(simon.brooks@daf.qld.gov.au)

Summary Weed infestations managed by the nationally cost-shared Tropical Weeds Eradication Program have been affected by the passage of two severe tropical cyclones: Severe Tropical Cyclone Larry in 2006 and Severe Tropical Cyclone Yasi in 2011. The timing of the cyclones coincided with vegetative phases of most target weeds' life-cycles. Destructive winds severely damaged the tropical forest vegetation and produced masses of vegetative debris, impeding site access and slowing ground survey operations. This resulted in immediate delays to survey and weed control schedules. Changes in seedling recruitment patterns may only be apparent in program reporting after one or two years while other effects may arise over 3 or 4 years. There are immediate and longer term effects of cyclones particularly on the staff required to meet annual surveillance targets and protect the cost-share investment to date.

Keywords Surveillance, disaster, Miconia, Mikania.

INTRODUCTION

The nationally cost-shared 'National Tropical Weeds Eradication Program' (NTWEP) commenced in late 2003 and currently targets *Limnocharis flava* (L.) Buchenau, *Miconia calvescens* DC., *Miconia racemosa* (Aubl.) DC., *Miconia nervosa* (Sm.) Triana and *Mikania micrantha* Kunth. Most infestations occur between the towns of Daintree (16°S) and south of Ingham (19°S) along the far north Queensland tropical coast and adjacent Atherton Tablelands. BOM (2018a) expect an average of 4.7 cyclones a year to form in the Coral Sea and Gulf of Carpentaria area. These cyclones tend to form between 5 and 15°S and reach maximum intensity between 10 and 20°S. Single infestations of *M. nervosa* and *M. racemosa* were not near the paths of recent severe cyclones. The effects of two cyclones on the target weeds and the eradication program are summarised below to provide guidance for responding to future extreme natural events.

Severe Tropical Cyclone Larry Cyclone Larry crossed the north Queensland coast with the centre of the 25 km-diameter eye passing near Innisfail on the morning of the 20 March 2006, and continued over the Atherton Tableland while weakening (BOM 2018b, CTS 2018). This was a severe category 4 cyclone with gusts estimated up to 230 km h^{-1} (CTS 2018). Destructive winds (>170 km h^{-1}) extended approximately 50 km from the centre between Babinda and El Arish (Figure 1). The core of this cyclone passed over *M. calvescens* and *L. flava* infestations; other *M. calvescens*, *L. flava* and *M. micrantha* locations were affected by destructive winds.

Severe Tropical Cyclone Yasi Cyclone Yasi crossed the north Queensland coast near South Mission Beach on the 3 February 2011 (Figure 1). With estimated sustained winds of 218 km h^{-1} and gusts up to 285 km h^{-1} this category 5 cyclone produced destructive winds extending over 130 km south of Innisfail (BOM 2018c). The most destructive winds passed over *M. calvescens*, *M. micrantha* and *L. flava* infestations.

EFFECTS ON WEEDS

The native range of all the eradication target species includes some tropical areas of Central, South America and the Caribbean where species have evolved with selective pressures from large scale disturbances, such as landslides and cyclones, and smaller scale disturbances, such as tree and branch falls. The effects observed on the three eradication target species with multiple infestations in the areas directly impacted by the two cyclones vary depending on their biology. Heavy rainfall is seasonally common in the tropics so any effects are not exclusively cyclonic, and could include erosion and exposure of buried seed. All species form a persistent soil seedbank but seed dispersed into an unsuitable habitat can be exhausted.

Limnocharis flava Both cyclones passed over known and (at the time) unknown infestations but there have been no large or direct consequences observed on this anchored aquatic herb. *Limnocharis flava* can reproduce by seed and vegetatively all year round (Figure 2). However, known infestations were visited by

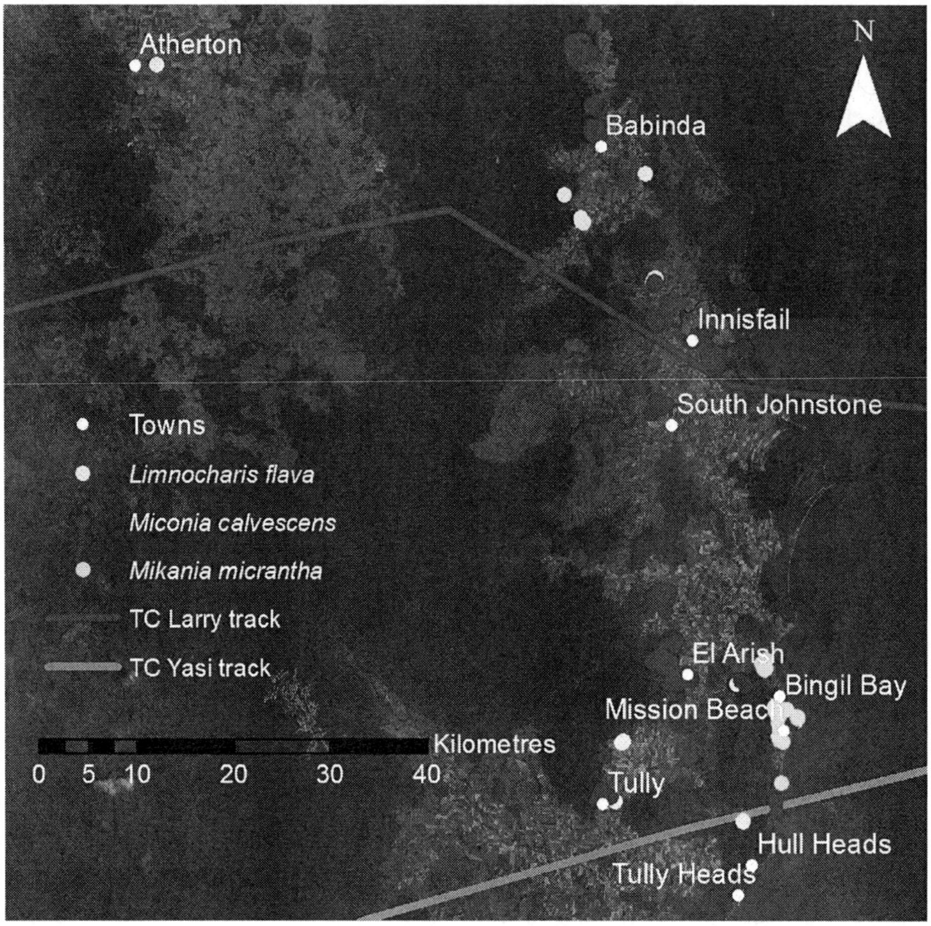

Figure 1. Tracks of Severe Tropical Cyclones Larry (2006) and Yasi (2011) over all known infestations as of December 2017. Points are not to scale and include non-naturalised locations on the far north Queensland coast.

field crews every four weeks, which prevents flowering stalks from maturing and negates the formation of vegetative plantlets on the flowering stalk (Weber and Brooks 2013). Unmanaged infestations are unlikely to produce wind dispersed seed into a narrow aquatic habitat, as flowering stalks bend towards the water as papery follicles mature (Weber and Brooks 2013). This species favours margins of still, shallow water and appears to be limited by faster flowing creeks and rivers with rockier substrates. Soil or plant-borne seeds have to reach a suitable habitat and occasional pre- and post-cyclone downstream surveys have not revealed seed spread specifically related to cyclones. New infestations are usually linked to cultivation. This species forms a persistent seed bank so heavy rainfall events could lead to local seed spread through slow moving water in suitable habitats, such as dam or drain overflow, flood back up and coastal inundation.

Miconia calvescens The response of *M. calvescens* plants to the cyclones may vary depending on the life stage. Cyclones can produce a large volume of leaf litter, but small seeded Miconia species are considered to be 'litter gap demanders' and thus prefer areas with less litter (Murphy *et al.* 2008, and references therein), or it may take some time for the litter to break down to allow seed germination. Murphy *et al.* (2008) recorded *M. calvescens* seedlings six months after cyclone Larry, with peak establishment nine months after the cyclone and higher seedling growth rates in more-damaged plots. However, larger tree fall gaps are less suitable habitat (Meyer 1998). Established seedlings that survive the cyclone may respond quite quickly to increased light levels (Murphy *et al.* 2008), while larger saplings can recover from trunk damage. Field crews measure the height and diameter of large plants encountered in the field (Brooks and Setter 2014) and 3.6% of 2413 records are badly damaged stems (from a variety of causes), these plants have a distinctly high ratio of basal diameter (cm) to height (m) (S. Brooks, unpubl. data). Although saplings and trees can recover the defoliation, a height reduction may influence their detection from the ground or air, with plants taking years to regrow to the height of the recovering rainforest canopy and maturity.

The native tropical vegetation has evolved with cyclones as a selective pressure, however a subset of invasive species including *M. calvescens* are more persistent in the rainforest environment after cyclones (Murphy *et al.* 2014). Meyer (1998) considers a series of six cyclones across Tahiti in 1982–1983 to have revealed the underlying *M. calvescens* incursion and promoted plants into the damaged native forest canopy. This 'revelation' was considered to be due to the favourable growth conditions rather than cyclone induced spread of *M. calvescens*. The low frequency of fruit production (7 cases) in February and March (Figure 2) has restricted cyclone related dispersal with spread patterns consistent with frugivore dispersal. New *M. calvescens* infestations between 2012 and 2014 are thought to have resulted from cultivation: the last new infestation was in 2014.

Mikania micrantha This species has a strongly seasonal flowering behaviour, commencing in April and May, so seed was not present on vines at the times of either cyclone (Figure 2). *Mikania micrantha* also readily reproduces from stem fragments with a single node. The destructive core of cyclone Yasi passed over *M. micrantha* infestations at Mission Beach and near Bingil Bay. This area was surveyed several times after cyclone Yasi and 20 new management areas were recorded in the six months to the end of December 2012, but this is thought to have resulted from ground searches expanding northwards. Some *M. micrantha* infestations are accessible to vehicles so there are opportunities for dispersal of soil borne seed and vegetative fragments during this event and subsequent clean-up operations.

Cyclone Yasi created a lighter environment for germination from the soil seed bank and *M. micrantha* seedling growth, and there was 274 ha of additional survey of cyclone disturbed infestations in 2011–2012 financial year. Unfortunately there was also a large increase in mature *M. micrantha* plants between August and December 2012, these plants were found across 17 known management areas. Although *M. micrantha* is renowned as a rapidly growing species, it still took 18+ months for the vines (from seed or stem fragments) to reach a detectible size and maturity. It is possible that more recruitment commenced in 2011–2012 wet season than immediately after the cyclone as mature plants were recorded in the second half of 2012. There were increases in emergence and growth for two years after cyclone Yasi and some of these plants were hard to detect and prevent maturing.

EFFECTS ON PROGRAM RESOURCES

The eradication program builds annual budgets, usually three years in advance, on the basis of search rates, frequency and area, summed for each target species. Operational scheduling for the available workforce is then based on the survey frequency for each species. There are annual 'wet season' disruptions due to heavy rain, preventing surveys for up to a week, but not dramatically altering the vegetation and search rates like cyclones do. The NTWEP has some ability to re-allocate some resources in the event of a cyclone, however there is only a small amount of flexibility built into the field survey schedule. Large problems with the search rate, extent or frequency of one species can influence the resources available for other species.

Cyclones bring an immediate disruption to field survey schedules, beyond the normal wet season delays. Cyclone Yasi bought at immediate loss of approximately 70 work days or 2 weeks operations for

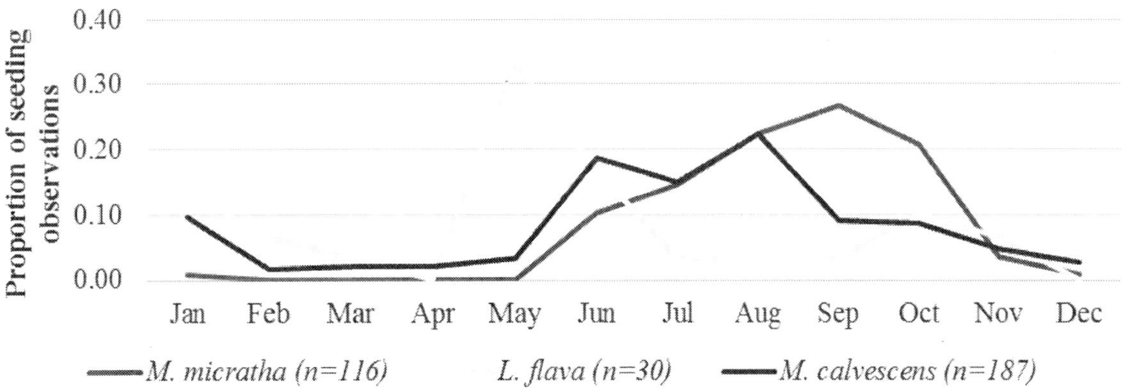

Figure 2. Proportion of seeding observations by NTWEP field crews, per month pooled for all infestations (2003–2017), including discovery records.

staff based at South Johnstone. Both cyclones caused damage to roads and property, with some staff unable to work immediately. There is also damage and disruptions to offices and program coordination activities. The eradication program had well-equipped, local field staff and vehicles that were diverted to assist with broader recovery activities. With infestations primarily occurring on privately owned land, contact with rural landholders affected by the cyclone needs to be carefully managed to maintain long term access. These factors particularly affect the schedule and the frequency of program activities. Even if the survey frequency for *L. flava* (4 weeks) and other species (*M. micrantha*, *M. racemosa* and *M. nervosa*) 3–4 times a year is not badly affected, these species collectively account for 20% of programs recorded annual survey effort (S. Brooks unpubl. data). *Miconia calvescens* infestations are surveyed at lower frequency (16–24 months) but to cover an annual search area target over 4000 hectares, different portions of infestations need to be surveyed throughout the year, which consumes 80% of the programs search effort. Cyclone Yasi impacted over a quarter of the *M. calvescens* survey area.

Destructive winds severely damaged the tropical forest vegetation and produced masses of vegetative debris. While landscape features can become more visible, the field crew's familiarity with infested areas and site accessibility is dramatically altered. Significant field resources were required to re-establish access to survey the rainforest around infestations. After cyclone Yasi this was estimated at two months' work for six staff or contractors. There is also a reduction in ground survey rates, after cyclone Larry crew search rates for *M. calvescens* dropped from 0.7 ha $person^{-1}$ day^{-1} to 0.3 ha $person^{-1}$ day^{-1} in cyclone affected vegetation. Beyond the immediate impedances of woody trunk and branch debris, regrowth of plants including the native palm *Calamus australis* Mart. (lawyer cane) and the exotic shrub *Rubus alceifolius* Poir. (giant bramble) slowed surveys for up to four years.

Cyclones have bought a reduction in in-kind contributions from program stakeholders and more focus on activities such as public access (road and track) clearing and feeding endangered wildlife, with less resources available for weed control beyond public access tracks. After cyclone Larry, three additional field crew were employed for a few months for track cutting. The NTWEP received an extra $300,000 over 2 financial years after cyclone Yasi, primarily for field workers, vehicles and equipment. Both cyclones also bought the need for increased weed identification and spread prevention training and information for workers on clean-up crews. After receiving ID training after cyclone Larry, a local contractor recognised a new *M. micrantha* patch several years later. After cyclone Yasi, NTWEP staff spent time producing and sharing maps of infestations for agencies coordinating clean up contracts.

FUTURE CYCLONES

The program will be managing infestations of target species into the 2030s (Jeffery and Brooks 2016) so it is highly likely that one or more cyclones will impact some infestations over this time frame. It is only possible to predict the timing, location and intensity of cyclones a day or two in advance (certainly not years), which limits effective forward planning. The broad impacts, identified above, will vary across species, infestations and within infestations due to the orientation of the land and the wind strength and direction(s). With field crews based in three locations (Mossman, Cairns and South Johnstone) the effects of a cyclone may directly impact one or two crews. With many unknown factors it is impossible to pre-determine the specific costs to the program. However, these past events provide guidance as to the types of impacts and time frames likely to arise from future cyclonic events as summarised below.

Cyclones Larry and Yasi severely disrupted the NTWEP workforce, scheduling, reduced site access and lowered the field search rates. There will be a short to medium term need to catch up or keep up the visit schedules and re-establish site access. This can only be done through additional field workers as soon as is practicable. The effect of vegetation disturbance on the search rates can continue and have direct and flow-on influences up to four years later, particularly for *M. calvescens*. The first few years after the cyclone may also be accompanied by increases in the search frequency or area, and control costs. Cyclones will bring a need to supply weed identification and spread prevention information and training, and maps. There is the potential for propagule movement during clean-up activities that need to be managed as part of the post cyclone weed seed spread prevention activities.

The natural spread of target weeds associated with cyclones has been limited by the time of the year and the requirement for viable seed to arrive in a suitable habitat. Future events should focus on locally-suitable habitats for *L. flava* that could be reached by slow moving shallow flood waters, as well as maintaining survey frequencies. Future cyclones will slow the *M. calvescens* searches and influence growth and detection rates. Similar effects would be expected at the *M. calvescens* and *M. racemosa* infestations near Kuranda and the *M. calvescens* and *M. nervosa* infestations near Whyanbeel. There was *M. micrantha* recruitment noted around known infestations after both cyclones, so there could be an increase in search area if mature plant discoveries extend the survey buffer.

The dramatic changes in tree canopies and light and provided enhanced opportunities for weed seed germination and seedling growth and while this can impact program reporting, it is useful for accelerating progress towards eradication. Similarly seed rendered unviable or dispersed into an unsuitable habitat also acts to deplete the population. Cyclones are a normal part of the tropical environment for native and exotic plants but have dramatic short to medium term impacts on both the vegetation and the people that survey it. Therefore the effects of cyclones on weed eradication programs need to be taken into account by policy makers in terms of adjustment to milestones and their timeframes. Flexible funding models are needed to cope with increased expenditure to re-establish site access, account for slower surveys and changes in survey frequency over several years. Eradication programs will need allowances to change performance criteria to account for the impacts of cyclones or other natural disasters.

ACKNOWLEDGMENTS

The data presented has been collected by numerous dedicated field workers in the National Tropical Weeds Eradication Program and stakeholder agencies.

REFERENCES

BOM, Bureau of Meteorology (2018a). Tropical cyclones in Queensland. http://www.bom.gov.au/cyclone/about/eastern.shtml (accessed 17 April 2018).

BOM, Bureau of Meteorology (2018b). Severe Tropical Cyclone Larry description. http://www.bom.gov.au/cyclone/history/larry.shtml (accessed10 April 2018).

BOM, Bureau of Meteorology (2018c). Severe Tropical Cyclone Yasi description. http://www.bom.gov.au/cyclone/history/yasi.shtml (accessed 10 April 2018).

Brooks, S.J. and Setter, S.D. (2014). Issues and solutions for researching weed eradication target species. Proceedings of the 19th Australasian Weeds Conference, ed. M. Baker, pp. 255-8. (Tasmanian Weed Society, Hobart).

CTS Cyclone testing station – School of Engineering, James Cook University Townsville (2018). Newsletter October 2006. https://www.jcu.edu.au/cyclone-testing-station/publications/newsletters (accessed 10 April 2018).

Jeffery, M. and Brooks, S.J. (2016). Eradication in the tropics: constantly changing and adapting. Proceedings of the 20th Australasian Weeds Conference eds R. Randall, S. Lloyd and C. Borger. pp. 23-7. (Weeds Society of Western Australia, Perth).

Meyer, J-Y. (1998). Epidemiology of the invasion by *Miconia calvescens* and reasons for a spectacular success. Proceedings of the First Regional Conference on *M. calvescens* control, eds J.Y. Meyer and C.W. Smith, pp. 72-7. (Papeete, French Polynesia).

Murphy, H.T., Brooks, S.J., Bradford, M.G., Metcalfe, D.J. and Westcott, D.A. (2008). Recruitment and growth dynamics of *Miconia calvescens* (Melastomataceae) in tropical forest impacted by Cyclone Larry. Proceedings of the 16th Australian Weeds Conference, eds R.D. van Klinken, V.A. Osten, F.D. Panetta and J.C. Scanlan (Queensland Weeds Society, Brisbane).

Murphy, H.T., Metcalfe, D.J., Bradford, M.G. and Ford, A.J. (2014). Community divergence in a tropical forest following a severe cyclone. *Austral Ecology* 39, 696-709.

Weber, J. and Brooks, S.J. (2013). The biology of Australian weeds 62. *Limnocharis flava* (L.) Buchenau. *Plant Protection Quarterly* 28, 101-13.

The eradication of weed species in New Zealand's National Interest Pest Responses Programme

Frances G. Velvin
Ministry for Primary Industries, PO Box 2526, Wellington 6140, New Zealand

Summary The National Interest Pest Response (NIPR) Programme is a nation-wide government-led programme that focuses on the eradication of pests that have been established for many years in New Zealand. The Ministry for Primary Industries is the lead agency. It provides coordination across the many regions and organisations involved in the NIPR programme, facilitates the use of best operational practices, and provides funding for pest management activities and operational research.

There are currently nine NIPR weed species under eradication. *Bryonia cretica* (white bryony), *Ehrharta villosa* (pyp grass), *Eichhornia crassipes* (water hyacinth), *Hydrilla verticillata* (hydrilla), *Phragmites australis* (phragmites), *Salvinia molesta* (salvinia), and *Zizania latifolia* (Manchurian wild rice) are internationally recognised as major environmental weeds; *Sorghum halepense* (Johnson grass) and *Moraea flaccida* (Cape tulip) are serious agricultural weeds. Significant progress toward eradication has been achieved since operations under the NIPR programme began 10 years ago, including the eradication of *Ceratophyllum demersum* (hornwort) from the South Island, and there have been no detections of hydrilla plants and Johnson grass for over 2 years and 6 years respectively.

Keywords Environmental weeds, eradication, National Interest Pest Response Programme.

Machine vision systems for robotic weed sensing in field environments

Cheryl L. McCarthy, Steven R. Rees and Craig P. Baillie
National Centre for Engineering in Agriculture, University of Southern Queensland, Toowoomba, Queensland 4350
(cheryl.mccarthy@usq.edu.au)

Robotic sensing systems for weed control are required to be developed that are compatible with current commercial farm operations, while also having potential to be integrated with autonomous weeding platforms in the future, to realise weed control cost savings in both the short and long term. Commercially available weed detection sensors in Australia are designed for 'green-from-brown' situations and are not capable of discriminating between weeds and crop when they are growing next to each other. Weeds growing in-crop are presently controlled by tillage operations, manual spot spraying, costly selective herbicides or over-reliance on chemicals in herbicide-tolerant crop.

Practical requirements for robotic weed sensing in commercial conditions are cost-effective and robust weed sensors, real-time weed detection at commercial groundspeeds, low calibration requirements, and operation in daylight and in typical crop growth stages. The National Centre for Engineering Agriculture, a research centre within the University of Southern Queensland, has developed a range of camera-based technologies for detecting weeds in-crop for site-specific herbicide application. These systems have been developed for the grains, sugarcane, cotton and horticulture industries in projects since 2007, for ground-based vehicles and more recently in application to machine vision sensors on aerial drones.

Low-cost consumer-style colour cameras have been demonstrated to extract plant size, and leaf shape and texture, using deterministic algorithms to discriminate weed from crop. A system developed for the sugar industry discriminated the leaves of Guinea grass (a grass weed) growing amongst sugarcane (a grass crop) and achieved a weed hit rate of 85% with less than 1% overspray onto crop, in a trial in which Guinea grass was automatically detected and spot-sprayed in real-time from a tractor. Enhanced discrimination was achieved through the addition of plant height segmentation using colour/depth sensing in the pyrethrum industry. Novel methods for operation in daylight without a shade hood, and real-time processing whilst travelling at up to 15–20 km h^{-1}, have been developed. Subsequently, weed detection systems from ground-based vehicles, suitable for 'green-from-green' commercial field conditions in major cropping industries, are currently undergoing commercialisation.

Aerial drone imagery provides the potential benefits of rapidly generating a weed map for a field without requiring multiple weed detection sensors to be mounted on a boom sprayer. A trial using conventional aerial drone imagery techniques demonstrated rapid quantification of the extent of weed populations in a fallow field. The technique is also potentially applicable to monitoring of plant species populations in various landscapes and for generating areas of interest for scouting on-foot. Current research for broadacre cropping is developing technologies for an automated and real-time workflow for drone flight to prescription map generation and green-from-green spot spray operation.

Keywords Spot spray, colour cameras, image analysis, prescription map, drones.

Needle in a haystack – detecting hawkweeds using drones

Mark Hamilton[1], Robert Matthews[2] and Jo Caldwell[3]
[1] Pest and Weeds Unit, NSW National Parks and Wildlife Service (NPWS), Hurstville, New South Wales 1481, Australia
[2] Helisurveys Pty Ltd, Jindabyne Airport, Jindabyne, New South Wales 2627, Australia
[3] Southern Ranges Branch, NSW National Parks and Wildlife Service, Tumut, New South Wales 2720, Australia
(mark.hamilton@environment.nsw.gov.au)

Summary Orange hawkweed (*Hieracium aurantiacum* L.) and mouse-ear hawkweed (*H. pilosella* L.) are perennial daisies native to Eurasia and have proven highly invasive in several countries outside their native range. Small incursions are present in New South Wales in Kosciuszko National Park and proximate to the reserve (only orange hawkweed). They pose a significant threat to south eastern Australia and are the subject of a New South Wales (NSW) and Victorian joint eradication program, coordinated in NSW by NPWS and partners.

A key challenge in the NSW Hawkweed Eradication Program is delimiting the infestation, particularly across the remote and rugged area where the infestation occurs. To supplement existing ground surveillance, NPWS sought to utilise drone surveying in the orange hawkweed operation as large and remote areas can be surveyed at low cost. Drones capture colour imagery and a specifically designed algorithm automates the detection process. In the 2017/2018 season, during peak flowering, 608 ha was surveyed by drone over 17 survey days. Eight confirmed infestations were detected, including a new location on Happy Jacks Plain.

Drone surveys have been incorporated into the program as an ongoing component, and are proving to be a highly cost-effective surveillance method. It is expected drone survey during flowering times will contribute significantly to delimitation of the orange hawkweed infestation and be pivotal in reaching eradication rapidly.

Keywords Delimitation, UAV, eradication, *Hieracium aurantiacum*, orange hawkweed.

INTRODUCTION

Orange hawkweed (*Hieracium aurantiacum* L.) and mouse-ear hawkweed (*H. pilosella* L.) are stoloniferous perennial herbs of the Asteraceae family. Native to Eurasia, they are highly invasive in many countries outside their native range, including New Zealand, United States of America, Canada and Japan. Orange hawkweed has naturalised in Australia in New South Wales (NSW), Victoria and Tasmania. In NSW, small infestations of orange hawkweed are found across eleven locations in the Jagungal Wilderness Area of Kosciuszko National Park (KNP) and at two locations on private land east of the reserve. One known NSW incursion of mouse-ear hawkweed is found in a separate part of KNP on the Main Range, proximate to Blue Lake.

In Australia, hawkweeds are at the early stages of establishment but have the potential to occupy large tracts of south-east Australia. Beaumont *et al.* (2009) estimate 27 million ha are susceptible to orange hawkweed invasion and economic modelling indicates potential losses to the grazing sector in excess of $68 million AUD per annum (CPI adjusted to present) (Brinkley and Bomford 2002). Due to the small infestation area and significant potential impacts, the objective is to eradicate both hawkweed species from NSW; similar efforts are occurring in Victoria.

The NSW Hawkweed Eradication Program is one of the largest weed eradication programs in Australia. Eradicating an organism from an invaded range is a deceptively difficult task: the infestation must be delimited; reproduction halted; above ground biomass controlled; and the seedbank exhausted. Owing to this difficulty, there have been no documented cases of state-wide eradications in NSW and very few examples exist in Australia (Panetta 2009).

Since detection in 2003, NPWS has coordinated an orange hawkweed control program, with the objective of eradication since 2009/2010. To be successful, any eradication must quickly and reliably detect the target organism, preferably before reproduction occurs, to ensure timely eradication. A key obstacle to successful eradication is the ability to detect all plants such that control can occur. Delimitation efforts often represent the single largest investment in a weed eradication.

The Jagungal Wilderness Area is remote and rugged, and thus an extremely difficult area to survey. It has very few roads or vehicle trails (Figure 1); the area is steep, being dissected by the Tumut River and its tributaries; and dense heath and Alpine ash forest are common. Within KNP, orange hawkweed infestations are found across an extent of 7080 ha, and including the two off-park locations the full extent of potential area is 31,721 ha (minimum convex polygon method). It is important to note that though this is a large extent, hawkweed is not expected to occur with equal probability across this area. Like other plants, hawkweed is limited by environmental factors that determine where it can disperse and establish. Most high-likelihood areas have been searched multiple times and it is not expected or necessary for all areas to be searched for hawkweed, at least not in one season, or even over several seasons.

Detection of orange hawkweed, a small herb with rosettes 5–20 cm diameter and flowers to 20 mm diameter and up to 40 cm high, is difficult, especially when not in flower. Until recently, the eradication program relied solely on humans undertaking ground surveillance, which has achieved impressive results (Hamilton *et al.* 2015). However, the large areas to be searched and the increasing difficulty reaching and surveying areas, combined with the declining detection certainty of human-assisted surveillance as weed populations become sparser (Chandler 2014), necessitated a method to survey large areas at low cost. Remotely piloted aerial systems (RPAS) or drones have the potential to do this. The benefits of RPAS for weed surveillance are well documented (Clements *et al.* 2014, Dehaan *et al.* 2014, Merz *et al.* 2016). Due to their light weight and small size, and the ability to automate the detection process through algorithms, drones can collect data at high resolution, and can do so at low cost, particularly when compared to manned aircraft. For these reasons, the use of this technology for hawkweed eradication was explored.

In 2014/2015 and 2015/2016 detection of orange hawkweed using drones was trialled by the Australian Centre for Field Robotics with NPWS (Hung and Sukkarieh 2015). The deep orange-red flowers are distinctive, and were considered sufficiently unique to explore the use of colour (RGB) imagery captured aerially by drones. Following this initial work, with assistance from Helisurveys Pty Ltd, drone surveillance and detection mechanisms have been improved and are now operational and integrated into the eradication program. This paper outlines the progress made to date, the process of orange hawkweed drone survey and detection, and the successes achieved in the 2017/2018 season. This paper largely deals with orange hawkweed drone detection, but mouse-ear hawkweed drone and remote sensing options are briefly discussed.

MATERIALS AND METHODS

The challenge at hand was to detect a small herb across a large and remote area with a limited budget.

Automating detection of orange hawkweed Due to the large survey areas and small target size, approximately 400 photos are generated per hectare surveyed. Manual photo processing would reduce any benefit gained from the rapid and low-cost survey. As such, an algorithm was developed by Helisurveys to identify pixels that corresponded to the colour of orange hawkweed flowers. Using the unique colour of the flower as the main predictor, from a small sample of images containing known hawkweed flowers, the upper and lower values of the red, green and blue colour channels were determined. A test version of the software provided an RGB colour detection algorithm that was improved upon once more data was collected, but was robust enough for initial drone survey. When more training data was collected and the algorithm improved, previous images and data sets were re-analysed to confirm no orange hawkweed flowers were missed using the test version of the algorithm.

The RGB values and ranges of orange hawkweed flowers were determined by manually viewing and logging the pixel values. Following this, a period of trial and error testing was performed to narrow pixel values down to an acceptable range. A program was developed to help log pixel values of orange hawkweed flowers, but no machine learning was used.

As more images were collected and true detections were found, the algorithm was refined to make it more likely to detect genuine orange hawkweed flowers. At the same time, the false positives were reduced by adding in shape, graduation of hues away from the centre of the flower, and size. Several filters that are run after the detection is logged were also developed. These filters help to reduce the false positives created by certain types of vegetation and flowers with similar colours.

Sensor and aircraft The camera used was a Canon 70D with an 18–55 mm lens, which was fixed at 48 mm during survey. When over the survey area, the camera was triggered via remote control to take photos at a rate of 3 each second with a shutter speed of 1/1250th of a second.

The RPAS used for the 2017/2018 season was a DJI M600 with a Ronin MX Gimbal. The M600 comes as standard with a LightBridge 2 system that provides the flight control link, telemetry and a realtime HD video downlink. The gimbal can be controlled via the LightBridge 2, and it was configured to point directly down and follow the direction of travel of the M600. The ground control software used for flight planning and execution of the flight plan was UGCS (Universal Ground Control System). This was chosen for its ability to create flight paths that follow an imported

Digital Elevation Model that maintains the drone at a constant 30 m above terrain level.

Image capture Using information from previous trials (Hung and Sukkarieh 2015), it was determined that the ideal Ground Sample Distance (GSD) was 2.6 mm. This was increased from 3.0 mm in 2016/2017 to help confirm detections during the human image analysis phase. Extra initial processing was done in the software to help account for variations in the image brightness. This allowed flights to not be restricted to only sunny, cloudless days. Orange hawkweed flowers were successfully detected on both full overcast and full sun days and between the hours of 8:30 and 17:00.

To attain the required GSD of 2.6 mm, the drone was flown at 30 m elevation with the camera lens set to 48 mm focal length. The speed was set at 8 m per second. This was determined to be the optimum compromise between being able to cover more area per flight and minimising pixel blur.

Field algorithm Initially, the image analysis software was run after the surveys were completed. This could take 6–8 hours to process 2 hectares worth of imagery. With further development, the time was reduced by a factor of 20. Because some of the survey sites are only accessible by helicopter, the inability to run the algorithm in the field was cost prohibitive and meant that ground truthing – and critically, herbicide treatment of hawkweed plants – could be delayed. A 'field' version of the software was developed to quickly process images, but at the possible expense of rigour. To date, the field version has not missed any detections the full version has detected, but the risk exists. It is used *in addition* to the full version to be able to rapidly verify detections on the day of survey.

False positives and manual culling Different survey sites and vegetation types provide different challenges for the algorithm: Some flights produce 10–20 false positives (out of 2500–3000 images), while others could produce up to 600–800 false positives. Filters were developed to reduce the number of false positives, but at the cost of potentially missing genuine detections. To date, the filters have not filtered out any genuine detections. Once the algorithm and filters are run, the remaining images must be viewed by a human to determine if ground truthing is required. The detection is highlighted by a red square overlayed on the image to help the viewer quickly identify the area for close inspection. These images are geotagged using information from the onboard GPS. Together with the images, a KML and CSV file are created with the coordinates of the image to aid with ground truthing.

Survey area selection As per other hawkweed surveys, priority drone survey sites were selected based on a simple surveillance prioritisation model. The model considers: 1) proximity to known infestations; 2) dispersal modelling from wind speed and direction data collected at hawkweed locations; 3) areas of past surveillance; and 4) vegetation type (M. Hamilton unpubl. data). Additional factors to consider are specific site requirements for drone surveying, such as large areas with no sudden changes in topography, and with a vegetation height lower than approximately 20 metres.

RESULTS

Since 2015/2016 the area surveyed by drones has dramatically increased, from 17 ha in 2015/2016, to 85 ha in 2016/2017, and currently to 608 ha in 2017/2018 (Figure 1). Drones now survey almost double the area of humans on the ground do. 2017/2018 saw the first detection of previously unknown patches of orange hawkweed. Prior to this, imagery was captured over known hawkweed infestations and hawkweed was successfully detected (to test the process) but no hawkweed was detected in survey areas, only absences were recorded.

In the 2017/2018 season, 608 ha was surveyed using drones, 515 ha in KNP and 93 ha off park. Surveys occurred during peak flowering period over 17 days from 18 December 2017 to 26 January 2018. A total of 290,277 images were captured and eight detections were confirmed as orange hawkweed. Detections included one infestation detected through broken canopy (Figure 2), a sole orange hawkweed flower amongst 4 small rosettes, and a small infestation in the western part of Happy Jacks Plain. The latter was particularly significant given that parts of this large plain have been surveyed for many years as it is ideal habitat for the weed, and it is a priority for survey as it is closest to the reserve boundary.

Area surveyed varied from 25–50 ha day^{-1}, averaging 35 ha. This was influenced by suitable drone survey areas, inclement weather, and transit time between sites, if multiple sites were surveyed in a day. Later in the survey program, a new drone with longer flying time extended surveying ability to 50 ha day^{-1} in suitable areas. For comparison, a survey of 40 ha using human surveyors would require a team of 5 staff for 16–20 days. This is over one order of magnitude more expensive when compared to drones. Including salaries, vehicle, equipment and accommodation expenses, on-ground survey using staff costs \$556–695 ha^{-1}, compared to the drone rate of approximately \$50 ha^{-1}. Future work will include assessments of drone versus human detection accuracy per unit cost.

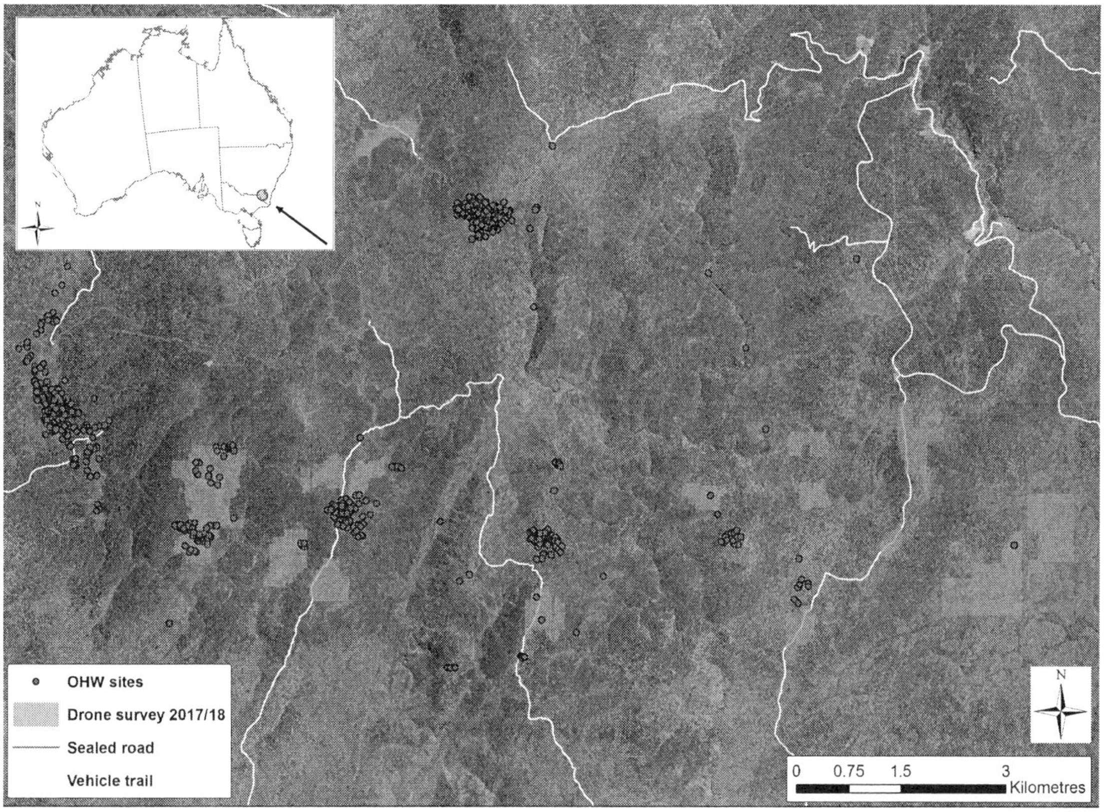

Figure 1. Orange hawkweed (*Hieracium aurantiacum*) infestations (blue dots) in Kosciuszko National Park and location of drone detection surveys (pink areas) in 2017/2018.

Figure 2. Example of a drone detection (through broken canopy) of orange hawkweed (*Hieracium aurantiacum*), Farm Ridge, Kosciuszko National Park. The four red boxes represent an area with one or more orange hawkweed flowers present. Total image area is approximately 17 × 11 m. Image taken from 30 m above ground.

DISCUSSION

This study outlines an example of where drone survey with automated detection (algorithm) can reliably detect a sparse weed across a large, remote landscape, and significantly contribute to delimitation in a large weed eradication program. Drones have commonly been used for detection or mapping of more abundant weeds or agricultural weeds in more controlled settings such as agriculture. In our work, drones are used to detect a rare weed in a remote and rugged landscape in a cost-effective manner.

Notwithstanding the benefits of drone survey and automated detection based on flower colour, the technology as it is applied here, is clearly limited to being useful only when flowers are present (~6–8 weeks each season). Cost comparisons made with human ground surveillance illustrate the much greater value for money from drone surveys, but mask the higher detection ability of humans (i.e. non-flowering plants are detected). Orange hawkweed flowers from mid-December to February, with a peak around late December. Optimal survey time is a period of 6 weeks and 100% detection certainty cannot be achieved via drone, as plants may not be in flower during survey. In addition, the small flower size requires high-resolution imagery for detection. This requires low altitude flights, thus compromising flight efficiency as smaller areas are covered in each flight.

To increase detection certainty and survey efficiency, NPWS is working with Macquarie University to explore use of hyper- or multi-spectral imagery, sourced from satellite or drone, to detect orange and mouse-ear hawkweed *plants*. The whole plant or plant clusters will be targeted. Being larger than the small flowers, this will allow the use of lower resolution imagery and higher elevation drone survey or satellite imagery. Analysis is under way to determine the reflectance of both hawkweed species in the non-visible spectrum relative to surrounding vegetation, and then test the effectiveness of this detection method.

Into the future, NPWS and partners will utilise drones to increase the area of surveillance to delimit the orange hawkweed infestation. A program for 2018/2019 is already being developed. Greater efficiencies including larger survey areas, more rapid detection verification and synchronisation with control programs will be explored. Drone survey using colour imagery and possibly hyperspectral imagery can contribute significantly to delimitation of hawkweed infestations and may be pivotal in more rapid delimitation, and ultimately successful eradication.

ACKNOWLEDGMENTS

The authors would like to thank Hillary Cherry for comments on the manuscript and the many staff and volunteers that dedicate their time to the NSW Hawkweed Eradication Program. The program in NSW is funded in part by the NSW Weeds Action Program, the Office of Environment and Heritage and NSW Local Land Services.

REFERENCES

Beaumont, L., Gallagher, R., Downey, P., Thuiller, W., Leishman, M. and Hughes, L. (2009). Modelling the impact of *Hieracium* spp. on protected areas in Australia under future climates. *Ecography* 32, 757-64.

Brinkley, T.R. and Bomford, M. (2002). Agricultural sleeper weeds in Australia: What is the potential threat? (Bureau of Rural Sciences, Canberra). 33 pp.

Chandler, B. (2014). Innovation vs. invasion: innovative techniques to assist the eradication of weed populations. Proceedings of the 19th Australasian Weeds Conference, ed. M. Baker, pp. 259-62. (Tasmanian Weed Society, Hobart).

Clements, D., Dugdale, T., Hunt, T., Fitch, R., Hung, C., Sukkarieh, S. and Xu, Z. (2014). Detection of alligator weed using an unmanned aerial vehicle. *Plant Protection Quarterly* 29, 84-9.

Dehaan, R.L., Weston, L.A. and Rumbachs, R. (2012). The design and the development of a hyperspectral unmanned aircraft mapping system for the detection of invasive plants. Proceedings of the 18th Australasian Weeds Conference, ed. V. Eldershaw, pp. 103-7. (Weed Society of Victoria, Melbourne).

Hamilton, M.A., Cherry, H. and Turner, P.J. (2015). Hawkweed eradication from NSW: Could this be 'the first'? *Plant Protection Quarterly* 30, 110-5.

Hung, C. and Sukkarieh, S. (2015). Using robotic aircraft and intelligent surveillance system for orange hawkweed detection. *Plant Protection Quarterly* 30, 100-Merz, T., Hrabar, S., Kendoul, F. and Jeffery, M. (2016). Unmanned helicopter system for miconia weed surveys. Proceedings of the 20th Australasian Weeds Conference, eds R. Randall, S. Lloyd and C. Borger, pp. 191-4. (Weeds Society of Western Australia, Perth).

Panetta, F.D. (2009). Weed eradication – an economic perspective. *Invasive Plant Science and Management* 2, 360-8.

Multispectral imagery and weed biocontrol: A useful pairing?

A.J. McConnachie[1], P. Jones[2], R. Shilpakar[3] and P. Dawson[3]
[1] Weed Research Unit, New South Wales Department of Primary Industries, Biosecurity and Food Safety, Locked Bag 6006, Orange, New South Wales 2800, Australia
[2] Department of Agriculture and Fisheries, Biosecurity Queensland, Ecosciences Precinct, GPO Box 267, Brisbane, Queensland 4001, Australia
[3] North West Local Land Services, PO Box 500, Tamworth, New South Wales 2340, Australia

Summary High resolution multispectral imagery is playing an increasingly important role in assessing agricultural crop health and vigour. In recent studies, such imagery has been trialled to assess the impact of weed biocontrol agents on their target plants. However, acquisition of this imagery can be limiting, due to flight costs. This study explored the use of unmanned aerial vehicles (UAVs) as a cost-effective, user-friendly tool in acquiring imagery to not only detect an invasive weed target, but also to monitor the dispersal and impact of a released biocontrol agent. *Cylindropuntia* spp. are Cactaceae of Mexican and southern USA origin. They have become weeds of arid and semi-arid regions of mainland Australia, with eight species currently recorded as naturalised. A biocontrol program was initiated in Australia in 1925 for the control of *C. imbricata*, through the introduction of a cochineal, *Dactylopius tomentosus*. More recently, different biotypes of *D. tomentosus* (which have specific impacts on the different *Cylindropuntia* spp.) have been imported, tested and approved for release. Two *Cylindropuntia* species, *C. fulgida* var. *mamillata* (boxing glove cactus) and *C. pallida* (Hudson pear), which have had *D. tomentosus* biotypes released against them were considered in this study. For *C. fulgida* var. mamillata, a UAV (equipped with a 12MP sensor) was used to acquire aerial imagery (RGB) of a long-term monitoring site to visualise cochineal crawler dispersal over time, as well as the overall impact of the cochineal on the weed. For *C. pallida*, UAVs fitted with multi-spectral sensors were used to assist in mapping the species in areas of its invaded range, as well as assessing the impact of the *D. tomentosus* biotype on target weed vigour. The results of these studies will be discussed, in particular reference to UAVs and their future use in monitoring and evaluation.

What's over the horizon for weed management in Australia – science outcomes to benefit Australian growers from the Bayer/GRDC partnership

Marco Busch
Bayer AG, Crop Science Division, Industriepark Hoechst, 65926 Frankfurt / Main, Germany

Summary Weeds are the single most important reason for crop losses globally, causing high management costs and threatening food security. Increasing weed resistance adversely affects crop production in various parts of the world, decreasing average yields for most of the crops significantly. In Australia, herbicide resistance started around 30 years ago with the rapid evolution of resistant weeds such as ryegrass, wild radish and others threatening Australian wheat production.

In order to cope with this problem screening and monitoring needs to reflect regional requirements and as such adapted strategies regarding greenhouse testing mirrored by a corresponding strategy for field testing. New tools such as novel herbicidal molecules with resistance-breaking properties are therefore urgently required. In June 2015 the Grains Research & Development Corporation (GRDC) and Bayer signed their *Herbicide Innovation Partnership* (HIP), an innovative cooperation model to jointly discover next-generation weed control solutions.

In order to meet the challenges of increasing productivity and sustainability in crop production and to best match Australia's need for innovative weed control solutions the herbicide screening cascade for new active ingredients at Bayer's global Competence Center for Weed Control in Frankfurt, Germany has been revised accordingly. This refers to the infrastructure use, set-up of individual test levels as well as the representation of Australian key weeds. The new cascade from Early Phase to Profiling covers all nominated Australian weeds, including specific resistant biotypes. Provision of high quality seed material matching specification (batch quantity and germination rate fulfilling automation needs) partly proved to be of utmost importance. Furthermore greenhouse follow-up tests with focus on e.g. resistance-breaking properties, formulations, selectivity etc. have been adapted. New herbicidal actives are being tested through the modified testing cascade allowing early assessment of delivery on Australian farmer needs.

These greenhouse activities are complemented by tailor-made field screening in Australia. The field screening in Australia with its unique models (e.g. with resistant weeds) strongly improved our ability of recognising the fit of new candidates for the Australian markets at a very early stage of our R&D process.

As a third pillar of this integrated approach a total of 33 post-doctoral researchers from Australia and New Zealand are participating over the entire period of the partnership in this global effort at Bayer's global Competence Centre for Weed Control in Frankfurt allowing for deep education in cutting-edge life-science research in a truly cross-functional set-up. As fully integrated members in multidisciplinary project teams they are contributing to various promising new research projects in order to identify new active ingredients for weed control.

Research priorities for weed suppression by crops in Australia

Michael Widderick[1], Deirdre Lemerle[2], Cameron Taylor[3], Chris Johansen[4], Abul Hashem[5], Vikki Osten[6], Tony Cook[7], Kerry Bell[1] and Andrew Storrie[8]
[1] Queensland Department of Agriculture and Fisheries, Leslie Research Facility, PO Box 2282, Toowoomba, Queensland 4350, Australia
[2] Graham Centre for Agricultural Innovation, Faculty of Science, Charles Sturt University, Wagga Wagga, New South Wales 2650, Australia
[3] Birchip Cropping Group Inc., PO Box 85, Birchip, Victoria 3484, Australia
[4] Consultant in Agricultural R&D, Leeming, Western Australia 6149, Australia
[5] Western Australia Department of Primary Industries and Regional Development, PO Box 75, Northam, Western Australia 6401, Australia
[6] 9 Burns Avenue, Emerald, Queensland 4720, Australia
[7] NSW Department of Primary Industries, 4 Marsden Park Road, Calala, New South Wales 2340, Australia
[8] AGRONOMO, PO Box 5944, Albany, Western Australia 6330, Australia
(michael.widderick@daf.qld.gov.au)

Summary Crop competition is an important tactic for suppressing weed growth and maintaining crop yield, especially with the increasing levels of herbicide resistance in Australia. In 2015, a team of weed scientists from the western, southern and northern grain production regions conducted a comprehensive review of research on crop competition in Australia, funded by the Grains Research and Development Corporation (GRDC). The aim of the study was to enable a 20% increase in grower adoption of this tactic for weed suppression by 2020. The project team: identified priority weed species; conducted a meta-analysis of pooled data to identify common trends; conducted a knowledge gap analysis based on the literature review and the meta-analysis; defined an approach for improved future crop competition experimentation; and identified and provided recommendations for future Research, Development and Extension (RD&E).

An updated assessment of priority and key weed species was provided. The need for more studies on agronomic means of enhancing weed competition were identified, especially for weakly competitive crops such as pulses. These included crop rotation, fertiliser manipulations and determination of traits conferring competitiveness that can be used in varietal selection and breeding programs. Further, it is necessary to guide farmers on understanding the benefits of growing a competitive crop as a viable alternative to chemical weed control. Plant breeding for crop traits associated with strong competitive ability will provide an alternative and longer-term solution to weed management.

Keywords Crop competition, review, meta analysis, research protocol, recommendations.

INTRODUCTION

Across all Australian grain growing regions, herbicides have been heavily relied upon for in-crop weed management. This reliance has resulted in widespread herbicide resistance in a range of common in-crop weed species. In Australia, there have been 49 weed species identified as having herbicide resistance across 16 sites of action (Heap 2018). Of these, 32 were confirmed as resistant in Australian grain cropping systems. As herbicide-resistant individuals survive herbicide treatments, other non-herbicide weed management tactics are required, including crop competition.

Competitive crops have been shown to have a positive impact on in-crop weed suppression and maintaining crop yield. Several agronomic practices can influence the competitiveness of a crop (Lemerle *et al.* 2001, Borger *et al.* 2015). These include crop species and cultivar, row spacing, density, seeding methods and row orientation. Furthermore, the competitive ability of a crop can be influenced by fertiliser placement, depth and time of sowing, effective control of insects and diseases, and a complex interaction of weather and soil conditions (Palta and Peltzer 2001).

Crop competition was used in Australia to increase the yield and thereby profitability of crops prior to the advent of selective herbicides and in organic crops. However, with the spread of herbicide resistance, there has been a renewed interest in competitive crops for weed management.

There has been much research during the last 20 years on crop competition across all Australian grain growing regions and published in refereed scientific journals, (Lemerle *et al.* 1996, Bajwa *et al.* 2017), while some more recent research has been reported only in GRDC Crop Updates.

GRDC identified a need to review all available literature to identify current knowledge and required future Research, Development and Extension (RD&E) needs.

A team of weed scientists from the western, southern and northern GRDC regions was sponsored by GRDC to:

- identify priority weed species;
- conduct a comprehensive review of literature on crop competition as a cultural weed management tactic;
- conduct a meta-analysis of pooled data from the literature to identify common trends; and
- conduct a knowledge gap analysis based on the literature review and meta-analysis.

Through this process, we identified future RD&E needs and defined a consistent approach for future experimentation.

The review is a basis for prioritising future RD&E investment. It also provides a standard protocol for a nationally consistent approach to experimentation on crop competition, thereby providing more reliable and comparable research results for use in predictive models of the benefits and costs of crop competition for weed control. This knowledge will likely increase adoption of this weed management tactic.

MATERIALS AND METHODS

Priority weed species The project team identified the most important key weed species for each region, and prioritised these based on potential impacts and current and future threats of herbicide resistance. The project team used their knowledge of industry weed issues and research in this process.

Literature review A comprehensive review was undertaken of reported research on crop competition across the wheat belt of Australia from the early 1980's to 2015 for each region (western, southern and northern agro ecological zones). The studies included in the literature reviews examine either individual or multiple crop competition factors across various crops, weeds, rotations, seasons and sites, and research approaches.

Meta analysis The meta-analysis was designed to pool data sourced from the literature and to identify common trends. For individual crop competition factors, like-measurements have been compared to identify trends. In both the southern and northern regions, mimic weeds have commonly been used in place of target weed species.

The meta-analysis consisted of three components:

1. a summary of the number of trials conducted in each region and for each weed species;
2. a total of 64 graphs of data, based on data presented in papers, to explore trends between crop competition factors and measures of weed growth; and
3. a summary of the pooled data across each region showing general trends.

Knowledge gaps In order to determine the knowledge gaps for crop competition, the project team firstly 'brainstormed' an importance rating for each key crop competition factor (crop species, cultivar, crop density, row spacing, row orientation, sowing time, depth of sowing, nutrition, soil characteristics, rainfall) and each key crop (wheat, barley, oats, canola, lupins, field pea, chickpea, faba bean, triticale, maize, soybean, sunflower, mung bean, soybean) on a 1 to 5 scale for each region. The relative importance for each combination of crop competition factor and key crop was calculated by multiplying this product by 4. This meant the relative importance value could range from 0 to 100 (100 = highly important). Priority was taken into account when identifying key knowledge gaps.

Secondly, following the review of literature, an intuitive rating (0% = nothing available, 100% = completely covered) was made as to how completely each crop competition by crop factor had been researched for each region.

Thirdly, to identify the knowledge gaps, the relative importance percentages (0–100%) were subtracted from the intuitive percentages (0–100%). Using these differences, priority levels were assigned. When the relative importance percentage was greater than the intuitive percentage of coverage by the literature, this indicated a potential knowledge gap. The reverse indicated there was sufficient knowledge for that crop competition by crop factor.

Finally, a list of national and regional priorities was developed, taking into account the importance of the crop competition factor and the crop species.

Experimental design protocol One of the key findings of this study is the need for an improved and consistent approach to experimentation on crop competition. The project team developed an experimental protocol to guide future planning and experimentation.

RESULTS

Priority weed species Common weeds across the regions were annual ryegrass (*Lolium rigidum* Gaudin), wild radish (*Raphanus raphanistrum* L.), wild oat (*Avena* spp.) and fleabane (*Conyza* spp.), while windmill grass (*Chloris truncata* R.Br.) and feathertop Rhodes grass (*Chloris virgata* Sw.) were seen as emerging problems.

Weeds specific to regions include brome grass (*Bromus* spp.) and prickly lettuce (*Lactuca serriola* L.) for the western and southern regions, and sweet summer grass (*Brachiaria eruciformis* (Sm.) Griseb.) for the northern region.

There is a more limited range of species in the western region reflecting the simpler farming rotations and limited ecological zones compared with the southern and northern regions.

Literature review A total of 88 references were included in the study. However, only 46 of these were used in the meta-analysis because these papers had data in a form that was accessible.

The most common focus of crop competition research was on crop productivity with a lesser focus on this tactic for weed suppression.

More research was done in the western and southern regions compared with the northern, possibly due to the earlier onset of herbicide resistance in annual ryegrass and wild radish.

Most of the research has been on wheat agronomy by manipulating factors such as crop density, row spacing and cultivar. Work on other crops varied with region, as expected.

Some experiments were confounded by crop rotation and other weed management practices, including herbicide application, therefore making it difficult to extract and interpret main crop competition effects.

Meta analysis Research to date has focused on a small number of key weeds in the western region, mainly annual ryegrass and wild radish. Similarly, research in the southern region was mainly on annual ryegrass. Research in the northern region was even more limited, but most has been on wild oat or other weed species present at the experimental site.

Key findings include that:

- an increase in crop density commonly resulted in a decrease in weed growth. This was consistent across crops and regions and for different weed species;
- cultivars can differ in their competitive ability across crop species with more competitive cultivars resulting in a decrease in weed growth. However, not all studies show a difference between cultivars and the effects of different cultivars can be inconsistent due to seasonal conditions;
- a wheat row orientation of east-west, as opposed to north-south, generally results in a reduction in weed growth, especially in the western region. However, results are inconsistent, with one southern region study having the opposite result;
- a decrease in row spacing generally resulted in a decrease in weed growth. This effect was consistent across crop species and regions. However, in several studies, wider row crops resulted in a reduction in weed growth due to interaction with in-crop control measures such as herbicide application;
- skip row research has only been conducted in the northern region. Generally, solid planting reduces weed growth compared with skip rows; and
- an early sowing time generally reduced weed growth. However, there has been little research on this crop competition factor.

Knowledge gaps The study identified RD&E needs at both national and regional levels. National RD&E needs include:

- breeding for, and evaluation of competitive ability of the most important crops and their cultivars. Also, for understanding of the traits and genetics that influence competitive ability (e.g. allelopathy, early vigour);
- whether gains can be made in the competitive ability of weakly competitive crops (e.g. pulses)? Such crops represent a weak phase in the crop rotation (in terms of crop competition) and therefore gains should be pursued;
- fertiliser rates and more efficient fertiliser placement (horizontally and vertically) to favour the crop over the weed, measuring impact of crop competitive ability on weed control;
- that there are insufficient data across all regions on many priority weed species;
- the effect of the timing of weed emergence on the efficacy of crop competition;
- whether specific crop rotations seem to out compete weeds compared to others?;
- that farmers need more guidance and demonstration/extension as to how to use crop competition and how to best implement it in the farming system;
- if crop and cultivar competitiveness information is available then it should be included in crop production agronomy guides; and that
- it may be feasible to model different crop competition by crop species interactions once there is sufficient data that has been measured with a consistent procedure. This information could be used in the extension of crop competition as a weed management tactic.

Regionally specific RD&E needs include:

For the western region:

- canola across all crop competition factors;
- field pea for cultivar and crop density; and

- chickpea for cultivar, crop density and row spacing.

For the southern region:

- lupins, faba bean, chickpea for cultivar, crop density and row spacing; and,
- canola for row spacing.

For the northern region:

- chickpea, faba bean, mungbean and maize for cultivar, crop density and row spacing; and, wheat and barley, for cultivars.

Experimental design protocol Recommendations for future crop competition research have been made to ensure reliable and useful data for change of practice, including experimental design, establishing the weed population, what to measure, statistical analysis and reporting.

Weed:crop competition is complex. The research needs to be kept relatively 'simple' to tease out the main effects. Much of the existing work combines several crop competition and non-crop competition factors and clear messages are lost.

DISCUSSION

To enable a 20% increase in grower adoption of crop competition by 2020, there is a need for both agronomic research (short-term) and plant breeding research (long-term) to improve the competitive ability of crops. In addition, Development and Extension is required regionally to provide farmers with knowledge as to how best to use crop competition in their farming system.

Future crop competition research requires national funding, coordination and communication for national breeding for competitive ability traits, regional packages for agronomy, and research consistency to improve the quality of research outputs. Bio-economic modelling is required to provide farmers and advisors with cost-effective and reliable predictions of the benefits of crop competition to ensure future adoption of this tactic for weed management where appropriate.

Since crop competition is only one aspect of weed management and crop agronomy, it needs to be investigated as part of an integrated package for weed control. For example: the interaction between crop competition and in-crop herbicides; the effect on weed seed retention for harvest weed seed control; and effect on in-crop disease levels.

We recommend a National Crop Competition Program to improve communication, coordination and quality of RD&E on crop competition for the grains industry of Australia.

ACKNOWLEDGMENTS

This work was funded by a Grains Research and Development Corporation Grant (DAQ00197).

REFERENCES

Bajwa, A., Walsh, M. and Chauhan, B. (2017). Weed management using crop competition in Australia. *Crop Protection* 95, 8-13.

Borger, C., Hashem, A. and Powles, S. (2015). Manipulating crop row orientation and crop density to suppress *Lolium rigidum. Weed Research* 56, 22-30.

Heap, I. (2018). The International Survey of Herbicide Resistant Weeds. www.weedscience.org (accessed 13 April 2018).

Lemerle, D., Verbeek, B., Cousens, R. and Coombes, N. (1996). The potential for selecting wheat varieties strongly competitive against weeds. *Weed Research* 36, 505-13.

Lemerle, D., Gill, G., Murphy, C., Walker, S., Cousens, R., Mokhtari, S., Peltzer, S., Coleman, R. and Luckett, D. (2001). Genetic improvement and agronomy for enhanced wheat competitiveness with weeds. Invited review. *Australian Journal of Agricultural Research* 52, 527-48.

Palta, J.A. and Peltzer, S. (2001). Annual ryegrass (*Lolium rigidum*) reduces the uptake and utilisation of fertiliser-nitrogen by wheat. *Australian Journal of Agricultural Research* 52, 573-81.

Managing clethodim-resistant annual ryegrass (*Lolium rigidum*) in canola with herbicides and competition

Samuel G.L. Kleemann, Gurjeet S. Gill and Christopher Preston
School of Agriculture, Food & Wine, University of Adelaide, PMB 1, Glen Osmond, South Australia 5064, Australia
(christopher.preston@adelaide.edu.au)

Summary Clethodim has been the main herbicide used for the control of annual ryegrass in canola. However, increasing clethodim resistance in recent years has been impacting on growers' ability to control annual ryegrass. In the absence of alternative post-emergent herbicides for annual ryegrass control in canola, more emphasis is being placed on pre-emergent herbicides. Two field trials were conducted at Roseworthy in South Australia in 2015 and 2016 to investigate the effectiveness of increased crop competition with hybrid canola in combination with pre-emergent herbicides to reduce seed set of clethodim-resistant annual ryegrass. In both years, annual ryegrass establishment was the same in both hybrid canola (Hyola 559TT) and open pollinated canola (ATR Stingray). The herbicide strategies reduced annual ryegrass establishment in both years, but there was no significant difference between cultivars. However, greater early vigour in hybrid canola was reflected in a significantly lower annual ryegrass head density in Hyola 559TT compared to ATR Stingray in both 2015 and 2016. Hyola 559TT without any herbicide treatment reduced ryegrass seed set by 14% in 2015 and by 37% in 2016 compared to ATR Stingray. However, combination of Hyola 559TT with the most effective herbicide treatment, reduced annual ryegrass seed set by 48% in 2015 and 57% in 2016 compared to the same herbicide treatments in ATR Stingray. In 2016, Hyola 559TT produced higher seed yields than ATR Stingray, but there was no difference in yield in 2015.

Keywords Crop competition, hybrid canola, clethodim resistance, annual ryegrass.

Reduction in sowthistle biomass and seed production through manipulation of pulse population and row spacing

Adam B. McKiernan[1], Greg Harvey[1], Linda Heuke[2], Michael Walsh[2] and Michael Widderick[1]
[1] Queensland Department of Agriculture and Fisheries, 13 Holberton Street, Toowoomba, Queensland 4350, Australia
[2] I.A. Watson Grains Research Centre, University of Sydney, 12656 Newell Highway, Narrabri, New South Wales 2390, Australia

Summary *Sonchus oleraceus* L. (sowthistle) is an annual broadleaf weed of northern winter crops including pulses. Several sowthistle populations have been confirmed resistant to Group B and Group M (e.g. glyphosate) herbicides. An integrated approach to sowthistle control including non-chemical methods such as growing a competitive crop will slow the further development of herbicide resistance. In this study we tested the competitive ability of faba bean and chickpea against sowthistle, to reduce sowthistle growth and seed production. Field trials consisted of five replicates of 13 treatments (row spacing [25 cm and 50 cm] × crop population [low, medium, high] × ±weeds + one crop free plot). Sowthistle seed was spread over the soil in +weed plots and the crop free plot after crop planting. All plots were irrigated to encourage weed and crop germination. At crop maturity, sowthistle biomass and seed production data were collected and the crop harvested. Overall, sowthistle reduced faba bean yield by 10% and chickpea yield by 7%. In the absence of crop competition, crop free plots had significantly higher sowthistle biomass and seed production than cropped plots. In both crops, increased crop population combined with narrow row spacing resulted in the greatest reduction in sowthistle biomass and seed production compared to low population and wide rows. In-crop sowthistle biomass and seed production can be significantly reduced through manipulation of faba bean and chickpea planting configuration. Growing a competitive pulse crop is an important component of an integrated weed management strategy in the face of herbicide resistance.

Keywords *Sonchus oleraceus,* common sowthistle, chickpea, faba bean, crop competition.

Glyphosate and 2,4-D amine resistance in common sowthistle (*Sonchus oleraceus*) and fleabane (*Conyza bonariensis*) in the Northern Grain growing region of Australia

Adam Jalaludin[1], Michael Widderick[1], John Broster[2] and Michael Walsh[3]
[1] Department of Agriculture and Fisheries, Toowoomba 4350, Queensland, Australia
[2] Graham Centre, Wagga Wagga, New South Wales, Australia
[3] University of Sydney, Narrabri, New South Wales, Australia
(adam.jalaludin@daf.qld.gov.au)

Summary Common sowthistle (*Sonchus oleraceus*) and fleabane (*Conyza bonariensis*) are common and problematic broadleaf weeds in the northern grains region, infesting fallows and crops throughout much of the year. A heavy reliance is placed on knockdown herbicides such as glyphosate and 2,4-D for the control of these weeds. As a result, resistance to glyphosate has been confirmed in both species and both are at risk of developing resistance to 2,4-D. In response, a GRDC-funded field survey has been conducted in winter and summer grain crops from late-2016 to early-2018 to determine the extent of glyphosate and 2,4-D resistance across the northern grain region. In total, more than 500 sites have been surveyed in New South Wales and Queensland, with more than 200 populations of sowthistle and 60 populations of fleabane collected. To date, screening of those populations with glyphosate and 2,4-D have identified 5 and 18 common sowthistle and fleabane populations, respectively, to have evolved resistant to glyphosate. No populations of either weed species have been identified as resistant to 2,4-D. Results from the survey will be provided to individual land managers to inform their weed management programs. Maps of the distribution of resistance will be produced to help guide and prioritise RD&E in relation to these weeds and their management.

Keywords Sowthistle, fleabane, glyphosate, 2,4-D amine, herbicide resistance.

Microwave: A novel approach to tackle herbicide resistance in no-till farming systems

Muhammad J. Khan, Graham I. Brodie and Dorin Gupta
School of Food and Agriculture, Dookie Campus, The University of Melbourne, Nalinga Road, Dookie, Victoria 3647, Australia
(mkhan3@student.unimelb.edu.au)

Summary Herbicide resistance has constrained sustainable crop productivity. Consequently, interest in chemical-free weed management has increased. We examined the effect of pre-sowing microwave (MW) soil heating to debilitate the weed seedbanks of confirmed herbicide-resistant weeds of the Australian rice and wheat production systems. The MW energy was projected through a horn antenna with aperture dimensions of 5.5 × 11 cm into the top soil (0–6 cm) horizon. An energy range of 500–700 J cm^{-2} gave a temperature gradient of 80–90°C. This temperature achieved approximately 70–80% reduction in weed establishment in both rice and wheat crops. Therefore, in the case of rice, a 34% increase in the grain yield (9.0 t ha^{-1}) was achieved compared to the non-MW treated scenario (6.7 t ha^{-1}). Similarly, a substantial increase in wheat grain yield of 39.2% was achieved through MW energy application (7.8 t ha^{-1}), compared to non-MW treated plots (5.6 t ha^{-1}). In summary, investment in this new technology can effectively minimise the weed pressure and promote sustainable crop production.

Keywords Microwave soil heating, seedbank, wheat, rice.

INTRODUCTION

Globally, 471 unique cases of herbicide resistance (species × site of action) have been documented in 86 crops, with 250 weed species being resistant to 160 herbicides (Heap 2016). This is a significant threat to the sustainable productivity of agricultural commodities in industrialised countries. Globally, 51 weed species are resistant to various chemical herbicides in rice crops and 73 weed resistance cases have been reported in wheat crops (Heap 2016). In Australia, no-till farming practices have increased the use of chemicals for weed management, which has ultimately shifted the weed flora to evolve herbicide resistance. In no-till farming system, early weed seed shedding favours weed seedbank accumulation in top soil layers. Controlling seed bank viability promotes long-term weed suppression in agriculture systems (Burnside *et al.* 1986). Physical methods of weed control can potentially manage herbicide resistant weed populations. Among those physical control methods, MW soil treatment has recently emerged as an alternative to tackle herbicides-resistant weeds in the Australian agricultural system (Brodie *et al.* 2017, Khan *et al.* 2017). It has been reported that pre-emergence MW irradiation of soil can potentially minimise weed establishment (Nelson 1996, Sartorato *et al.* 2006, Brodie *et al.* 2017). It can also destroy weed plants' reproductive parts and their seeds that are covered up by soil to a depth of several centimetres (Brodie *et al.* 2007). Considering all these findings, therefore, the efficacy of microwave for weed control in crops, under field conditions, is not well studied. Hence, the objective of this study was to evaluate the effect of microwave soil treatment for weed management in the Australian dryland wheat and rice farming systems, under field conditions.

MATERIALS AND METHODS

A rice crop field experiment was conducted between October 2015 and April 2016 at Dookie Campus, The University of Melbourne, Victoria, Australia. An area of 73.5 m^2 was excavated and manually levelled into a turkey-nest pond so the area could be flood irrigated to grow rice. The individual experimental plot sizes were 2.0 × 2.0 m, which were arranged in a randomised complete block design with five replications, and with a 0.5 m untreated buffer zone between each experimental plot. The experiment consists of two treatments: an untreated control (MW_0 = 0 sec) and microwave treated (MW_1 = 120 sec). A MW prototype was developed for soil irradiation. It was based on the magnetron of a commercial microwave oven (EMS8586V; Sanyo; Tokyo, Japan) operating at 600 W with a frequency of 2.45 GHz, which was assembled with a rectangular waveguide of 86 × 43 mm channelling the MW energy into a pyramidal horn antenna with aperture dimensions of 110 × 55 mm. After microwave soil treatment, the plots remained undisturbed until the next day. The pre-germinated seeds of the rice variety Opus were broadcast at a seeding rate of 125 kg ha^{-1}. A similar experimental protocol was adapted for a dryland wheat crop experiment, under field conditions between May–December 2016. The number of tillers, fresh biomass, and dry biomass, for both rice and wheat crops, were measured for randomly selected 0.25 m^2 quadrats, drawn from each plot, for early growth assessment in response to MW soil heating, and weights were converted to tonnes per hectare. At physiological maturity, the crops were manually harvested from whole plots for yield assessment.

After harvesting, the crop biomass was dried in an air oven (TR1050 27124, Nabertherm, Lilienthal, Germany) for 24 h at 65°C, and final crop yield was converted to tonnes per hectare. Weed density and biomass accumulation were assessed from whole plot areas (2.6 m^2) at maximum tillering stage, and dry biomass of weeds was assessed at crop harvest.

Furthermore, two further rice field trials were conducted from October 2016 to April 2017 in a randomised complete block design with five replications at two different locations – Dookie, Victoria and Jerilderie, New South Wales. A MW weed killer has been designed to eradicate the weeds between crop rows and for pre-emergence soil treatment. It has four independently controlled, 2 kW microwave generators operating at 2.45 GHz. The trailer is powered by two onboard 7 kVA, 3-phase electrical generators (Brodie *et al.* 2017).

The MW energy is channeled to the ground via waveguides (390 mm long) and horn antennae with aperture dimensions of 55 × 110 mm. However, in this study, treatment of up to 60 sec duration occurs while the trailer was stationary. The selection of 60 sec was based on the results of a previous field trial, where a treatment duration of 120 sec with MW power source of 0.7 kW was used (Khan *et al.* 2017). The trailer was then moved forward by about 8 to 10 cm and treatment was done again in the next small section of soil. At both locations, after MW soil treatments, the plots remained undisturbed for a few hours, after which the rice variety YRM70 was sown with a small seed plotter at the seed rate of 80 kg ha^{-1} with a row-to-row spacing of 25 cm. All the agronomic practices were maintained, and final data were collected at crop harvesting.

For all the trials, the obtained data were subjected to normality assessment and single factor Analysis of Variance was performed using MATLAB® R2015b (The Mathworks Inc., Natick Massachusetts, USA) to determine statistically significant means, and a least significant difference (LSD) test was used to compare the treatment means at a 5% probability level.

RESULTS

Pre-emergence MW weed management relies on the temperature distribution in the soil. The applied MW energy density in the treated plots, accounting for MW energy reflections from the soil in the absence of using any wave-guide tuning, was approximately 560 J cm^{-2}. This increased the temperature profile of top soil (0–6 cm) to approximately 75 ± 5°C, evidenced through thermal images (Figure 1), and a thermometer.

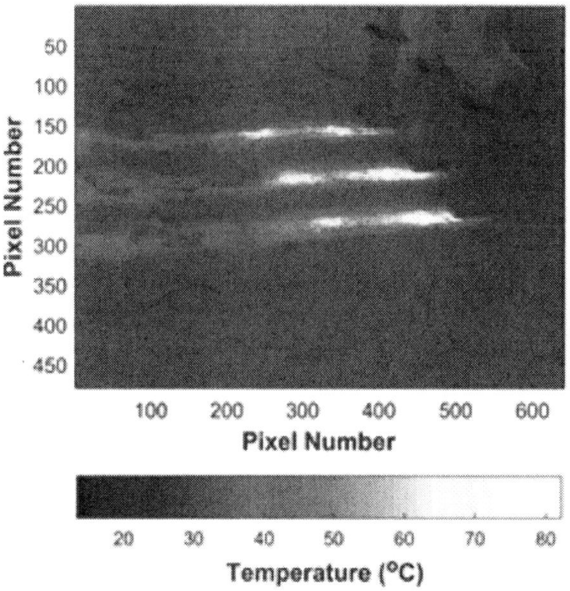

Figure 1. Infrared thermal image of microwave treated plot for reducing weed seedbank viability.

In the case of the first rice crop, the MW soil treatment significantly (P<0.001, Figure 2) reduced the weed density (17.6 plants $plot^{-1}$), weed fresh biomass (156.4 g $plot^{-1}$) and weed dry biomass accumulation (21.6 g $plot^{-1}$) compared to the untreated control plots; 94.8 plants $plot^{-1}$, 612.8 g $plot^{-1}$ and 122.6 g $plot^{-1}$ respectively. Overall, the pre-emergence MW irradiation of soil gave a 70–80% reduction in weed establishment.

Similarly, in the dryland wheat field experiment, the pre-emergence application of MW energy, fed through a horn antenna into the soil, significantly (P<0.001, Figure 2) reduced weed density, weed fresh weight, and dry weight in the wheat crop (Table 2) compared to the untreated control plots. The overall reductions in weed density, weed fresh weight, and dry weight, achieved with MW application in the dryland wheat crop, were 62.9%, 89%, and 90%, respectively.

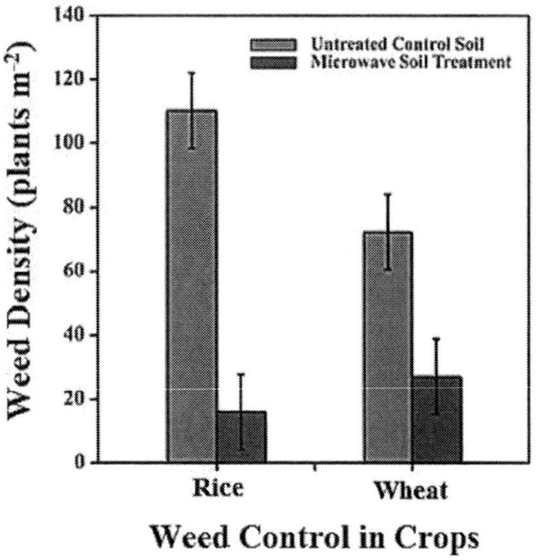

Figure 2. Effect of pre-cropping microwave soil heating on the weed density of rice and wheat crop measured at crop harvesting at Dookie, Victoria.

During the first year's rice experiment, the projection of MW energy into soil, for pre-emergence weed management, significantly ($P<0.001$, Table 1) increased the tiller density (419 m^{-2}) and grain yield (9.0 t ha^{-1}) of rice, compared to the untreated control (292 m^{-2}, and 6.7 t ha^{-1}, respectively). A significant difference was observed in the case of dry biomass at maximum tillering stage (Table 1), but no statistically significant difference was observed at harvesting between microwave treated plots (27.8 t ha^{-1}) and untreated control plots (22.8 t ha^{-1}).

Dryland wheat production is strongly influenced by weed management practices. The application of MW into soil for pre-emergence weed control significantly increased the dryland wheat productivity (Table 1). The 50.9% gain in fresh weight and 42.4% gain in dry weight, compared with the control plots, were acquired from the MW treated plots. Tiller density is the main contributor of final grain yield. The maximum number of tillers was developed in the MW treated plots (387 m^{-2}) compared to the untreated control plots (268 m^{-2}). In the present investigation, the final incremental increase of 33.1% in dry biomass production and 39.2% in grain yield was attained through MW energy application into the soil for weed management.

DISCUSSION

The higher crop productivity could be attributed to a 70–80% reduction in weed establishment achieved through MW irradiation of soil, ultimately leaving more room for crop growth. Thermal devitalisation of weed seedbanks in the top soil layer may be the possible cause of minimum weed interference with the various crops. This was evidenced by (Vidotto *et al.* 2013), who explored the effectiveness of high temperature on seed viability of six weed species including *Echinochloa crus-galli* (L.) P.Beauv., a problematic weed of rice regions globally. They stated that 80–100% germination reduction was achieved through raising the soil temperature to 79.6°C. This

Table 1. Comprehensive view of microwave weed control on the productivity of rice and wheat crop under field conditions at Dookie, Victoria. Different superscripts across the row depicting a significant difference at 5% probability level.

Agronomic crop characteristics	Rice: Soil treatments – Microwave treated	Rice: Soil treatments – Untreated control	Rice: $LSD_{5\%}$	Rice: % change	Wheat: Soil treatments – Microwave treated	Wheat: Soil treatments – Untreated control	Wheat: $LSD_{5\%}$	Wheat: % change
Number of tillers (m^{-2})	419^a	292^b	113.9	43%	387^a	268^b	62.0	44.4%
Fresh biomass weight (t ha^{-1}) [a]	46.3^a	25.1^b	9.4	84.6%	30.8^a	20.4^a	10.4	50.9%
Dry biomass weight (t ha^{-1})	27.8^a	22.8^a	6.04	21.9%	19.7^a	14.8^b	4.8	33.1%
Grain yield (t ha^{-1})	9.0^a	6.7^b	2.04	34.38%	7.8^a	5.6^b	2.3	39.2%
Partial factor productivity of nitrogen (kg grain kg N^{-1}) [b]	72	53.6	–	–	65	46	–	–

[a] Data collection at maximum tillers establishment for both crops (60 days after sowing).

[b] Partial factor productivity of nitrogen (PFP) $= \frac{Y_0}{N_r} + \frac{\Delta Y}{N_r}$, change in crop yield with nitrogen application was calculated based on work done by (Cassman *et al.* 1998).

Note: Applied nitrogen during cropping period of rice was 125 kg N ha^{-1} and for wheat 120 kg N ha^{-1}.

is the same temperature range (70–80°C) that was acquired by MW irradiation of the soil in the present study. This effectively induced an inhibitory effect on the weed population and therefore increased the rice crop yield. These results are strongly supported with findings of other work (Brodie 2016).

An increase in crop dry biomass was also reported by Gibson *et al.* (1988) who demonstrated that shoot dry biomass of birch significantly increased in MW irradiated soil. Their experiment was to evaluate the effect of MW treatment of soil supplemented with two mycorrhizae on birch seedlings; the shoot dry biomass progressively increased with irradiation duration of soil, with the highest dry shoot biomass being 84 mg for MW irradiation.

In addition to weed suppression, numerous studies have reported the supplementary effect of MW irradiation of soil on crop growth. However, the exact reason for good crop growth and development in MW treated soil is still in question. It has been demonstrated that MW irradiation markedly alters the physical and chemical properties of soil organic matter (SOM; Hur *et al.* (2013)) and enhances the humification of SOM (Kim and Kim 2013). Relevant to this, SOM is aggregates of organic residues in soil at different stages of humification. Thermal denaturation of these biopolymers, induced by MW irradiation, could increase the concentrations of free amino acids for succeeding turnover to CO_2 and the ammonia pool (NH_4^+), which might have led to higher yield of the crops in MW treated soil, in the present investigations.

ACKNOWLEDGMENTS

This research work was executed with the support of the Australian Rural Industry Research and Development Corporation and the Grains Research and Development Corporation. We thank Rice Research Australia Pty. Ltd. – SunRice™ for use of their trial site at Old Coree, Jerilderie, New South Wales.

REFERENCES

Brodie, G. (2016). Derivation of a cropping system transfer function for weed management: Part 2. Microwave weed management. *Global Journal of Agricultural Innovation, Research & Development* 3, 1-9.

Brodie, G., Hamilton, S. and Woodworth, J. (2007). An assessment of microwave soil pasteurization for killing seeds and weeds. *Plant Protection Quarterly* 22, 143.

Brodie, G., Khan, J., Gupta, D., Foletta, S. and Bootes, N. (2017). Microwave Weed and Soil Treatment in Agricultural Systems. AMPERE Newsletter

Burnside, O.C., Moomaw, R.S., Roeth, F.W., Wicks, G.A. and Wilson, R.G. (1986). Weed seed demise in soil in weed-free corn (*Zea mays*) production across Nebraska. *Weed Science* 34, 248-51.

Gibson, B.F., Frances, M.F. and Deacon, J.W. (1988). Effects of microwave treatment of soil on growth of birch (*Betula pendula*) seedlings and infection of them by ectomycorrhizal fungi. *New Phytologist* 108, 189-204.

Heap, I. (2016). International Survey of Herbicide Resistant Weeds. http://www.weedscience.org/ (accessed 1 August 2018).

Hur, J., Sung-Won, P., Min Chan, K. and Han, S.K. (2013). Enhanced binding of hydrophobic organic contaminants by microwave-assisted humification of soil organic matter. *Chemosphere* 93, 2704-10.

Khan, M.J., Brodie, G. and Gupta, D. (2017). The effect of microwave soil treatment on rice production under field conditions. *Transactions of the American Society of Agricultural and Biological Engineers* 60, 517-25.

Kim, M.C. and Kim, H.S. (2013). Artificial and enhanced humification of soil organic matter using microwave irradiation. *Environmental Science and Pollution Research* 20, 2362-71.

Nelson, S.O. (1996). A review and assessment of microwave energy for soil treatment to control pests. *Transactions of the American Society of Agricultural and Biological Engineers* 39, 281-9.

Sartorato, I., Zanin, G., Baldoin, C. and De Zanche, C. (2006). Observations on the potential of microwaves for weed control. *Weed Research* 46, 1-9.

Vidotto, F., De Palo, F. and Ferrero, A. (2013). Effect of short-duration high temperatures on weed seed germination. Annals of Applied Biology 163, 454-65.

Comparing the weed seed collection efficacy of stripper and draper type harvester fronts during harvest

Annie Rayner[1], John Broster[2], Alison Chambers[2] and Michael Walsh[1]
[1] Sydney Institute of Agriculture, School of Life and Environmental Sciences, University of Sydney, Narrabri, New South Wales 2390; Australia
[2] Graham Centre for Agricultural Innovation (Charles Sturt University and NSW Department of Primary Industries), Charles Sturt University, Locked Bag 588, Wagga Wagga, New South Wales 2678, Australia
(aray8265@uni.sydney.edu.au)

Summary The wide-spread reliance on herbicides for weed control has resulted in high frequencies of herbicide-resistant weed populations in many Australian agricultural regions. Harvest weed seed control (HWSC) practices provide growers with an end-of-season weed control option, by targeting the seed of weeds maturing at the same time as the crop. This targeting of weed seeds helps to lower the seed bank and reduces the reliance on herbicides for in-crop weed control. In some Australian production regions growers have adopted stripper fronts, which collect just the grain heads or pods allowing the retention of upright and intact stubble. With reduced crop residues collected the speed and efficiency of harvest is dramatically increased. The aim of this study was to compare the effectiveness of stripper and draper type harvester fronts in collecting annual ryegrass seed during the harvest of cereal crops. In December 2017, at two locations near Wagga Wagga, 50 annual ryegrass plants were marked and their seed-bearing tillers counted prior to harvest. At the same time 20 representative plants were collected to determine the average seed production per tiller. After harvest the remaining tillers were collected by cutting at the soil surface and the remaining seed were counted. Stripper fronts collected up to 87% of the seeds, while a draper front would have collected up to 91%. Stripper fronts appear to be an excellent option for collecting weed seeds at harvest. They would provide an opportunity to achieve both high stubble retention without negatively impacting on HWSC.

Keywords Annual ryegrass, harvester front, HWSC, weed seed collection.

The extent of herbicide resistance in ryegrass and wild oats in New South Wales and Queensland

John C. Broster[1], Allison J. Chambers[1], Adam Jalaludin[2], Michael J. Widderick[2] and Michael J. Walsh[3]
[1] Graham Centre for Agricultural Innovation (Charles Sturt University and NSW Department of Primary Industries), Charles Sturt University, Locked Bag 588, Wagga Wagga, New South Wales 2678, Australia
[2] Queensland Department of Agriculture and Fisheries, PO Box 2282, Toowoomba, Queensland 4350, Australia
[3] I.A. Watson Grains Research Centre, University of Sydney, 12656 Newell Highway, Narrabri, New South Wales 2390, Australia
(jbroster@csu.edu.au)

Summary Herbicide resistant weeds are a major impediment to grain growers across the Australian cropping regions. Since resistance was first reported in ryegrass and wild oats in Australia in the 1980s the extent of resistance in these weeds has increased. Knowledge of the extent of resistance in a region is important in designing suitable extension messages to manage resistance. Random seed collection surveys for annual ryegrass and wild oats were conducted across the northern cropping region (NSW and QLD) with the aim of establishing the frequency of herbicide resistance in these species to commonly used herbicides. Approximately 1000 paddocks were surveyed with annual ryegrass and wild oats occurring in 68 and 56% of paddocks, respectively. Approximately 60% of collected annual ryegrass populations were resistant to diclofop and sulfometuron. There were lower levels of resistance to clethodim (10%), trifluralin (11%) and glyphosate (3%) herbicides. In contrast to annual ryegrass, the frequency of herbicide resistant wild oat populations was much lower with only approximately 40% resistant to clodinafop and less resistant to clethodim and idosulfuron. There were regional differences herbicide resistance with 10% of annual ryegrass populations from northern NSW found to be resistant to glyphosate compared to 2% in the remainder of the state. However, there were lower frequencies of resistance to all tested selective herbicides in this region compared to the rest of the NSW. Despite the high frequencies of resistance, some herbicides (e.g. pyroxasulfone and prosulfocarb + metolachlor) remain effective across all collected annual ryegrass and wild oat populations.

Keywords Herbicide resistance, ryegrass, wild oats.

Metabolic profiling for benzoxazinoids in weed suppressive and early vigour wheat genotypes

James M. Mwendwa[1], Paul A. Weston[1], Inge Fomsgaard[2], Bente B. Laursen[2], William B. Brown[1], Hanwen Wu[1,3], Jane C. Quinn[1], Jeffrey D. Weidenhamer[4] and Leslie A. Weston[1]
[1] Graham Centre for Agricultural Innovation, Wagga Wagga, New South Wales 2678, Australia
[2] Department of Agroecology, Aarhus University, Forsøgsvej 1, DK-4200 Slagelse, Denmark
[3] Department of Primary Industries, Plant Biosecurity, Wagga Wagga, New South Wales 2650, Australia
[4] Department of Chemistry, Geology and Physics, Ashland University, Ashland, Ohio 44805 USA
(jmwendwa@csu.edu.au)

Summary Replicated and randomised wheat (*Triticum aestivum* L.) cultivar trials were conducted in moderate to low rainfall zones at Wagga Wagga (572 mm) and Condobolin (449 mm) NSW, respectively in 2014–2016. At each experimental site, crop and/or weed growth were monitored at selected growth stages including tillering, vegetative, grain filling, harvest and after crop harvest. In addition, shoots, roots, rhizoplane and bulk rhizosphere soil samples were collected for metabolomics profiling and biomass evaluation. Plant tissue samples were extracted in methanol using an automated Buchi high pressure extractor while soil samples were extracted using a rotary shaker. Extracts were filtered and specifically analysed for unique secondary metabolites or allelochemicals associated with weed suppression, specifically benzoxazinoids (BXs), using liquid chromatography coupled to quadrupole time-of-flight mass spectrometry (UPLC-MS QToF). Metabolic profiling of wheat shoots, roots, and soils resulted in detection of up to 14 individual BXs including BX glycosides, lactones and hydroxamic acids of interest. Both qualitative and quantitative differences in BXs were observed and were cultivar-, growth stage- and location-dependent. Plant part and rhizosphere location (distance from root) also impacted BX concentration. Further metabolic profiling provided crucial information regarding crop metabolism, as well as the biosynthesis and release of metabolites associated with weed suppression in currently available commercial wheat cultivars, in contrast to weed suppressive rye (*Secale cereale* L.) and a heritage wheat cultivar Federation, both recognised for their potent ability to suppress weeds. We conclude that certain commercial wheat cultivars maintained high yield potential and were significantly more weed suppressive, depending on year and location, due to both their early growth habit and canopy architecture as well as the release of BX metabolites into the rhizosphere over time.

Keywords Weed suppression, metabolomics, residue, competition, resource allocation.

Opportunities for weed management using virtual fencing technology

Rick Llewellyn[1], Danila Marini[2], Marta Monjardino[1], Sue Belson[2], Jackie Ouzman[1] and Caroline Lee[2]
[1] CSIRO, Waite Campus, Urrbrae, South Australia 5064, Australia
[2]CSIRO, Armidale, New South Wales 2351, Australia
(rick.llewellyn@csiro.au)

Summary Grazing is a well-recognised tool for the management of many major weeds. However, fencing costs and the limitations of temporary electric fencing constrain its potential application and effectiveness. New fenceless spatial grazing technology offers the opportunity for grazing to be applied more effectively and more often by Australian land managers. After more than a decade of development, the GPS-based technology to facilitate spatially targeted grazing using 'virtual fencing' is nearing commercial availability for use with cattle. The technology essentially locates animals in relation to a user-defined 'virtual fence' and trains animals to respond to an audio cue when they approach the virtual fenceline. This can be used to focus grazing pressure within a targeted area, exclude livestock from sensitive areas and/or to manage livestock 'herding'. In this paper we present results from recent trials in Southern Australia testing the potential for virtual fencing to be used to target sheep grazing. Results demonstrate the ability to use virtual fencing within a paddock to focus grazing pressure on a weedy area while successfully excluding livestock from areas where high stocking density would cause environmental damage and usually prevent the use of high grazing pressure as a weed control tool. The potential economic and weed control benefits from being able to target increased grazing pressure on sub-paddock areas is also shown using bio-economic modelling. The results demonstrate the potential benefits of the development of virtual fencing technology for use with sheep. Further opportunities for environmental and agricultural production benefits are presented.

Keywords Virtual fencing, spatial grazing, weed control, livestock, precision agriculture.

Ecology of threehorn bedstraw (*Galium tricornutum*): Implications for management and harvest weed seed control

Samuel G.L. Kleemann and Gurjeet S. Gill
School of Agriculture, Food and Wine, University of Adelaide, PMB 1, Glen Osmond, South Australia 5064, Australia
(gurjeet.gill@adelaide.edu.au)

Summary Threehorn bedstraw is an important broadleaf weed of crops in southern Australia but its behaviour in field crops has not been adequately investigated. Populations of bedstraw (n = 10) were collected from the southern cropping region to investigate their seed dormancy and seedling emergence pattern. Seeds of these populations were placed outdoors in seedling trays in autumn and watered as needed until spring to prevent water stress. There was no variation in the pattern of seedling emergence in these populations but their cumulative germination ranged from 14–46% of the initial seedbank. Presence of such high levels of ungerminated seedbank in bedstraw is consistent with previous reports of its high seedbank persistence. Seedling emergence in these populations only occurred after the onset of cold conditions in late winter (July-August), which is consistent with stimulation of seed germination by cold-stratification. Competitive effects of varying bedstraw densities were investigated in wheat (cv. CLF-Hatchet and Sceptre) and lentils (cv. Hurricane XTA). There was a hyperbolic relationship between bedstraw density and its seed production ($R^2 = 0.82–0.93$) in both crops, however seed production was much higher in lentils (>20,000 seeds m^{-2}) than wheat (3800–5900 seeds m^{-2}). Despite its late seedling emergence, there was an exponential relationship between bedstraw density and yield loss in wheat and lentil. Maximum yield loss relative to the weed-free control was 10–14% in wheat and ≥30% in lentils. These results highlight the need to develop effective bedstraw management tactics, especially in lentils. Bedstraw showed no seed dispersal prior to the harvest-ready stage of wheat and lentil (100% retained). Therefore, harvest weed seed control practices including seed catchers, weed seed destructor and narrow windrow burning could be highly effective in its long-term management provided seeds could be separated into the chaff fraction in the harvester.

Keywords Ecology.

INTRODUCTION

Threehorn bedstraw (*Galium tricornutum* Dandy) is an important broadleaf weed of winter crops in southern Australia. This weed has become more prevalent across the Wimmera district of Victoria, and mid north and Yorke Peninsula of South Australia where lentil area has increased considerably over the last 10 years. Lentils are weak competitors with weeds (McDonald *et al.* 2007), and at present there are few herbicide options for the selective control of this weed in lentils.

Despite increasing concerns about threehorn bedstraw in southern Australian farming systems, there are significant knowledge gaps in its seed biology. Chauhan *et al.* (2006a, b) reported that germination of threehorn bedstraw was inhibited by light, but stimulated by exposure to cold-stratification or treatment with gibberellic acid. According to Mennan (2003), seed persistence in catchweed bedstraw (*Galium aparine*) was limited to 2–3 years in the field but could be increased by deep burial. Catchweed bedstraw is a competitive weed of field crops, which can reduce cereal yields by 30–60% (Rola 1971). Seeds of catchweed bedstraw were found to be a significant contaminant of cereal grain, and it was shown to increase crop lodging by smothering crops, which made harvest more difficult (Defelice 2002).

Even though threehorn bedstraw is an important weed in southern Australian, there is little information available on its seedbank persistence and crop-weed interference. At present there is considerable interest in the grains industry in harvest weed seed control (HWSC) tactics including weed seed catchers, weed seed destructor and narrow windrow burning. The effectiveness of these practices depends largely on the timing of weed seed dispersal and the amount of weed seed present above the cutter bar height at crop harvest (Walsh and Powles 2014). However, there is no information available on seed production, retention and shedding behaviour of threehorn bedstraw, which will strongly influence prospects of HWSC for its management.

More information on seedling emergence pattern, seed persistence, seed production, retention, and shedding behaviour of threehorn bedstraw is needed to facilitate development of effective control strategies. Studies were therefore undertaken to address these knowledge gaps.

MATERIALS AND METHODS

A field study was undertaken in 2017 at Roseworthy (34°32′ S, 138°41′ E), South Australia, to investigate field seedling emergence and seedbank recruitment of 10 populations of bedstraw randomly collected from cropping fields in Victoria and South Australia. Seeds were collected before crop harvest in October to November 2016 from more than 200 mature plants at each site, and were immediately stored outdoors in naturally lit storage units to maintain ambient temperature conditions until used for experiments. In the autumn after seed collection (March 2017), 100 seeds of each population (n = 10) were sown in plastic trays (33 cm by 28 cm by 5 cm) containing potting mix and were maintained outside during the normal growing season at Roseworthy. There were four replicate trays of each population. Trays were watered as required to maintain good soil moisture conditions for seed germination and seedling emergence. Weed seedlings were counted (at first leaf appearance) from April to the end of October (7 months or 190 d). Initially, emergence was recorded at weekly intervals but after 1.5 months, because of low emergence, it was recorded at 2-week intervals. Seedlings were counted and removed until no further emergence was recorded in three consecutive measurements. The counts were expressed as cumulative seedling emergence (i.e. percentage of the total emergence). Cumulative seedling emergence data were fitted using a sigmoidal logistics model (GraphPad Prism v.6.0; GraphPad Software, San Diego, California) to compare the time taken (d) by populations to reach 50% maximum seedling emergence (t_{50}).

Additional field studies were undertaken at Roseworthy to determine seed production, crop yield loss and pattern of seed dispersal of bedstraw in wheat (2016 and 2017) and lentils (2017). Plots of nil, low (10–20 plants m^{-2}), medium (50 plants m^{-2}), and high (≥100 plants m^{-2}) weed density of bedstraw were established by hand-broadcasting weed seeds prior to sowing wheat and lentils. In 2016, the experiment was planted with wheat cv. CLF-Hatchet at 100 kg ha^{-1} seed rate on 6 June, whereas wheat cv. Sceptre at 100 kg ha^{-1} and lentil cv. Hurricane XTA at 45 kg ha^{-1} seed rates were planted on 7 and 8 June in 2017. Diammonium phosphate required to supply 22.5 kg N ha^{-1} and 25 kg P ha^{-1} was banded below the crop seed at sowing. Plots (5 m long and 1.5 m wide) were sown using a knife-point and press wheel seed drill on 0.25 m row spacing, and both crops were managed according to the district practice. The experiments were established as a randomised complete block design with four replicates. Weed seed production of bedstraw was assessed within four weeks of crop senescence in November 2016 and 2017. Plant density and seed production was assessed in two quadrats (0.5 m by 0.5 m) in each plot. A hyperbolic model was found to best fit the relationship between weed density and weed seed production. High density plots of bedstraw were also used to measure temporal pattern of weed seed dispersal. At seven weeks before the expected harvest-ready stage of the crop (≤12% grain moisture content), three circular seed traps (15 cm diam. by 10 cm deep) were randomly placed between crop rows in each plot. A wire screen was fitted to the bottom of each trap to capture seed rain. Eight holes at the bottom of the seed trap allowed rainwater drainage. Weed seeds were collected weekly from the traps until just before crop harvest. Collected seeds were identified and counted to quantify seed dispersal. Grain yield was determined using a Kew plot harvester, and a one-phase exponential decay model was fitted to the data to assess the impact of weed density on grain yield. Contamination of grain sample by weed seeds at harvest was determined by counting the number of bedstraw seeds in a grain subsample (~500 g) from each plot.

RESULTS

All populations exhibited no seedling emergence until 28 DAS (Figure 1). Furthermore, the time taken to reach 50% of maximum seedling emergence was similar across the populations and ranged from 95–106 days (Table 1). The start of seedling emergence coincided with minimum air temperatures dropping below 5°C in the field (Figure 1). Cumulative seedling emergence from the seedbank did not exceed 50% for any of the populations, and ranged from 14–46% (Table 1). These results indicate large persistence of bedstraw seedbank from one year to the next.

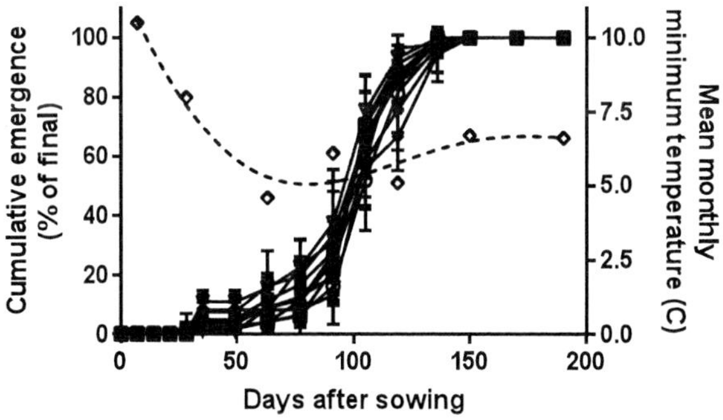

Figure 1. Cumulative seedling emergence pattern of bedstraw populations (n = 10) collected from Victoria and South Australia. Dashed line represents mean monthly minimum temperature, and vertical bars are the SE of the mean.

Table 1. Cumulative seedling emergence from the seedbank and time taken (days) to reach 50% of maximum seedling emergence (t_{50}) of bedstraw populations from Victoria and South Australia. Values in parenthesis represent SE of mean, and t_{50} values were calculated by fitting cumulative emergence data to a sigmoidal logistic model.

Population	Cumulative seedling emergence (% of seedbank)	Time taken (d) to reach 50% seedling emergence (t_{50})	r^2
BE2 YP	32 (3.6)	98.1	0.95
BE5 SE	38 (4.0)	99.9	0.99
BE9 WD	14 (3.8)	99.2	0.97
BE14 YP	45 (2.6)	99.5	0.99
BE15 YP	31 (2.6)	105.6	0.98
BE16 YP	36 (4.1)	105.1	0.99
BE17 MN	46 (3.7)	99.6	0.99
BE18 SE	27 (3.7)	99.8	0.98
BE19 SE	28 (1.7)	94.9	0.98
BE21 MN	24 (4.8)	105.1	0.97

There was a hyperbolic relationship between bedstraw density and seed production ($R^2 = 0.82–0.93$) in wheat and lentil (Figure 2). Despite delayed seedling emergence of bedstraw relative to both wheat and lentils, plants were still capable of significant seed production, which was considerably higher in lentils (>20,000 seeds m^{-2}) than wheat (3800 seeds m^{-2}) (Figure 2b). Seed production was also influenced by the growing season (Figure 2a), and was strongly correlated ($R^2 = 0.78–0.96$) to rainfall and biomass production (data not presented). In response to above-average spring rainfall in 2016 (179 mm) at Roseworthy, bedstraw seed production in wheat was more than 1.5-fold higher (5900 seeds m^{-2}) than in 2017 (3800 seeds m^{-2}), which received below average spring rainfall (88 mm).

Bedstraw caused a reduction in wheat and lentil yield at low to moderate weed infestations (Figure 3). Maximum yield loss relative to the weed-free control for lentils (1937 kg ha^{-1}) was estimated to be more than 30%, whereas the losses were much smaller (10–14%) in wheat where weed-free yield in both growing seasons exceeded 5 t ha^{-1} (data not presented).

Even though bedstraw reached physiological maturity ~40 d before the harvest-ready stage of both crops, it did not shed any seeds into the seed traps (100% retention, Table 2). The contamination of crop grain with bedstraw seeds was significantly higher in lentils (5092 seeds kg^{-1} of grain) than wheat (12–234 seeds kg^{-1} of grain). This result indicates that a large

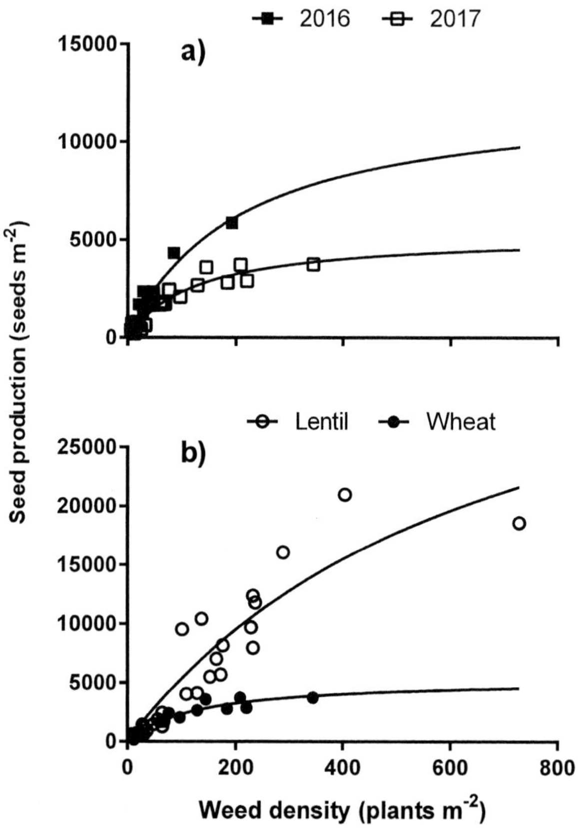

Figure 2. Seed production of bedstraw in wheat in 2016 and 2017 growing season a), and in competition with wheat and lentils in 2017 b). The hyperbolic model was fitted in GraphPad Prism.

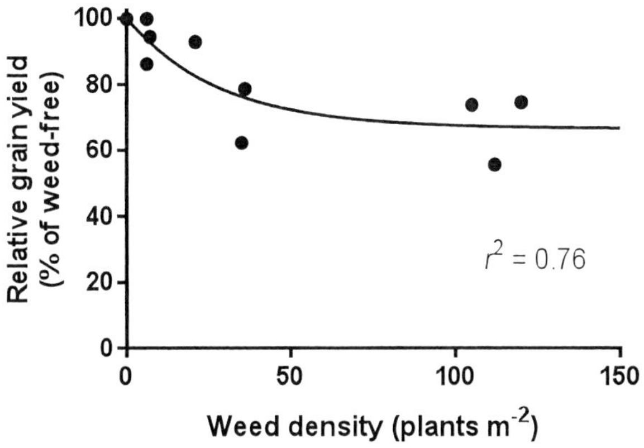

Figure 3. Relationship between density of bedstraw and relative grain yield of lentil (Hurricane XTA) in 2017. Relative yield = (weed density yield/weed free yield) × 100; Mean weed-free yield = 1937 kg ha^{-1}. The non-linear curve represents a one-phase exponential decay model (GraphPad Prism).

Table 2. Bedstraw seed retention (%) in the inflorescence at the harvest-ready stage of wheat and lentils and grain contamination in 2016 and 2017. Values in parenthesis represent the SE of mean.

	Bedstraw seeds	
Crop (season)	Retention (%)	Grain contamination seeds kg^{-1} of grain
Wheat 2016	100	234 (73)
Wheat 2017	100	12 (4)
Lentil 2017	100	5,092 (1,477)

amount of bedstraw seed ends up in the grain fraction during crop harvest and does not exit the harvester in the chaff fraction as is the case with weeds such as ryegrass.

DISCUSSION

Seedling emergence behaviour was similar between the populations of bedstraw, however seedling recruitment was slow with most populations requiring 95–106 days to reach 50% emergence. In contrast, much faster rates of seedling recruitment (<30 days) have been observed for other common local weeds such as Indian hedge mustard and sowthistle, which are known to germinate rapidly and possess little or no innate dormancy (Chauhan 2006). Cumulative seedling emergence of populations ranged from 14–46% of the initial seedbank. Such high levels of ungerminated seedbank in bedstraw are consistent with previous reports of high seedbank persistence (Chauhan *et al.* 2006b). Seedling emergence in bedstraw populations only occurred after the onset of cold conditions in late winter (July-August), which is consistent with stimulation of seed germination in this species by cold-stratification (Chauhan *et al.* 2006a). Late seedling emergence observed in bedstraw would enable many plants to escape pre-sowing weed control tactics.

Despite its late seedling emergence relative to wheat and lentils, bedstraw was capable of large seed production, particularly in lentils where seed production exceeded 20,000 seeds m^{-2}. Relative to wheat, lentils are known for their poor weed competitive ability (McDonald *et al.* 2007), and seed production of this magnitude may provide some explanation for the increasing prevalence of this weed in areas of lentil production. Even at moderate weed infestations (150 plants m^{-2}), bedstraw reduced lentil grain yields by ≥30%, whereas the yield losses in wheat were more modest (10–14%). Previously Rola (1971) reported that catchweed bedstraw was capable of reducing wheat yields by more than 30–60%, and attributed its high competitive ability to its climbing habit and ability to smother the crop.

In two growing seasons of contrasting spring rainfall, bedstraw did not shed any seeds into the seed traps (100% retained). This is in stark contrast to other weed species such as barley grass which shed more than 95% of its seed by crop harvest (Kleemann and Gill 2018). Complete seed retention in bedstraw was reflected in the high level of grain contamination with bedstraw seeds, especially in lentils (Table 2). The level of crop grain contamination was 25 to 400-fold higher in lentils than wheat, which could be partly related to greater weed seed production in lentils than wheat (Figure 2), but could also be associated with greater observed susceptibility of bedstraw to lodge in wheat. Catchweed bedstraw was also reported to easily lodge (Defelice 2002) because of its high biomass accumulation, weak stems, and tangled growth habit. In contrast to wheat, the short growth habit of lentils resulted in more vigorous bedstraw plants, which were less prone to lodging. Lower harvest height used in lentils than wheat (5 cm versus 15 cm) could have also contributed to greater grain contamination by bedstraw seeds in lentils. These results suggest that HWSC practices could assist in the management of bedstraw provided its seeds could be effectively captured and separated into the chaff fraction.

ACKNOWLEDGMENTS

We are grateful to Grains Research and Development Corporation (GRDC) for funding the research, and Jerome Martin and Malinee Thongmee for providing technical assistance.

REFERENCES

Chauhan, B.S. (2006). Ecology and management of weeds under no-till in southern Australia. PhD Thesis (Ph.D). http://hdl.handle.net/2440/37881 (accessed 1 August 2018).

Chauhan, B.S., Gill, G.S. and Preston, C. (2006a). Factors affecting seed germination of threehorn bedstraw (*Galium tricornutum*) in Australia. *Weed Science* 54, 471-7.

Chauhan, B.S., Gill, G.S. and Preston, C. (2006b). Seed germination and seedling emergence of threehorn bedstraw (*Galium tricornutum*). *Weed Science* 54, 867-72.

Defelice, M.S. (2002). Catchweed bedstraw or cleavers, *Galium aparine* L. – A very "sticky" subject. *Weed Technology* 16, 467-72.

Kleemann, S.G.L and Gill, G.S. (2018). Investigation of post-maturity weed seed dispersal as an indicator of susceptibility of weed species to harvest weed seed control. Proceedings of the 18th European Weeds Research Society Symposium, ed. A. Simončič, pp. 115. (European Weed Research Society, Ljubljana, Slovenia).

McDonald, G.K., Hollaway, K.L. and McMurray, L. (2007). Increasing plant density improves weed competition in lentil (*Lens culinaris*). *Australian Journal of Experimental Agriculture* 47, 48-56.

Mennan, H. (2003). The effects of depth and duration of burial on seasonal germination, dormancy and viability of *Galium aparine* and *Bifora radians* seeds. *Journal of Agronomy and Crop Science* 189, 304-9.

Rola, J. (1971). Causes and effects of weed compensation in crops. *Weed Abstracts* 20, 425.

Walsh, M.J. and Powles, S. B. (2014). High seed retention at maturity of annual weeds infesting field crops highlights the potential for harvest weed seed control. *Weed Technology* 28, 486-93.

Lime improves control of wild radish and annual ryegrass in acid soils of Western Australia

Abul Hashem and Catherine Borger
Department of Primary Industries and Regional Development, 75 York Road, Northam, Western Australia 6401, Australia
(abul.hashem@dpird.wa.gov.au)

Summary Most sandy soils of Western Australia (WA) are acidic, which may influence the competitive ability of crops against weeds like wild radish (*Raphanus raphanistrum* L.) and annual ryegrass (*Lolium rigidum* Gaudin). We examined the effect of lime applied to acidic soils on weed control and crop yield over four seasons at three locations (Eradu, Wongan Hills and Merredin) of the WA wheatbelt. All experiments were conducted in a split-plot design with lime in the main plots replicated four times and herbicides in the sub-plots. Limesand was applied in the first year (2010) on the soil surface and incorporated by sowing. Lime treatments increased soil pH by approximately one unit in the top soil (0 to10 cm) but the sub-soil pH (20–30 cm) remained at or below 4.8. Lime treatments had no effect on the density of weeds in the first three years after application, but in the fourth year lime at 2.5 and 5.0 t ha^{-1} reduced the in-crop density of wild radish by 45% at Eradu and annual ryegrass by up to 60% at Merredin. Herbicides applied in the sub-plots increased barley grain yield by 54–58% in the fourth year (2013) after lime application at Merredin and Eradu. Lime applied to the soil surface can reduce weed density but it may take three to four years after application to show the effect.

Keywords Soil acidity, lime, wild radish, annual ryegrass, weed control.

INTRODUCTION

The most important causes of soil acidification on agricultural land are the application of ammonium-based fertilisers and urea, elemental sulphur fertiliser and the growth of legumes (Goulding 2016). Crop growth is reduced in acidic soils, which is likely to make the crop less competitive against weeds (Scott *et al.* 1997). This may compound the effects of weed competition resulting in further reductions in grain yields. Wheat yield can be reduced by 50% due to annual ryegrass competition (Hashem *et al.* 1998). The presence of 10–75 wild radish plants m^{-2} can reduce wheat yields by 7–56% and lupin yields by 28–92% (Hashem *et al.* 2006). A potential option of increasing the competitive ability of crops against annual ryegrass in low pH soil is to apply lime (Goulding 2016). Gazey and Andrew (2010) demonstrated that the application of lime increased the biomass of a barley crop and decreased the biomass of annual ryegrass with markedly different soil pH profiles in a long-term lime trial at Kellerberrin, Western Australia (WA).

Application of lime may also influence herbicide efficiency. There is little data available on the influence of lime on the performance of herbicides to control annual ryegrass or wild radish. The aim of this study was to examine the impact of lime and herbicides on the control of annual ryegrass or wild radish in low pH soils. The study hypothesised that application of lime would improve weed suppression and increase grain yield in a wheat-wheat-lupin-barley rotation.

MATERIALS AND METHODS

Four sites were established in WA, to investigate annual ryegrass (Merredin and Wongan Hills) and wild radish (Eradu and Wongan Hills). The crop rotation at all four sites was wheat-wheat-lupin-barley, grown from 2010–2013. Trials at each site were set up in a split plot design. Four rates of limesand (0, 1.25, 2.5 and 5 t ha^{-1}) were allocated in the main plots and five rates of selective herbicide (Table 1) in the sub-plots. The unit plot size was 20 m by 2 m, replicated four times. Measurements included pH ($CaCl_2$) of soils collected just before sowing crop and application of herbicide each year, weed and crop density (from two 50 cm by 50 cm quadrats per plot) and crop yield.

RESULTS

Lime effect on soil pH In 2010–2012, lime increased soil pH by about one unit at 0–10 cm depth, but the pH in the sub-soils remained below 4.8 (the desirable pH for optimum crop plant growth) at all sites. In 2013 (the fourth year), lime at 2.5 and 5 t ha^{-1} increased soil pH ($CaCl_2$) by 0.67−1.23 units at 0–10 cm depth, 0.22–0.87 units at 10–20 cm and 0.08–0.35 units at 20–30 cm.

Lime effect on weed density and crop Lime did not have a significant impact on weeds in first, second and third year after lime application (from 2010–2012). In the fourth year (2013), initial annual ryegrass density was reduced with the increases in lime rate (compared to no lime); from 28 plant m^{-2} at 0 t ha^{-1} to 16 plant m^{-2} at 5.0 t ha^{-1} (41% reduction) at Wongan Hills and from 139 plant m^{-2} on no-lime to 56 plants m^{-2} at 5.0 t ha^{-1} (59% reduction) at Merredin site (Table 2). Likewise, lime reduced the initial density of wild radish in 2013 by 1% (at 1.5 t ha^{-1}) to 48% (at 5.0 t ha^{-1}) at Eradu where wild radish density was 120 plants m^{-2} in the no-lime control (Table 2). Wild radish density at Wongan Hills was too low to show a significant impact of lime, with an average of 2 plants m^{-2} in the no-lime control (Table 2).

Likewise, lime had no significant impact on crop yield from 2010–2012. However, in the fourth year (2013), lime significantly increased barley grain yield by 6% (at 5.0 t ha^{-1}) at the Wongan Hills wild radish site and 7% (at 2.5 t ha^{-1}) at the Wongan Hills annual ryegrass site (Table 3). At Eradu and Merredin, ryegrass site, there was no significant increase in grain yield resulting from lime (Table 3).

Herbicide effect on weed control and crop growth
In the wheat crops grown in 2010–2011, annual ryegrass control at Wongan Hills improved with increasing

Table 1. Herbicides applied across lime treatments to control annual ryegrass (Merredin and Wongan Hills) and wild radish (Eradu and Wongan Hills) in different crops from 2010–2013.

Site and weed species	Herbicide treatments
Annual ryegrass at Merredin and Wongan Hills	Wheat crop in 2010 and 2011: Sakura® (pyroxasulfone 850 g kg^{-1}) at 0, 60, 90, 120, 150 g ha^{-1} at pre-plant.
	Lupin crop in 2012: Simazine (simazine 500 g L^{-1}) 2000 mL ha^{-1} at post-plant across all plots except in the no herbicide control.
	Barley crop in 2013: Boxer Gold® (prosulfocarb 800 g L^{-1} + s-metolachlor 120 g L^{-1}) at 0, 1000, 1500, 2000, and 2500 mL ha^{-1}at pre-plant.
Wild radish at Eradu and Wongan Hills	Wheat crop in 2010 and 2011: Velocity® (bromoxynil 210 g L^{-1} + pyrasulfotole 375 g L^{-1}) at 0, 250, 350, 500 and 670 mL ha^{-1} at post-emergent.
	Lupin crop in 2012: Simazine® (simazine 500 g L^{-1}) 2000 mL ha^{-1} at post-plant across all plots except in the no-herbicide control.
	Barley crop in 2013: Velocity® (bromoxynil 210 g L^{-1} + pyrasulfotole 375 g L^{-1}) at 0, 250, 350, 500 and 670 mL ha^{-1} at post-emergent.

Table 2. Cumulative effect of lime applied in 2010 and herbicides applied each year from 2010–2013 on the initial density of weeds recorded after crop emergence and before application of post-emergent herbicides in 2013 season at each site. For herbicide rates, refer to Table 1.

		Wild radish (plant m^{-2})		Annual ryegrass (plant m^{-2})	
Lime/ Herbicide	Rate	Eradu	Wongan Hills	Wongan Hills	Merredin
Lime (t ha^{-1})	0	120	2	28	139
	1.25	105	2	26	114
	2.5	63	1	19	82
	5	66	1	16	56
	LSD (P<0.05)	39.4	NS[1]	11.1	35.3
Herbicides	Rate 1	163	5.1	86	235
	Rate 2	63	1.4	9	114
	Rate 3	65	0.5	7	87
	Rate 4	72	0.5	5	34
	Rate 5	78	0.1	3	19
	LSD (P<0.05)	39.5	1.27	14.6	82.6

[1] NS= Non-significant.

rates of Sakura®, leading to increased grain yield. At Merredin, increasing rates of Sakura® improved control of annual ryegrass and improved wheat grain yield by 4–8% in 2011 (data not presented). No clear interaction of lime and herbicide was found on weed control or yield at either site of ryegrass. At the wild radish sites, increasing rates of Velocity® improved weed control at Eradu and Wongan Hills. Velocity® at lower rates was more effective on wild radish at Eradu than at Wongan Hills, and increased wheat grain yield by 17–18% in 2011 (data not presented). Again, there was no interaction of lime and herbicide at either site of wild radish.

In the 2012 lupin crops, simazine, together with the cumulative effect of herbicides applied in 2010 and 2011 seasons, significantly increased control of wild radish and annual ryegrass compared to the no herbicide control plots.

In the 2013 barley crop, Velocity® was effective on wild radish at both sites. Further, the broadleaf herbicides sprayed on wild radish in 2010, 2011 and 2012 had a cumulative effect on the reduction of wild radish in 2013. Prior to spraying Velocity® in 2013, the density of wild radish counted at the tillering stage of the barley crop was reduced from 163 to 63 plants m^{-2} (64% reduction) at Eradu and 5.1 to 0.1 plants m^{-2} (98% reduction) at Wongan Hills (Table 2).

Herbicide effect on grain yield In 2012, herbicide treatments increased lupin grain yield by 62–227%, with the exception of the Merredin annual ryegrass site where lupin growth was very poor due to low rainfall in 2012 season (data not presented).

Increases in rates of Boxer Gold® controlled annual ryegrass by 51–91% and increased the yield of barley by 6–16% at Wongan Hills (Table 3). Likewise, herbicides increased ryegrass control by 51–92% and barley grain yield by 79–116% at Merredin (Table 3). Velocity® increased barley grain yield by 132–138% at Erdau, 7–9% at Wongan Hills (Table 3).

DISCUSSION

Application of lime increased soil pH in the top 10 cm. However, it took four years for the lime to leach down, to alter pH at a depth of 10–30 cm. Further, sub-soil pH was still lower than is optimal for crop growth (pH ≥4.8). At all sites except Eradu, the pH in the sub-surface layers (20–30 cm) with or without lime remained at or below 4.5. This low pH in the sub-soil suggests that soil acidity at the sub-soil layers was restricting barley crop root growth even three years after lime application.

Prior to 2013 there were only small increases in soil pH and consequently no observed effect on the density of annual ryegrass and wild radish populations. Surface-applied limesand took a long time to impact on sub-soil pH; therefore growers need to consider incorporation of limesand via deep ripping or other methods of soil renovations. A change in the pH at 2.5 and 5.0 t ha^{-1} in the upper soil layers in 2013, increased the barley crop competitiveness against weeds, as indicated by reduced weed density and increased grain

Table 3. Effect of increasing rates of lime and herbicides on the grain yield of barley at Eradu and Merredin, 2013. For herbicide rates used, refer to Table 1.

		Barley grain yield (t ha^{-1})			
		Wild radish		Annual ryegrass	
Lime/herbicide	Rate	Eradu	Wongan Hills	Wongan Hills	Merredin
Lime (t ha^{-1})	0	3.28	3.96	3.36	1.11
	1.25	3.30	4.09	3.36	1.11
	2.50	3.35	4.18	3.59	1.13
	5.00	3.33	4.21	3.56	1.16
	LSD (P<0.05)	NS	0.123	0.194	NS
Herbicides	Rate 1	1.60	3.84	3.13	0.62
	Rate 2	3.71	4.09	3.43	1.11
	Rate 3	3.81	4.18	3.62	1.20
	Rate 4	3.72	4.14	3.56	1.33
	Rate 5	3.72	0.138	3.60	1.34
	LSD (P<0.05)	0.197	0.138	0.217	0.184

yield. A competitive crop is a highly effective form of weed control. There was no interaction of herbicide and lime in this short term study. However, annual ryegrass prefers soil at the same pH as crops. Therefore, liming should make ryegrass less stressed, and less stressed weeds are more responsive to herbicide. Some evidence indicates that wild radish prefers acidic soils (Willis 2006) and so liming may be detrimental to the growth and development of wild radish. More research is required on the long term effect of lime on wild radish and ryegrass.

These results reinforce the fact that surface applied lime takes a long time to alleviate the sub-surface acidity problem and suggests the necessity of continuing to monitor the limed plots for a longer period of time.

ACKNOWLEDGMENTS

We are grateful to GRDC (UWA 000146) for funding the project. Research Collaboration of Australian Herbicide Resistance Initiative (AHRI) is greatly appreciated. Thanks are due to Barb Sage, Dave Nicholson, and DPIRD Research Support units at Geraldton, Merredin and Wongan Hills for providing technical services. Thanks are also to Liebe Group for presenting the trials at Wongan Hills to the growers on the field walks and field days.

REFERENCES

Gazey, C. and Andrew, J. (2010). Long-term effect of lime application on soil pH, crop yields and annual ryegrass competition. Proceedings of 2010 Agribusiness Crop Update, pp. 229-33. (DAFWA and GRDC, Perth, Western Australia).

Goulding, K.W.T. (2016). Soil acidification and the importance of liming agricultural soils with particular reference to the United Kingdom. *Soil Use and Management* 32, 390-9.

Hashem, A., Cheam, A., Bowran, D. and Piper, T. (1998). Annual ryegrass control in wheat by chemical and non-chemical options. Proceedings of Crop Protection Technical Symposium: Highlights of weed research and development in Western Australia, eds D. Bowran and T. Piper, pp. 32-3. (Perth, Western Australia).

Hashem A., Pathan, S. and French, B. (2006). Wild radish-lupin competition: Difference in the competitive ability of lupin cultivars. Proceedings of the 15th Australian Weeds Conference, eds C. Preston, J.H. Watts and N.D. Crossman, pp. 391-4. (Weed Management Society of South Australia, Adelaide).

Scott, B.J., Conyers, M.K., Poile, G.J. and Cullis, B.R. (1997). Subsurface acidity and liming affect yield of cereals. *Australian Journal of Agricultural Research* 48, 843-54.

Willis, M. (2006). Does liming limit the growth and development of wild radish (*Raphanus raphanistrum*)? BS (Agriculture) Dissertation, The University of Western Australia, Perth.

Developing biological control options for the integrated management of *Conyza bonariensis* and *Sonchus oleraceus*

L. Morin[1], K. Nagalingam[2], K.B. Ireland[1], V. Lesieur[3,4], M. Ollivier[4], M. Rafter[2], G.C. Hunter[1], T. Thomann[3], M. Jourdan[3], A. Sheppard[1,3], J.F. Martin[4], M.S. Tixier[4], M.D. Vitorino[5] and S. Raghu[2]
[1] CSIRO Health & Biosecurity, GPO Box 1700, Canberra, Australian Capital Territory 2601, Australia
[2] CSIRO Health & Biosecurity, GPO Box 2583, Brisbane, Queensland, Australia
[3] CSIRO Health & Biosecurity, European Laboratory, Campus International de Baillarguet, Montferrier-sur-Lez, Montpellier 34980, France
[4] CBGP, Montpellier SupAgro, INRA, CIRAD, IRD, Univ. Montpellier, Montpellier, France
[5] FURB, Blumenau, Brazil
(louise.morin@csiro.au)

Summary Flaxleaf fleabane (*Conyza bonariensis*; native range – South America) and common sowthistle (*Sonchus oleraceus*; native range – Eurasia) are major weeds of cropping systems in northern New South Wales and southern Queensland. Fleabane and sowthistle are estimated to cause revenue losses in excess of $43 million and $8 million respectively for grain producers. Both weeds are predominant in the fallow period where they use vital stored soil moisture and act as alternate hosts for pest insects. Development of resistance to herbicides in both weeds is making populations increasingly difficult to manage in agricultural environments. Given the success of classical biological control in the management of other weeds (including ruderal species) in the Asteraceae (e.g. *Onopordum* sp., *Cirsium* sp., *Parthenium hysterophorus*, *Chromolaena odorata*), we have commenced a project to develop biological control solutions for these two weed species. Our work includes (a) determination of the invasion history of these species using genomic data; (b) physiological and bioclimatic modelling of their distributions in their native range and in Australia; (c) exploration for natural enemies in the native range informed by the results from (a) and (b); and (d) undertaking risk assessment of candidate agents in their respective native ranges and under quarantine conditions in Australia. This talk will provide an overview of the project, and the results achieved to date.

Keywords Common sowthistle, flaxleaf fleabane, integrated weed management, classical biological control, fungi, arthropods.

Recruitment and fecundity of annual ryegrass, great brome grass, barley grass, doublegee and sowthistle

Catherine P.D. Borger and Abul Hashem
Department of Primary Industries and Regional Development, PO Box 483, Northam, Western Australia 6401, Australia
(catherine.borger@dpird.wa.gov.au)

Summary An experiment was conducted at Wongan Hills Western Australia (WA) in 2016 and 2017, to investigate the biology of ten weed species. The trial investigated recruitment, germination time, survival over the cropping season, seed production and time of shedding. This paper reports the results of annual ryegrass (*Lolium rigidum* Gaudin), great brome grass (*Bromus diandrus* Roth), barley grass (*Hordeum leporinum* Link), doublegee (*Rumex hypogaeus* T.M.Schust. & Reveal, formerly *Emex australis* Steinh) and sowthistle (*Sonchus oleraceus* L.).

Great brome grass seedling recruitment was higher than annual ryegrass recruitment in 2017, even though annual ryegrass had much higher seed set in 2016. Further, brome grass had the greatest impact on crop yield in 2017. Brome grass, barley grass, doublegee and sowthistle all shed their seed earlier than annual ryegrass. Harvesting as early as possible or swathing would be necessary to capture seeds from these species for harvest weed seed destruction.

Keywords *Lolium rigidum*, *Bromus diandrus*, *Hordeum leporinum*, *Rumex hypogaeus*, *Emex australis*, *Sonchus oleraceus*, germination, seed production, shedding.

INTRODUCTION

Annual ryegrass is the worst cropping weed in Australia in terms of area infested, impact on crop yield and cost of herbicide resistance management (Llewellyn *et al.* 2016). However, changes to the agronomic system and environment have ensured that other winter weed species are increasingly detrimental. For example, brome grass species are now the fourth most problematic weed in Australia (Llewellyn *et al.* 2016). A trial at Wongan Hills, Western Australia (WA) aimed to investigate the biology of eleven weed species, including Afghan melon, annual ryegrass, barley grass, button grass, caltrop, doublegee, great brome grass, roly poly, sowthistle, windmill grass and wireweed. In this paper we compare the emergence, survival, seed production and shedding time of annual ryegrass, great brome grass, barley grass, doublegee and sowthistle growing in a wheat crop.

MATERIALS AND METHODS

An experiment was conducted at the Department of Primary Industries and Regional Development (DPIRD) Wongan Hills Research Station in 2016 and 2017 on yellow sand. In 2016, a site with few weeds was selected and prior to the trial, existing weeds were cleared using non-selective herbicide and cultivation (Table 1). Prior to seeding wheat cv. Mace, plots were sown with one of 11 different weed species; Afghan melon, annual ryegrass, barley grass, great brome grass, button grass, caltrop, doublegee, roly poly, sowthistle, windmill grass and wireweed. Seeds of each weed species were harvested from elsewhere on the Research Station or in the surrounding district in the prior year. The trial also included weed free control plots. The trial was established as a randomised block design with two dimensional balance (i.e. 12 weed species comparisons and 3 replications were blocked in two directions, Coombes 2008).

No in-crop herbicides were applied during the 2016 growing season. Any additional weed species in the trial were removed by hand and other pests were controlled as required. In 2017, the trial was resown to wheat cv. Mace. While a non-selective herbicide was applied, low summer and autumn rainfall at the site ensured that winter weeds did not germinate prior to seeding the crop. Again, non-target weeds were removed by hand and other crop pests were controlled where necessary.

Harvest was conducted at a height of 15 cm, and all chaff was spread back on the plot. Therefore, those weed seeds captured in the grain sample were removed from the plot but any weed seeds in the chaff were retained.

Each year, ten weeds per plot were marked with pegs (after seeding) to monitor growth and seed production of individual plants in the initial weed cohort. For all weed species, later cohorts emerged during the season, but the ten marked plants were the earliest weeds to appear in each plot as these early weeds were the largest/most competitive in crop. The seed heads of the marked weeds were bagged and checked every two weeks to monitor total seed production and time of shedding. Height of each seed head was also measured.

Weed density was assessed at multiple times during the season (time of assessment varied between weed species) from 50 cm by 50 cm quadrats. Weed density and seed production per plant were used to estimate weed seed production m^{-2}. This paper will discuss growth of the winter weeds; great brome grass, barley grass, annual ryegrass, doublegee and sowthistle.

RESULTS

Emergence and fecundity Emergence from the initial 100 seeds m^{-2} in 2016 ranged from 50 plants m^{-2} for sowthistle to 68 m^{-2} for barley grass (Table 2). Seed production was greatest in annual ryegrass, followed by great brome grass. However, plant density in 2017 was greatest for great brome grass, followed by annual ryegrass. These two species had increased seed production in 2017 compared to 2016. Barley grass, sowthistle and doublegee had reduced plant density in 2017 compared to 2016.

In each year, the first cohort of great brome grass and annual ryegrass (i.e. the marked plants in each plot) emerged at the same time as the crop and these marked plants remained at a similar growth stage to the crop throughout the year (data not presented). The marked plants in each plot had high survival rates during the growing season, with 97% survival of great brome grass and 100% survival of annual ryegrass until senescence prior to crop harvest. By comparison, the barley grass, doublegee and sowthistle all had the initial cohort germinate after the crop was established. The initial 10 barley grass plants per plot emerged from June to September and many plants died prior to seed set. All barley grass had naturally senesced by crop harvest, but some of the doublegee and sowthistle plants were still green in November 2016 and 2017.

Weed seed shedding In 2016 and 2017, the earliest possible harvest date occurred in mid-November (14/11/2016 and 9/11/2017, Table 3). In each year annual ryegrass shed 24% and 13% of seed prior to harvest in 2016 and 2017. All annual ryegrass seed heads were above harvest height. Compared to annual ryegrass, great brome grass seed shed rapidly over November. In mid-November (14/11/2016 and 15/11/2017), great brome grass had shed 64% of seed in 2016 and 70% of seed in 2017. Harvest in 2016 did not occur until December and 100% of seed was shed prior to harvest. In 2017 the crop was ready for harvest in November (9/11/2017) when great brome grass had only shed 28% of seed. All great brome grass seed heads formed above harvest height.

Barley grass started shedding seed in October, but retained 62% of seed at crop maturity in 2016 and 96% in 2017 (Table 3). However, in this particular trial, the barley grass was poorly competitive due to late emergence and the plants did not grow taller than 10–12 cm. Shedding was delayed because the seed heads on the short plants were protected from wind by

Table 1. Agronomic details for each operation involved in the experiment in 2016 and 2017.

Date	Operation
8/2/2016	Non-selective herbicide: glyphosate 810 g ai ha^{-1} (Roundup PowerMAX®) + 2,4-D ester 330 g ae ha^{-1} + triclopyr 36 g ai ha^{-1} (Garlon®).
7/4/2016	Cultivation: to 5 cm (autumn tickle to stimulate germination and emergence).
14/4/2016	Non-selective herbicide: glyphosate 1080 g ai ha^{-1} + carfentrazone-ethyl 6 g ai ha^{-1} (Hammer®).
9/5/2016	Cultivation: to 5 cm to kill remaining weeds.
26/5/2016	Sow weeds: manually spread 100 seeds m^{-2} of Afghan melon, annual ryegrass, barley grass, great brome grass, button grass, caltrop, doublegee, roly poly, sowthistle, windmill grass or wireweed over each plot. A weed free control treatment was also included.
26/5/2016	Seeding: Wheat cv. Mace (knife points with press wheels), at 80 kg ha^{-1}, 22 cm crop row spacing, 3–4 cm depth. MacroPro Plus® at 80 kg ha^{-1}, approximately 3 cm below the seed.
16/8/2016	Pest control: prothioconazole/tebuconazole 31.5/31.5 g ai ha^{-1} (Prosaro®) + alpha-cypermethrin 25 g ai ha^{-1} (Alpha-Scud®) + 1% wetter
1/12/2016	Harvest: at a height of 15 cm.
25/5/2017	Non-selective herbicide: paraquat/diquat 337.5/287.5 g ai ha^{-1} (Spray.Seed®).
25/5/2017	Seeding: The crop was sown as for 2016. No additional weed seeds were sown.
17/7/2017	Fertiliser: 50 L ha^{-1} Flexi-N®
12/9/2017	Pest control: dimethoate 160 g ai ha^{-1} (Dimethoate®)
23/11/2017	Harvest: at a height of 15 cm.

the crop. All seed heads were below the harvest height of 15 cm. As the barley grass plants had senesced, crop harvest caused sufficient disturbance to shed the remaining seed heads, and seeds were not captured by the harvester.

Sowthistle and doublegee plants started shedding seed in October, in both years. Sowthistle shed 94–95% of seed and doublegee shed 100% of seed prior to harvest in 2016 and 2017.

Note that the heavy spring rainfall in 2017 allowed many doublegee to remain green and continue growing throughout harvest and set seed during summer (Bureau of Meteorology 2017). Table 3 only includes doublegee seed from those plants that had senesced at harvest. Most sowthistle plants were below harvest height, with less than 1% of the marked plants taller than 15 cm in 2016 and 2017. All doublegee plants were below harvest height.

Crop yield In 2016, weed density was low and none of the weed species had a significant impact on crop yield (Table 2, Table 4). In 2017, the great brome grass plots had lower yield than the plots with other weed species ($P<0.001$, LSD: 407.6).

Table 4. Average yield of wheat in each year.

Species	2016 yield (kg ha^{-1})	2017 yield (kg ha^{-1})
Annual ryegrass	3961	1867
Great brome grass	3933	1251
Barley grass	4690	2062
Doublegee	5121	2230
Sowthistle	3856	2297
Weed free control	4008	1934

Table 2. Average density and seed production (m^{-2}) of each weed species in the 2016 and 2017 growing season. *Note that 2017 seed production is not available for doublegee as the plants remained green over summer and so assessment of seed production was delayed.

	2016		2017	
Species	Plant density	Seed production	Plant density	Seed production
Annual ryegrass	65	9366	203	29496
Great brome grass	60	3484	349	29129
Barley grass	68	889	42	884
Doublegee	55	336	7	*
Sowthistle	50	3207	8	2009

Table 3. The cumulative percent of seed shed by each weed species from October in each year. In 2016, the crop was harvested on 1/12/2016 but was ready for harvest on 14/11/2016. In 2017, the crop was harvested on 23/11/2017 but was ready for harvest on 9/11/2017. All species had 100% shedding at harvest as the harvester either caught seeds or caused sufficient disturbance that the seeds fell to the ground.

Species	7/10/2016	24/10/2016	7/11/2016	14/11/2016	25/11/2016	1/12/2016
Annual ryegrass	0	1	3	9	24	100
Great brome grass	0	0	1	64	100	
Barley grass	0	8	13	38	100	
Doublegee	0	63	82	82	100	
Sowthistle	0	27	32	87	95	100
Species	**9/10/2017**	**24/10/2017**	**9/11/2017**	**15/11/2017**	**23/11/2017**	
Annual ryegrass	0	1	3	13	100	
Great brome grass	0	6	28	70	100	
Barley grass	0	4	4	8	100	
Doublegee	25	25	100			
Sowthistle	0	0	72	94	100	

DISCUSSION

Great brome grass did not produce as many seeds as annual ryegrass in 2016, but still had higher seedling recruitment in 2017. Annual ryegrass seeds are small, with a very thin seed coat. Spafford Jacob *et al.* (2006) found that seed predation (by ants, mice etc.) of small seeds such as annual ryegrass can be close to 100% over summer in WA, whereas larger seeds like wild oats are a less desirable food source. Seed predation is the most likely reason for the low recruitment of small seeded species like annual ryegrass compared to large seeded species like great brome grass in the 2017 season. The great brome grass used in the current trial germinated at the same time as the crop, and was highly competitive with wheat, as indicated by the reduced crop yield in 2017. Therefore, it is very important to control great brome grass through an integrated weed management program (Storrie 2014). Harvest weed seed destruction may be an effective control measure for great brome grass. Walsh and Powles (2014) noted great brome grass seed retention of 77% at crop maturity, and Hashem *et al.* (2017) found 74% of great brome grass seed could be captured if harvest height was 10 cm. This is similar to the 72% seed retention at crop maturity in 2017 and much higher than the 36% retention in 2016 in the current study. It is clear that great brome grass seed retention at harvest varies between years due to seasonal conditions and the time at which the crop reaches maturity. Lodging of brome grass seed heads may also occur in some locations or seasons, which would influence the success of harvest weed seed destruction. For successful harvest weed seed destruction, harvest should be completed as soon as possible in those fields where great brome grass is a problem. Alternatively, swathing may be used to reduce weed seed shedding. Further research is required to determine the success of swathing or harvest weed seed destruction for great brome grass.

The barley grass used in this trial had delayed emergence, which has also been observed in barley grass in South Australia (Gill and Fleet 2012). Late emergence allows the seedlings to avoid non-selective herbicides, but would reduce their ability to compete with the crop (Gill and Fleet 2012). Barley grass retained seed more effectively than great brome grass, but plants in the current study were very short and so were protected from the wind. This would reduce shedding. Other populations may have earlier shedding and further research is required to determine if harvest weed seed destruction is effective for barley grass.

Doublegee and sowthistle also germinated well after the crop and were not highly competitive in this trial. These species are unlikely to be highly competitive in wheat crops, although further research is required in other locations and crop species. Prior research has indicated that doublegee can grow up into the crop canopy, particularly in short pulse crops, and can contaminate yield in some locations (Peltzer and Douglas 2017). Sowthistle tends to be more common in the southern wheatbelt and may be more competitive in this region than in Wongan Hills (Borger *et al.* 2018).

ACKNOWLEDGMENTS

The research was part of GRDC project 'Seed bank biology of emerging weeds' (UA00156). The trial was conducted by DPIRD staff Shari Dougall and Bruce Thorpe at the Wongan Hills Research Station, Nerys Wilkins and Dave Nicholson.

REFERENCES

Borger, C., Hashem, A., Amjad, M., Wild, C., Nicholson, D., Douglas, A., Peltzer, S., Swift, B., Rossi, A., Delroy, J., Cockburn, E., Butler, A., Clarke, E., Chambers, K. and D'Antuono, M. (2018). Summer weeds within the Western Australian wheatbelt – a GRDC survey. Proceedings of the Grains Research Updates. (Grains Industry Association of Western Australia, Perth, Western Australia).

Bureau of Meteorology (2017). Climate data online. http://www.bom.gov.au/climate/data/ (accessed 4 January 2017).

Coombes, N. (2008). DiGGer (licensed freeware). http://www/austatgen.org/files/software/downloads/ (accessed 15 January 2017).

Gill, G.S. and Fleet, B. (2012). Seed dormancy and seedling recruitment in smooth barley (*Hordeum murinum* ssp. *glaucum*) populations in southern Australia. *Weed Science* 60, 394-400.

Hashem, A., Amjad, M. and Borger, C. (2017). Effect of harvest height and windrow burning on brome grass seed bank. Proceedings of the Grains Research Updates. (Grain Industry Association of Western Australia, Perth, Western Australia).

Llewellyn, R., Ronning, D., Clarke, M., Mayfield, A., Walker, S. and Ouzman, J. (2016). Impact of weeds on Australian grain production: the cost of weeds to Australian grain growers and the adoption of weed management and tillage practices. https://grdc.com.au/Resources/Publications/2016/03/Impact-of-weeds-on-Australian-grain-production (accessed 20 March 2018).

Peltzer, S. and Douglas, A. (2017). Doublegee. https://www.agric.wa.gov.au/grains-research-development/doublegee (accessed 12 January 2017).

Spafford Jacob, H., Minkey, D.M., Gallagher, R.S. and Borger, C.P. (2006). Variation in postdispersal weed seed predation in a crop field. *Weed Science* 54, 148-55.

Storrie, A.E. (2014). 'Integrated weed management in Australian cropping systems' (Grains Research and Development Corporation, Australia).

Walsh, M.J. and Powles, S.B. (2014). High seed retention at maturity of annual weeds infesting crop fields highlights the potential for harvest weed seed control. *Weed Technology* 28, 486-93.

Impact of chaff lining/tramlining on the seed persistence and emergence of weeds

Annie van der Meulen[1], Annie Rayner[2], Michael Widderick[1], Adam Jalaludin[1], John Broster[3], Alison Chambers[3], Linda Heuke[2], Shona Robilliard[2] and Michael Walsh[2]
[1] Department of Agriculture and Fisheries, PO Box 2282, Toowoomba, Queensland 4350, Australia
[2] University of Sydney, 12656 Newell Highway, Narrabri, New South Wales 2390, Australia
[3] Charles Sturt University, Locked Bag 588, Wagga Wagga, New South Wales 2678, Australia
(annemieke.vandermeulen@daf.qld.gov.au)

Summary Chaff lining and chaff tramlining are forms of Harvest Weed Seed Control (HWSC) that, owing to their low cost and ease-of-implementation, have potential for wide-spread adoption. The chaff environment is likely to be suboptimal (compacted soil and chaff covering) for seed persistence and seedling establishment, therefore, this practice has the potential to be as effective as other forms of HWSC in driving down weed seed banks. To examine the efficacy of chaff lining and chaff tramlining treatments, a series of field experiments were set up at three locations (Wagga Wagga, Narrabri and Irongate) during the 2017 harvest. The aim of these studies was to investigate the survival of 100 seed lots of annual ryegrass, brome grass, wild oats, wild radish, turnip weed, and common sowthistle placed under chaff lines and tramlines of wheat, barley, chickpea, canola and lupin. Using chaff collected from two of the field sites, pot trials will be conducted to examine the emergence of annual ryegrass under different chaff types (wheat, barley, canola and lupin) and increasing amounts of chaff (equivalent to 0, 3, 6, 12, 18, 24, 30 and 42 t ha^{-1}). In the face of worsening herbicide resistance, chaff tramlining, as a form of HWSC, is likely to be an effective tactic for improved control of key weeds of Australian cropping systems.

Keywords Chaff lining, chaff tramlining, harvest weed seed control, annual ryegrass, brome grass, wild oats, wild radish, turnip weed, common sowthistle.

Newly-developed RIM models for integrated management of brome and barley grass

Marta Monjardino and Rick Llewellyn
CSIRO Agriculture and Food, Locked Bag 2, Glen Osmond SA 5064, Australia
(marta.monjardino@csiro.au)

Summary Managing the weeds brome grass (*Bromus rigidus* Roth) and barley grass (*Hordeum* spp.) in the cropping systems of southern Australia is becoming increasingly problematic due to evolving weed populations including the emerging risk of herbicide resistance. We use the new Brome RIM and Barley Grass RIM bio-economic models to demonstrate how an integrated weed management approach can be evaluated. As an example, we show the effect of increased crop competition on reducing reliance on selective herbicides, lowering weed seedbanks and increasing profitability over time.

Keywords Brome RIM, Barley Grass RIM, bio-economic modelling, IWM, crop competition.

INTRODUCTION

Brome grass (*Bromus rigidus* Roth) and barley grass (*Hordeum* spp.) are important weeds in the no-till farming systems of southern Australia (Llewellyn *et al.* 2015). Managing these grass weeds is an increasingly challenging task requiring integration of several practices, often in low-rainfall, low-cost cropping environments, with relatively limited herbicide options within cereal crops, and emerging herbicide-resistance risk (Kleeman and Gill 2009, Boutsalis *et al.* 2014, Owen *et al.* 2015).

The evolution of herbicide resistant populations of annual ryegrass (*Lolium rigidum* Gaudin) led Australian weed researchers to promote an integrated weed management (IWM) approach that combines chemical, physical and biological practices with the aim to kill existing weeds, prevent seed set and deplete the seed bank (e.g. Powles *et al.* 1997). Brome and barley grass populations with delayed germination are now demanding a more strategic IWM approach be applied (Kleeman and Gill 2009).

Decision support tools can help analyse the long-term biological and economic effects of combinations of alternative strategies over a cropping sequence. The aim of this study is to provide an overview of the newly developed Brome RIM and Barley Grass RIM tools, and present a simple example of model results demonstrating the value of increasing crop competition for more sustainable management of brome and barley grasses in a low rainfall region. We discuss the long-term benefit of technologies that could improve crop establishment on sandy soils, and hence increase crop competition against weeds that are rapidly becoming resistant to selective herbicides.

MATERIALS AND METHODS

The RIM models The new bio-economic models were adapted from the original RIM (Ryegrass Integrated Management) model that was developed in the early 2000s (Pannell *et al.* 2004) and later improved (Lacoste and Powles 2014, 2015) as a decision support tool for testing the biological and economic performance of IWM practices to control herbicide-resistant annual ryegrass in the grain belt of Western Australia. Since then, RIM has been adapted to several weed species and cropping systems worldwide (Monjardino *et al.* 2003, Torra *et al.* 2010, Beltran *et al.* 2012, Bagavathiannan *et al.* 2015, Spammer 2018).

The Brome RIM and Barley Grass RIM models represent the population dynamics of brome grass and barley grass over a period of up to 10 years. There a range of enterprise choices, including cereals, oilseed and grain legume crops, as well as pastures/fodder for livestock production. Weed control includes sowing-related options, selective and non-selective herbicides, grazing, spring and harvest options, as well as user-defined practices.

The model includes a detailed representation of the biology of weeds, crops and pasture, as well as the financial costs and returns of agricultural production and management in the context of the typical dryland mixed system of southeastern Australia. The user specifies the crop/pasture and management sequences for up to 10 years and the model calculates the consequences with respect to crop yields, weed population dynamics and profitability. Crop yield may be improved by good agronomy such as suitable crop rotations and effective weed control, but significantly reduced by weed competition, delayed sowing, disease build up or through phytotoxic damage by herbicides applied in-crop.

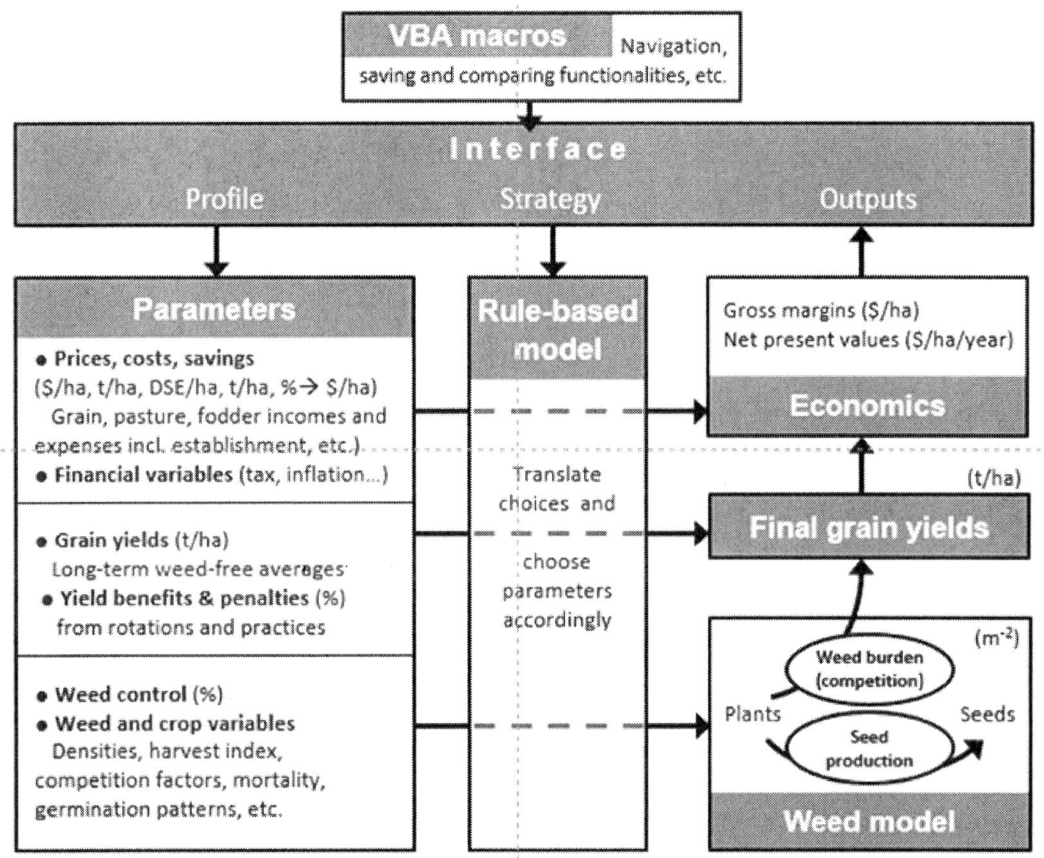

Figure 1. Key relationships between the main components of the RIM model (Lacoste and Powles 2015).

Weed management scenarios The Brome RIM and Barley Grass RIM models were used to assess the value of improved crop establishment and competition as means of non-chemical control of these weed species.

Two scenarios were simulated with each RIM model for a default common crop sequence of Wheat–Barley–Wheat–Grain Legume in the South Australian-Victorian Mallee cropping region (Moodie 2017). A key assumption was a starting low weed density of one (1) brome and barley grass plant m^{-2} in the spring prior to the first year of the analysis. The two scenarios compared were:

1. ***standard crop competition*** **(baseline)**
   Sowing: delayed in cereals; dry in legumes; standard seeding rate; no-till system.
   Knockdown: glyphosate in all cereal phases, except double-knock in first wheat.
   Pre-emergent: trifluralin in cereals, trifluralin + metribuzin in barley, simazine in legumes.
   Post-emergent: clethodim in legume crops;
   Spring: crop–topping in legume crops.
2. ***high cereal crop competition***
   Baseline + high competition of cereal crops.

Parameterisation The underlying parameters of Brome RIM and Barley Grass RIM can be adjusted by the user to reflect local conditions. Key biological factors that drive the pattern of weed population change over time include weed and seedbank population dynamics (i.e. germination rates during the growing season, and natural mortality rates of seedlings, dormant seeds during season and of seeds over summer), as well as weed–crop competition effects for each crop type.

In the analysis we used model default values, including some key ones that describe the main setting (Table 1). While not shown here, brome grass is a more aggressive germinator than barley grass (Kleeman and Gill, 2009), although the latter has a shorter growing season which allows it to set seed even in the driest of seasons (e.g. Mudge 2016).

Brome grass is more competitive than barley grass against all crops, especially cereals. High crop competition is assumed in both Brome RIM and Barley Grass RIM to be equivalent to the High Seeding Rate option at the lower cost of the Standard Seeding Rate option available in the models (Table 1).

The herbicide efficacy values used in this analysis are based on a range of regional trials (e.g. Kleeman and Gill 2008, Boutsalis *et al.* 2014, Mudge 2016, Fleet *et al.* 2017).

RESULTS

For both weeds, the high crop competition scenario returned a higher long-term average gross margin (annuity) than that simulated for the standard baseline scenario, resulting in an average annual net benefit of increasing crop competition of \$24 ha^{-1} and \$16 ha^{-1} to control brome grass and barley grass, respectively (Table 2).

An average net profit gain of 12% and 9% for more effective crop competition against brome (Figure 2a) and barley grass (Figure 2b), respectively, resulted from better crop yields due to a significant reduction in the number of mature weed plants setting seed (less 94% and 187% for brome and barley grass, respectively), as well as less weed seeds in the soil (less 114% and 239% for brome and barley grass, respectively) (Table 2, Figures 3a and 3b), without the extra cost of weed control.

DISCUSSION

The new Brome RIM and Barley Grass RIM models were used to illustrate the potential for improving crop establishment and thus competition against weeds that are becoming increasingly difficult to manage in the southeastern dryland cropping regions of Australia.

The results indicate the potential for greater crop competition to reduce weed seedbanks and improve profit, at least for the default rotation and management strategy used in the models. This is particularly relevant in the case of brome grass which is often found on sandy soils where achieving strong crop establishment can be difficult (McBeath *et al.* 2017).

Table 1. Key enterprise-related model parameters used in the analysis.

Parameter	Wheat	Barley	Legume
Brome grass competition factor	0.80	0.73	0.58
Barley grass competition factor	0.50	0.45	0.36
Std seed rate (kg ha^{-1})	60	60	90
High seed rate (kg ha^{-1})	90	90	120
Std seed costs (\$ ha^{-1})	21	21	34
Weed-free yield (t ha^{-1})	1.5	1.7	1.0
Farm grain price (\$ t^{-1})	240	198	365
% of Brome / Barley grass plants controlled			
Trifluralin	30/60	30/60	
Trifl + Metribuzin	45/65	45/65	
Simazine			60
Clethodim			98
Spray-topping			60

Table 2. Average 10-year annual gross margins (annuity), mature plants setting seed and seeds in soil in autumn results (year 10) generated for the standard and high crop competition scenarios for both brome grass and barley grass.

	Brome grass		Barley grass	
Model output	Standard	High	Standard	High
Annuity (\$ ha^{-1} $year^{-1}$)	177	201	173	189
Mature plants m^{-2}	93	33	78	14
Seeds in soil m^{-2}	870	292	1,109	178

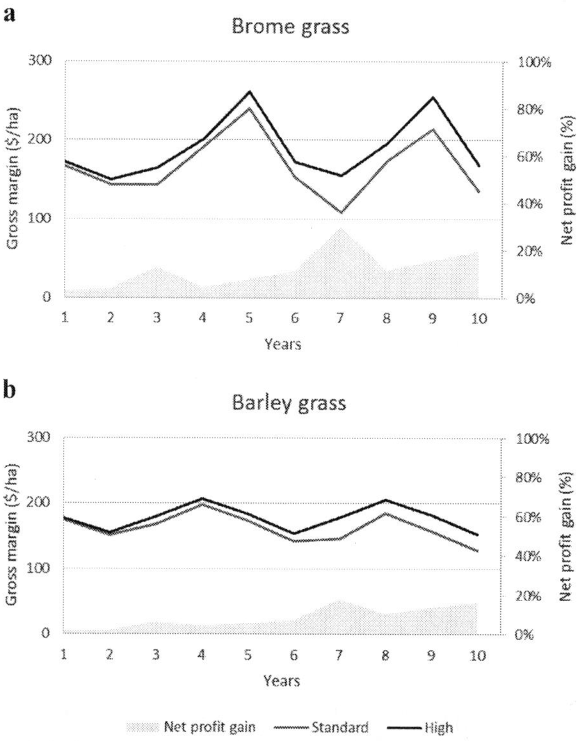

Figure 2. Annual gross margin over a 10-year WBWL crop sequence simulated with a) Brome RIM and b) Barley Grass RIM for the standard and high crop competition scenarios.

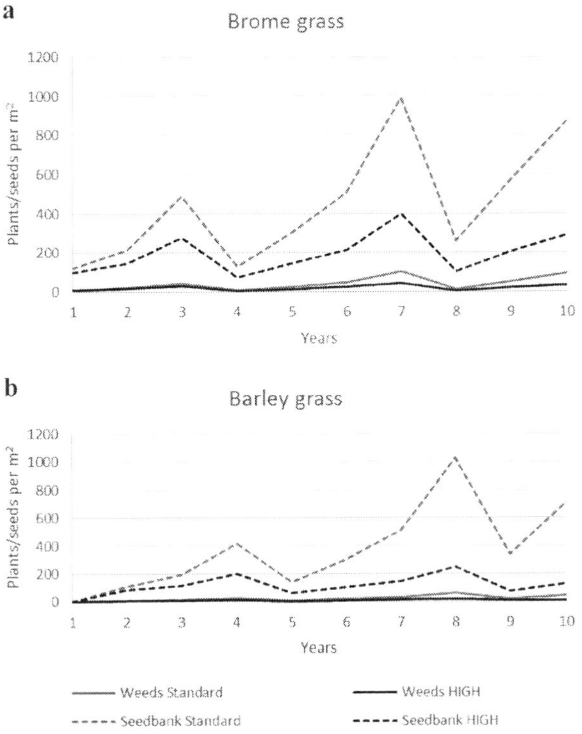

Figure 3. Weed and seedbank dynamics over a 10-year WBWL crop sequence simulated with a) Brome RIM and b) Barley Grass RIM for the standard and high crop competition scenarios.

This brief analysis highlights the long-term biological and economic benefits of potential innovations that could improve crop competition, demonstrating potential for analyses around specific technologies, such as high vigour cultivars (e.g. Mudge 2016), soil wetter agents and on-row sowing in non-wetting soils (McBeath *et al.* 2017) that may increase the relative advantage of the crop against weeds.

Overall, the new RIM models provide a practical platform to effectively test a range of weed management options for weeds that are increasingly demanding a strategic integrated management approach.

ACKNOWLEDGMENTS

This work was funded by GRDC and CSIRO with generous collaboration from the Australian Herbicide Resistance Initiative (AHRI) at the University of Western Australia. We thank G. Gill and S. Kleemann from the University of Adelaide and M. Moodie from Mallee Sustainable Farming for their expert input into the Brome and Barley Grass RIM models. We acknowledge the original RIM team, and M. Lacoste from AHRI who developed the 2013 RIM version with support from GRDC.

REFERENCES

Bagavathiannan, M., Lacoste, M., Powles, S., Steckel, L., Popp, M. and Norsworthy, J. (2015). Palmer amaranth integrated management model (PAM): demonstration of the framework and seeking stakeholder inputs. Proceedings of the Beltwide Cotton Meetings, San Antonio, United States of America.

Beltran, J., Pannell, D., Doole, G. and White, B. (2012). A bioeconomic model for analysis of integrated weed management strategies for annual barnyard grass (*Echinochloa crus-galli* complex) in Philippine rice farming systems. *Agricultural Systems* 112, 1-10.

Boutsalis, P., Kleemann, S., Gill, G. and Preston, P. (2014). A hidden threat: widespread Group B herbicide resistance in brome across south-eastern Australia. Proceedings of the 19th Australasian Weeds Conference, ed. M. Baker, pp. 202-5. (Tasmanian Weeds Society, Hobart).

Fleet, B., Cook, A., Richter, I., Shepperd, W., Preston, C. and Gill, G. (2017). Management of group A herbicide resistant barley grass in pasture phase. GRDC Eyre Peninsula Farming Systems 2016. https://eparf.com.au/wp-content/uploads/2017/04/39.-Managment-of-group-A-herbicide-resistant-barley-grass-in-pasture-phase.pdf (accessed 11 April 2018).

Kleeman, S. and Gill, G. (2008). Applications of metribuzin for the control of rigid brome (*Bromus rigidus*) in no-till crops of Southern Australia. *Weed Technology* 22, 34-7.

Kleeman, S. and Gill, G. (2009). Population ecology and management of rigid brome (*Bromus rigidus*) in Australian cropping systems. *Weed Science* 57, 202-7.

Lacoste, M. and Powles, S.B. (2014). Upgrading the RIM model for improved support of integrated weed management extension efforts in cropping systems. *Weed Technology* 28, 703-20.

Lacoste, M. and Powles, S.B. (2015). RIM: anatomy of a weed management decision support system for adaptation and wider application. *Weed Science* 63, 676-89.

Llewellyn, R.S., Ronning, D., Ouzman, J., Walker, S.R., Mayfield, A. and Clarke, M. (2015). Impact of weeds on Australian grain production and adoption of tillage practices. (GRDC and CSIRO, Australia). 160 pp.

McBeath, T., Llewellyn, R., Gupta, V., Davoren, B., Shoobridge, W. and Kroker, S. (2017). On-row sowing for brome competition on non-wetting sands. Mallee Sustainable Farming. http://www.

msfp.org.au/row-sowing-brome-competition-non-wetting-sands (accessed 11 April 2018).

Monjardino, M., Pannell, D.J. and Powles, S.B. (2003). A multi-species bio-economic model for integrated management of *Lolium rigidum* and *Raphanus raphanistrum* in Australian farming systems. *Weed Science* 51, 798-809.

Moodie, M. (2017). Crop sequencing considerations for the Mallee. Mallee Sustainable Farming (MSF).

Mudge, B. (2016). Over dependence on agrichemicals – Barley grass trial. Unpublished report. Barry Mudge Consulting for Upper North Farming Systems.

Owen, M.J., Martinez, N.J. and Powles, S.B. (2015). Herbicide resistance in *Bromus* and *Hordeum* spp. in the western Australian grain belt. *Crop and Pasture Science* 66, 466-73.

Pannell, D.J., Stewart, V., Bennett, A., Monjardino, M., Schmidt, C. and Powles, S.B. (2004). RIM: a bio-economic model for integrated weed management. *Agricultural Systems* 79, 305-25.

Powles, S.B., Preston, C., Bryan, I.B. and Jutsum, A.R. (1997). Herbicide resistance: impact and management. *Advances in Agronomy* 58, 57-98.

Spammer, Z. (2018). The financial and managerial implications of herbicide resistance of annual ryegrass in the Central Swartland. Master of Commerce, Stellenbosch University, South Africa.

Torra, J., Cirujed, A.A., Recasens, J., Taberner, A. and Powles, S.B. (2010). PIM (Poppy Integrated Management): A bio-economic decision support model for the management of *Papaver rhoeas* in rain-fed cropping systems. *Weed Research* 50, 127-39.

Technical manuals and training on the soil behaviour of pre-emergent herbicides and modes of action (MOA's) of post-emergent herbicides

Mark Congreve and John Cameron
Independent Consultants Australia Network, PO Box 718, Hornsby, New South Wales 1630, Australia
(john@icanrural.com.au)

Summary With the rapid and widespread increase in herbicide resistance in Australian cropping landscapes, it is increasingly important to include greater diversity in weed management programs. Inclusion of non-herbicide weed management tactics such as harvest weed seed control, crop competition, or other non-herbicide techniques, has become a mandatory requirement for sustainable farming systems. Similarly, herbicides when used must be effective and perform to their optimum level. In environments where post-emergent herbicides have developed resistance at levels rendering them no longer effective, the inclusion of pre-emergent herbicides has become common place. Technical manuals and training workshops have been developed, detailing the science that underpins how herbicides work. Greater knowledge of how herbicides work will enable grain advisors to better optimise their advice to growers on herbicide use. Manuals and training cover the following key topics: the soil behaviour of pre-emergent herbicides – volatility, photodegradation, influence of organic matter, soil binding coefficients and cation exchange capacity of the soil; water solubility, breakdown pathways and DT50 values; how these product features coalesce to inform how different herbicides behave in the soil; and modes of action of post-emergent herbicides – how different modes of action work, herbicide entry through the leaf, translocation and metabolism and; the complexities of herbicide resistance and the implications these factors have in optimising performance of particular modes of action.

Keywords Binding coefficient, water solubility, DT50, breakdown pathway, translocation, leaf entry, herbicide metabolism, herbicide resistance.

INTRODUCTION

This paper reviews an extension program targeted to tertiary trained grains advisors. The program was developed to enhance industry knowledge of herbicides and their key performance criteria. This Grains Research and Development Corporation (GRDC) initiative was developed and run by Independent Consultants Australia Network (ICAN), with training activities conducted across the northern grains region of Australia between 2014 and 2018.

The genesis of this project was a desire to arm decision-makers with a better technical understanding of the key factors influencing herbicide performance.

With herbicide resistance now very common in Australian farming systems, it is increasingly important to ensure effective herbicide application. Several of the weaker resistance mechanisms are dose responsive within the constraints of registered application rates. With the objective of achieving maximum levels of control to drive the weed seedbank down over time, this translates to farming systems with a low tolerance of weed survivors to re-supply the weed seedbank for following seasons. In systems with low tolerance of weed survivors, there is increased pressure to ensure that there is a diversity of weed management tools used and that each and every time herbicides are used, they are targeted to optimise performance and minimise risk.

Understanding how herbicides work within the plant and behave in the environment will assist advisors to correctly target their use in accordance with label directions.

MATERIALS AND METHODS

What was done? To improve industry knowledge and understanding of herbicide modes of action and their key performance criteria, it was determined that technical resources needed to be developed. These resources give users a practical, science-based understanding of the key factors influencing field performance of herbicides. These resources consolidated information from a wide range of global sources into a single resource designed for agronomists and weed specialists in the Australian grains industry.

While herbicide technical information is available from a range of global sources, the location and format of much of this material can make it difficult for agronomists to find and interpret this information for provision of advice to growers. The manuals and associated videos developed in this project curate this body of knowledge into language and formats relevant to the needs of Australian field facing agronomists and weeds advisers.

In addition to the manuals, a series of face to face small group training workshops were conducted. These workshops targeted grains agronomists and key influencers. The factors dictating herbicide performance were explained and then contextualised to show how differences in chemistry can impact field performance. Case examples relevant to local experiences were used to bring to life the background science.

Who did we target? In broadacre grains production, the majority of growers utilise the advice of an agronomist for their crop protection advice. By engaging an agronomist, growers benefit from specialised expertise and experience gained from the provision of similar advice to other growers in the region. It is increasingly common for grain growers to rely heavily on the advice of their agronomist. This is particularly evident in areas related to crop protection and weed management where mistakes can be costly and specialist knowledge reinforced through regular use is often required. This is driving an increasing number of growers to delegate many of the crop protection decisions to their agronomist.

For this reason, the primary target audience for the resources developed and associated training, was grains agronomists and key influencers. While most of the promotion for training workshops was directed to this group, approximately 10% of participants were growers.

Another reason for targeting agronomists and key influencers was that each individual typically influences 10–100 growers. There are significant multiplying factors at play when technical training is delivered to this group.

RESULTS

What did we achieve? There were two primary deliverables in this extension project:

1. development of technical resources; and
2. workshop training of agronomists.

Two technical manuals were produced. The first manual was titled 'The soil behaviour of pre-emergent herbicides' which focuses on their interaction with the soil and the environment. This manual covers herbicide volatility, photo degradation, herbicide binding to stubble, solubility and binding properties influences on mobility of the herbicide and herbicide degradation. These influence the length of residual control and the crop rotation constraints. The manual explains the understanding of how these factors influence herbicide availability and then seeks to collate technical information by individual herbicide to explain how that herbicide is likely to perform under specific soil types and environmental conditions.

This manual was first produced in 2014, with a second edition published in 2018 and is available for download from the GRDC at: www.grdc.com.au/SoilBehaviourPreEmergentHerbicides. Two associated videos were also produced in collaboration with Anvil Media. These videos can be found at: https://www.youtube.com/watch?v=s63GYYyflzw&t=2s https://www.youtube.com/watch?v=LJNjuMWS57U&t=0s

The second technical manual covers post-emergent herbicides, including their place in the management of grass and broadleaf weeds, their mode of action (how, where, when and why they kill weeds), plant uptake and translocation to the site of activity and the role of metabolism in herbicide breakdown and crop selectivity. This understanding then leads into detailed explanation of herbicide resistance and the mechanisms involved, and the key factors in maximising the performance of different modes of action. (At the time of writing this paper a specific URL for this manual was not yet available. The manual will be available at www.grdc.com.au).

Associated training workshops have been conducted across Queensland and New South Wales grain growing regions. Between 2014 and 2018, 30 half-day pre-emergent herbicide workshops were completed. In addition, 27 half-day post-emergent workshops have now been completed during 2017 and 2018. Exit survey ratings of the workshops demonstrated most participants found the workshops to be of very high value (Table 1).

While the fundamental knowledge of how herbicides work is generic to the herbicide and/or mode of action, herbicide examples used in each workshop are typically those most relevant to the geographic location.

Generally, the intent was to keep an individual workshop series to a particular time of year and grouped to locations running a similar farming system. This allowed workshop materials and herbicide examples selected to be of most immediate value to participants (e.g. a workshop series targeting summer fallow use of residuals in the northern region would focus on different herbicide examples to a series of workshops run in March in southern New South Wales where examples might be focused on ryegrass management).

Participant feedback, via exit surveys and questions raised during the workshops provided insights into the key learnings and take-away messages obtained by participants and their unmet needs.

DISCUSSION

What did we learn? Having completed 57 workshops (with more than 1000 participants) across Queensland and New South Wales, several observations can be made. The workshops attracted all levels

Table 1. Workshop exit survey ratings 2014 to November 2017 (a more detailed exit survey separately evaluating the individual workshop components was used for the post-emergent workshops).

		1 Waste of my time	2 Not very helpful	3 Workshop needs improvement	4 Some good stuff here	5 Excellent	Weighted average
27 pre-emergent workshops	Overall workshop value	–	–	0.4%	33%	66%	4.7
20 post-em workshops	Leaf entry, translocation and metabolism	–	–	0.7%	29%	71%	4.7
	Understanding herbicide resistance	–	–	1%	28%	71%	4.7
	Maximising the performance of each MOA	–	–	0.7%	30%	69%	4.7

of agronomists – from highly experienced through to new graduates.

Through exit surveys, the key areas of benefit most frequently cited were:

- troubleshooting product complaints;
- understanding herbicide resistance in the context of what is seen in the paddock;
- matching herbicide application with herbicide properties; and,
- in the context of pre-emergent herbicides, matching herbicide selection with soil type, stubble, weed seed location and environment.

There was a significant trend for agronomists to prefer morning workshops over afternoon workshops (where both were offered).

Average workshop attendance was approximately 20 participants. Workshop sizes of 15–20 appears to be the ideal 'sweet spot', having enough diversity of shared experience in the room while not being too large. In almost all workshops, it was noticed that as attendance increased above 20, discussion from the audience decreased proportionally. As a rule of thumb, workshops were closed when registrations approached 25 participants (however, up to 30 individuals participated in some workshops).

The half-day workshop format was appreciated by many, allowing agronomists to get back in the paddock after lunch. This format was well suited to the pre-emergent workshop content. However, the depth of content required to be covered in the post-emergent workshop indicates that future workshops would be better suited to a three-quarter day format.

Some agronomists indicated a preference to participate in both workshops back to back, (e.g. post-emergent herbicides in the morning and pre-emergent herbicides in the afternoon), while another significant cohort of participants appeared to prefer a 'short and sharp' stand-alone half-day workshop format.

A small, but significant, percentage of participants have repeated a workshop, with feedback that their understanding was improved the second time around.

While growers were not the primary target of these workshops, the growers who did attend reported that they found the workshops highly beneficial. Many of these growers indicated that 'all' growers should be exposed to this level of detail of herbicide understanding and should participate in the workshop 'in full'. However, it should be noted that these participating growers are highly interested, and therefore invested, in this topic (which can be observed by them actively seeking out participation in these workshops), and are likely to place a higher value on the content of these workshops.

In the first few years of delivery of the pre-emergent workshop series, workshop participants were asked an exit survey question "Does this information need to be delivered to growers? If so, in what format?" The grouped responses can be seen below (Table 2).

These data suggest that there is an opportunity to further enhance grower knowledge (at least for pre-emergent herbicides where this data was collected).

Areas for further research, development and extension Through questions asked during the workshops and exit surveys, we have been able to compile a list of knowledge gaps that would benefit from either further research or extension. These are as follows:

Table 2. Should the information contained in these technical workshops be delivered to growers? If so, in what format?

Delivery to growers	No. of mentions
YES = 157 (55%)	
Absolutely. Growers need to know this	23
Yes. Same workshop format (at least for interested growers)	42
Yes. But interactive format / case studies / not as technical	52
Yes. Via small group grower meetings (focusing on application in particular)	19
Yes. But delivered by their agronomist	7
Yes. Via demo plots and spray application day	11
Yes. Needs to be mandatory for younger / newer agros	3
MAYBE = 61 (21%)	
Probably. But general guidelines only. Product names only	14
Possibly. Via development of an App / Decision support tool where they enter their paddock information and it tells them what is important. Or video.	4
Possibly at some level by fact sheets / press article / manual	27
Maybe. But needs to be very simple. Updates?	16
NO / NOT SURE = 66 (23%)	
Not sure / maybe. Some info is very technical & important bits might be missed	22
No. Leave this to advisers. Too complex / not interested	44

- there were numerous requests to extend the workshops to other grains regions outside of New South Wales and Queensland (and a few requests to extend the training to non-grains industries, specifically horticulture and sugar cane); and
- that while the current post-emergent workshop focuses on herbicide activity from leaf entry through to activity at the target site, adjuvants and application are only covered in minimal detail. While there are other detailed workshops dedicated to herbicide application, there have been numerous requests to develop a similar technical training workshop for adjuvants.

Specific to pre-emergent herbicides

- Continual on-going extension is required regarding the use of winter cereal grass pre-emergent herbicides, specifically in relation to:
  - different seeder types (i.e. use patterns with various disc seeders);
  - use in stubble retained systems;
  - quantifying interactions with stubble burning;
  - for northern New South Wales and Queensland, understanding the impact of no rainfall at planting (crops and weeds germinating on stored soil water only), or large rainfall events at planting to emergence (autumn storms in the north).
- Additional herbicide registration opportunities including:
  - pre-emergent herbicides in fallow systems; and
  - pre-emergent herbicide(s) for use at planting of mungbeans and adzuki beans to control *Chloris* species.
- Improved plant back guidance for many older herbicides, especially when used in fallow situations, including regionally specific information where applicable e.g. summer *vs* winter dominant rainfall regions.
- Potential for the development of robust, cost effective in-paddock test kits to measure residual herbicide levels, including sampling protocols.
- Enhanced microbial degradation is known to occur with many herbicides when used relatively frequently, however practically no data is available under Australian broadacre farming systems. Enhanced degradation will lead to reduced longevity of weed control and possibly shorter rotational constraints. Microbial breakdown is the primary dissipation pathway for pre-emergent modes of action B, C, D, J and K, so leading candidates from these modes of action should be targets for investigation.

Specific to post-emergent herbicides

- Hydrophilic herbicides (particularly glyphosate) are particularly sensitive to conditions affecting spray drying and resulting leaf cuticle penetration. Different plant species have different leaf cuticle properties, which alters the rate of hydrophilic herbicide entry, before the spray has dried. Practically no data exists to show different leaf uptake speed by species, and what little data does exist is old and generated with historical formulations and adjuvants. This issue is particularly relevant for summer spraying and will be exacerbated with the onset of resistance that will require higher levels of herbicide penetration to achieve control. Studies are required to demonstrate level of leaf penetration x difficult to control species x formulation x climatic conditions x adjuvant.

- The surfactant package (in the formulation, or tank mix) is extremely important in aiding herbicides (in particular lipophilic herbicides) to penetrate the leaf. Better information is required to show which surfactant systems are optimal for individual species, especially when applied under challenging climatic conditions (e.g. summer spraying), or when herbicide resistance compromises performance.
- Frosts often occur at the time of season when it is desired to use Group A herbicides in winter crops. More information is sought on how to maximise performance in these situations i.e. apply during frosts and accept poorer control, or delay application and target larger weeds?
- Herbicide 'safeners' are often included in the formulation of some Group A, B and H herbicides. These safeners increase metabolic activity, increasing safety to the crop. However, these safeners are likely to be having the same effect on the weeds. Questions that arise include:
  - what is the magnitude of increased safety to key weeds?;
  - is the use of safeners leading to increased selection for metabolic resistance in weeds?; and
  - is it possible to develop some form of safener delivery that only treats the crop and not the weed?
- Tank mixing 'dissimilar' herbicides is poorly understood and requires further attention. Most users appear primarily concerned with physical tank mix compatibility, with available resources more focused on tank mixing procedure. In addition to physical compatibility, further information is required:
  - explaining biological compatibility / incompatibility between different herbicides;
  - understanding different application requirements for various herbicides and what compromises are being made e.g. preferred application setup (droplet size, spray volume) is very different for glyphosate and Group A or Group G, so will be a compromise if these herbicides are tank mixed; and
  - understanding the correct choice of adjuvant for tank mixes (and that tank mixing may require a compromise in the choice of adjuvant).
- Better understanding of the implications of tank mixing foliar fertilisers. Most available information focuses on physical spray tank compatibility, with little information available on biological interactions. Some foliar fertiliser formulations are known to impact on herbicides when tank mixed. Effects can be negative (foliar fertiliser cations acting like 'hard' water) or possibly positive (overcoming herbicide induced micro nutrient unavailability).
- Additional herbicide registrations opportunities include:
  - conversion of remaining optical sprayer permit claims to full label claims;
  - registration of isoxaflutole for post-emergent control of *Chloris truncata* R.Br. in fallow (in conjunction with paraquat); and
  - development of Group G 'contact' herbicides as additional second knocks in a double-knock strategy.
- Development of some form of in-field resistance testing to enable results prior to herbicide application.
- Enhanced understanding of the resistance mechanisms in individual populations. Certain resistance mechanisms are known to be dose responsive, so understanding the mechanism, or more importantly the dose responsiveness, is important for development of management strategies.

CONCLUSION

Outstanding feedback in terms of participation attendance at workshops and commentary on the associated technical material that was produced in this extension project indicates that it has met, or possibly exceeded, expectations of the target audience: that is broadacre agronomists across Queensland and New South Wales.

Participants have been exposed to the underlying factors that are important in optimising field performance of their selected herbicide tools, and now have access to technical reference materials that will assist in furthering this understanding, especially as herbicides at our disposal come under increasing challenge with herbicide resistance.

There is further opportunity to enhance knowledge by additional investment to address knowledge gaps identified across the delivery of this training.

While a significant percentage of agronomists across Queensland and New South Wales have now participated in this training, feedback indicates that there is a similar need to extend learning to other grain production areas across Australia. In addition, there continues to be an on-going need to facilitate training to those entering the broadacre grains agronomy (recent graduates or those transferring from other regions or cropping systems).

ACKNOWLEDGMENTS

ICAN wishes to acknowledge the GRDC for providing project scope and material development and extension funding (GRDC Project Code ICN00016), Dr. Christopher Preston for technical review of manuals and Anvil Media for video production.

How distinguishable are hawkweeds from co-occurring vegetation species in the eye of hyperspectral sensor?

Hsing-Chung Chang[1], Kerrie Tomkins[1], Rene Heim[1], Will Farebrother[1], Chad Ajamian[1], Mark Hamilton[2] and Hillary Cherry[2]
[1] Department of Environmental Sciences, Macquarie University, New South Wales 2109, Australia
[2] Pest and Weeds Unit, NSW National Parks and Wildlife Service (NPWS), Hurstville, New South Wales 1481, Australia

Summary Hawkweeds are an eradication target in New South Wales, Australia, and infestations occur in remote and mountainous areas of Kosciuszko National Park. Traditional field-based surveys are critical to detect plants, but are very time consuming and labour intensive. Therefore, there is a strong demand to use remote sensing to detect hawkweeds over remote terrain to assist eradication.

This study collected hyperspectral profiles (ranging from 350 nm to 2500 nm of the spectrum of electromagnetic radiations) of orange and mouse-ear hawkweeds and co-occurring vegetation species in Kosciuszko National Park. Samples were collected using both an optic lens and a leafclip. The optic lens measures the overall spectral information of the plant and includes any disturbance/noise interference. The leafclip, with its own illumination source, measures the reflectance of the leaf directly. The spectral ratios of hawkweeds compared to other co-occurring species were analysed to show the wavelengths at which there were significant differences between the spectral profiles of selected target species. A classification method called 'Random Forest' was used to test the accuracy of identifying hawkweeds and differentiating them from co-occurring species. The ability to detect hawkweed plants using remote sensing will assist program managers to search large expanses of inaccessible land in a cost effective manner, allowing delimitation of the entire infestation, and ultimately, eradication.

Use of GIS technology in weed management

Paras Acharya
Greater Sydney Local Land Services (NSW Department of Industry), GPO Box 4515, Penrith, New South Wales 2750, Australia
(paras.acharya@lls.nsw.gov.au)

Summary Greater Sydney Local Land Services (GS LLS) is a regional organisation that works with local and state government organisations, land managers and the community through its role on the Regional Weed Committee to assist with strategic planning and coordination of weed management in the Greater Sydney region. GS LLS has always been proactive and instrumental in the use and adaptation of new technologies that assist in the weed management in the region. This presentation focuses on the use of spatial technology and applications for the collection of weed surveillance and management data to support weed management at a regional scale. Several applications have been developed to assist Local Control Authorities to meet reporting requirements to the Biosecurity Information System, identify and manage high risk pathways and sites, and to manage a cross jurisdictional weed eradication program.

These applications have been developed by GS LLS in collaboration and liaison with various stakeholders involved with weed management in the Greater Sydney region. The presentation aims to demonstrate the use of GIS technology to develop applications that provide a common platform to collect, store and analyse weed data in a regional context. The adaptation of GIS technology also enables a real time data collection, on the fly reporting and access of information for non GIS users.

Keywords GIS, weed management, collaboration.

Unmanned aerial vehicles used to control giant reed (*Arundo donax* L.)

Philip Milling
6 Rosin Place, Bolwarra Heights, New South Wales 2320, Australia
(phil@skylandmanagement.com.au)

Summary Unmanned Aerial Vehicles (UAVs) continue to grow in popularity to improve efficiency and effectiveness in many industries. Most of the uses to date have involved mapping, land surveying, photography, site monitoring and recording. However, their use for weed control, has been restricted.

Using the UAV as a tool to apply herbicide from above has enabled deep and full application of herbicide. Results to date have shown 80–90% brownout of giant reed after a single treatment using glyphosate. Therefore, understanding this technology and its applications can provide weed managers with efficient and effective options which have not been available in the past. Combining the latest of technologies with tried and true methods can significantly enhance our weed management efforts.

Keywords UAV, glyphosate, spot treatment.

INTRODUCTION

Arundo donax L., commonly known as giant reed (sometimes called elephant grass) is a weed primarily of riparian areas. It is found in every state of Australia and the Northern Territory. In New South Wales (NSW) it is predominantly found along the central and northern coast. In the Hunter valley of NSW, giant reed is prevalent along several major waterways. These include the Pages, Paterson, Allyn, Williams and Hunter Rivers.

Giant reed is a bamboo like plant that can grow up to 10 m tall. It has hollow woody canes that support blue/green leaves that attach to the stem by a sheath wrapping around the stem. It produces a tall plume like flower head, up to 60 cm, which produce seeds. However, in many cases, including in the Hunter Valley, these seeds are unviable. The Hunter population has been studied and found to be sterile clones (Azadeh Haddadchi and Gross, n.d.). Instead, the plant reproduces vegetatively from spreading rhizomes or from fragments of rhizomes that break off and establish at new locations. Fragments of fresh stems can produce roots from the nodes if in contact with soil.

Giant reed forms impenetrable thickets (several hundred stems m^{-2}), creating a monoculture, where no other vegetation can establish or persist. If left unabated, these thickets can spread along entire reaches of a river. Although it is thought that they provide protection for riverbanks, they can also cause large sections to break away because their root systems, although dense, will only grow to a depth of up to one metre. This means the root system can be undercut and easily dislodged. This dislodged giant reed then causes further issues downstream as a flood rack of vegetation, which can cause significant infrastructure and riparian damage. This mobilised giant reed may eventually lodge on a riverbank or bed, where it will set root and establish as a new giant reed colony. Giant reed can block or significantly alter the course of a river.

Giant reed can also establish in areas away from riparian areas and has been seen establishing on soil and gravel stock piles and along areas where roadworks have been undertaken. These most likely have established from plant matter that has been transported in these materials, or possibly on machinery associated with these activities.

Other issues that giant reed presents include its high water use. It can use up to 2000 L m^{-2} (Csurhes 2009), significantly higher than that of native riparian vegetation. Evidence also shows that giant reed has no habitat or feed value to native fauna. In fact, the plant has toxins in it which make it unpalatable to native fauna and is also suspected of releasing toxins into the water, which deters other plants while favouring its own growth (Bell 1997, Dudley 1998, CRC 2005)

The advance of giant reed along our waterways is having a significant detrimental impact on their function. One of the main reasons it has been able to establish so extensively has been due to the time consuming and costly control of traditional methods. Traditional methods of control are varied, with different rates of success. The more proven methods such as cut and paint, although generally effective, are very labour intensive and, hence, costly. Other methods like root removal by mechanical means is costly, and causes a high amount of ground disturbance, which is not desirable, particularly in riparian areas. Foliar spraying is a quicker and more cost-effective method, however the ability to get the coverage required is impeded by the sheer size and density of most thickets.

In all these cases, as with most weed control, follow up is required. This is because giant reed has a high resilience to control due to its extensive root mass. Therefore, in most cases the use of a variety of control methods is required.

MATERIALS AND METHODS

The trials comprise a combination of control methods, based on the stage in the control timeline as well as the location of the giant reed on the riverbank. The giant reed at the first trial site on Paterson River at Tocal in the lower Hunter valley NSW stretched from the river bank up onto the flood plain, with the thickest vegetation being up to 70–80 m long and up to 30–40 m wide. The second site, on the Pages River in the upper Hunter valley, near the township of Murrurundi, was composed of more isolated clumps on the river banks, ranging in diameter from a few meters up to 15–20 m. To overcome the difficulty in accessing the giant reed an Unmanned Aerial Vehicle (UAV) was used to apply herbicide as the primary treatment. The aircraft used was a Yamaha® Rmax rotary wing UAV. The first trial site along the Paterson River, Tocal was broken into six adjacent sections, each of which received different herbicide treatments.

The treatments used were glyphosate 360 g L^{-1} (Roundup® Biactive) and glyphosate 570 g L^{-1} (Roundup® UltraMax), both with and without an acidifier (LI700®). The initial application was done in autumn 2016 on four of the six plots only, using a low rate, then followed up with treatment of all six sites at a higher rate in spring 2016 (Table1).

The next stage was to burn the plots to remove a significant proportion of the biomass. The burn was initially scheduled for autumn 2017 however circumstances delayed this to late winter. The burn was unfortunately hampered on the day by some light rain and no wind. Despite these conditions, those areas that did burn showed a significant reduction in the amount of biomass.

Following the burn, a follow up herbicide application on the regrowth was done using the UAV, in spring 2017. This application was done when the canes had reach approximately one m in height.

Once the giant reed is mostly under control the site will be revegetated through aerial seeding using the UAV. The seed mix will include a sterile cover crop and suitable native riparian vegetation for the site. Note that a buffer strip along the water edge was not treated, to provide some stability to the river bank. Once vegetation is well established behind this narrow strip of Giant reed, it can be manually removed using cut and paint methods.

The second trial site at Pages River, Murrurundi, was first sprayed using the Yamaha® Rmax UAV in Autumn 2017. The results from the first trial site at Tocal indicated that the higher rate (20%) of glyphosate 360 g L^{-1} without acidifier would most likely achieve the best result, thus it was included.

RESULTS

While the trials are currently in progress, the results to date appear to be quite successful. One of the benefits anticipated from using the UAV was its ability to force the herbicide through the full stratum of the giant reed. Water sensitive paper was placed from ground level up to 2400 mm off the ground to establish if this was achieved. All aerial paper had droplets on them, with decreasing coverage closer to the ground with some ground coverage found. This showed that the herbicide was getting right through the plant canopy.

At Paterson River, the autumn treatment at the lower rate resulted in a biomass brownout rate of 80–90% of the thickets. It resulted in no new shoots in the following spring, as was evident in the non-treated adjacent giant reed. New giant reed shoots off existing canes showed signs of herbicide damage, that is they were stunted and had yellowed margins, (Figure 1). This was generally only found less than half way up the canes, with the top half of the canes having no re-growth.

The benefit of the acidifier was indistinguishable when comparing plots that did not have it added, and hence it was decided that this would not be used in the next herbicide treatment. The lower concentration

Table 1. Herbicide treatments trialled for giant reed.

Plot	Herbicides	Autumn 2016 – rate	Spring 2016 – rate
1b	Glyphosate 360g L^{-1} (Roundup® Biactive)	2%	20%
1ba	Glyphosate 360g L^{-1} (Roundup® Biactive)	2% + acidifier (LI 700®)	20%
2u	Glyphosate 570g L^{-1} (Roundup® UltraMax)	1.4%	13%
2ua	Glyphosate 570g L^{-1} (Roundup® UltraMax)	1.4% + acidifier (LI 700)	13%
3b	Glyphosate 360g L^{-1} (Roundup® Biactive)	**No treatment**	20%
3u	Glyphosate 570g L^{-1} (Roundup® UltraMax)	**No treatment**	13%

of glyphosate 360 g L^{-1} was slower acting than the glyphosate 570 g L^{-1} however the final result was the same. The weaker formulation achieved the 80–90% brownout at 12–14 weeks, whereas the other formulation achieved the same milestone at 8–10 weeks.

The single application of the higher dilution rate of glyphosate had a significant result with plot 3u achieving close to 100% mortality within 6 weeks and plot 3b achieving the same result in 8–10 weeks. The four plots previously treated in autumn showed further decline resulting from the secondary application. A further 50% of the live giant reed was killed with this second treatment. This result was not as successful as anticipated. The reason for this may be due to the fact that these plants were still showing signs of the impacts from the previous treatment and may not have had enough foliage on them for the second application.

The burn of the giant reed at Paterson River took place in winter 2017. It was initially planned for autumn 2017, however this was not possible due to several factors outside of the control of the trial. Burning of hayed off giant reed was patchy, with some denser areas burning well. Regrowth in these areas was minimal through the following dry spring and summer. Since significant rainfall fell in early autumn 2018 recovery of the weed has increased, however the density is greatly reduced.

Figure 1. Glyphosate symptoms on new shoots off existing canes of giant reed (Paterson River).

The findings from the Tocal (Paterson) trial influenced the approach to be taken at the Murrurundi site, that is the higher rate of herbicide was required to achieve the desired result. The giant reed treated at this site achieved a greater than 95% brownout rate (Figure 2). This could have been as a result of the seasonal conditions at the time which appeared to be more favourable than those at Tocal.

The giant reed at the Murrurundi site was burned in the winter of 2018, with great success. The burns eliminated the majority of the above ground biomass. This burn was far more successful than that at Tocal. This is because the conditions where right at the time as well as the high percentage of dead canes, which burned more readily. Figure 3 below shows the results of the burn (same clump as in Figure 2).

Figure 2. Higher dilutions of glyphosate worked well on giant reed at Murrurundi, with >95% brownout with near perfect control (6 months after treatment, November 2017).

Figure 3. Burned giant reed clump at Murrurundi, as shown in Figure 2 (August 2018).

DISCUSSION

Based on the results, the recommended application rate of herbicide using this application method is the higher rate (glyphosate 360 g L^{-1} at the dilution rate of 20% (e.g. 200 mL product in 1 L water and glyphosate 570 g L^{-1} at a dilution of 13–14%). These rates achieved the best result with a single treatment and left the giant reed in a suitable state to burn. Giant reed has an extensive root mass and is a large plant. The key is to apply sufficient volume of herbicide onto the plant, which the UAV was able to achieve. Although the glyphosate 570 g L^{-1} (Roundup® UltraMax) worked quicker, the glyphosate 360 gL^{-1} (Roundup® Biactive) provided almost the same result after a longer time.

As giant reed tends to grow adjacent to or in waterways, the only option is to use a herbicide such as (Roundup® Biactive), which is permitted for use in aquatic zones. Despite these interim results, the use of the rotary winged Unmanned Aerial Vehicle (UAV) has proven effective and efficient in the application of herbicide on giant reed. The UAV presents several advantages over traditional methods in the treatment of giant reed. Firstly, the aircraft efficiently accesses the entire thicket and secondly, the rotor wash created by the UAV accurately pushes the herbicide through the entire stratum of the giant reed.

The ability to get over the entire thicket and achieve the necessary herbicide coverage has resulted in a significant time and cost saving as well as having zero impact on the ground. This also applies to other weeds that would otherwise be difficult to access or to gain full coverage using ground-based methods.

The burn is an important step in the control of giant reed as it eliminates the above ground biomass which has the potential in a high river flow event to form flood racks, which pose a significant risk to downstream infrastructure. Burning was the chosen method for biomass reduction in this case, however if this is not possible them mulching may be done to achieve the same outcome.

As with any weed control program follow up treatment will be necessary. It is anticipated that monitoring and required treatment will be necessary for approximately 5 years post primary treatment. This can be achieved by spot spraying small regrow before it reaches more than one m tall.

ACKNOWLEDGMENTS

The trials were a collaborative effort between the private and public sector: Sky Land Management Pty Ltd; Hunter Local Land Services; and NSW Department of Primary Industries, Tocal College. Burning at Murrurundi was undertaken by the landholder.

REFERENCES

Bell, G.P. (1997). Ecology and management of *Arundo donax*, and approaches to riparian habitat restoration in southern California. *In* 'Plant Invasions: Studies from North America and Europe', eds J.H. Brock, M. Wade, P. Pysek, P. and D. Green pp. 103-13. (Blackhuys Publishers, Leiden, The Netherlands).

CRC for Australian Weed Management (2005). Press Release: 28 December 2005. Giant musical reed threatens Yarra, Torrens.

Csurhes, S. (2009) Giant reed. *Arundo donax* – Weed Risk Assessment. Queensland Primary Industries and Fisheries. https://www.daf.qld.gov.au/__data/assets/pdf_file/0006/59973/IPA-Giant-Reed-Risk-Assessment.pdf (accessed 23 August 2018).

Dudley, T.L. (1998). *Arundo donax*. *In* Noxious wildland weeds of California, eds C. Bossard, J. Randall, and M. Hoshovsky, pp. 53-8. (University of California Press, California).

Haddadachi, A. and C.L. Gross (n.d.). An assessment of the mechanisms of seed infertility in giant reed: *Arundo donax* L. (Poaceae). Ecosystem Management, University of New England Armidale.

Collector app mapping to assess effectiveness of invasive grass control

Steve Taylor, Jenny Conolly and Renee Brawata
ACT Parks and Conservation Service, 500 Cotter Road Weston Creek,
Australian Capital Territory 2611, Australia
(steve.taylor@act.gov.au)

Summary The Collector app has been used for several years in the Australian Capital Territory (ACT) for mapping infestations of naturalised plants and invasive plants and their control. The app is used on both Smartphones and tablets. There is now sufficient data to reveal trends in distribution and density of targeted species after control work. African lovegrass, serrated tussock and Chilean needle grass are the most commonly controlled invasive grasses in the ACT. All three of these invasive species have massive impacts on native plant diversity, so understanding the resources required for control work is important. The data shows that with stable resourcing over a number of years, land managers can gain control of invasive grass infestations. Direct control involves using herbicides such as glyphosate and/or flupropanate, but other management tools also play an important part in successful control. The analysis also revealed how to improve both the accuracy of polygons and visual estimates of density.

Keywords Invasive grasses, African lovegrass, serrated tussock, Chilean needle grass.

INTRODUCTION

Invasive alien grasses are regarded as an increasing landscape threat (Godfree *et al.* 2017). They impact upon landscape connectivity, biodiversity and productivity.

The invasiveness and impact of *Nassella trichotoma* (Nees) Hack. ex Arechav. (serrated tussock) and *Nassella neesiana* (Trin. & Rupr.) Barkworth (Chilean needle grass) in south eastern Australia is well documented (Osmond *et al.* 2008, Snell *et al.* 2007, respectively). These two invasive grasses are well established in urban Canberra and are increasing in distribution in rural areas, including Tidbinbilla Nature Reserve and Namadgi National Park (Taylor and Conolly 2017).

Eragrostis curvula (Schrad.) Nees (African lovegrass) has also proven to be highly invasive and threaten grassy ecosystems (Downey *et al.* 2010, Firn *et al.* 2010, Firn *et al.* 2012, Firn *et al.* 2018). It is also well established in urban and peri-urban areas of the Australian Capital Territory (ACT) and is spreading along river and road corridors into Namadgi National Park and Tidbinbilla Nature Reserve.

Hyparrhenia hirta (L.) Stapf (Coolatai grass) established in northern parts of NSW, has also been shown to threaten biodiversity (McArdle *et al.* 2004). This species has recently started to spread into Southern NSW (e.g. Southern Highlands and Monaro) and a small infestation was recently detected in the ACT.

Anthoxanthum odoratum L. (sweet vernal grass) has spread considerable distances in Kosciuszko National Park and is starting to spread rapidly in nearby Namadgi National Park. It has a relatively lower impact in natural or undisturbed ecosystems but will dominate disturbed areas lacking perennial grass cover.

Nassella leucotricha (Trin. & Rupr.) Pohl (Texas needle grass) may also be present in the ACT but it could be concealed by the more abundant Chilean needle grass, so its status is uncertain.

This paper focuses on results of control efforts for serrated tussock, Chilean needle grass and African lovegrass because there is sufficient mapping data for analysis. All three of these species meet the spatial and impact based definitions of invasive plants: spatial (naturalised plants that produce reproductive offspring, often in very large numbers, at considerable distances from parent plants, and have the potential to spread over long distances) and impact (naturalised plants that threaten biological diversity) (Hui and Richardson 2017).

Managing the spread and impact of invasive grasses requires accurate mapping of infestations and control effort. This helps with location of follow-up control sites and with planning and reporting. Measuring successful invasive grass management involves three variables: (i) reduction in area of infestations; (ii) reduction in density of infestations; and (iii) replacement of invasive grasses with local native plant species. This paper examines recent mapping data with respect to (i) and (ii).

MATERIAL AND METHODS

Naturalised plants and invasive plant control have been recorded on ArcGIS On-line and its associated field app, Collector, since 2014 (Esri 2018). Staff, contractors and volunteers all edit the one unique feature layer (spatial data file), so there is no confusion about where control work has been undertaken.

This data has revealed trends in the area and density of invasive grass infestations.

Approximately 182 ha of Chilean needle grass, 266 ha of African lovegrass and 699 ha of serrated tussock were controlled multiple times in the lowlands grasslands of Canberra Nature Park reserves, between the years 2014–2017. The total area of each species treated was stratified by grassland type: Natural Temperate Grassland (NTG) (396 ha); native grassland other than NTG (514 ha); and exotics dominated grassland (237 ha). Data from weed control sites was categorised by treatment number. To determine the success or otherwise of control work, only sites with follow-up control were included in the analysis.

Each treatment involved spot spraying using glyphosate and/or flupropanate. The herbicide rates were those listed in permit PER9792 (APVMA 2018). The lower rates in this permit were used (e.g. 150 mL flupropanate 100 L^{-1} water for African lovegrass) to minimise off-target damage. There was usually only one treatment a year in the serrated tussock and Chilean needle grass infestation sites, whereas African lovegrass infestation sites often required more than one treatment per year. Weed control operators searched for and destroyed target species, with thoroughness of control achieved by a grid search and/or zig-zag search methods.

Once control work was complete the Collector app was used on smartphones or tablets to create a polygon around the distinct infestations (streaming mode gave greater accuracy) so the area of the control work could be calculated. The area controlled for each target species was calculated by grassland type and treatment number to compare changes in area treated over time. The proportion of area in each grassland type was also calculated by treatment number to demonstrate prioritising of high value conservation areas (NTG) for invasive weed control.

A density rating for the target species was assigned to each polygon: <1% cover (density 1), 1–10% cover (density 2), 11–25% cover (density 3), 26–50% cover (density 4), >50% cover (density 5). For each targeted species the number of sites (polygons) in each density category was shown by treatment number.

Visual estimates of density or cover are prone to error depending on the size of the area it applies to – smaller polygons tend to have less error. There is also the issue of observer bias and therefore data consistency when different weed control operators are undertaking the work.

Field workers estimate density in one of three ways:

- estimate the area covered by a target plant and multiply by the number treated plants and then divide by the polygon area (can be very accurate if plants are a similar size);
- visual sample of quadrats within the polygon and produce an average density estimate for the polygon (good for larger polygons); or
- visual clumping of plants within the polygon to determine density (works well with small polygons).

RESULTS

The type of mapping data that was available from the Collector app for the analysis is shown (Figures 1 and 2). The reserves are located within urban Canberra. The tan coloured polygons in Figure 1 are serrated tussock control. The green coloured polygons in Figure 2 are Chilean needle grass control. The search areas in both cases were the entire reserves (green shaded areas).

Figure 1. Serrated tussock control (tan/brown polygons) at Jerrabomberra Grasslands West. Pink/purple dots are rare native plant locations. Reserve size is 290 ha.

Figure 2. Chilean needle grass control (green polygons) at Crace Grasslands. Pink/purple dots are rare native plant locations. Reserve size is 145 ha.

Data collected during mapping of invasive plants including area and density sprayed (Figures 3–8) showed successful reduction in area and density of infestations for all three invasive grasses studied. They also show that high value grasslands such as native dominated and natural temperate grasslands, have been prioritised for treatment.

The first treatment for all species led to a significant drop in infestation area. Reasons for this are: (i) when larger – older plants are killed the polygons are considerably reduced in size; (ii) outliers are often younger plants that produce less seed; (iii) the seed bank at the edge of the infestation range front is smaller; and (iv) in polygons where flupropanate was used there is a suppressing effect on further germination for up to two years. Once the infestations start being broken down then polygons reduce in size. There are often more polygons mapped after the first treatment but the cumulative area is less than the large initial polygons.

Figures 6–8 show persistence at the various sites. Denser initial infestations require more treatments before no follow-up control was required during the study period. This is no surprise because seed banks of the target species would be larger and there is less interspecific competition from neighboring plants so a higher level of re-infestation is expected. Infestations are broken up once treatment begins, but smaller polygons remain that have a relatively high density requiring further follow-up control. This is an artifact of mapping with polygons.

African lovegrass, unlike serrated tussock and Chilean needle grass, continuously flowers and seeds through its growing season in the ACT (September to

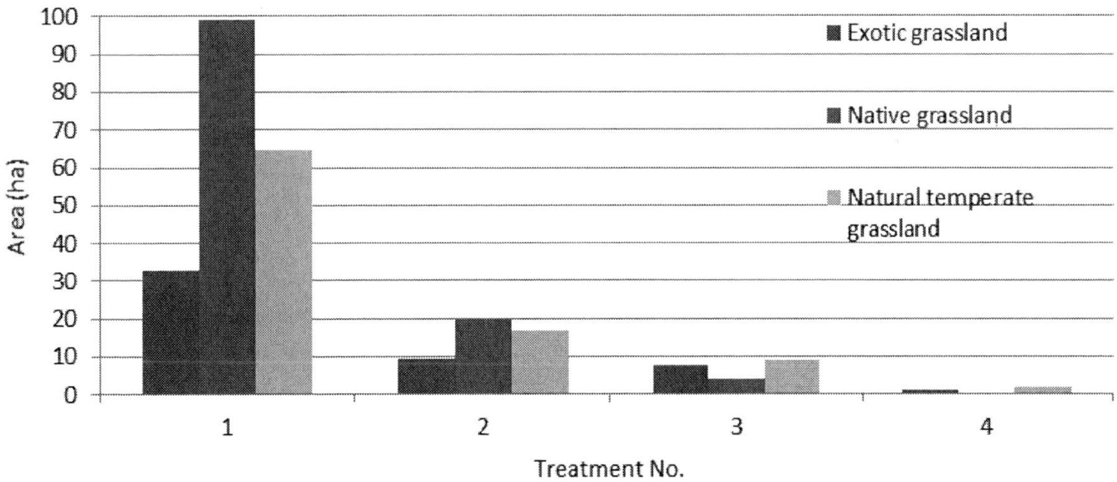

Figure 3. Area of African lovegrass controlled and treatment number. Each treatment involved spot spraying with glyphosate and/or flupropanate.

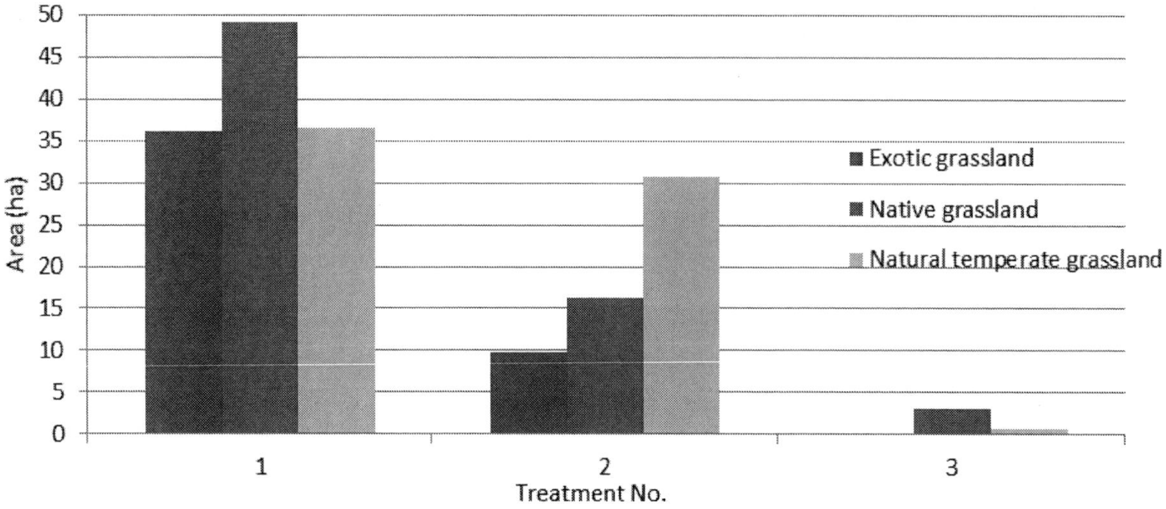

Figure 4. Area of Chilean needle grass controlled and treatment number. Each treatment involved spot spraying with glyphosate and/or flupropanate.

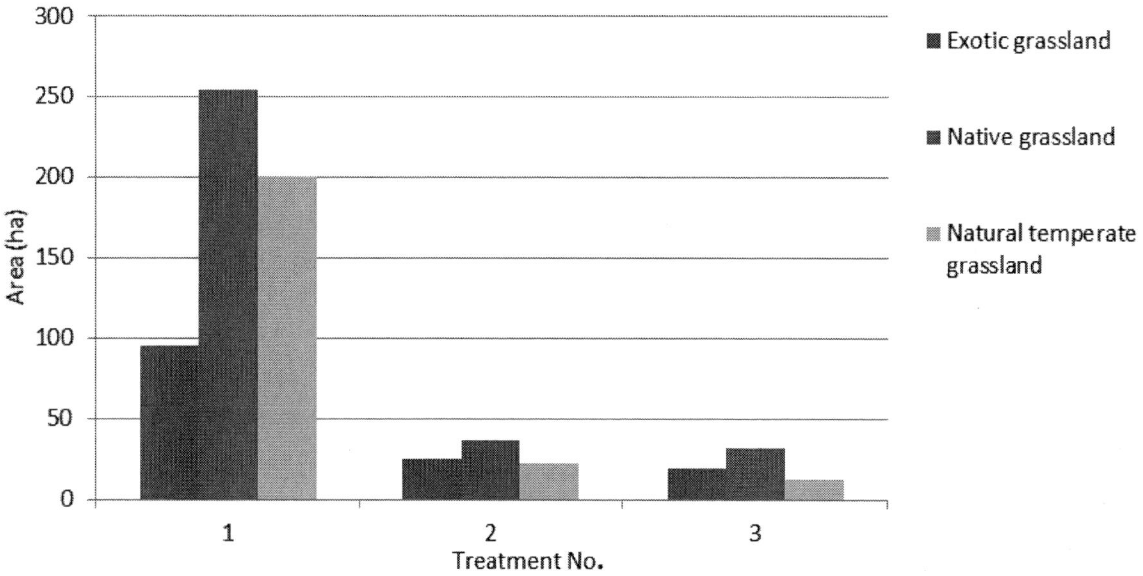

Figure 5. Area of serrated tussock controlled and treatment number. Each treatment involved spot spraying with glyphosate and/or flupropanate.

April). The flowering corresponds with rainfall, particularly for infestations that have been slashed, burnt or grazed before the rain. The resulting propagule pressure can therefore be relative high compared to other invasive grasses. This may explain why relatively more treatments are required for African lovegrass compared to the other invasive grasses over the study period (Figure 6 c.f. Figures 7 and 8).

Accuracy and consistency of data is important in interpretation of these results. This was a recognised problem in visual estimates of density (Morrison 2016). Factors that help with accuracy of visual estimates of treated invasive grasses are: small polygons; easily accessible terrain; treatment season that makes the target species more visible; and spray dyes. Consistency is very important for interpretation of trends in data. This can be an issue for density estimates if different spray operators map the different treatments. Fortunately this source of consistency error is lower than it otherwise would be because spraying contractors are often allocated to the same reserves over a number of years.

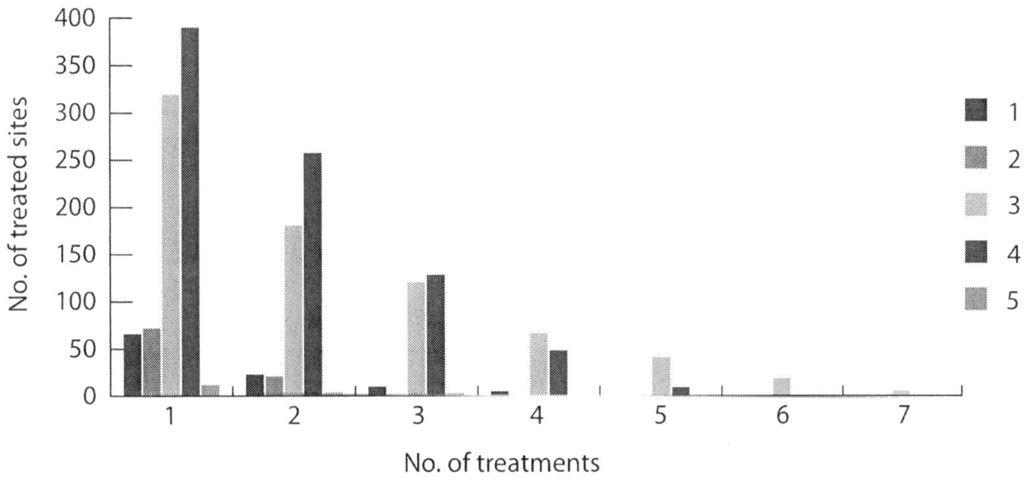

Figure 6. No. of sites in each density category (1 lowest, 5 highest) by treatment – African lovegrass. Each treatment involved spot spraying with glyphosate and/or flupropanate.

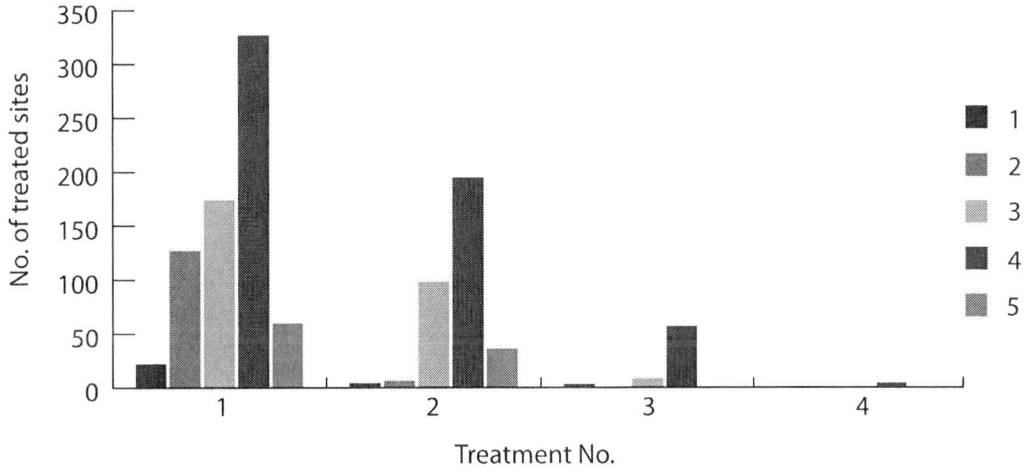

Figure 7. No. of sites in each density category (1 lowest, 5 highest) by treatment – Chilean needle grass. Each treatment involved spot spraying with glyphosate and/or flupropanate.

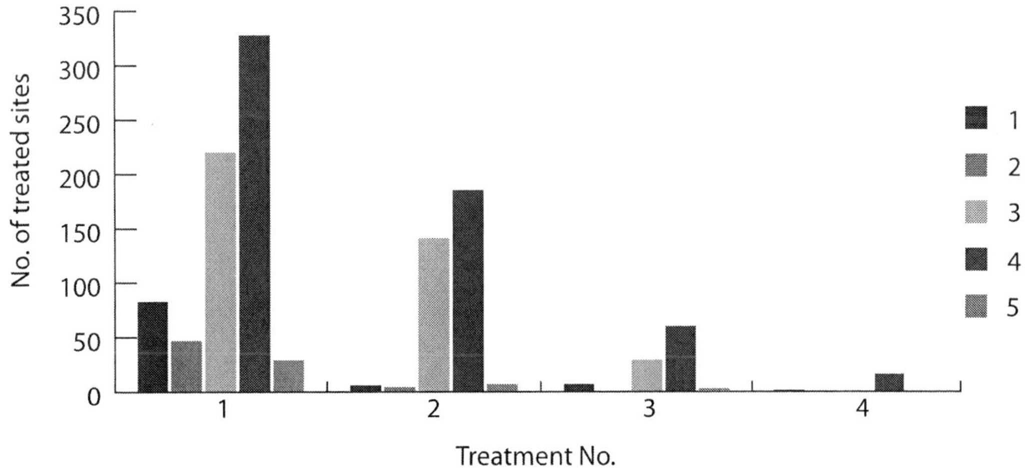

Figure 8. Number of sites in each density category (1 lowest, 5 highest) by treatment – serrated tussock. Each treatment involved spot spraying with glyphosate and/or flupropanate).

Feedback from staff was that an additional field of spray mix volume should be recorded on the Collector app maps for larger polygons. This will be added to the feature layer in 2018/2019. It will be a good method of verifying the accuracy of visual estimates.

DISCUSSION

The Collector app-ArcGIS On-line mapping system records all the treated invasive plant infestations and ensures follow up control sites are not overlooked. The Collector app data, which is synced to ArcGIS On-line, produced the summary data for the bar charts in this study. The data from the bar charts shows:

- on-going follow-up control is essential to deal with re-infestation from seed banks and adjacent un-treated infestations;
- denser infestations require more follow-up control, so on-going resourcing is essential to bring such infestations under control (the use of the residual herbicide flupropanate can partially offset this); and
- relatively more control effort is required for African lovegrass (a species that seeds a number of times in its growing season, making re-infestation more likely) as opposed to invasive grasses that seed once a year (Chilean needle grass and serrated tussock in this study).

The introduction of spray mix volume as a new field on the Collector app maps will increase the accuracy and consistency of visual density estimates for larger polygons.

ACKNOWLEDGMENTS

Thank you to ACT Government staff, contractors and volunteers for making the effort to map their invasive plant control work. Extra thanks to Nic Mikhailovich and Kirsten Tasker for feedback on improvements to the fields used in mapping with the Collector app. Thank you also to Alexi Williams, Allan Bendall and Josh Thomson for help trialing the Collector app when it was first introduced. And also thanks to Harley Baker, Grant Woodbridge, Dave Whitfield, Bernadette Brown, Ellyse Sheridan, Craig Wainwright, Glenn Charnock, Liam Banyer, Nic Daines, Mal Copelin, Waltraud Pix and other field workers for promoting the use of the Collector app with their colleagues, which contributed to good uptake of this mapping technology.

REFERENCES

APVMA, Australian Pesticides and Veterinary Medicines Authority. http://permits.apvma.gov.au/PER9792.PDF (accessed 1 August 2018).

Downey, P.O., Scanlon, T.J. and Hosking J.R (2010). Prioritizing weed species based on their threat and ability to impact on biodiversity: a case study from NSW. *Plant Protection Quarterly* 25, 111-26.

Esri (2018). Collector for ArcGIS http://www.esri.com/products/collector-for-arcgis (accessed 1 August 2018).

Firn, J., MacDougall, A.S., Schmidt, S. and Buckley, Y.M. (2010). Early emergence and resource availability can competitively favour natives over a functionally similar invader. *Oecologia* 163, 775-84.

Firn, J., Prober, S.M. and Buckley, Y.M. (2012). Plastic traits of an exotic grass contribute to its abundance but are not always favourable. PLoS ONE 7(4):e35870.

Firn, J., Ladouceur E. and Dorrough J. (2018). Integrating local knowledge and research to refine the management of an invasive non-native grass in critically endangered grassy woodlands. *Journal of Applied Ecology* 55, 321-30.

Godfree, R., Firn, J., Johnson, S., Knerr, N., Stohl, J. and Doerr, V. (2017). Why non-native grasses pose a critical emerging threat to biodiversity conservation, habitat connectivity and agricultural production in multifunctional rural landscapes. *Landscape Ecology* 32, 1219-42.

Hui, C. and Richardson D.M. (2017). 'Invasion dynamics'. (Oxford University Press, England).

Morrison, L.W. (2016). Observer error in vegetation surveys: a review. Journal of Plant Ecology 9, 367-79.

McArdle, S.L., Nadolny, C. and Sindel B.M. (2004). Invasion of native vegetation by Coolatai Grass Hyparrhenia hirta: impacts on native vegetation and management implications. Pacific Conservation Biology 10, 49-56.

Osmond, R., Veebeek, M., McLaren, D.A., Michelmore, M., Wicks, B., Grech, C. J. and Fullerton, P. (2008). Serrated tussock – National best practice manual. Victorian Department of Primary Industries.

Snell, K., Grech, C. and Davies, J. (2007). Chilean needle grass – National best practice manual. Victorian Department of Primary Industries.

Taylor, S. and Conolly, J. (2017). Spread of invasive grasses into Namadgi National Park, Proceedings of the 19th NSW Weeds Conference. (Weed Society of New South Wales, Armidale).

Weed surveillance using search theory

Oscar Cacho[1], Dane Panetta[2], Hillary Cherry[3], Susie Hester[1,4], Sue Bower[5] and Hank Bower[6]
[1] UNE Business School, University of New England, Armidale New South Wales 2351, Australia
[2] School of Agriculture and Food, The University of Melbourne, Parkville, Victoria 3010, Australia
[3] Pest and Weeds Team, National Parks and Wildlife Service, Hurstville, New South Wales 1481, Australia
[4] Centre of Excellence for Biosecurity Risk Analysis (CEBRA), The University of Melbourne, Parkville, Victoria 3010, Australia
[5] LHIB Flora Management Officer,
[6] Manager Environment/World Heritage,
Lord Howe Island, New South Wales 2898, Australia
(ocacho@une.edu.au)

Summary The success of a weed incursion response ultimately depends on the ability to find and kill individual plants. Understanding detectability will allow development of operational models to guide eradication efforts by allocating resources as efficiently as possible. We present the basics of search theory to show why obtaining a measure of detectability can help us develop effective tools for allocating surveillance resources. This is particularly important in the context of new methods of searching for weeds becoming available. The use of detector dogs, remote sensing, drones and other tools provide new opportunities to improve efficiency, but the right mix of methods will depend on the cost to achieve a given probability of detection. We use examples from existing weed eradication responses to show how detectability is affected by environment, search mode and life stage of the plant. We conclude by proposing ways of making these decision-analysis principles more accessible to practitioners.

Keywords Detection, eradication, decision analysis.

INTRODUCTION

Invasive plant managers are required to design and implement strategies that use available budgets as efficiently as possible. Budgets are typically distributed between the treatment of known infestations, research to improve control methods and future management decisions, active surveillance, and community engagement to encourage the public to report infestations—passive surveillance. The decision problem faced by invasive plant managers is illustrated (Figure 1).

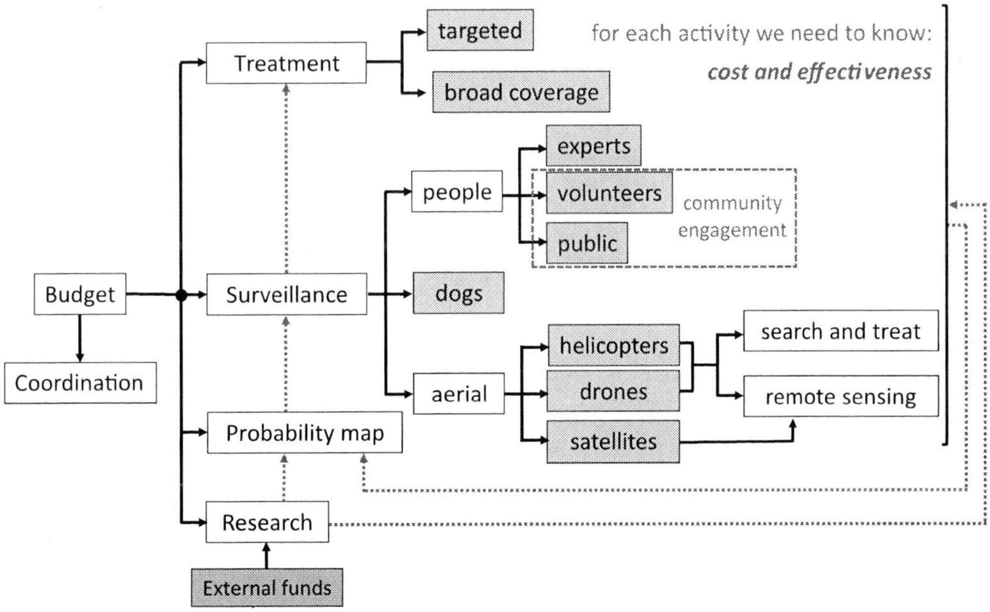

Figure 1. Conceptual model of the decision problem facing managers of eradication and containment programs. Dotted lines indicate transmission of information.

Within the expected budget, an agency must decide how to allocate resources to different forms of surveillance, treatment and research. Each of these actions can involve different activities. For example, in the case of treatment, targeted application of herbicide may be necessary in natural ecosystems whereas broad treatment is possible in cropping systems. Surveillance informs treatment (illustrated by a broken line in Figure 1) by assigning target areas. For surveillance in public lands, the typical case involves field officers (experts) searching for the weed in areas known or suspected to be invaded. Targeted treatment may occur simultaneously with surveillance, but it can also occur separately, as when contractors are sent later to spray. This form of surveillance may be supplemented by volunteers working under the guidance of field officers and, in some cases, informed by members of the public reporting sightings of the weed. Community engagement activities influence the effectiveness and cost of surveillance by attracting volunteers, and by enabling members of the public to recognise and report weeds (Hester and Cacho 2017). For large eradication efforts, a dedicated coordinator is required, and this is a significant fixed cost.

The use of detector dogs can be beneficial in situations where the weed is present at very low densities and/or is difficult to distinguish visually from the surrounding vegetation. Remote detection using satellites, drones or helicopters to collect photos or imagery can be used to pinpoint areas for treatment. Aerial surveillance is valuable when large search areas are involved, and for gaining access to inaccessible or remote sites. Aerial surveillance can also be combined with treatment when helicopters or drones apply herbicide directly to weeds as they are found (e.g. Leary *et al.* 2014).

The incursion response is guided by a 'probability map' (Figure 1) based on the best information available. Ideally, the map would show likelihood of weed presence in space, based on habitat suitability, proximity to known infestations and other factors (Hauser and McCarthy 2009). This information would guide surveillance and treatment efforts. The probability map can be improved over time as new surveillance and treatment information is incorporated. The feedback from surveillance to the probability map can be improved through additional research that combines data on the incursion with other information (e.g. biology/ecology of the target weed).

DETECTABILITY AND SEARCH THEORY

To solve the allocation problem, we need to know the effectiveness and cost of each mode of surveillance and treatment. Effectiveness is measured as the probability that a weed will be found when present (P_D) and the probability that a treated weed will be killed (P_K), in the case of surveillance and treatment respectively. The probability that a weed will be extirpated from an area can be expressed as:

$$P_E = P_D \times P_K. \quad (1)$$

For a random search path:

$$P_D = 1 - \exp(-RSW) \qquad (2)$$

Where R = detectability (m); S = speed of search (m/s); W = search effort (s m^{-2}). The term within brackets is also called coverage (c). The variables that determine coverage are related to costs, so expressing the problem in this way allows us to come up with realistic estimates of what is possible with a given budget or, conversely, to justify budget requests for a target outcome (Cacho *et al.* 2007).

The detectability of a target can be calculated based on the detection profile, a representation of the searcher's performance showing the probability that the target will be detected as a function of its lateral distance from the searcher (Figure 2). The efficiency of search per unit of distance covered is given by the area under the lateral range curve, which is equivalent to the area of the rectangle in Figure 2.

In practice, the two key probabilities (P_D and P_K) are dependent on the resources available and their allocation. The WeedSearch model (Cacho and Pheloung

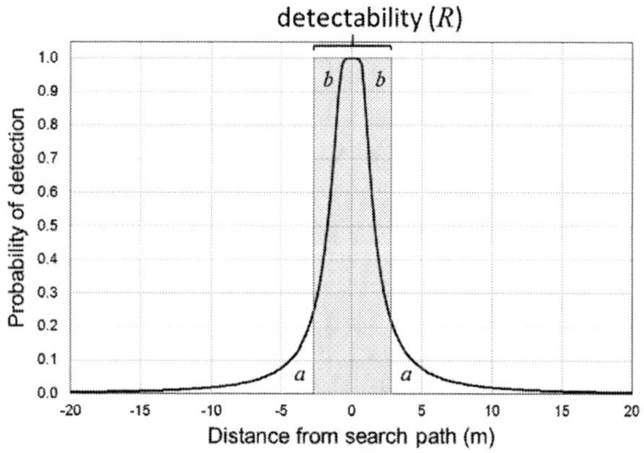

Figure 2. Lateral range curve (LRC) showing the probability that a target will be detected depending on its distance from the search path. Detectability (R) is calculated as the width of a rectangle with area equal to the area under the LRC. The areas labelled a equal the areas labelled b. See Cacho *et al.* (2006) for details.

2007) provides a simple tool to assess eradication feasibility based on a small number of parameters related to biological and logistical aspects of the weed and the environment it invades (Panetta *et al.* 2011), but more work is needed to understand how detectability is affected by surveillance method.

APPLYING THE MODEL

To solve the allocation problem we need estimates of weed detectability (*R*). The ability to detect a weed for a given search effort depends on the weed's visibility, experience of the searcher, and search method. Weed visibility is as much a function of the vegetation and terrain in which it occurs as of the characteristics of the target weed, which vary with phenology and growth stage. This means our detectability parameter needs to account for search mode, environment, and plant characteristics as they change seasonally and with life stage. These factors are combined to illustrate how the problem could be tackled (Table 1). Numerical values are not available, but we indicate the relative detectability of selected weeds depending on life stage and surveillance method.

The visibility of weeds across different life stages has a significant impact on appropriate surveillance techniques. For example, mature bitou bush (*Chrysanthemoides monilifera* subsp. *rotundata* (DC.) Norl. is easy to spot, especially when flowering, but a seedling or juvenile is not. The pink bark of cherry guava trees (*Psidium cattleianum* Afzel. ex Sabine), more prevalent on mature plants, and the new growth which flushes with a pink/reddish tinge contribute to its visibility. Cherry guava in a shrubland/heathland situation occurs on the upper slopes and edges of cliffs of Lord Howe Island (LHI). In these cases, the plant will form part of the canopy; combined with the flushing foliage this is a key to spotting, and treating, the target from a helicopter (AH, Table 1). Orange hawkweed (*Hieracium aurantiacum* L.) resembles a number of other native and introduced species, and is most detectable at flowering. However, it flowers and sets seed within several weeks, meaning that timely detection is critical (Constantine *et al.* 2016). The duration of maximum visibility during the year influences the temporal allocation of search effort, requiring periods of intense activity by staff and volunteers.

Table 1. Detectability of weeds depending on growth form, habitat type and growth stage. Estimates are approximations based on Harris et al. (2001) and authors' experience. Life stages: seedlings (S); juveniles (J); mature (M); and mature flowering or with flushing foliage (MF). Surveillance methods applicable for each life stage (see Figure 1): people: PE (experts); PV (volunteers-trained); PP (public -untrained); D (detection dogs)a; aerial: AH (helicopter – human visual sighting); AD (drones – remote imagery/video/photos); and AS (satellites – remote imagery/photos)b. For these methods L, M, H and N refer to low, medium, high and no potential for detection.

Weed growth form (*examples*)	**Habitat type**	**Life stage**	**Surveillance mode and level of detectability**						
			PE	PV	PP	D	AH	AD	AS
Tree or tall shrub *Psidium cattleianum* (cherry guava)	Forest	S	H	L	L	M	N	N	N
		J	H	L	L	M	L	L	L
		M	H	M	L	M	M	M	?
		MF	H	H	M	M	M	M	?
Shrub *Chyrsanthemoides monilifera* (bitou bush)	Shrubland – including coastal dune habitat	S	M	M	L	M	N	N	N
		J	M	M	L	M	L	L	L
		M	H	H	L	M	H	H	M
		MF	H	H	H	M	H	H	M
Annual or herbaceous perennial *Hieracium aurantiacum* (orange hawkweed)	Open habitat*	S	M	L	L	M	N	N	N
		J	H	M	L	M	L	L	L
		M	H	M	L	M	L	L	L
		MF	H	H	M	M	M	M	L

* Open habitat includes grassland, river beds, sand dunes and other natural and induced open habitats.
[a] Detectability estimates for dogs are based on olfactory (rather than visual) detection.
[b] Detectability estimates for aerial surveillance (AH, AD and AS) are based on scenarios where the weed has emergent foliage either above or in canopy).

Use of dogs in detection of weeds (D, Table 1) can increase our ability to detect plants before flowering, which reduces the risk of reproductive escape. McLean and Sargisson (2017) show that dogs can be effectively trained to discriminate specific weed species, and indeed trials in Kosciuszko National Park on detection of hawkweeds by dogs show promise. The reliance on olfactory detection could be particularly useful in the understory where dense vegetation hinders visual detectability.

An interesting case of public detections (PP, Table 1) occurs in LHI, where 80–90% of the local population (350 inhabitants) would be able to detect a cherry guava. Long-term inhabitants of LHI have grown up with this plant, and have been involved in the campaign to eradicate the plant, but detectability could be lower among others.

Regarding aerial surveillance, drones (AD, Table 1) are proving useful in some situations. For example, the visibility of mature bitou bush by drones can be high, as the plant has a distinct leaf shape and colour, and it is highly detectable using aerial photographs when in full flower. Drones in combination with automatic image recognition are being used for orange hawkweed detection in Kosciuszko National Park (Hamilton *et al.* 2018). As this technology improves it will increase the effectiveness and reduce costs of remote area surveillance and monitoring. While remote sensing may pick up sizeable infestations of weeds, it is unlikely to detect very small numbers of plants, particularly if these occur in the understory (Panetta and Timmins 2004). However, multispectral imagery collected from drones and satellites may enable detection of small patches (~ 1 m^2) if the target weed has a unique spectral signature to other species in the invaded environment.

Research plays a critical role in the management of the weeds considered (Table 1). In the case of hawkweeds, considerable investment has enabled research on modelling and prioritising surveillance, developing effective herbicide techniques, and understanding the ecology and biology of the weed (Moore *et al.* 2011, Caldwell and Wright 2014). It should be possible to create tables of detectability parameters for some of the major weeds based on search theory concepts. This would be the first step towards making the allocation model outlined in Figure 1 operational.

There has been considerable progress since Panetta and Timmins (2004) noted the lack of quantitative approaches available to assess the feasibility of eradication. The probability that a plant invasion will be eradicated is strongly correlated to its detectability. It follows that understanding detectability is the key to developing operational models that can guide eradication efforts by allocating resources as efficiently as possible.

REFERENCES

Cacho, O.J., Spring, D., Pheloung, P. and Hester, S. (2006). Evaluating the feasibility of eradicating an invasion. *Biological Invasions* 8, 903-17.

Cacho, O.J., Hester, S.M., and Spring, D. (2007). Applying search theory to determine the feasibility of eradicating an invasive population in natural environments. *Australian Journal of Agricultural and Resource Economics* 51, 425-33.

Cacho, O. and Pheloung, P. (2007). WeedSearch: Weed Eradication Feasibility Analysis Software Manual. CRC for Australian Weed Management Project 1.2.8.

Caldwell, J. and Wright, G. (2014). Orange hawkweed control program report 2013/14. Office of Environment and Heritage, NSW National Parks and Wildlife Service, Hurstville.

Constantine, A., Hauser, C.E., Primrose, K. and Smith, N. (2016). Hawkweed (*Hieracium* spp.) surveillance: Development of a targeted and robust plan for the Victorian Alps. *Plant Protection Quarterly* 31, 28-32.

Hamilton, M.A., Matthews, R. and Caldwell, J. (2018). Needle in a haystack – detecting hawkweeds using drones. Proceedings of the 21st Australasian Weeds Conference, 9–13 September, Sydney, New South Wales, eds S. Johnson, L. Weston, H. Wu and B. Auld, pp. 126-30. (Weed Society of New South Wales).

Harris, S., Brown, J. and Timmins, S.M. (2001). Weed surveillance–how often to search. Science for Conservation 175. Department of Conservation, Wellington, New Zealand. 27 pp.

Hauser, C.E. and McCarthy, M.A. (2009). Streamlining 'search and destroy': cost-effective surveillance for invasive species management. *Ecology Letters* 12, 683-92.

Hester, S.M. and Cacho, O.J. (2017). The contribution of passive surveillance to invasive species management. *Biological Invasions* 19, 737–48.

Leary, J., Mahnken, B.V., Cox, L.J., Radford, A., Yanagida, J., Penniman, T., Duffy, D.C. and Gooding, J. (2014). Reducing nascent Miconia (*Miconia calvescens*) patches with an accelerated intervention strategy utilizing herbicide ballistic technology. *Invasive Plant Science and Management* 7, 164-75.

McLean, I.G. and Sargisson, R.J. (2017). A dog as a generalist plant detection tool. *Weed Research* 57, 287-92.

Moore, J.L., Hauser, C.E., Bear, J.L., Williams N.S.G. and McCarthy, M.A. (2011). Estimating detection–effort curves for plants using search experiments. *Ecological Applications* 21, 601-7.

Panetta, F.D. and Timmins, S.M. (2004). Evaluating the feasibility of eradication for terrestrial weed incursions. *Plant Protection Quarterly* 19, 5-11.

Panetta, F.D., Csurhes, S., Markula, A. and Hannan-Jones, M. (2011). Predicting the cost of eradication for 41 Class 1 declared weeds in Queensland. *Plant Protection Quarterly* 26, 42-6.

The state (and territory!) of data collection on national priority weeds

Bo Raphael, Jessica Evans, Katherina Ng, Luke Maloney and Sandra Parsons
Australian Bureau of Agricultural and Resource Economics and Sciences (ABARES), GPO Box 858, Canberra, Australian Capital Territory 2601, Australia
(bo.raphael@agriculture.gov.au)

Summary A role of the Australian Government in invasive species management is to provide national coordination to foster improved outcomes across Australia. The Australian Government also has a role to support the collection and collation of national weed data and information. To better understand the current national picture of data collection for agricultural and environmental priority weeds, ABARES developed an online survey aimed at Australian and state/territory governments and regional natural resource management organisations. The survey provides a national picture of the data collected and shared on priority weeds (Weeds of National Significance, National Environmental Alert, agricultural sleeper weeds, regional priority weeds) across Australia including distribution, impacts, resource allocation, emerging weed threats and data sharing. This paper highlights key results from the 2017–2018 survey to inform policy and decision-making, providing a baseline for similar future surveys.

Keywords Priority weeds, survey, data, WoNS.

The Victorian Weed Spotters: Recruitment and training of citizen scientists

Zachariah Munakamwe[1], Angela Constantine[2] and Catherine McInerney[3]
Agriculture Victoria
[1] PO Box 3100, Bendigo, Victoria 3552, Australia
[2] PO Box 5646, Cranbourne, Victoria 3977, Australia
[3] 70a Camp Road, Anglesea, Victoria 3230, Australia
(Zachariah.munakamwe@ecodev.vic.gov.au)

Summary The Weed Spotters program is a Victorian Government initiative delivered by Agriculture Victoria, aimed at engaging a passive surveillance network to identify and report Victoria's highest priority, State prohibited weeds (SPW). The program involves training volunteer Weed Spotters to identify and report target weeds, thereby allowing early detection and treatment with the ultimate aim of eradication. These citizen scientists are strategically recruited from various agencies identified as most likely to encounter, correctly identify and report the target weeds.

When compared to the general public, Weed Spotters have been found to be more likely to make accurate reports of SPWs. This is likely a result of strategic recruitment from key organisations, and importantly, due to the training they receive in the identification of target species. A total of 222 Weed Spotters were recruited and trained in the 2016/2017 and 2017/2018 financial years. These new recruits now form part of a network of over 2588 registered Weed Spotters.

A participatory approach was used to deliver the training, utilising live weed sample demonstrations and group discussions. An evaluation of the training was conducted which found that it was well received by participants, with some improvements to the training, such as increased time to examine live plant specimens, identified and incorporated for future training sessions.

Keywords Agriculture Victoria, early detection, citizen scientists, participatory training, State prohibited weeds, passive surveillance.

INTRODUCTION

A combination of prevention, early detection and intervention is an effective biosecurity approach for the management and eradication of high risk invasive species (Agriculture Victoria 2018), while engagement and training of citizen scientists is a proven effective biosecurity strategy (Laidlaw *et al.* 2015). Weed Spotters is a surveillance network of volunteers, first conceptualised back in 2001 (McInerney and Smith 2008) that combines both of these approaches.

Weed Spotters are volunteers who are trained to identify and report State prohibited weeds (SPWs). SPWs are the highest category of declared noxious weeds in Victoria and are managed with the aim of eradicating them from the State (*Catchment and Land Protection Act 1994* (Vic)). Weed Spotters are trained to focus on a small, target group of species, enabling them to confidently identify and report the highest priority weeds in Victoria. This provides a valuable surveillance tool in the management of SPWs (McInerney and Smith 2008), contributing to their prevention, early detection and ultimately, eradication. At the time of writing there are 2588 registered Weed Spotters in Victoria.

THE WEED SPOTTERS PROGRAM (2016/2017–2017/2018)

Recruitment Strategic Weed Spotter recruitment was first introduced in 2008, prior to which recruitment was undertaken on an ad-hoc basis, generally targeting people interested in weeds. The strategically adopted approach ensures that there is a competent surveillance network covering the State, with the necessary skills, opportunity and motivation to detect and report SPWs.

This process began with a systematic stakeholder analysis (Molony 2014) that identified potential key target agencies from which to recruit Weed Spotters. This considered ability to detect the weeds, interest, relationship with the Department, and the potential likelihood of encountering SPWs during daily operational activities. These key stakeholders were largely other government agencies whose staff spent substantial time undertaking outdoor work, in areas where these target species were most likely to be found. Target agencies included water management authorities, local governments, road managers, Catchment Management Authorities and Landcare groups. The analysis ensured that the most appropriate people were targeted for recruitment and training to maximise the number of accurate reports received.

Training Weed Spotter training consists of a two- to three-hour session featuring information about targeted

SPWs. The main objectives are to train participants to identify and report SPWs, distinguish them from similar species, identify common places to detect them and to increase understanding of weed related biosecurity principles and responsibilities.

Training sessions are delivered in spring and summer when most SPWs are actively growing and most easily identified. Groups of no more than 15 participants are preferred to ensure effective interaction and full engagement. An online version of the Weed Spotter training is currently being developed for those who are unable to attend a face-to-face session with the opportunity to participate.

Weed Spotter training sessions are free and each participant receives a package of weed identification resources including a Weed Spotter training manual, SPW calendar, SPW identification cards, SPW brochures and a Weed Spotter newsletter subscription. This is fully funded by Agriculture Victoria. The network is maintained at a size that can be adequately supported by Agriculture Victoria to ensure regular contact and support materials for all participants.

Training evaluation The 2016/2017–2017/2018 training was evaluated by means of a short survey. The purpose of the evaluation was to provide an indication of the effectiveness of the training and to identify areas for improvement for the delivery of future sessions. At the end of each training session, participants were asked to complete a questionnaire with seven closed-ended statements. A section for additional comments and observations was also included at the end of the statements.

RESULTS AND DISCUSSION

Recruitment During 2016/2017–2017/2018, 15 sessions were delivered across Victoria, training and recruiting 222 new Weed Spotters. State government organisations represented the largest number of recruits (37%), drawn from VicRoads, Department of Environment, Land, Water and Planning, Agriculture Victoria, and Parks Victoria (Figure 1). The next highest number of recruits was from rural city councils (28%). The remaining participants were from various agencies and groups including water management authorities, city councils, Catchment Management Authorities and Landcare groups.

Training The training that Weed Spotters receive has proven to be effective, resulting in a high number of accurate reports of target species. Approximately one third (34%) of all the SPW reports received (both negative and positive) in the 2016/2017 and 2017/2018 financial years were from registered Weed Spotters. Forty-eight percent of the confirmed accurate reports were also from Weed Spotters (Table 1). Only 21% of the inaccurate reports were from Weed Spotters, with the remaining 79% of the inaccurate reports from the general public (Table 2).

Training evaluation The response to the training evaluation survey was extremely high and overall positive. A total of 191 out of 222 (86%) training participants completed the survey. Nearly all respondents (>99%) reported that their knowledge of SPWs and the negative impact they can have on Victoria was increased as a result of the training. Only one

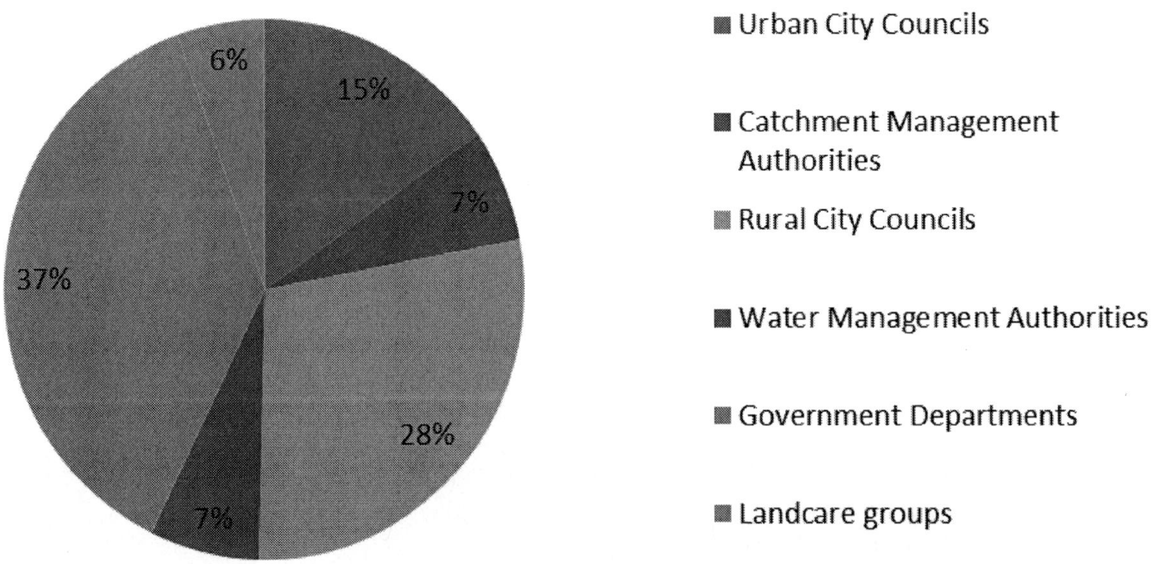

Figure 1. Agencies from which Weed Spotters were recruited and trained in 2016/2017 and 2017/2018.

Table 1. State prohibited weeds: Accurate reports in 2016/2017 and 2017/2018.

Year	Accurate reports	From registered Weed Spotters	From general public	% from Weed Spotters
2016/2017	21	9	12	43%
2017/2018	8	5	3	63%
Total	29	14	15	48%

Table 2. State prohibited weeds: Inaccurate reports in 2016/2017 and 2017/2018.

Year	Accurate reports	From registered Weed Spotters	From general public	% from Weed Spotters
2016/2017	16	2	14	13%
2017/2018	13	4	9	31%
Total	29	6	23	21%

participant (out of 191) responded negatively by strongly disagreeing with that statement. The opportunity to examine live weed samples was very popular across all participants. SPW identification cards and SPW calendars were the most popular hard copy resources provided, with many participants asking for additional copies to distribute to colleagues.

The strategic recruitment and training of 222 new Weed Spotters that have the ability, opportunity and motivation to detect and report SPWs has been a significant achievement. Having an extensive network increases the probability of detecting Victoria's highest risk weed species at an early stage of invasion, making control and management more effective and increasing the chance that infestations can be eradicated from Victoria.

There is constant fluctuation in the number of registered Weed Spotters, as people change employment, move to a new house or decide to discontinue their participation in the program. In addition to building capacity in the network to detect and report SPWs, training new participants also helps maintain the critical mass and distribution of the network.

Comparison of the accuracy of reports received from Weed Spotters and the general public show that trained, strategically recruited Weed Spotters are more likely to make accurate reports of target species than the general public. By recruiting from targeted groups and providing them with training and resources, the number of accurate reports increases, which decreases wasted staff time in following up on false reports. This increases the time available for managing new and existing infestations, rather than chasing false leads.

An example of the benefits of Weed Spotters is demonstrated by a recent report received of water hyacinth (*Eichorrnia crassipes* (Mart.) Solms), a SPW in Victoria. A registered Weed Spotter reported an advertisement for water hyacinth on a 'Buy, Sell and Swap' Facebook group in 2017. The newly trained Weed Spotter made the report just days after attending a Weed Spotter training session. Agriculture Victoria Authorised Officers responded to the report, leading to the seizure of 31 water hyacinth plants from a backyard pond on private property. The trace back investigations led to the discovery of an additional 21 water hyacinth plants that were removed from a suburban property. Both infestations were located near major waterways, posing a serious risk of the plants spreading. These two water hyacinth infestations would not have been detected without the contribution from the Weed Spotter who was a member of the private Facebook group where the plants were advertised.

The results of the training evaluation survey show that the training was very well received by participants and that the resources provided were of value. The results specifically highlighted the value the participants gained from the opportunity to examine live weed specimens. This feedback has been incorporated into the training delivery by ensuring that sufficient time is allowed for participants to closely examine the plants, making the training engaging and exciting for participants. Evaluation of future training sessions will continue in the interests of continual improvement.

Victoria is recognised nationally and internationally for leadership in prevention and preparedness of invasive species (Agriculture Victoria 2018). Early detection and rapid response are essential to successfully eradicate SPWs. The role of Weed Spotters in support of these government interventions cannot be overemphasised. They are trained to be the 'eyes and ears' of the Department (Smith 2006) and provide a valuable contribution to preventing the establishment and spread of new high risk invasive plants. Over the past two years, Agriculture Victoria have strategically recruited over two hundred new Weed Spotters from targeted organisations, which are more likely to accurately report target species compared to the general public. By evaluating the Weed Spotter training, we have ensured that Weed Spotters are engaged and responsive to the training.

ACKNOWLEDGMENTS

Special thanks to all Weed Spotters for their valuable contributions to protecting Victoria from State prohibited weeds.

REFERENCES

Agriculture Victoria (2018). Invasive Plants and Animals Policy Framework. http://agriculture.vic.gov.au/agriculture/pests-diseases-and-weeds/protecting-victoria-from-pest-animals-and-weeds/invasive-plants-and-animals/invasive-plants-and-animals-policy-framework (accessed 2 March 2018).

Molony, L. (2014). Evaluation of Weed Spotters. Internal Report prepared for the Department of Environment and Primary Industries, Victoria. Unpublished document.

Laidlaw, M.J., Holland, A. and Guymer, G. (2015). Many eyes on the prize: The role of citizen scientists in active weed surveillance. Proceedings of the 20th Australasian Weeds Conference, eds R. Randall, S. Lloyd and C. Borger, pp. 17-9. (Weeds Society of Western Australia, Perth).

McInerney, C.J. and Smith, N. (2008). The Victorian weed alert program: Weed spotters an integral surveillance tool. Proceedings of a symposium held on Surveillance for Biosecurity: Pre-Border to Pest Management, eds K.J., Froud, I.A. Popay, and S.M. Zydenbos, pp. 175-82. (New Zealand Plant Protection Society, Paihia, New Zealand).

Smith, N. (2006). The Victorian weed alert rapid response plan. Weed spotters and the weed alert contact officers. Proceedings of the 15th Australian Weeds Conference, eds C. Preston, J.H. Watts and N.D. Crossman, pp. 743-6. (Weed Management Society of South Australia, Adelaide).

Weeds at the early stage of invasion: pilot project evaluation

Kate Blood[1] and Bec James[2]
[1] Department of Environment, Land, Water and Planning, PO Box 7, Beaufort, Victoria 3373, Australia
[2] Department of Environment, Land, Water and Planning, Box 3100, Bendigo Delivery Centre, Victoria 3554, Australia
(Kate.Blood@delwp.vic.gov.au)

Summary The Weeds at the Early Stage of Invasion (WESI) project focuses on high risk early invaders that threaten biodiversity. The project team works with public land managers anywhere in Victoria.

Various tools and a decision making framework help public land managers make informed decisions about the effective management of early invaders.

Between 2014 and 2017, the project team established a series of pilot projects to test, improve and assist adoption of the early invader tools. Over this period, a total of 16 pilot projects were conducted of which seven were chosen to be evaluated. A total of 52 individuals were involved in the pilot projects including 13 agencies/groups.

In 2017 a pilot project evaluation report was developed, summarising the findings of the evaluation. The report concluded that the early invader pilot projects had been a valuable way of testing, refining and gaining adoption of the early invader process and tools in Victoria. This model could be beneficial for the development and adoption of other invasive species projects.

Keywords Early invader pilot projects, Weeds at the Early Stage of Invasion.

INTRODUCTION

The Weeds at the Early Stage of Invasion (WESI) project focuses on high risk, early invaders that threaten biodiversity. We work with Department of Environment, Land, Water and Planning (DELWP) and Parks Victoria staff looking after public land anywhere in Victoria. By investigating the barriers that prevent action on early invaders, WESI has created a process and tools to assist public land managers.

The WESI project's framework leads public land managers through a decision making process, which is supported by a set of detailed guides and other tools. The focus of the WESI project is to enable the early intervention and localised eradication of high risk invasive weed species on public land in Victoria at any scale. It aims to support, enable and build the capabilities of DELWP and Parks Victoria staff working on early invaders that threaten biodiversity.

The project team set up a series of pilot projects to assist the adoption of the WESI process and tools by local public land managers. It also provided an opportunity to test and refine the tools out in the field with real life scenarios.

The purpose of the evaluation plan was to evaluate the WESI pilot projects that aim to:

- to cover a range of scenarios and weed types to test and improve different aspects of the WESI processes and tools; and
- promote the WESI Project within the regions and aid its adoption.

The evaluation looked at:

- the appropriateness and delivery of the pilots;
- the distribution of the package of WESI tools to the pilot groups;
- whether the pilots were successful?;
- how useful the package of WESI tools was?; and
- how the package of WESI tools could be improved?

Audiences and stakeholders for the evaluation The WESI pilot evaluation plan informed the primary investor, the Weeds and Pests on Public Land (WPPL) program, the WESI Steering Group, project team and pilot participants.

A single report (James and Blood 2017) was produced to inform the participants of the outcomes of the pilot evaluation.

The report participants included nine people representing seven different pilot projects.

MATERIALS AND METHODS

Evaluation scope and methods The pilot evaluation took the form of low-key interviews to limit the burden of the evaluation on the pilot participants, while at the same time providing adequate results for a meaningful evaluation. The interviews were conducted by the WESI project officer, Bec James, face-to-face or over the phone with the pilot participants. The evaluation informed future project direction and improved the existing processes and tools.

Evaluation approach and key questions Outlined below are the key evaluation questions asked by the WESI project team to pilot project participants as part of the evaluation. The specific audience for each question is indicated with brackets.

Activities

1. Were the pilot projects designed and implemented appropriately? [pilot project participants] Were the pilot projects implemented the right way?
2. Was the pilot project implementation strategy implemented as planned? [WESI project team] If not, why not?

Outputs

3. Was the planned Decision Making Framework (DMF) and the tools created and distributed to the pilot project groups, provided as planned? [WESI project team] What was planned? What was delivered? What changes were made along the way? Why?
4. To what extent did the pilot project groups use the WESI process and tools, and apply them to their situations? [pilot project participants]
5. How suitable was the WESI process and tools for the target audience and intended users? [pilot project participants]

Outcomes (Results)

6. To what extent was the package demonstrated to work, in a variety of settings? [pilot project participants] What worked? What didn't? What contextual (other related) factors influenced the results e.g. budget, time? What lessons were learnt about the use of the package and how it could be improved into the future?
7. To what extent do pilot project participants rate the success of the pilot project they were involved in? [pilot participants] Provide a rating on the overall success of the pilot: 5 – Highly successful, 4 – Successful, 3 – Partly successful, 2 – Unsuccessful, 1 – Very unsuccessful. Please provide a reason for the rating provided.
8. How do pilot project participants rate the usefulness of the most recent WESI package? [pilot participants] Provide a rating on the overall usefulness of the pilot project: 5 – Highly useful, 4 – Useful, 3 – Partly useful, 2 – Not useful, 1 – Useless. Please provide a reason for the rating provided.

Improvement

9. How can the package be improved, to increase its usefulness and effectiveness? [pilot project participants]

Data collection methods, sample sizes and timing Over the 2014–2017 WESI project period, a total of 16 pilots projects were conducted and some are still underway. A focused selection of seven of these pilot projects were chosen for evaluation, representing a variety of scales and situations. Pilot projects were undertaken in each of six DELWP regions.

The pilots were undertaken at different scales (e.g. small reserves to statewide) and tested all or different components of the WESI package of tools. The tools included a decision making framework, a set of six guides, weed risk database, templates and other items.

Data was captured from one-on-one interviews (in person or via phone), document reviews, collection of other specific data, including the outcomes achieved by the pilot projects.

RESULTS

The pilots were implemented mostly as planned in an appropriate manner. All participants interviewed thought the pilot projects were successful; and of the participants interviewed after the WESI guides were published, 100% thought the package of tools was useful.

Lessons learnt

- *Lesson A:* The WESI process and tools can be effectively applied at a range of scales, for new and established programs, and for a range of likely situations to improve investment decisions for early invaders. There are additional situations where it would be valuable to pilot.
- *Lesson B:* Without WESI team support, it could be challenging to gain adoption of the WESI process and tools.
- *Lesson C:* The existing WESI tools need to be simplified and, with additional tools, made more accessible to people in the field.
- *Lesson D:* There is a demand for a weed knowledge centre of information (including traits and risk ratings) to support the effective and efficient use of the WESI process and tools.
- *Lesson E:* Without resources and tools to achieve biodiversity outcomes, it is difficult to undertake planning and on-ground activities to implement the WESI process.

Interpretation of findings From the WESI project team's point of view at the end of this round of pilot projects, it is clear that the range of developed tools could be further refined and manipulated into a more succinct and field-friendly format.

Ongoing support for regional staff is required to learn about and effectively use the WESI process and tools. A weed knowledge centre of information and risk ratings will significantly help the use and adoption of the WESI tools.

The provision of resources to assist public land managers with planning, eradication activities and monitoring will give the WESI project wings and accelerate its adoption.

RECOMMENDATIONS

The evaluation report recommended that the WESI project be continued with additional resources for at least a further three years to undertake the following recommendations:

- *Recommendation A:* Pilot the WESI process and tools across further situations to further improve investment decisions for early invaders.
- *Recommendation B:* Continue the WESI project to support biodiversity and public land managers with improved early invader management.
- *Recommendation C:* Refine existing tools and develop additional WESI tools which are more accessible to people in the field.
- *Recommendation D:* Develop a weed knowledge centre, accessible to all, containing information (including trait and risk ratings) to support the effective and efficient use of the WESI process and tools.
- *Recommendation E:* Provide resources to biodiversity and public land managers to undertake planning and on-ground activities to implement the WESI process and tools to achieve better biodiversity outcomes.

CONCLUSIONS

The early invader pilot projects undertaken from 2014–2017 were a valuable way of testing, refining and gaining adoption of the WESI process and tools in Victoria. One hundred percent of participants interviewed thought the pilot projects were successful; and of the participants interviewed after the WESI guides were published, 100% thought the package of tools was useful.

The pilot projects have reinforced the usefulness of the WESI package and suggested further refinements and additional tools to increase their effectiveness, including a weed knowledge centre.

The WESI decision making framework is logical and easy to understand. Pilot project participants were able to quickly understand and follow its intent. The set of six WESI guides that support the decision making framework is considered a great resource. Summarising the set of guides into one document will help its adoption, especially with operations staff.

Continued support from the WESI team for public land managers is important for further adoption of the WESI package. Future pilot projects will test additional scenarios. Further resources are necessary to assist planning and on-ground eradication of early invaders to protect biodiversity – the ultimate aim.

ACKNOWLEDGMENTS

The WESI team would like to thank all the pilot project participants, the Weeds and Pests on Public Land Program and the WESI Steering Group for supporting the WESI project. The WESI project is funded with support of the Victorian Government.

REFERENCES

James, B. and Blood, K. (2017). Weeds at the Early Stage of Invasion: Pilot Projects. Evaluation Report 2014–2017. Department of Environment, Land, Water and Planning, Victoria.

First incursion, weed policy implementation and eradication of *Boerhavia erecta* (Nyctaginaceae) in northern Australia

John O. Westaway[1], Bert Lukitsch[2] and Roni Opden[2]
[1] Department of Agriculture and Water Resources, PO Box 37846, Winnellie, Northern Territory 0821, Australia
[2] Department of Environment and Natural Resources, Government, PO Box 496, Palmerston, Northern Territory 0831.
(john.westaway@agriculture.gov.au)

Summary The herbaceous weed, erect tar vine, *Boerhavia erecta* was reported to Government weed management authorities in Darwin in 2017, the first record for Australia. This American species is a widespread and sometimes serious weed in Australia's neighbouring tropical countries, particularly in annual crops. As such *B. erecta* had been included on the Northern Australia Quarantine Strategy (NAQS) target weed list as a species that is likely to arrive in Australia and potentially impact domestic agriculture and the environment. The species' presence on a Darwin rural property was verified using morphological and molecular identification. Commonwealth and State jurisdictions were notified of the detection.

Weeds Management Branch of the NT Department of Department of Environment and Natural Resources (DENR) declared the species and arranged to collaborate with NAQS to assess population(s), and manage this new incursion. DERN prepared an Incident Action Plan aimed at preventing the growth and spread of *B. erecta* which comprised a communications strategy and operational response plan. Communications included a weed alert pamphlet and letters to residential, farmer and nursery stakeholders. Operational response included traceback, surveillance of properties and repeated chemical treatment of the infestation with consent of affected landholder.

B. erecta appears to be confined to a single rural property near Darwin and prospects of eradication of this fast growing weed are promising due to the early response and collaborative effort between government departments and affected landholder. This case has also provided an opportunity for DENR to refine its weed rapid response policy.

Keywords *Boerhavia*, weed, policy, eradication.

Iris pseudacorus L. (Irideae, Iridaceae) in South Africa: Feasibility of eradication

Thulisile P. Jaca, Reshnee Lalla and Menzi Nxumalo
South African National Biodiversity Institute, Private Bag X 101, Pretoria 0001, South Africa
(t.jaca@sanbi.org.za)

Summary *Iris pseudacorus* is a perennial herb in the Iridaceae family, native to North Africa, Eurasia and the Mediterranean region. It has been introduced to several countries in the world including United States, Canada, and New Zealand where it is now considered invasive. It is an aggressive invader of wetlands and mostly found along river banks, stream banks and edges of the dams. In South Africa the species is listed as category 1a on the National Environmental Management Biodiversity Act 2004 (Act No. 10 of 2004) [NEM:BA] Alien Invasive Species List 2014, meaning that it is an eradication target. As part of a long-term management plan for the species in South Africa, its extent of invasion is currently being assessed. Here we report on the current extent of invasion of the species, seed germinability, the feasibility of eradication, and provide recommendations for management. The first record of this species in South Africa was in 2004 in one province. We have since discovered that *I. pseudacorus* has increased in abundance and range especially after 2010. At present, the species has been recorded from five provinces as 64 isolated populations in total. The species spreads mostly by rhizomes, however, in our seed germinability tests we found that seeds are highly viable in South Africa. In laboratory conditions, 90% of seeds germinated at 25°C. For feasibility of eradication, three herbicides were tested (imazapyr 100 g L^{-1}, glyphosate 480 g L^{-1} ae, and imazapyr 300 g kg^{-1} + glyphosate 216 g kg^{-1}). Imazapyr was effective in controlling *Iris pseudacorus* at concentrations of 0.5% and 1%. Regrowth was only observed in ~2% of the population towards the centre, 11 months after treatment with 0.5% imazapyr (100 g L^{-1}). Given its known distribution, we recommend that the species be reclassified as category 1b on the NEM: BA, and that further research is conducted on the possibility of biocontrol as a suitable management option.

Keywords: Iridaceae, *Iris pseudacorus*, invasive, South Africa.

Frogbit (*Limnobium laevigatum*) detection and eradication in New South Wales

Charlie Mifsud[1] and Terry Inkson[2]
[1] NSW Department of Primary Industries, Grafton, New South Wales 2460, Australia
[2] MidCoast Council, Forster, New South Wales 2428, Australia
(charles.mifsud@dpi.nsw.gov.au)

Summary Frogbit (*Limnobium laevigatum*) is an emerging aquatic weed in New South Wales (NSW) and is listed as Prohibited Matter under the *NSW Biosecurity Act 2015.* On 12 July 2017 a member of the public reported to MidCoast Council a suspicious plant growing on a pond in bushland located at the village of Green Point south of Forster. MidCoast Council staff inspected the site on 13 July and suspected the plant to be frogbit. A sample was submitted to the NSW Herbarium and the plant was confirmed as frogbit on 18 July. This was the first known naturalised infestation in NSW and posed an environmental threat to waterways within the adjoining national park. A rapid response plan was initiated involving staff from MidCoast Council, NSW Department of Primary Industries and National Parks. This plan involved:

1. treatment of the known infestation undertaken on 17 July;
2. surveillance of all 312 residential properties in Green Point on 8–9 August and high risk sites identified within 10 kilometres of the known infestation between August–October 2017; and
3. a communication plan involving advertising to alert the public undertaken by 31 July.

During surveillance the probable source of the infestation was located on a neighbouring residential property in Green Point where frogbit and other aquatic weeds were discovered in ponds. No other infestations of frogbit were discovered during the surveillance. A management plan was initiated on the residential property to eradicate the frogbit. A trace-back was initiated on the source of the frogbit to the residential property however the original supplier was no longer in business. Other sources of frogbit were located for sale online through Ebay and Gumtree. These sellers were contacted, their properties inspected and the plants confiscated.

Keywords Frogbit, prohibited matter, rapid response, eradication.

Australian weed biocontrol: A look at the past, present and future

Kerinne Harvey and Andrew McConnachie
Weed Research Unit, Department of Primary Industries, Locked Bag 6006, Orange, New South Wales 2800, Australia
(kerinne.harvey@dpi.nsw.gov.au)

Summary For over a century, Australia has successfully implemented the applied discipline of biocontrol as an approach to alleviate issues caused by problematic plant species (weeds). The benefits of this applied ecological approach lie not only in the efficacy in controlling the weed targets, but also because of the environmentally-friendly and self-sustaining attributes of many programs, with these benefits very often translating into a natural successive process of ecological restoration. As a result, biocontrol often forms a vital component of integrated weed management strategies. However, because these programs require significant long-term financial investment and expertise to match the time scales of the issue, Australia has seen a major decline in capability for weed biocontrol research, extension and on-ground management over the past 15 years. This has been recognised at a state and federal level, with several significant recent funding commitments being made towards existing and new weed biocontrol programs. This talk will look at a couple of the historically successful biocontrol of weeds programs in Australia, focussing on the role these programs have had in restoring ecological balance to the ecosystems that were once invaded. Here we further present in its early stage the development of a co-dependent partnership model to sustainably fund the future needs of research, development and extension in weed biocontrol. Collectively, this model encourages communication and consultation in bringing together Australia's leading weed biocontrol RD&E agencies for rebuilding capability and capacity, to fast-track and maximise the on-ground delivery of biocontrol agents.

Keywords Weed biocontrol, co-dependent partnership funding model.

Impacts of a pre-dispersal seed predator on achene production in the aquatic macrophyte, *Sagittaria platyphylla*

Raelene M. Kwong[1,2], Jean Louis Sagliocco[1], Nathan E. Harms[3] and Julie G. Nachtrieb[4]
[1] Agriculture Victoria, AgriBio, Centre for AgriBioscience, Bundoora, Victoria 3083, Australia
[2] Department of Ecology, Environment and Evolution, La Trobe University, Bundoora, Victoria 3086, Australia
[3] US Army Engineer Research and Development Center, Vicksburg, Mississippi 39180, USA
[4] US Army Engineer-Lewisville, Aquatic Ecosystem Research Facility, Lewisville, Texas 75057, USA
(rae.kwong@ecodev.vic.gov.au)

Summary To aid prioritisation of potential biological control agents of *Sagittaria platyphylla* (Engelm.) J.G.Sm. for use in Australia and South Africa, we studied the effects of a pre-dispersal seed weevil, *Listronotus appendiculatus* (Boheman) (Coleoptera: Curculionidae) on achene production and plant performance.

Levels of achene production and insect damage were compared between insecticide-treated and control plants over one flowering season in an experimental study pond. Based on its potential to significantly reduce achene production under field conditions in the native range, *L. appendiculatus* may make a suitable candidate for the biological control of *S. platyphylla* by assisting in slowing the rate of spread or reinvasion back into herbicide-treated areas. However, it is unlikely to provide population-level impacts on well-established infestations.

Keywords *Sagittaria platyphylla,* delta arrowhead, weed biological control, *Listronotus appendiculatus,* pre-dispersal seed predation.

INTRODUCTION

Sagittaria platyphylla (Engelm.) J.G.Sm. (Alismataceae) is an aquatic emergent herb indigenous to North and Central America that has become invasive in Australia and South Africa (Adair *et al.* 2012). Initially valued as an aquatic ornamental for use in ponds and aquaria, *S. platyphylla* has become a serious aquatic weed, particularly in shallow water bodies such as irrigation channels, drains, creeks and wetlands where it can rapidly form dense and extensive infestations that impede water flow and displace native species. Classical biological control is considered a desirable option for the management of *S. platyphylla* due to the difficulty and expense of controlling the weed in irrigation systems as well as the need for alternative approaches for use in sensitive aquatic habitats.

During surveys for natural enemies in North America (Kwong *et al.* 2014), the weevil, *Listronotus appendiculatus* (Boheman) was identified as a promising candidate for the biological control of *S. platyphylla* for three reasons. Firstly, it was found across most of the plant's native distribution from temperate to subtropical climates and in open and shaded habitats. Secondly, it could reach very high densities causing high levels of damage to reproductive structures (Kwong *et al.* 2017). Thirdly, it appears to have a limited host range and is not known to attack other species outside *Sagittaria* (Muenchow and Delesalle 1992, Harms and Grodowitz 2009).

Towards refining our understanding of the potential of *L. appendiculatus* as a candidate biological control agent for *S. platyphylla*, we conducted an herbivore-exclusion experiment in the USA to: (1) measure the seasonal impact of adult and larval feeding on plant reproductive performance; and (2) to determine if plants compensate for fruit herbivory through increasing sexual or asexual reproduction.

MATERIALS AND METHODS

The herbivore-exclusion study was conducted in an earthen, clay-lined pond (28 m × 75 m), located at the U.S. Army Engineer Lewisville Aquatic Ecosystem Research Facility (LAERF), Lewisville, Texas, USA (33.067016°, −96.954724°).

On 6 June 2012, 24 plots (0.5 × 0.5 m) were randomly placed approximately 3–5 m apart along the margins of the study pond within *S. platyphylla* stands so that each half of the pond contained 12 plots. Plots were randomly assigned a treatment of either non-herbivore (treated) or herbivore (control). Herbivores were eliminated from non-herbivore plots by foliar application of a mixture of Karate® insecticide (Syngenta Crop Protection, Greensboro, NC) at the rate of 0.1 mL L^{-1} in a water solution (0.030 mL insecticide applied per plot) and Thoroughbred® non-ionic-organosilicone surfactant (Estes Incorporated, Wichita Falls, TX) at the rate of 1 mL L^{-1}. The insecticide application was extended 0.5 m beyond the plot boundaries to ensure full coverage and applied when there was no breeze to ensure insecticide did not drift onto control plots. Plots were treated with insecticide every two weeks beginning 8 June and ending 13 September 2012. Herbivore plots were sprayed with water in the same manner and frequency as non-herbivore plots. No differences in initial plant height or water depth were detected among treatments (ANOVA, $F_{1,21}$ = 3.28, P = 0.08 for plant height; $F_{1,21}$ = 0.46, = 0.5 for water depth). Initial plant density in herbivore plots was marginally greater than in non-herbivore plots (Table 2). A small-scale greenhouse trial on potted *S. platyphylla* determined that the insecticide/surfactant mixture elicited no phyto-toxic or phyto-stimulation responses (R. Kwong unpubl. data).

Fruit herbivory Study plots were examined approximately every four weeks for fruit production and herbivore damage. The total number of undamaged and damaged fruit (surface or internal damage) per inflorescence was recorded. An additional assessment of internal damage was measured as the proportion of fruit per plot with damaged receptacles, which was assessed following the removal of achenes from each fruit (see *Achene herbivory* section below). The number of immature fruit per plot that had been completely destroyed by feeding activity was also recorded.

Achene herbivory Following fruit herbivory assessments, fruit from each plot were pooled and air dried. Achenes from mature fruit were dislodged from their receptacles and all extraneous plant material and insect larvae removed. To estimate mean achene weight per fruit, the pooled achenes per plot were weighed and averaged across the number of mature fruit per plot. Individual achene weight was determined by counting the number of achenes (intact + eaten + unfertilised) present in a 0.025 g subsample of the pooled achenes for each plot. The number of achenes per fruit was then calculated as the ratio of the achene weight per fruit and the individual achene weight. Weevil larvae targeted the achene embryos and did not feed upon unfertilised achenes. Therefore, when estimating the proportion of damaged achenes per fruit, unfertilised achenes were subtracted from the total number of achenes.

Plant response To evaluate end of growing season differences in plant density and biomass due to herbivore-plant interactions, all 24 plots were harvested at 3.5 months after commencement of the trial (22 September 2012). The total number of *S. platyphylla* plants and tubers per plot were counted. Foliage and tuber material were dried separately to constant weights in an oven at 55°C for a minimum of 48 h.

Post-trial herbivory To assess the continued activity of weevils on fruit damage and achene production, *S. platyphylla* was sampled on two occasions (22 October and 21 November 2012) in an adjacent stand occurring within the same pond. This stand had not previously been sprayed with insecticide or water. On each sampling occasion, 12 quadrats were haphazardly placed within the stand and were assessed for the following measures: number of fruit per plot, proportion of fruit per plot with internal damage, number of achenes per fruit and proportion of damaged achenes per fruit. Estimates for number of achenes and percentage of damaged achenes per fruit followed the same procedures outlined in *Achene herbivory* above.

Statistical analysis All statistical analyses were conducted using GenStat 16th Edition (VSN International Ltd., www.vsni.co.uk). The exclusion experiment was a randomised block design with blocking based on two halves of the pond. Estimated means for non-herbivore (insecticide) and herbivore (control) treatments and for the different sampling dates were compared using analysis of variance (ANOVA) or linear mixed models. The choice of statistical technique depended on the complexity of the structure of the data and the degree of balance. For some response variables, missing values such as the absence of fruit on a particular sampling date, created an unbalanced design (Quinn and Keough 2002). Proportion data were analysed using generalised linear mixed models (GLMMs), with the link functionlogit(p) = log(p/1−p) applied to the expected proportions, thus necessitating the results being presented as back-transformed means on the original scale. The analysis of plant density at the end of the trial used pre-trial plant density as a covariate.

RESULTS

The removal of herbivores from plots using an insecticide spray treatment had no discernible influence on the number of fruit produced by plants at the individual or plot level throughout the duration of the trial (Table 1, Figure 1a,b). Nor did the application of insecticide appear to deter insect pollinators from visiting flowers as the percentage of unfertilised achenes between herbivore and non-herbivore plots were comparable (Table 1.).

Fruit herbivory The insecticide treatment reduced fruit herbivory on *S. platyphylla*, measured as piercing or chewing surface damage and internal weevil larva feeding damage. Temporal variation in herbivory occurred and treatment differences appeared rapidly (13 June 2012) for both surface and internal damage (Table 1, Figure 1c,d). Surface damage was highest at the start of the trial irrespective of treatment and was likely due to superficial feeding by adult *L. appendiculatus* and the pentatomid bug *Euschistus ictericus* (Linnaeus), which was also observed on the fruits of *S. platyphylla* at the site. Across the duration of the trial, on average 30 and 34% of fruit in herbivore plots showed signs of surface and internal damage respectively and insecticide treatment reduced these levels of damage by 39% for surface damage and 83% for internal fruit damage (Table 1). Weevil larvae tunnelled into and consumed tissue within the receptacle. Overall, receptacle damage in herbivore plots occurred in almost half of the fruit, but ranged from 46.7% (95% CI = 31.2, 62.8) at the start of the trial and increased to 65.9% (95% CI = 52.6, 77.1) by the completion. In contrast, no receptacle damage was recorded in the insecticide plots (Table 1). When immature fruit were heavily attacked, they were completely destroyed, while no immature fruit were destroyed in the non-herbivore plots (Table 1).

Achene herbivory Inflorescence-feeding insects decreased the total weight of achenes per fruit across the duration of the trial (Table 1). This was due to an overall 22% reduction in the mean total number of achenes (fertilised and unfertilised) per fruit from 518 in non-herbivore plots to 403 in herbivore plots. Of total achenes, the number that remained intact and viable (i.e. not damaged by larvae) was 38% lower in herbivore plots compared to non-herbivore (i.e. 291 compared to 467) (Table 1, Figure 1e,f). In addition to feeding on receptacle tissue, larvae also consumed the embryos of fertilised achenes. Overall, larvae destroyed an average of 14.3% of achenes in herbivore plots compared to only 0.06% of damage in

Table 1. Comparisons of fruit and achene production and levels of herbivory by the pre-dispersal seed predator, *Listronotus appendiculatus* between herbivore (control) and non-herbivore (insecticide-treated) plots of *Sagittaria platyphylla*. Values are means with 95% confidence intervals [minimum, maximum] shown in parentheses.

Outcome measures	Comparisons between treatment groups		
Trial with repeated measures [T1-T5]	Herbivore (control)	Non-herbivore (insecticide-treated)	p-value
Fruit herbivory measures			
Number of fruit per inflorescence [means]	5.84 [5.42, 6.25]	5.37 [4.97, 5.76]	0.168
Number of fruit per plot [back-transformed means]	23.76 [17.50, 32.26]	26.50 [19.64, 35.75]	0.689
% Fruit per inflorescence with surface damage [back-transformed mean]	29.92 [23.19, 37.65]	18.12 [13.68, 23.61]	0.012
% Fruit per inflorescence with internal damage [back-transformed mean]	34.00 [27.93, 41.14]	5.80 [4.02, 8.32]	<0.01
% Fruit per plot destroyed at immature stage [back-transformed means]	9.28 [5.27, 13.30]	0.00	<0.01
% Damaged receptacles per plot [back-transformed means]	49.32 [38.36, 60.36]	0.00	<0.001
Achene production measures per plot on mature fruit			
Achene weight per fruit (mg) [means]	0.0578 [0.0492, 0.0664]	0.0747 [0.0663, 0.0830]	0.004
Total number of achenes (fertilised and unfertilised) per fruit [means]	403 [356, 449]	518 [472, 563]	0.001
Number of intact (undamaged) achenes per fruit [means]	291 [240, 343]	467 [417, 517]	<0.001
% of damaged achenes per fruit [back-transformed means]	14.18 [9.63, 20.41]	0.062 [0.030, 0.130]	<0.001
% of unfertilised achenes per fruit [back-transformed means]	9.94 [7.30, 13.39]	7.22 [5.29, 9.78]	0.086

non-herbivore plots (Table 1; $F = 166.15$; $df = 34.6$; $P = <0.001$, time $F = 4.15$; $df = 59.8$; $P = 0.005$; interaction $F = 6.48$; $df = 136.4$; $P < 0.001$), although two periods of greater damaged were observed in herbivore plots on the 11 July (22%) and the 21 September (39%).

Plant response Over the duration of the trial, there was around a three-fold increase in plant density in non-herbivore plots and a four-fold increase in herbivore plots, however no treatment effect was evident in the number of plants per plot at the end of the trial when corrected for initial plant density (Table 2). Tuber abundance and total dry weight of tubers per plot did not differ between herbivore and non-herbivore plots, however above ground plant biomass was 35% less in non-herbivore than in herbivore plots, although the effect was marginally significant (Table 2).

Post-trial herbivory Fruit production continued at the site for the two consecutive post-trial assessment dates, although the percentage of fruit attacked by weevil larvae reduced by 74% (Table 3) and very little feeding damage was observed to the achenes.

Table 2. Comparison of plant density and above and below-ground plant biomass between insecticide-treated (non-herbivore) and control (herbivore) plots of *Sagittaria platyphylla*. Values are means with 95% confidence intervals [minimum, maximum] shown in parentheses.

Outcome measures	Comparisons between treatment groups		
	Herbivore (control)	Non-herbivore (insecticide-treated)	p-value
Plant density per plot pre and post			
Pre-trial number of plants [means]	15.2 [11.4, 19.0]	20.7 [16.9, 24.5]	0.05
End of trial number of plants [means]	61.9 [52.3, 71.5]	59.3 [49.7, 68.9]	0.69
End of trial plant response per plot			
Number of tubers [means]	50.1 [12.1, 88.1]	113.1 [22.5, 203.7]	0.12
Dry tuber biomass (g) [back-transformed means]	8.16 [3.87, 17.23]	21.3 [10.1, 45.0]	0.07
Above ground biomass (g) [means]	67.7 [52.0, 83.4]	44.1 [28.4, 59.8]	0.04

Table 3. Post-trial herbivory assessments on an untreated (non-insecticide) stand of *Sagittaria platyphylla*. Comparisons of fruit and achene production and levels of herbivory by the pre-dispersal seed predator, *Listronotus appendiculatus* between two dates. Values are means with 95% confidence intervals [minimum, maximum] shown in parentheses.

Outcome measures	Comparisons between treatment groups		
Trial with repeated measures [T1–T5]	22 October 2012	21 November 2012	p-value
Fruit herbivory measures			
Number of fruit per plot [back-transformed means]	23.00 [15.9, 30.1]	17.00 [8.73, 25.3]	0.24
% Fruit per inflorescence with internal damage [back-transformed mean]	20.63 [14.00, 29.40]	5.39 [2.05, 13.42]	0.01
Achene production measures per plot on mature fruit			
Number of intact (undamaged) achenes per fruit [means]	351 [263, 440]	435 [331, 539]	0.21
% of damaged achenes per fruit [back-transformed means]	3.8 [1.8, 11.0]	0.05 [0.04, 0.23]	<0.001

DISCUSSION

Impact on reproductive performance Our investigations demonstrated that the pre-dispersal seed predator *L. appendiculatus* negatively affected the reproductive performance of *S. platyphylla* in three ways. Firstly, when larvae attacked immature fruit they often completely destroyed the fruit leaving behind a mass of frass and the remnants of achenes. Hence, an average of 10% of all fruit produced per plant was destroyed prior to achene maturation. Secondly, feeding damage of receptacle tissue disrupted the development of achenes resulting in an overall 22% reduction in the number of achenes produced per fruit. In herbivore plots, an average of 50% of fruit across the 3-month trial sustained insect damage to the receptacle. Finally, direct consumption of the fertilised achene embryos resulted in a further 14% destruction of achenes per fruit. Therefore, the remaining number of achenes produced per fruit was reduced from 467 in non-herbivore plots to 291 in herbivore plots, which represents an average overall reduction of 38%. During our field surveys conducted at 38 sites from 2010 to 2012 (Kwong *et al.* 2017), achene destruction varied markedly from site to site which may be associated with a variety of factors including: (a) habitat or climate factors operating at a local or regional scale; (b) episodic rather than chronic impacts; and/or (c) levels of parasitism or predation of the pre-dispersal seed predators or of the plant's pollinators. While parasitism levels were not quantified during our insect-exclusion experiment, we have documented a number of parasitic wasp species from *L. appendiculatus* larvae (N. Harms unpubl. data). In addition, the predatory reduviid bugs (Heteroptera: Reduviidae) were observed at some sites, piercing fruit and feeding on *L. appendiculatus* larvae. In contrast, weevil populations were in such high abundance at other sites, such as Yazoo Wildlife Management Reserve in Mississippi, that few of the sampled fruit produced achenes.

Plant response to herbivory Herbivores and pathogens affect plants in different ways and as such, plants exhibit a diversity of responses that can vary from susceptibility (the inability of the plant to resist

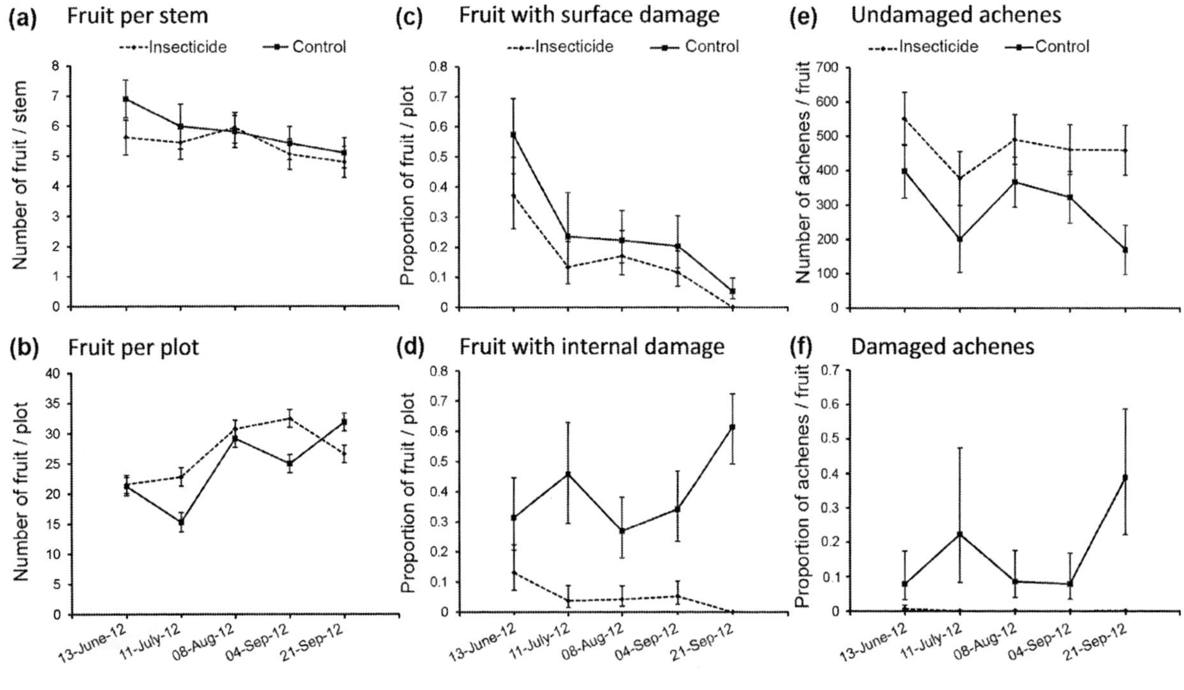

Figure 1 (a–f). Effects of insecticide treatment to exclude the curculionid pre-dispersal seed predator, *Listronotus appendiculatus* on *Sagittaria platyphylla* over a 14-week period. Fruiting head (fruit) production: (a) the number of fruit per stem; and (b) the total number of fruit per plot fruit. Fruit damage: proportion of fruit per plot with (c) surface and (d) internal damage. Achenes: (e) the number of undamaged achenes per fruit, and (f) the proportion of achenes damaged per fruit. Insecticide treatment commenced on 8 June 2012. Dashed line represents non-herbivore plots (insecticide-treated), solid line represents herbivore plots (control, treated with water). Means (a) and back-transformed means (b) to (f) are plotted with 95% CI.

herbivore damage) to resistance (plant traits or life-history mechanisms that enable plants to avoid, defend or tolerate herbivore damage), or overcompensation (plants that have higher fitness when they are damaged compared with related plants that are undamaged (Agrawal 2000). In terms of seed predation, plants may avoid or minimise losses of seed to predators in at least two ways (Atlan *et al.* 2010): phenological escape and predator satiation. Phenological escape results from asynchrony between predator and prey populations. In this case, fruit production occurs at a time when herbivore populations are minimal, either before or after peak seed predation. Predator satiation occurs where plants produce an overabundance of fruit thereby reducing the probability of an individual fruit being attacked (Janzen 1971). We found evidence for both of these strategies. *Sagittaria platyphylla* used a phenological avoidance strategy by continuing fruit production for up to two months following inactivity by *L. appendiculatus*. Secondly, *S. platyphylla* produced an abundant supply of fruit across the season with the proportion of fruit attacked never reaching more than 60%. However, rates of fruit predation were found to be highly variable in natural populations, ranging from no attack to almost 100% of fruit attacked (Kwong 2016), suggesting that predator satiation may not occur at all sites or at all times.

In addition to avoidance and satiation strategies, another response to herbivory may be through compensation, by replacing tissue lost to herbivory or reallocating resources to other plant parts. In our study, we expected to find a compensation for fruit herbivory through the production of more fruit, yet we found no evidence for an increase in fruit production at either the plant or plot level. As with other studies where seeds from non-attacked fruit were heavier than from insect-attacked fruit (English-Loeb and Karban 1992), we expected to find an increase in weight of intact achenes, yet we found no evidence for this.

The reallocation of resources to vegetative biomass is seldom reported in the literature as a response to herbivory of reproductive structures, therefore we expected to find no effect of this in our study. Yet in our study, aboveground plant biomass was 1.5 times greater in herbivore compared to non-herbivore plots. A likely explanation for the possible reallocation of resources in our study might be associated with insect herbivore damage to leaves and petioles, which would have been reduced following insecticide application. At the study site, leaf damage in the form of small shot holes was caused by adults of the weevil, *Listronotus lutulentus* (Boheman), while leaf petioles were damaged by larvae of both *L. lutulentus* and *L. appendiculatus* boring into and pupating at the base of petioles. The increase in aboveground plant biomass in herbivore plots was not associated with plant density as plant density levels were similar between herbivore and non-herbivore plots. We therefore hypothesise that leaf and petiole herbivory may have induced a plant growth compensatory response.

To assess the potential of *L. appendiculatus* for the biological control of *S. platyphylla* in introduced countries we need to determine what influence seed herbivory has on the population dynamics of the weed. *Sagittaria platyphylla* is a perennial, long-lived species due to its prolific clonal growth via stolons and tubers. While we do not yet understand the importance of seed production on population growth rate, we suspect that seedling recruitment is unlikely to play an important role in the population dynamics of established populations. *Sagittaria platyphylla* has proven difficult to control with conventional methods due to submersed rosettes and subterranean tubers that are protected from contact herbicide application. However, populations can rapidly recolonise from the existing seed bank when conditions become favourable, such as during draw-down events in warmer months (Flower 2004). Hence, the reduction in seed banks caused by *L. appendiculatus* may assist in reducing the amount of seed available for re-colonisation following herbicide application.

From a biocontrol perspective, we predict that *L. appendiculatus* on its own is unlikely to reduce existing stands of *S. platyphylla* in the long term, but rather should be seen as one of a complementary suite of control agents that attack different life history stages of the weed.

ACKNOWLEDGMENTS

We are grateful to the Rural Industries Research and Development Corporation (Project No. PJR-007053), Murrumbidgee Irrigation Ltd and the Victorian Department of Economic Development, Jobs, Transport and Resources (formerly Department of Environment and Primary Industries) (Australia) for project funding.

REFERENCES

Adair, R., Keener B., Kwong, R., Sagliocco, J-L. and Flower, G. (2012). The biology of Australian weeds 60. *Sagittaria platyphylla* (Engelmann) J.G. Smith and *Sagittaria calycina* Engelmann. *Plant Protection Quarterly* 27, 47-58.

Agrawal, A.A. (2000). Overcompensation of plants in response to herbivory and the by-product benefits of mutualism. *Trends in Plant Science* 5, 309-13.

Atlan, A., Barat, M., Legionnet, A.S., Parize, L. and Tarayre, M. (2010). Genetic variation in flowering phenology and avoidance of seed predation in native populations of *Ulex europaeus*. *Journal of Evolutionary Biology* 23, 362-71.

English-Loeb, G.M. and Karban, R. (1992). Consequences of variation in flowering phenology for seed head herbivory and reproductive success in *Erigeron glaucus* (Compositae). *Oecologia* 89, 588-95.

Flower, G.E. (2004). The biology and control of arrowhead. Report to Goulburn-Murray Water, Tatura, Victoria. http://www.g-mwater.com.au/downloads/arrowhead/giles flower_arrowhead_report.pdf (accessed 18 April 2016).

Harms, N.E. and Grodowitz, M. (2009). Insect herbivores of aquatic and wetland plants in the United States: a checklist from literature. *Journal of Aquatic Plant Management* 47, 73-96.

Janzen, D.H. (1971). Seed predation by animals. *Annual Review of Ecology, Evolution and Systematics* 2, 465-92.

Kwong, R., Sagliocco, J-L., Harms, N., Shearer, J.F., Keener, B. and Green, P. (2014). Prospects for the biological control of delta arrowhead (*Sagittaria platyphylla*), an invasive aquatic species in Australia. Proceedings of the XIV International Symposium On Biological Control of Weeds, eds F.A.C. Impson, C.A. Kleinjan and J.H. Hoffmann, pp. 53-67. (University of Cape Town, South Africa).

Kwong, R.M. (2016). The invasive aquatic macrophyte *Sagittaria platyphylla* (Alismataceae): Is it a suitable target for classical biological control? PhD dissertation, La Trobe University.

Kwong, R.M., Sagliocco, J-L., Harms, N.E., Butler, K.L., Green, P.T. and Martin, G.D. (2017). Biogeographical comparison of the emergent macrophyte, *Sagittaria platyphylla* in its native and introduced ranges. *Aquatic Botany* 141, 1-7.

Muenchow, G. and Delesalle, V. (1992). Patterns of weevil herbivory on male, monoecious and female inflorescences of *Sagittaria latifolia*. *American Midland Naturalist* 127, 355-67.

Quinn, G.P. and Keough, M.J. (2002). 'Experimental design and data analysis for biologists'. Cambridge University Press.

Project summary of biocontrol agent *Eueupithecia cisplatensis* for *Parkinsonia aculeata* in Queensland by Queensland DAF

Kelli Pukallus, Joshua Nicholls, Centaine Ferris and Judy Clark
Queensland Department of Agriculture and Fisheries, Biosecurity Queensland, Tropical Weeds Research Centre, Charters Towers, Queensland 4820, Australia.
(kelli.pukallus@daf.qld.gov.au)

Summary The leaf-feeding looper, *Eueupithecia cisplatensis* Prout (Lepidoptera: Geometridae) commonly referred to as UU, is a biological control agent for the invasive Weed of National Significance, *Parkinsonia aculeata* L. (parkinsonia). With funding support from Meat and Livestock Australia (MLA), UU was imported from Argentina to Brisbane by Commonwealth Scientific and Industrial Research Organisation (CSIRO) in 2010. Mass-rearing commenced in 2013 at the Queensland Government's Department of Agriculture and Fisheries (Qld DAF) Tropical Weeds Research Centre (TWRC) in Charters Towers. Field releases of UU have occurred at 105 sites throughout Queensland, with establishment observed at 50% of monitored sites in central Queensland and 75.5% in North Queensland.

Keywords Queensland Department of Agriculture and Fisheries, UU, parkinsonia.

INTRODUCTION

Parkinsonia aculeata L. (parkinsonia) is a small, thorned tree originating from the Americas, and found mainly in riparian areas and flood plains. Parkinsonia impedes grazing due to forming dense thickets, reduces grass production and creates feral animal reserves (Qld DAF 2018a).

Three biocontrol agents have been previously released to help control parkinsonia; a sap-sucking bug (*Rhinacloa callicrates* Herring) and two seed feeders (*Mimosestes ulkei* Horn and *Penthobruchus germaini* Pic). All are established and causing various degrees of impact throughout Queensland (van Klinken and Heard 2012). Native range surveys commenced in central and South America in 2002 by CSIRO for new biological control options. *Eueupithecia cisplatensis* Prout (UU) was imported from Argentina in 2010, with approval granted from Australian Department of Agriculture for release in 2013. Mass-rearing of UU commenced at the Queensland DAF's Tropical Weeds Research Centre (TWRC) in March 2013, providing in-kind services to CSIRO. Funding from Meat & Livestock Australia Ltd through Australian Government's Rural Research and Development (R&D) for Profit Program, was received in March 2016 via an arrangement with CSIRO supporting an increase in the mass-rearing and release of the UU project already underway at the TWRC.

The larval stage of *E. cisplatensis* consumes leaflets, damages the leaf and causes defoliation of parkinsonia. This reduction in photosynthetic capabilities impacts the health of the plant, making it more susceptible to fungal pathogens and reduces the overall vigour and reproductive capability of the plant.

METHODOLOGY AND RESULTS

UU biology The life cycle of UU is completed in 23–27 days on average (25°C, 65% RH, 14:10 day length) under glasshouse conditions, with life stages completed in: egg to larva 5–7 days; larva to pupa 13–15 days; and pupa to adult 5 days. Adults live for up to one week, mating within 24 hours of emergence and females start laying eggs shortly after mating.

Eggs are ovoid, a rusty-brown or green colour and laid individually, or in chandelier formation on the leaflets (Figure 1a). Females lay an average of 60–80 eggs. Larvae are cryptic and camouflaged when in situ on a leaf, laying along the midrib (rachis) or at a 45 degree angle (Figure 1b). Larval colour ranges from lime green, rusty red/brown through to light green/yellow.

Pupae are brown, 10 mm in length, and occur encased in a silken cocoon, which has hydrophobic properties to assist UU to survive in flooded conditions (Figure 2a). They are attached to the stem/branch or along the rachis in the field. Adults are a mottled grey/black colour, have a 15 mm wingspan, and at rest, position their wings along the leaf (Figure 2b).

Rearing methodology Adults were collected and sexed from gauzed rearing cages (150 cm × 120 cm × 140 cm) and placed (15 adults maximum) into décor plastic containers lined with paper towel. The adult boxes were kept at 25°C and opportunistically moistened at normal day/night conditions. Once the eggs hatched, the paper towel was removed and placed, eggs facing down, on a parkinsonia plant in a rearing cage, and removed after four days to allow for total eclosion. Plants were replaced when leaflets were near completely consumed, with most cages going through

Figure 1. *Eueupithecia cisplatensis* (UU) (a) eggs laid on leaflets and (b) larva at rest (45 degree angle) on a parkinsonia leaf.

Figure 2. *Eueupithecia cisplatensis* (UU) (a) pupae encased and removed and (b) female adult moth.

7–8 host plants with 250 larvae, over 21 days. Cage floors were cleaned daily to remove fallen plant material and frass and were wiped with methylated spirits, to prevent bacterial and fungal colonies from developing. Larvae wandered to pupate, pupating on the plant, around the pot rim/base or corners/edges of the cage. Pupae were collected with tweezers after browning and hardening, and stored in takeaway containers at 9°C for up to 3 weeks, until released. Adults were collected from uncollected pupa from different cages to ensure genetic diversity within adult boxes.

Release methodology The initial release strategy was using larvae placed in clustered branches, attached with a lock-tie onto tips of parkinsonia branches (Figure 3a). Predation from wasps and ants was immediate, with limited larvae surviving. Eggs and adults from adult boxes were occasionally released, by wrapping and attaching the paper towel with the eggs attached onto branches, similar to larval releases. Pupal releases were the preferred method and are quicker, easier and less bulky for landholder/stakeholders and have reduced transport costs. Shallow plastic takeaway containers, lined with paper towel, containing 600–700 pupae were placed into a delta trap and attached to a parkinsonia branch (Figure 3b). The moths emerge over several weeks, mate and are able to fly to new trees or higher on the plant, compared to the limited mobility offered by larval or egg releases.

Release sites Release sites were on private grazing/agricultural land, mine sites, National parks, roadside verges or council reserves in riparian zones or on flood plains. *Eueupithecia cisplatensis* has been released at 105 sites within five local government areas within Queensland since April 2013. With the assistance of landholders, stakeholders, Queensland government and local council employees.

A total of 297,029 pupae, 504,600 larvae or eggs and 3,270 adult moths have been released (Table 1). Release frequency and quantity have varied at each release site, ranging from one to 20 separate releases over the years, depending on establishment and if they were classified a nursery site.

Figure 3. *Eueupithecia cisplatensis* (UU) (a) larval release method and (b) pupal release method.

Monitoring and establishment Release sites were monitored periodically and opportunistically from six months after release and were conducted using the beat-sheet method. A white sheet is placed under a section of foliage/tree and hit with a hoe handle several times, dispersing the larvae onto the sheet. The number and size range of the larvae were recorded, along with other Lepidopteran larvae or past agents, such as *R. callicrates*. Sites are given a present or non-present rating after monitoring, or unknown if the sites have not been revisited (Figure 4). Establishment (continued present rating) in monitored sites was 75.5% North Queensland and 50% Central Queensland (Table 1).

Monthly monitoring at a nursery site along the Burdekin River, north of Charters Towers (−19.80489° S, 146.07713° E), shows the larval population increases during summer and autumn. Averages are 4.3 and 5.42 larvae per tree for these seasons, coinciding with major rain events and new leaf growth, compared with spring (2.49) and winter (1.59). Three monthly monitoring was conducted at additional sites along the east-coast in the Burdekin region and in Central Queensland region since September 2017.

Establishment at non-release sites has been observed at several locations within Queensland, and has occurred more than 20 km from known release sites. At a non-release site near Brandon in North Queensland (−19.50228° S, 147.30742° E) in April 2016, over 40 larvae were counted from monitoring one section of a small tree (<2 m tall). Significant damage was noted to the tree, with limited leaflets, browned midribs and no flowers.

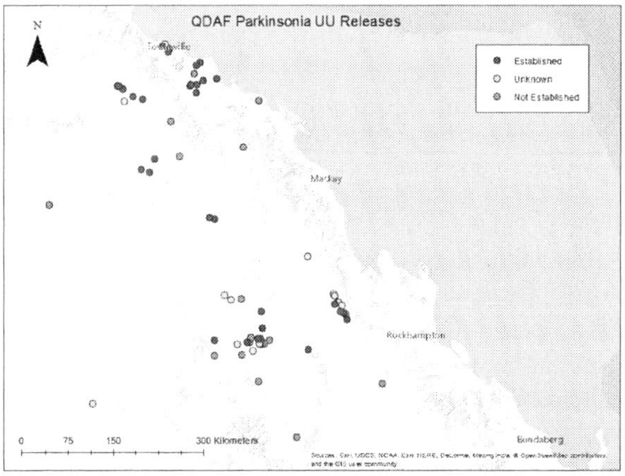

Figure 4. *Eueupithecia cisplatensis* (UU) release sites on parkinsonia throughout Queensland (April 2013–2018).

Table 1. *Eueupithecia cisplatensis* (UU) release numbers on parkinsonia in Queensland (April 2013–2018).

Region	# sites	Pupa	Larva/eggs	Adults	No. sites establishment noted	% established of monitored sites
Central Queensland (−24.8 to −22.2° S)	56	115,692	23,100	0	19	50
North Queensland (−21.6 to −19.2° S)	49	181,337	481,500	3,270	37	75.5
Totals	105	297,029	504,600	3,270	56	65.1

Impact The impact of biological control agents is often difficult to quantify in the field. Laboratory experiments on larval consumption of parkinsonia's bipinnate leaf structure have shown that a single UU larva can consume up to 118 leaflets (average 53.4) over a 15 day larval stage at 25°C (12:12h day/night). Consumption peaks when larval length reaches 20 mm, at around day 13. In the field the larvae are dispersed on the plant, rather than clumped, causing dispersed impacts across the plant. Larvae consume the leaflets, either whole or partly, and damage the rachis causing the leaf to eventually abscise.

Time lapse photographs taken every hour in rearing cages have shown daily larval movement throughout the plant architecture, and nocturnal feeding habitats. A UU larva can move up to 21 nodes on a single stemmed parkinsonia seedling (average height of 359.1 mm) over 13 days, travelling approximately 240 mm in stem length. Daily larval movement and leaflet consumption is sporadic and they rarely stay in the one location on a leaf.

DISCUSSION

With good field establishment, equating to seasonal persistence and agent dispersal, what does it mean in terms of impact and control of parkinsonia?

We know one larva consumes on average 53.4 leaflets within its development, and the average number of UU larvae sampled on parkinsonia in the field ranges from 1.59–5.42 per part tree, compared to the real impact that has been observed in the field from a population density of 40 UU on sampling a part tree. Accordingly, the current field density needs to increase significantly (by over 700%) to have an impact on the number of leaflets on a mature tree, conceivably to slow the reproductive potential and therefore the spread of parkinsonia, based on those numbers. With the completion of upcoming seedling exclusion and further cage trials, the impact of UU and the potential to reduce the population of parkinsonia seedlings will be better quantified.

Since 2014, release sites have been in local government drought declared or partly declared areas (Qld DAF 2018b), providing limited food resources at times for UU. Yet throughout these years, UU has persisted and spread to new locations. Seasonal peaks in population densities of UU are created, with a lag effect once rain events occur, as shown at the monitoring site at Burdekin River. The long term environmental events are unknown, but if UU can survive through the current conditions it will persist in the long term. This project has mass-reared and released the largest number of agents from Qld DAF's TWRC facility and has been a long-running program, due for completion shortly.

We are on the way to deeming the project a success due to the large numbers of UU released, establishment success, but whether a reduction in seedling recruitment and associated reduction of control activities for parkinsonia is observed remains to be seen in the years ahead.

ACKNOWLEDGMENTS

Funding support was received from MLA, CSIRO and the Australian Government. Thanks to technical support from CSIRO staff (S. Raghu, Andrew White and Gio Fichera). Release assistance from regional and city councils of: Burdekin; Central Highlands; Charters Towers; Isaac; and Townsville; Capricorn Catchments Inc, Fitzroy Basin Association Inc, the Central Highlands Regional Resources Use Planning Cooperative; and Queensland Parks and Wildlife Services staff. Thanks to numerous landholders across Queensland for access to their properties, release and monitoring assistance.

REFERENCES

Qld DAF, Queensland Department of Agriculture and Fisheries (2018a), Parkinsonia fact sheet. https://www.daf.qld.gov.au/__data/assets/pdf_file/0014/55040/IPA-Parkinsonia-PP36.pdf (accessed 15 April 2018).

Qld DAF Queensland Department of Agriculture and Fisheries (2018b), Queensland Drought situation maps. https://www.longpaddock.qld.gov.au/drought/archive/ (accessed 15 April 2018).

van Klinken, R.D and Heard, T.A (2012) *Parkinsonia aculeata* L. – parkinsonia. *In* 'Biological Control of Weeds in Australia', eds M. Julien, R. McFadyen and J. Cullen, pp. 437-47. (CSIRO Publishing, Australia).

Progress towards biological control of bellyache bush and prickly acacia

Dianne B.J. Taylor, Boyang Shi, Jason Callander and Kunjithapatham Dhileepan
Biosecurity Queensland, Department of Agriculture and Fisheries, Ecosciences Precinct, Dutton Park, Queensland 4102, Australia
(Di.Taylor@daf.qld.gov.au)

Summary Bellyache bush (*Jatropha gossypiifolia*) and prickly acacia (*Vachellia nilotica* subsp. *indica*), both Weeds of National Significance, are capable of dominating pasture and native grasslands, hampering stock management and expensive to control with herbicides. Bellyache bush is also toxic to stock and people. Biological control is the most economically viable management option for both weeds. Bellyache bush has been a target of biological control in Australia since 1997, with one agent released to date that failed to establish. Following field surveys in Peru, Paraguay and Bolivia, a small leaf-mining moth (*Stomphastis* sp.) was imported from Peru for host specificity testing in quarantine in 2014. Testing of 48 species has been completed, with the insect proving to be highly host specific. An application to release *Stomphastis* sp. is currently being considered by federal regulatory bodies. Prickly acacia has been a target of biological control since the 1980s, but with limited success to date. Based on plant genotype and climate matching, the search for new agents was redirected to Ethiopia and Senegal, focussing on gall-inducers. Several potential agents have been identified, two of which, a gall thrips (*Acaciothrips ebneri*) from Ethiopia and stem-galling fly (*Notomma mutilum*) from Senegal, have been imported into quarantine in Australia for host specificity testing. Imported in late 2015, the gall thrips has been tested on more than 50 plant species. Results are promising with gall induction only occurring on prickly acacia. The stem-galling fly was imported in April 2018. A culture is currently being established in quarantine and host testing will begin later in the year.

Overwintering habitat and the survival of the biocontrol beetle, *Cassida rubiginosa*

Michael G. Cripps[1], Sarah D. Jackman[1], Chikako van Koten[1], Cor J. Vink[2] and Michael B. Dodd[3]
[1] AgResearch, Lincoln, Private Bag 4749, Christchurch 8140, New Zealand
[2] Canterbury Museum, Christchurch, New Zealand
[3] AgResearch, Grasslands Research Centre, Private Bag 11008, Palmerston North 4442, New Zealand
(mike.cripps@agresearch.co.nz)

Summary The leaf-feeding beetle, *Cassida rubiginosa* Müller, was introduced to New Zealand in 2007 as a biological control agent against the pasture weed, *Cirsium arvense* (L.) Scop. (Californian thistle). The beetle is now established in most regions, but outbreak populations have only been observed in one region (Wairarapa). The overwintering habitat type might affect the survival of the beetle and its ability to achieve outbreak populations. We tested the overwintering survival of *C. rubiginosa* in three different habitats: radiata pine forest, native forest, and pasture. In autumn 2016, adult diapausing *C. rubiginosa* were placed in overwintering cages established in the three habitats, and in spring their survival was assessed. Temperature and relative humidity was recorded in each habitat, and predatory spiders were also sampled. The total recovery rate of overwintered beetles was greatest in the native forest (42 ± 7.9%), followed by ungrazed pasture (32.5 ± 7.5%), and pine forest (7.5 ± 4.2%). The total percentage recovered was not significantly different between the native forest and pasture habitats, but both of these habitats had greater recovery rates than the pine habitat. The native forest provided the best insulation against temperature and humidity extremes on a daily basis, and across the winter season, and is the likely explanation for the greater recovery rates in that habitat. This study has provided initial evidence that the particular overwintering habitat may contribute to the success of this biocontrol agent, and suggests that native forest is likely to provide a better habitat than pine forest.

Keywords Green thistle beetle, Californian thistle, biological control.

INTRODUCTION

Cassida rubiginosa Müller (Coleoptera: Chrysomelidae) is a leaf-feeding beetle of European origin that was introduced to New Zealand in 2007 as a biological control agent against *Cirsium arvense* (L.) Scop. (Californian thistle) (Cripps *et al.* 2011). Under controlled experimental conditions the beetle has been shown to reduce the density, size, and biomass of the weed, indicating that it has potential as a biocontrol agent (Cripps *et al.* 2010). However, in the field it seldom reaches population densities capable of exerting a regulating influence on its host plant. In Europe, and in North America (where the beetle was inadvertently introduced prior to 1901), this is primarily attributed to high rates of predation and parasitism (Ang and Kok 1995). However, in the novel ecosystem of New Zealand that has no native thistle species the beetle apparently experiences a high degree of enemy-free space, which allows for population outbreaks (Cripps 2013).

The beetle is now established in most regions of New Zealand, although its abundance is variable ranging from low, but persistent populations, to outbreak populations of hundreds of individuals per thistle shoot (M. Cripps, pers. obs.). One possible explanation for the variable abundance is the suitability of overwintering habitat. In the native range, the beetle is known to migrate from meadows, where it feeds on thistle species, to nearby broadleaf forests where it overwinters under leaf litter, which is thought to better insulate against environmental extremes (Kosior and Klein 1970). Life table studies in the native range have indicated that the overwintering survival rate of *C. rubignosa* is between 16 and 34%, and is a severe bottleneck in its population increase, although the causes for the low survival rate are unknown (e.g. predation, or environmental factors) (Koji *et al.* 2012). Therefore, understanding the overwintering conditions that enhance the survival of this beetle might help promote its effectiveness as a biocontrol agent. To this end, we tested the overwintering success of *C. rubignosa* in three different habitats typical of New Zealand agricultural landscapes: native forest, high-producing perennial pasture, and planted radiata pine forest.

MATERIALS AND METHODS

In autumn 2016 (18 April), 120 overwintering adult beetles were collected from two field cages located at the Lincoln campus of AgResearch (43°38′21″ S; 172°28′28″ E) and transported to the trial location at Tangihanga Station, near Gisborne (38°33′11″ S;

177°52′59″ E). The native forest was estimated to have 80% canopy cover and was composed of *Pittosporum eugenioides* A.Cunn. (lemonwood), *Alectryon excelsus* Gaertn. (tītoki), *Corynocarpus laevigatus* J.R.Forst. et G.Forst. (karaka), *Piper excelsum* G.Forst. (kawakawa), and *Melicytus ramiflorus* J.R.Forst. et G.Forst. (māhoe). The ground cover was a layer of leaf litter (4–6 cm deep) with no live vegetation. The pasture was completely open (i.e. no shading), and had 100% ground cover that was predominately composed of *Lolium perenne* L. (perennial ryegrass), with a small amount of *Agrostis capillaris* L. (browntop) and a *Crepis* sp. (hawksbeard). The pine forest was a mature plantation of *Pinus radiata* D.Don, visually assessed to have approximately 70% canopy cover. The ground cover was a layer of pine needle litter (2–3 cm deep), and the New Zealand native grass, *Microlaena stipoides* (Labill.) R.Br. (ca. 10% ground cover).

On 20 April, four replicate overwintering cages were established at each of the three habitats. The cages were made of Lumite® mesh material (BioQuip, USA) dug into the ground around the edges and supported by wire struts. The established cages were 50 × 90 cm, and approximately 30 cm high. A small rosette of *Cirsium vulgare* (Savi) Ten. (ca. 10 to 15 cm diameter) was collected from a nearby paddock and transplanted into each of the cages. This was planted as a food source for the beetle since in late winter to early spring the beetle enters a post-diapause quiescent state, and occasionally feeds when weather conditions are favourable (M. Cripps, pers. obs.). Ten adult beetles were placed inside each cage (40 individuals total per habitat). A data logger (Digitech QP-6013) contained inside a protective weather shield was placed inside each cage. Data loggers were placed on the ground surface to record temperature and relative humidity hourly. On 4 October 2016 the cages were examined for the presence of *C. rubignosa* first by physically searching through the vegetation and litter material in the caged area. Second, the caged area was sampled with a battery-powered vacuum (14.4V wet–dry Hoover Handivac) for any additional *C. rubiginosa* individuals. Potential predators were sampled (ca. 50 cm perimeter from the cage edge with a leaf-blower vacuum) in autumn at the time of cage set up, and again in spring at the time of assessment. Since nearly all predators collected were spiders we focused on identifying this group.

Unfortunately, the rosettes of *C. vulgare* in the forest habitats did not survive until spring, likely as a result of insufficient sunlight. The lack of a food source appeared to have resulted in the death of some beetles. Therefore, the recovery of beetles were grouped into two categories, 'alive' (active beetles), and 'recently dead', which were beetles that were dead or moribund (inactive), but still with normal colouration, and no indication of decay, such as fungal growth.

The relationship of the adult *C. rubignosa* recovery rate with each environmental variable (relative humidity (RH) and temperature) was analysed using a generalised linear model (GLM) with a binomial distribution through a logit-link function. Statistically significant relationships were assessed with the Akaike Information Criterion (AIC) in order to determine the relationship that had the closest fit to the data (smaller AIC values indicate a closer fit to the data.). Once the relationship with the closest fit to the data was determined, another variable was added, and assessed if it significantly improved the relationship. The assessment of these improvements was made using analysis of deviance. The environmental extreme variables included in the analyses were the time above 10°C and time below 70% RH since 10°C is the predicted developmental threshold for *C. rubignosa* (Ward and Pienkowski 1978), and since <10% of RH observations were below 70%.

RESULTS

The mean (±SE) percent recovery rate of alive beetles was greatest in the pasture (22.5 ± 6.7%), followed by native forest (10.0 ± 4.8%), and the pine forest (0%). However, the total recovery rate (alive plus recently dead) was greatest in the native forest (42 ± 7.9%), followed by pasture (32.5 ± 7.5%), and pine forest (7.5 ± 4.2%). The percentage of alive beetles recovered was not significantly different between the native and pasture habitats, but alive recovery in both of these habitats were significantly greater than the pine habitat (P=0.002). Similarly, the total percentage recovered was not different between the native and pasture habitats, but both of these habitats had greater recovery rates than the pine habitat (P=0.001 and 0.010, respectively). The recovery rate of recently dead beetles was significantly greater in the native habitat than either the pasture or pine habitats (P=0.019 and 0.010, respectively).

The total recovery rate of *C. rubignosa* decreased with increasing temperature and decreasing RH. The best models (AIC values <130) explaining *C. rubignosa* recovery rates were those based on:

1. maximum daily RH range; and
2. number of hours when temperature >1°C and RH <70%.

All of these models showed a negative relationship with the total recovery rate (Figure 1, P<0.001).

A total of ten different spider species were found, half of them introduced species (Table 1). In the native forest, three of the five spider species found were

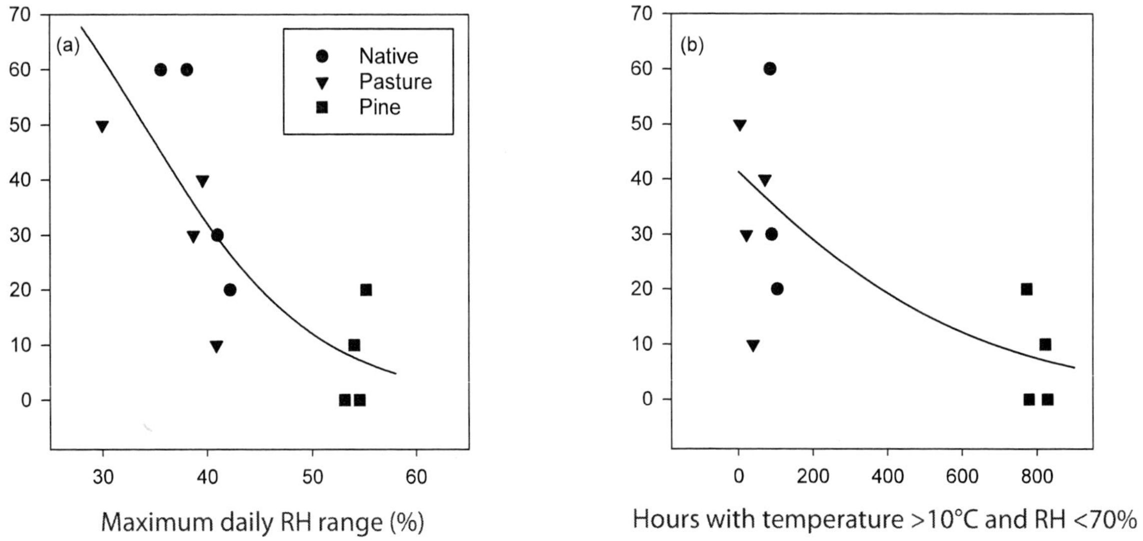

Figure 1. The two best models (AIC values <130) predicting the total recovery rate of overwintered *Cassida rubiginosa* from three different test habitats (native forest, pasture, or pine forest). The best predictors of recovery rate (%) from temperature and relative humidity (RH) recorded inside the cage over the study period (20 April to 4 October 2016) in order of the closest fit to the data are:
(a) maximum daily RH range: $Y = 100/(1 + \exp(-4.2181 + 0.1241*X))$; and
(b) hours with temperature >10°C and RH <70%: $Y = 100/(1 + \exp(0.3505 + 0.0027*X))$.

endemic, and comprised 68% of the individuals. In contrast, seven spider species were found in the pasture of which five were introduced species that comprised 80% of the individuals. The pine forest was depauperate in both species and abundance with only an individual each of two species collected. The introduced linyphiid, *Tenuiphantes tenuis* Blackw., was the only species common to all three habitats (Table 1).

DISCUSSION

Many insects complete diapause in mid-winter, but remain in a quiescent state until temperatures warm before becoming active (Danks 1991, MacMillan and Sinclair 2011). The quiescent state can be easily departed and returned to and it is possible for the overwintering beetle to temporarily become active and feed in small amounts before quickly returning to an inactive state when the temperature cools. The total recovery rate of beetles was influenced most strongly by temperature and RH extremes (time above 10°C and below 70% RH), and the native forest habitat provided the best insulation against these extremes on a daily basis, and across the winter season. As such, it is likely that the native forest habitat was more conducive for maintaining a quiescent state and therefore the beetle could survive longer there even without a food source. Leaf litter was also shown to be the preferred overwintering substrate in a multi-choice experiment (Spring and Kok 1999), conferring with the

Table 1. Spider species collected from vacuum samples taken in the three overwintering test habitats (native forest, pine forest and pasture) for *Cassida rubiginosa*. The total number of each spider species (#) and its endemic (E) or introduced (I) status to New Zealand are indicated.

Habitat	Species	#
Native forest	*Anoteropsis hilaris* (L. Koch) (E)	1
	Dunedinia denticulate Mill. (E)	2
	Rinawa pula Forster (E)	1
	Steatoda capensis Hann (I)	5
	Tenuiphantes tenuis (Blackw.) (I)	1
Pasture	*Anoteropsis hilaris* (L. Koch) (E)	2
	Cryptachaea blattea (Urquh.) (I)	1
	Erigone wiltoni Locket (I)	2
	Mermessus fradeorum (Berl.) (I)	1
	Steatoda capensis Hann (I)	2
	Tenuiphantes tenuis (Blackw.) (I)	1
	Theridiidae sp. (E)	2
Pine forest	*Tenuiphantes tenuis* (Blackw.) (I)	1
	Desidae sp. (E)	1

observed natural overwintering habitat, and the greater survival rates under leaf litter in native New Zealand forest found in our study. While the pasture habitat did not have a significantly different recovery rate of *C. rubignosa* compared to the native forest it is important to note that this was ungrazed pasture. At the time of cage establishment in the pasture habitat, the sward had been recently grazed to 3–4 cm, and by the time of assessment the sward had grown to approximately 30 cm in height. If the pasture had continued to be grazed it is likely that the recovery rate of *C. rubignosa* would have been reduced by this disturbance, but nevertheless this suggests that ungrazed habitats could also be suitable for overwintering. In contrast, the pine forest was the worst habitat for the overwintering of *C. rubiginosa*, and had the most extreme temperature and RH conditions (Figure 1).

To date, no predators of *C. rubiginosa* have been documented in New Zealand, although generalist predators such as spiders are certain to consume some, and have been observed feeding on adult *C. rubiginosa* in rearing cages and field trial sites (M. Cripps, pers. obs.). Pasture boundary habitats can influence the species composition and abundance of spiders in New Zealand and therefore could affect the overwintering survival of *C. rubiginosa* in different habitats (Bowie *et al.* 2014). However, the overwintering success of *C. rubiginosa* in our study did not appear to be influenced by spider predators, regardless of habitat.

This study provides initial evidence that the particular overwintering habitat available locally to *C. rubignosa* may contribute to the success of this biocontrol agent, and suggests that native forest with a leaf litter layer is likely to provide a better habitat than pine forest. Overall, the results suggest that the overwintering conditions for *C. rubignosa* are improved when insulated against extreme flucuating temperature and humidity. However, in order to generalise across New Zealand it would be necessary to repeat this experiment with increased replication in space and time with a greater number of beetles. Future studies should also include winter surveys for overwintering beetles in different habitats.

ACKNOWLEDGMENTS

We thank Derrick Wilson for assistance with spider collections, and Rose Greenfield for assistance with the cage set up. We also thank the managers of Tangihanga Station for allowing access to the property to carry out this work. This research was funded by the New Zealand Ministry of Business, Innovation, and Employment through AgResearch core funding of 'Pasture Weed Ecology' research and the 'Future Farm Systems' project.

REFERENCES

Ang, B. and Kok, L. (1995). Seasonal mortality of *Cassida rubiginosa* (Coleoptera: Chrysomelidae) due to incomplete development and parasitism in Southwestern Virginia. *Journal of Entomological Science* 30, 9-17.

Bowie, M.H., Klimaszewski, J., Vink, C.J., Hodge, S. and Wratten, S.D. (2014). Effect of boundary type and season on predatory arthropods associated with field margins on New Zealand farmland. *New Zealand Journal of Zoology* 41, 268-84.

Cripps, M.G. (2013). Observations on the thistle-feeding tortoise beetle, *Cassida rubiginosa* (Coleoptera: Chrysomelidae). *The Weta* 45, 5-13.

Cripps, M.G., Edwards, G.R., Bourdot, G.W., Saville, D.J., Hinz, H.L. and Fowler, S.V. (2010). Effects of pasture competition and specialist herbivory on the performance of *Cirsium arvense*. *Biocontrol Science and Technology* 20, 641-56.

Cripps, M.G., Gassmann, A., Fowler, S.V., Bourdôt, G.W., McClay, A.S. and Edwards, G.R. (2011). Classical biological control of *Cirsium arvense*: Lessons from the past. *Biological Control* 57, 165-74.

Danks, H. (1991). Winter habitats and ecological adaptations for winter survival. *In* 'Insects at low temperature', eds R.E. Lee and D.L. Denlinger, pp. 231-59. (Chapman and Hall, New York).

Koji, S., Kaihara, K. and Nakamura, K. (2012). Stage-specific mortality, fecundity, and population changes in *Cassida rubiginosa* (Coleoptera: Chrysomelidae) on wild thistle. *Applied Entomology and Zoology* 47, 457-65.

Kosior, A. and Klein, J. (1970). Migrations and wintering of cassids (Coleoptera, Chrysomelidae, Cassidinae) under the climatic conditions of the Ojców National Park. *Acta Zoologica Cracoviensia* 15, 315-41.

MacMillan, H.A. and Sinclair, B.J. (2011). Mechanisms underlying insect chill-coma. *Journal of Insect Physiology* 57, 12-20.

Spring, A. and Kok, L. (1999). Winter survival of *Cassida rubiginosa* (Coleoptera: Chrysomelidae), a biological control agent of Canada thistle. *Journal of Entomological Science* 34, 489-93.

Ward, R.H. and Pienkowski, R.L. (1978). Biology of *Cassida rubiginosa*, a thistle-feeding shield beetle. *Annals of the Entomological Society of America* 71, 585-91.

Molecular analysis of ecological interactions for optimising biocontrol of the invasive weed *Sonchus oleraceus* L. (Asteraceae) in Australia

M. Ollivier[1], V. Lesieur[1], M. Jourdan[2], T. Thomann[2], S. Raghu[3], L. Morin[4], A. Sheppard[4], M.S. Tixier[1] and J.F. Martin[1]

[1] CBGP, Montpellier SupAgro, INRA, CIRAD, IRD, Univ. Montpellier, Montpellier, France

[2] CSIRO Health and Biosecurity – European Laboratory, Montferrier sur Lez, France

[3] CSIRO Health and Biosecurity – Brisbane, Australia

[4] CSIRO Health and Biosecurity, GPO Box 1700, Canberra, Australian Capital Territory 2601, Australia

(melodie.ollivier@supagro.fr)

Summary Community ecology is a promising approach for optimising biological control of weeds. The characterisation of ecological interaction networks associated to the target plant in its natural context allows deciphering the complexity of interactions within arthropods community; and gives precious clues on potential undesirable effects induced by biocontrol agents' introduction. Based on this approach, several research questions are addressed: i) What is the diversity and specify within arthropods' community feeding on the plant? ii) Is there a top-down control of herbivores by natural enemies? iii) Are there differences in network structure between contrasted geographical areas? Our study targets the sowthistle, *Sonchus oleraceus*, that is native to Europe but invasive in Australia. The development of herbicide resistance makes it extremely difficult to manage in its invasive area. As an alternative to pesticide use, biocontrol solutions are explored via a collaborative research program between Montpellier SupAgro and CSIRO. First steps of the project focus on the molecular characterisation and comparison of the ecological networks among three different climatic regions in its native range. This is done using cutting-edge molecular approaches (*i.e.* multiplex metabarcoding through NGS) to reveal interactions from arthropod gut-content. Network metrics such as species richness, linkage density or connectance are computed and compared between sites to assess differences in levels of complexity among changing environments. The study will be further expanded to Australia in order to confront results between invasive and native ranges and contribute to a better assessment of risks and potential changes induced by such weed management strategy.

Keywords Ecological interactions, molecular analysis, sowthistle, biological control.

Biological control of *Sonchus oleraceus*: what is known, what is new and what is still missing?

V. Lesieur[1,2], T. Thomann[2], M. Jourdan[2], M. Ollivier[1], A. Sheppard[2], J.F. Martin[1], M.S. Tixier[1], L. Morin[3] and S. Raghu[4]

[1] CBGP, Montpellier SupAgro, INRA, CIRAD, IRD, Univ. Montpellier, Montpellier, France

[2] CSIRO Health & Biosecurity, European Laboratory, Montferrier sur Lez, France

[3] CSIRO Health & Biosecurity, GPO Box 1700, Canberra, Australian Capital Territory 2601, Australia

[4] CSIRO Health & Biosecurity, Brisbane, Queensland, Australia

Summary Common sowthistle, *Sonchus oleraceus* L. (Asteraceae), is a widespread agricultural and garden weed in its native range of Europe and northern Africa. Now widely distributed, this invasive species is an increasingly important weed in Australia. Development of herbicide resistance is making populations extremely difficult to manage in cropping systems. As an alternative method to herbicides, CSIRO in collaboration with Montpellier SupAgro (France) has initiated a classical biological control program for this weed. Characterising the natural enemy community associated with *S. oleraceus* is crucial to further select potential biological control agent(s). However, this community is surprisingly poorly understood across its native range. From a literature review and early field collections, two flies, the leaf-gall former, *Cystiphora sonchi* (Cecidomyiidae) and the bud-gall former, *Tephritis formosa* (Tephritidae), are considered as promising candidates. Their host specificity is currently being investigated by testing a series of key plants (*i.e.* economically important species and Australian native species). Moreover, native range surveys, guided with a climate-matching approach, have been carried out in Morocco and Western Europe in 2017 and 2018. To date, more than 40 arthropods species and, at least, five species of fungi have been collected. Potential new candidates (one rust and one hoverfly) have been identified and are currently maintained at the laboratory for host-testing.

Keywords Common sowthistle, weed, natural enemies' community, arthropods, fungi.

Matching expectations and reality: Potential impact of biological control on flaxleaf fleabane

Michelle A. Rafter[1], Kerri Moore[1], Nagalingam Kumaran[1], Gavin Hunter[2], Kylie B. Ireland[2], Louise Morin[2] and S. Raghu[1]
[1] Commonwealth Scientific and Industrial Research Organisation (CSIRO), Ecosciences Precinct, 41 Boggo Road, Dutton Park, Queensland 4102, Australia
[2] Commonwealth Scientific and Industrial Research Organisation (CSIRO), Black Mountain Science & Innovation Park, Clunies Ross Street, Acton, Australian Capital Territory 2601, Australia
(michelle.rafter@csiro.au)

Summary Flaxleaf fleabane (*Conyza bonariensis*) is a major weed of cropping in northern New South Wales and southern Queensland causing revenue losses in excess of $43 million for grain producers. Development of resistance to herbicides is making populations increasingly difficult to manage in these agricultural environments. Biological control is a potentially useful and valuable tool for the management of flaxleaf fleabane given the success of previous biological control programs against other weeds in the Asteraceae family such as *Parthenium hysterophorus* and *Chromolaena odorata*. In order to increase the likelihood of biological control being effective in the cropping context clear management expectations should be set. In addition to this, an agent which is determined to be safe for release ought to have a negative impact on the growth and fecundity of the plant. For these reasons while native range agent exploration is being undertaken in South America, an online survey of land managers was conducted to help define the management expectations of a potential biological control agent for flaxleaf fleabane. A simulated herbivory experiment involving removal of both above and below ground plant biomass was also performed to determine if a defoliating or root feeding insect will be an effective biocontrol agent. Defined management objectives and results of the simulated herbivory study will guide selection of potential biological control agents and evaluation of their effectiveness in the field, provided they are approved for release into the Australian environment.

Keywords *Conyza bonariensis*, biocontrol, stakeholder engagement, simulated herbivory, agent prioritisation.

Genomics, lineages and phenotypes: Investigations into the stability of the *Puccinia chondrillina–Chondrilla juncea* pathosystem in Australia

Gavin C. Hunter[1], Kylie B. Ireland[1], Mireille Jourdan[2], Jim Cullen[1] and Louise Morin[1]
[1] CSIRO Health and Biosecurity, Black Mountain Science and Innovation Park, Clunies Ross Street, Acton, Australian Capital Territory 2601, Australia
[2] CSIRO European Laboratory, Campus International de Baillarguet, 34980 Montferrier-sur-lez, France
(gavin.hunter@csiro.au)

Summary *Puccinia chondrillina* (Skeleton weed rust) was the first deliberate release of a fungal pathogen for biological control of an invasive weed. Over three decades four strains of *P. chondrillina* were released in Australia targeting the narrow and intermediate leaf forms of the agricultural weed *Chondrilla juncea* (Skeleton weed). Strain IT32, released in 1971 showed exceptional impact against the narrow-leaf form while strains TU21, IT36 and TU788 released in the 1980s and 1990s against the intermediate leaf-form exhibited poor to moderate impact. Unfortunately no strains of *P. chondrillina* have been identified which are pathogenic and specific to the broad-leaf form of *C. juncea* in the field. It is unknown which strains of *P. chondrillina* have persisted in Australia since their release or the possible effect that evolutionary forces may have exerted on their genotype to shape their populations or affect their pathogenicity. In this study we employ whole genome sequencing of *P. chondrillina* isolates collected in 2007 and 2016, and of stored urediniospores representing the original released strains, to determine the lineage of contemporary isolates and identify genetic changes that may have occurred over time. In addition, the pathogenicity phenotype of each contemporary isolate is determined through inoculations on the three forms of *C. juncea*. The hypothesis of little genetic change of *P. chondrillina* in Australia will be tested. These results will provide valuable information on the stability of the *C. juncea – P. chondrillina* pathosystem and possibly shed additional light on the lifecycle of skeleton weed rust in Australia.

Keywords Skeleton weed, Pucciniales, fungal genomics, pathogenicity phenotype.

Boxthorn rust (*Puccinia rapipes*): a potential biological control agent of African boxthorn (*Lycium ferrocissimum*)

Kylie B. Ireland[1], Gavin Hunter[1], Alan Wood[2] and Louise Morin[1]
[1] Commonwealth Scientific and Industrial Research Organisation (CSIRO), Black Mountain Science & Innovation Park, Clunies Ross Street, Acton, Australian Capital Territory 2601, Australia
[2] ARC-Plant Protection Research, Private Bag X5017, Stellenbosch, Western Cape Province 7599, South Africa
(kylie.ireland@csiro.au)

Summary Suitability of the rust fungus *Puccinia rapipes* as a biocontrol agent for African Boxthorn (*Lycium ferocissimum*), a major environmental and agricultural weed in regional Australia, was assessed. Studies were undertaken to elucidate the life cycle, genetic and phenotypic diversity and host specificity of *P. rapipes*. Laboratory and field surveys indicate that the rust is macrocyclic and autoecious – i.e. produces all five spore stages on *L. ferocissimum*. Two accessions of the fungus have been purified and cultured in quarantine in Australia, representing the Eastern and Western Cape distributions of the rust in South Africa. Preliminary testing shows that both accessions are capable of infecting at least three provenances of the weed, from the ACT, South Australia and Western Australia, while field surveys in South Africa indicate that the rust can infect multiple haplotypes of the weed. Further host testing of the native Australian Boxthorn (*L. australe*) and exotic goji berry (*L. barbarum*), alongside multiple haplotypes of *L. ferocissimum*, will be undertaken to elucidate the nature of the rust's host specificity. DNA-based molecular biology work will also be conducted to confirm the identify of all five spore stages of *P. rapipes* collected in the field in South Africa and to assess the genetic diversity of the fungus across its native range. Results of preliminary host-specificity testing and molecular characterisation will be used to make a preliminary assessment of the risk of significant impact on non-target plan species, before deciding if further testing working toward release of the fungus in Australia should proceed.

Keywords Biocontrol, plant pathogen, host specificity.

Water influences the impact of a native parasitic plant ('snotty gobble') on gorse

Robert M. Cirocco[1], José M. Facelli[1] and Jennifer R. Watling[1,2]
[1] DP 312 Benham Building, School of Biological Sciences, The University of Adelaide, Adelaide, South Australia 5005, Australia
[2] All Saints Building, Manchester Metropolitan University, Manchester M15 6BH, UK
(robert.cirocco@adelaide.edu.au)

Summary Weeds of National Significance such as *Ulex europaeus* (gorse) have devastating socio-economic and environmental impacts. At The University of Adelaide we have been researching the potential use of a native parasitic vine *Cassytha pubescens* otherwise known as 'snotty gobble' as a novel bio-control for major shrubby weeds such as gorse and blackberry. Native parasitic plants may have detrimental effects on performance of invasive weedy hosts by removing resources via suckers and thus contribute to their demise. Glasshouse and field studies have documented severe effects of parasites on invasive species, but the effects of parasites may be highly variable depending on environmental conditions. We conducted a glasshouse experiment investigating the effects of *Cassytha pubescens* on growth and physiology of gorse under high versus low water supply. The strong effect of *Cassytha* on growth of gorse was more severe in the well-watered treatment likely being due to improved parasite performance in these conditions. *Cassytha pubescens* also negatively affected nitrogen status of gorse and there was evidence of long term breakdown of the photosynthetic apparatus of the host in response to infection. The data indicate that the native parasite negatively affects photosynthetic performance and growth of gorse by removing large amounts of nitrogen from the host. Thus, *Cassytha* continues to show promise as an effective native bio-control against major invasive shrubby weeds in Australia, particularly in areas of high water availability and if successful may be used to help restore our native biodiversity.

Keywords Native bio-control, *Cassytha pubescens*, gorse, nitrogen, parasitic plant-host association, photosynthesis, *Ulex europaeus*, water availability, Weeds of National Significance.

This study has been published: Cirocco, R.M., Facelli, J.M. and Watling, J.R. (2016). High water availability increases the negative impact of a native hemiparasite on its non-native host. *Journal of Experimental Botany* 67, 1567-75.

Promising new biological control agents for Queensland

Elizabeth L. Snow, Peter Jones, Natasha Riding and Michael Day
Biosecurity Queensland, Department of Agriculture and Fisheries, PO Box 267, Brisbane 4001, Australia
(liz.snow@daf.qld.gov.au)

Summary A range of promising new biological control agents are being assessed and released in Queensland: (1) five different biotypes of the cochineal insect, *Dactylopius tomentosus* (Lamark), are being field-released on several cactus species (*Cylindropuntia* spp.). One biotype is achieving excellent results, killing coral cactus (*Cylindropuntia fulgida* (Engelm.) F.M. Knuth var. *mamillata* (Engelm.) Backeb.) around Longreach (central Queensland) and Hebel (south-western Queensland). This is an exciting development for a weed species that represents a serious long-term threat to pasture production across vast areas of arid and semi-arid rangelands; (2) the gall-fly, *Cecidochares connexa* (Macquart), has achieved excellent control of chromolaena (*Chromolaena odorata* (L.) R.M. King & H. Robinson) in Papua New Guinea. Testing for specificity has been completed and an application seeking field release in Australia was submitted to the Australian Government in 2015; (3) host-specificity testing has been completed for the fungal pathogen, *Puccinia spegazzinii* de Toni, for the control of *Mikania micrantha* Kunth, a major weed of wet tropical areas overseas and a target of eradication in Australia; (4) biology studies on a root-feeding beetle for mother-of-millions (*Kalanchoe delagoense* Eckl. & Zeyh), a widespread, toxic weed of pastures are currently being undertaken prior to host-specificity testing; and (5) research has commenced to find potential agents for giant rat's tail grasses (*Sporobolus* spp.). Recent prospecting in South Africa has identified several species of stem-boring insects that are now being assessed for possible release in Australia. A new project based in Vanuatu and funded by the New Zealand Government will enable exploration in South America and south-east Asia for biological control agents targeting *Solanum torvum* Sw. (devil's fig) and *Senna tora* (L.) Roxb. (sicklepod).

Keywords Biological weed control, invasive plants, *Chromolaena odorata*, *Cylindropuntia* spp., *Kalanchoe delagoense*, *Mikania micrantha*, *Sporobolus* spp.

INTRODUCTION

Weeds continually cause severe impacts to agricultural and natural ecosystems across Australia. In some cases, conventional control is not feasible and biological control is viewed as the only long-term, sustainable solution to their management. The approach is also cost effective, having an average benefit:cost ratio of 23:1 (across all projects) (Page and Lacey 2006). Using quarantine facilities at the Ecosciences Precinct at Dutton Park, Brisbane, one of only four such facilities in Australia, the Department of Agriculture and Fisheries has the capacity to undertake a number of projects researching imported biological control agents. A range of promising new biological control agents are being assessed for several invasive plant species, as follows: (1) the augmentation of biotypes of *Dactylopius tomentosus* (Lamark) available for the biological control for *Cylindropuntia* spp., which are serious emerging threats to arid and semi-arid rangelands; (2) *Chromolaena odorata* (L.) R.M. King & H. Robinson, considered one of the world's worst weeds, and a serious threat to agriculture and natural ecosystems; (3) *Mikania micrantha* Kunth, a target of a nationally cost-shared eradication program, is a vine that poses a long-term threat to a range of tropical tree crops and native plant communities in the World Heritage listed Wet Tropics of far north Queensland; (4) *Kalanchoe delagoense* Eckl. & Zeyh, a major weed in Queensland and northern New South Wales, found in a range of drier, sub-coastal habitats and toxic to cattle; (5) *Sporobolus* spp., aggressive and quick-spreading grass weeds of pastures; (6) *Solanum torvum* Sw. (devil's fig), an environmental and pasture weed and (7) *Senna tora* (L.) Roxb. (sicklepod), an agricultural weed (both invading Vanuatu). While the host range of biological control agents has already been tested and approved for release overseas for several of these species, further host-specificity testing is required in Australia to meet strict risk assessment criteria as set out by the Australian Department of Agriculture and Water Resources (DAWR) and Department of Environment and Energy (DoEE). The use of insects already tested overseas saves time and money as no exploration is required and testing elsewhere tends to provide data to support the probability of a positive outcome regarding host-specificity under Australian conditions. This paper outlines progress to date.

RESULTS AND DISCUSSION

***Cylindropuntia* spp.** Several different biotypes of *D. tomentosus* were tested in the quarantine facility at the Ecosciences Precinct (Dutton Park, Brisbane) for their efficacy in providing control of various *Cylindropuntia* spp. The 'cholla' biotype of *D. tomentosus* was imported from South Africa in 2012 to target *C. fulgida* var. *mamillata* (coral cactus). A further 16 biotypes were imported from the United States of America and Baja California, Mexico between 2011 and 2015 to target the remaining seven species of *Cylindropuntia* naturalised in Australia. Biological control success is dependent on matching the most effective biotype to the most susceptible target *Cylindropuntia* species. As *Cylindropuntia* species are morphologically variable and hence taxonomically challenging, particularly when the plants are young, molecular analysis was used to allow biotype matching. Designated photo-monitoring plots were also established at three release sites to measure the impact of the insect over time.

All biotypes of *D. tomentosus* have been approved for release and to-date five are currently being mass-reared for future releases, providing an effective management tool for stakeholders nation-wide. The remaining biotypes are still being assessed for their impact on naturalised *Cylindropuntia*. Three of these biotypes have been released at designated monitoring sites to assess their impact. The 'cholla' biotype was released in March 2016 at Longreach (central Queensland) and April 2016 at Hebel (south-western Queensland) and has been highly damaging (Figures 1 and 2). After two years, only six plants in a 1 ha monitoring plot at Longreach are still alive, while 86% of an initial population density of 3990 plants ha^{-1} are dead at the Hebel monitoring site. Due to the level of damage, there is a very low probability that plants will recover, as cladodes are completely desiccated.

There are high expectations for the '*californica* var. *parkeri*' biotype that targets Hudson pear (*Cylindropuntia pallida* (Rose) F.M. Knuth). A monitoring program has been set up at two sites near Lightning Ridge, New South Wales. Molecular analysis has demonstrated that two plastid markers and a nuclear region allow the differentiation of the eight naturalised *Cylindropuntia* species and enables researchers to match targeted *Cylindropuntia* species to the most effective cochineal biotype – a concept that has been described as 'genetic optimisation' (Jones *et al.* 2016). Stakeholders requiring a specific biotype to control their *Cylindropuntia* infestation are encouraged to submit samples of the cactus for molecular identification.

Figure 1. *Cylindropuntia fulgida* var. *mamillata* at 'Leander Station', Longreach, prior to release of biological control agents (2016).

Figure 2. *Cylindropuntia fulgida* var. *mamillata* at 'Leander Station', Longreach post-release (2017).

Chromolaena odorata The gall-fly *Cecidochares connexa* (Macquart) (Figure 3) has achieved excellent control of *C. odorata* in Papua New Guinea and several other countries. It was imported into Australia in 2012 for host-specificity testing against 18 plant species (Day *et al.* 2017). These were in addition to 122 plant species tested in seven other countries prior to release of the insect in those locations.

Host-testing of *C. connexa* has been completed and it has been found to be suitably host-specific. Based on these results, an application to release the gall-fly in Australia was submitted in 2015 and is still being assessed by the Australian Government.

Mikania micrantha The fungal pathogen *Puccinia spegazzinii* de Toni has been imported and tested pre-emptively in case the current eradication program for *M. micrantha* in Queensland is unsuccessful. This

Figure 3. *Cecidochares connexa* (Image courtesy C. Wilson.)

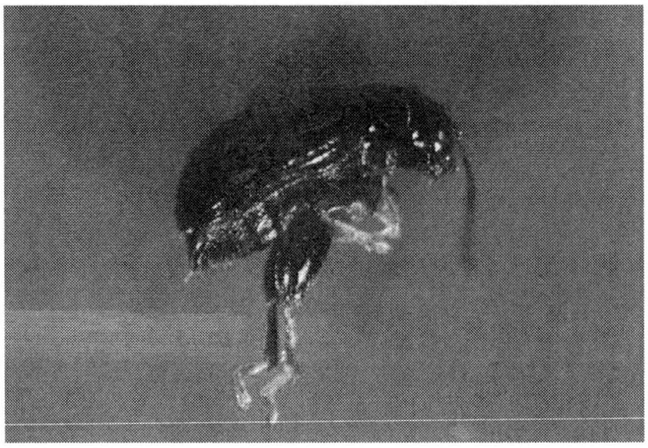

Figure 4. *Rhembastus* sp., a new agent for *Kalanchoe delagoense*.

pathogen has been tested in five countries against a total of 273 plant species with no non-target impacts (Day *et al.* unpublished data). It was imported from Papua New Guinea in 2015 and has been tested on 14 plant species in the tribe Eupatorieae and six species in the tribe Heliantheae.

Testing of the pathogen has been completed with host specificity being confirmed. An application for release in Australia will be lodged in 2018.

Kalanchoe delagoense Exploration was conducted in Madagascar to find suitable biocontrol candidates for *K. delagoense*. In 2017, the leaf and root-feeding beetle *Rhembastus* sp. (Figure 4), was imported into quarantine from Madagascar and its biology and host-specificity is being studied.

The adults of this insect feed on the leaves and plantlets produced at the end of the phyllodes and deposit their eggs within the root system of the plant. Upon emergence, larvae feed on the roots of the plant. Previous host-specificity testing in South Africa demonstrated that the insect could complete development on five non-target species in the Crassulaceae family. Despite this, the insect still has potential in Australia, as there are only eight native species in the Crassulaceae family (Witt *et al.* 2006). Once the biological requirements of the insect have been established under quarantine conditions, the project will move onto the host-specificity testing phase to assess suitability.

***Sporobolus* spp.** Researchers at Rhodes University in South Africa have been contracted to conduct field exploration to find suitable agents for the biological control of *Sporobolus* species. As Australia has a number of native congeners, particular attention is being given to native range field-host testing and molecular studies to help identify suitable candidates for detailed testing under laboratory conditions. Recent exploration in South Africa has found several species of stem- boring insects. More detailed host-specificity testing will be conducted pending field studies.

Solanum torvum* and *Senna tora Two new projects will undertake exploration in South America and south-east Asia for potential biocontrol agents for these two weed species. The work is being funded by the New Zealand Government. This project has not yet commenced, however any agents identified during prospecting may be of value to Australia's future biological control research. This is a good example of overseas work benefitting Australian biological control projects and providing a significant cost-saving for the exploration phase of the project.

ACKNOWLEDGMENTS

Australian Government Department of Agriculture and Water Resources, as part of its Rural R&D for Profit program (administered by Meat and Livestock Australia (round 1) and AgriFutures (round 2)), New South Wales Department of Primary Industries, New Zealand (NZ) Ministry of Foreign Affairs and Trade and NZ Landcare Research Ltd. Also Steve Csurhes for comments on the manuscript.

REFERENCES

Day, M., Riding, N. and Senaratne, W. (2017). Technical highlights, Invasive plant and animal research 2016–2017. (Queensland Government, Brisbane)

Jones, P.K., Oakey, J., Day, M.D. and Vitelli, J.S. (2016). Improving the effectiveness of the release strategy for the array of *Dactylopius tomentosus* biotypes for the biocontrol of *Cylindropuntia* species by using DNA analysis. Proceedings of the 20th Australasian Weeds Conference, eds R.

Randall, S. Lloyd and C. Borger, pp. 137-40 (Weeds Society of Western Australia, Perth).

Page, A.R. and Lacey, K.L. (2006). 'Economic impact assessment of Australian weed biological control'. Technical Series #10. 151 pp. (CRC for Australian Weed Management, Adelaide).

Witt, A.B.R., McConnachie, A.J., Palmer, W.A. and Grobbelaar, E. (2006). Distribution, biology and host range of *Rhembastus* sp. (Coleoptera: Chrysomelidae), a candidate for the biological control of *Bryophyllum delagoense* (Crassulaceae) in Australia, *Biocontrol Science and Technology* 16, 859-69.

Biological control of mistflower (*Ageratina riparia*) with a fungal agent facilitates native vegetation regeneration

Ben Gooden[1,2], Louise Morin[1], Andrew White[3], Melissa Piper[1], Jamie Fagg[2], Kris French[2], Alice Yeates[4] and Shon Schooler[3,4]
[1] CSIRO Health and Biosecurity, GPO Box 1700, Canberra, Australian Capital Territory 2601, Australia
[2] Centre for Sustainable Ecosystem Solutions, School of Biological Sciences, University of Wollongong, New South Wales 2522, Australia
[3] CSIRO, GPO Box 2583, Brisbane, Queensland 4001, Australia
[4] University of Wisconsin-Superior, PO Box 2000, Superior, WI 54880, USA
(Ben.Gooden@csiro.au)

Summary Biological control represents a sustainable long-term strategy for mitigating impacts of invasive alien plants on native ecosystems. It is often assumed, but rarely demonstrated, that a decline in alien plant abundance and competitive performance as a function of biological control will result in recovery of impacted native flora across invaded vegetation communities. Here, we examine patterns of native vegetation community responses to biological control of the invasive plant mistflower *Ageratina riparia* (Regel) R.M.King & H.Rob. (Asteraceae) with the white-smut fungus *Entyloma ageratinae*. Native vegetation (i.e. species richness, community composition, and percent cover of vascular plants) was surveyed in 24 permanent 1 m × 10 m belt transects at seven sites within wet sclerophyll forest in New South Wales and south-east Queensland. Surveys were conducted in areas of dense mistflower infestation (>70% cover) prior to or soon after the fungus colonised each site in 2011, and subsequently in 2012 and 2017. Surveys were conducted prior to agent release in 2011, then post release in 2012 and 2017. We found a significant 5-fold reduction in mistflower cover as a function of agent activity through time, which was in turn associated with a 4-fold increase in native plant richness, from an average of five species per transect in invaded areas in 2011 to an average of more than 20 species in 2017. Such positive effects of biological control on native plant richness were similar in New South Wales and Queensland, although vegetation composition varied by region. The reduction of mistflower abundance following biological control, and flow-on increases in native species richness, are likely to lead to benefits for ecosystem structure and function.

Keywords Biological control, ecosystem monitoring, fungal pathogen, invasive species impacts, vegetation community.

Biocontrol post-release: could earlier failures lead to overlooked successes?

Iain D. Paterson[1], Zezethu Mnqeta[1] and Paul O. Downey[2]
[1] Department of Zoology and Entomology, Rhodes University, PO Box 94, Grahamstown 6140, South Africa
[2] Institute for Applied Ecology, University of Canberra, Australian Capital Territory 2601, Australia
(paul.downey@canberra.edu.au)

Summary Whilst there have been several major successful biological control programs, in which the released agents devastate the target weed population, the vast majority of agents have been less successful, with the actual status of many released agents being either unknown or unconfirmed. Often this absence of evidence of the agent's establishment in the field is subsequently described as an actual 'establishment failure' which often results in no further evaluations or research on the agent being undertaken. Surprisingly given the large number of these 'failed' agents, or agents whose release status is unknown, there has been little impotence shown to reassess such agents and confirm and document their actual status, or determine and describe the reasons why they failed to establish and therefore seemingly discarded as an agent. Here we re-examined one such agent (*Phenrica guerini*) released on *Pereskia aculeata* in South Africa. *Phenrica guerini* was first released in 1991. Based on preliminary post-release evaluations from a subset of sites, it was concluded in 1997 that due to poor establishment rates that further releases of the agent be halted. Interestingly additional releases were subsequently made of the agent 12 years later, without any post-release evaluations. In 2014 a dedicated survey showed the early decision to halt the program may have been premature. Based on this example combined with an examination of the status of numerous other agents we believe that there may be many other overlooked successes, which if documented could have positive ramifications for the science and control of weeds.

Keywords Post-release evaluations, biological control, *Phenrica guerini*, establishment.

Biocontrol research for environmental weeds management in NSW

Louise Morin[1], Andrew McConnachie[2] and Peter Turner[2,3]
[1] CSIRO Health & Biosecurity, Canberra, Australian Capital Territory, Australia
[2] NSW Department of Primary Industries, Orange, New South Wales 2800, Australia
[3] Previous address: Office of Environment & Heritage, Hurstville, New South Wales 2220, Australia
(louise.morin@csiro.au)

Summary Classical biological control (biocontrol) involves the deliberate introduction of host-specific natural enemies of a target weed from its native to its invaded range. In 2016, the NSW Environmental Trust invested in biocontrol research for environmental weeds in NSW. Project scoping identified that the best approach was to take advantage of previous or current biocontrol projects to significantly reduce their cost, time and required resources. The project comprised two stages, with stage 1 now completed. In this first stage, a transparent prioritisation framework was devised, based on a matrix assessment system that involved:

1. initial screening steps to identify eligible weeds,
2. an assessment of the threat each weed poses to natural ecosystems,
3. the feasibility of undertaking a biocontrol project for each weed using the most promising candidate biocontrol agent, and
4. the likelihood of the selected agent being successful at controlling the weed if released in a new environment.

The framework was applied to generate a justified, priority list of biocontrol weed targets. Of the 67 eligible weeds identified and assessed, 18 were deemed to be priorities. These species were then allocated to a cell in a matrix of weed threat versus biocontrol prospects. Considering the first and second priority weeds identified and resources available for undertaking research in Stage 2, the Environmental Trust selected five weeds for investment: balloon vine (*Cardiospermum grandiflorum*), sea spurge (*Euphorbia paralias*), leaf cactus (*Pereskia aculeata*), broadleaved pepper tree (*Schinus terebinthifolius*) and yellow bells (*Tecoma stans*). Stage 2 has now begun with research on a candidate biocontrol pathogen agent of sea spurge.

Keywords Biological control, Environmental Trust, prioritisation framework, balloon vine (*Cardiospermum grandiflorum*), sea spurge (*Euphorbia paralias*), leaf cactus (*Pereskia aculeata*), broadleaved pepper tree (*Schinus terebinthifolius*), yellow bells (*Tecoma stans*).

Weed biocontrol in temperate Australia – a new manual for best management practice

Paul R. Sullivan
Weed Research Unit, NSW Department of Primary Industries, Tamworth Agricultural Institute, 4 Marsden Park Road, Calala, New South Wales 2340, Australia
(paul.sullivan@dpi.nsw.gov.au)

Summary Weed biocontrol using insects, mites and pathogens has a long and distinguished history in Australia. Successful long-term weed biocontrol requires not only the motivation, expertise and capacity, but also the latest best practice guidelines if practitioners are to effectively implement programs. Best practice biocontrol is an evolving field-based science. Sourcing quality data on the latest methodology is imperative and an ongoing process. For new programs, methodology is usually recommended by researchers, however over time, practitioners fine tune and develop the methodology which becomes best practice. Quality data on best practice may not exist, and even where data does exist, there is often poor communication of this data between researchers, extension staff and practitioners. Experience has shown that biocontrol is optimised when there is open and frequent communication between these role players. Developing and communicating best practice can be further complicated by geography which leads to varying methodologies. A best practice manual has been developed for weed biocontrol in temperate Australia. This practitioner friendly manual addresses common questions and methodology issues and provides the key steps for undertaking weed biocontrol programs. The manual gives biocontrol recommendations for 45 weeds and provides detailed information on their biocontrol agents. Information for biocontrol agents includes their lifecycle, impact and abundance; and how to collect, rear and monitor them. This manual will serve as a valuable tool for practitioners and will supplement the weed biocontrol course that is offered by NSW DPI.

Keywords Biological control, biocontrol, weed management, communication, extension, methodology.

Drivers of change and adaptation tools for reducing the impact of invasive plants under a changing climate

Michelle Leishman
Department of Biological Sciences, Macquarie University, New South Wales 2109, Australia
(michelle.leishman@mq.edu.au)

Summary Climate change is likely to facilitate invasion by exotic plant species via a range of processes, from changes in introduction pathways to invasive spread facilitated by changed environmental conditions. In addition, many exotic invasive plant species have traits that will make them highly successful invaders under a changing climate, resulting in a shift to more exotic-dominated vegetation. This talk examines the main drivers of vegetation change under climate change and how these may affect the relative success of native and exotic plant species. It also looks at current climate adaptation tools for reducing the impact of invasive plants under a changing climate.

Keywords Climate change, weed futures, invasion mechanisms.

Future weeds RD&E investment under a collaborative national biosecurity approach

Andreas Glanznig and Richard J. Price
Centre for Invasive Species Solutions, Building 22, University of Canberra, Belconnen, Australian Capital Territory 2617, Australia

Summary To reduce the huge impacts of weeds to manageable levels, governments and industry have committed to major national 2030 targets – for example, the Meat Industry Strategic Plan commits to invest in research and innovation to reduce weed and pest impacts by $150m a year, while governments have committed to a national target to reduce the impacts of invasives on Australia's biodiversity by 10%. To achieve these targets requires a new major collaborative effort to accelerate and transform weed management innovation.

In March 2018, at the request of the Commonwealth, the Centre for Invasive Species Solutions (CISS) began the process to develop its national plan for weed R&D investment over 2020–2030. This draft Plan proposes a major, decade long coordinated investment in public good weeds RD&E that would give effect to broad agreed priorities developed by Australia's governments and particular livestock industries. The aspiration places the planned investments firmly within National Biosecurity System arrangements, in particular the National Biosecurity RD&E Framework. Like the Framework, the Plan seeks to engender robust and integrated (systems) R&D capabilities and infrastructure that effectively and collaboratively support management of weed risks across the invasion curve. The Plan will prioritise investment in weed RD&E activities with the results feeding into continuously updated pipelines of best management practice. An exposure draft of the Plan will be released at the 21st Australasian Weeds Conference for ongoing consultation.

Keywords Weeds, invasive plants, investment frameworks, research, engagement.

Trifludimoxazin: A new PPO inhibitor that controls PPO resistant weed biotypes

G.R. Armel[1], I. Francis[2], R. Nielson[3], M. Witschel[4] and R. Liebl[1]
[1] BASF Corporation, Research Triangle Park, NC, USA
[2] BASF Australia Ltd, Southbank, Australia
[3] BASF Corporation, Limburgerhof, Germany
[4] BASF Corporation, Ludwigshafen, Germany
(gregory.armel@basf.com)

Summary Trifludimoxazin [1,5-dimethyl-6-thioxo-3-(2,2,7-trifluoro-3,4-dihydro-3-oxo-4-prop-2-ynyl-2*H*-1,4-benzoxazin-6-yl)-1,3,5-triazinane-2,4-dione] is a new inhibitor of protoporphyrinogen IX oxidase (PPO or Protox). This is the first PPO inhibitor containing a triazinone heterocycle. Trifludimoxazin is very active when applied PRE or POST on dicot/broadleaf weeds including PPO resistant *Amaranthus* biotypes which are not controlled by currently registered PPO inhibitors like the diphenylether herbicides (e.g., fomesafen, lactofen, etc.), sulfentrazone, or flumioxazin. Trifludimoxazin has also demonstrated activity on key monocot/grass weeds including *Lolium* spp. The combination of trifludimoxazin plus saflufenacil improved the burndown and spectrum of weed control over solo trifludimoxazin and therefore will be a key mix partner along with other non-PPO inhibitor chemistries as part of a resistance management strategy. Trifludimoxazin is expected to receive registration in key countries for use in multiple crops and total vegetation management by the early part of the next decade. This presentation will help explain the physiochemical properties of trifludimoxazin that make it such a potent inhibitor of the PPO enzyme in a broad range of weed species.

Keywords Trifludimoxazin, resistance, broadleaf weed, grass weeds.

Trifludimoxazin: A new novel herbicide for grass and broadleaf weed control in Australian winter cereals

Ian Francis, Marco Montagna, Russell Ison and Gavin Heard
BASF Australia Ltd, Level 12, 28 Freshwater Place, Southbank, Victoria 3006, Australia
(ian.francis@basf.com)

Summary Trifludimoxazin [1,5-dimethyl-6-thioxo-3-(2,2,7-trifluoro-3,4-dihydro-3-oxo-4-prop-2-ynyl-2*H*-1,4-benzoxazin-6-yl)-1,3,5-triazinane-2,4-dione] is a potent, novel inhibitor of protoporphyrinogen IX oxidase (PPO or Protox). Trifludimoxazin is very active when applied PRE or POST on dicotyledons/broadleaf weeds such as wild radish (*Raphanus raphanistrum* L.), and has also demonstrated activity on key monocotyledons/grass weeds including annual ryegrass (*Lolium rigidum* Gaud.). The combination of trifludimoxazin plus saflufenacil improved the spectrum of burndown and residual weed control over either herbicide applied alone. Trifludimoxazin is expected to receive its first registration in Australia for use in pre-plant burndown and pre-emergent residual weed control in winter cereals. This registration will provide Australia cereal growers a valuable new tool in management of weeds including species resistant to alternate modes of action.

Keywords Trifludimoxazin, cereals, broadleaf weed, wild radish, annual ryegrass, resistance.

Cinmethylin – a new herbicide for cereals: Mode of action and biokinetic properties

Ruth Campe[1], Eva Hollenbach[1], Lara Kämmerer[1], Janneke Hendriks[2], Wolfgang Höffken[3], Helmut Kraus[4], Jens Lerchl[1], Thomas Mietzner[3], Matthias Witschel[3], Johannes Hutzler[1], Laurent Picard[1] and Klaus Kreuz[1]
[1] BASF SE, Speyerer Straße 2, D-67117 Limburgerhof, Germany
[2] Metanomics GmbH, Tegeler Weg 33, D-10589 Berlin, Germany
[3] BASF SE, Carl-Bosch-Straße 38, D-67056 Ludwigshafen, Germany
[4] BASF Corporation, 26 Davis Drive, 27709-3528 NC, Research Triangle Park, United States
(klaus.kreuz@basf.com)

Summary The necessity of a new mode of action to combat herbicide resistant weeds raised the interest in the dormant molecule cinmethylin as a new solution for the control of weedy grasses in cereals. To elucidate its mode of action we applied a novel chemoproteomic pull-down approach. Three potential target proteins belonging to the same family of fatty acid thioesterases (FAT) bound to cinmethylin with high affinity. FAT enzymes play a crucial role in plant lipid biosynthesis by mediating the release of fatty acids (FA) from their site of synthesis in the plastids for subsequent formation of glycerolipids and very long chain fatty acids at the endoplasmic reticulum. By GC/MS analysis we could show that cinmethylin treatment leads to depletion of both saturated and unsaturated free FAs in the plant, emphasising that FA release for subsequent lipid biosynthesis is the primary target of cinmethylin. Furthermore, we demonstrated that physiological effects and downstream metabolic changes induced by cinmethylin differ substantially from other lipid biosynthesis inhibitor herbicides. Our results therefore suggest FAT inhibition by cinmethylin as a new mode of action and powerful tool for weed resistance management. The biokinetic properties of cinmethylin i.e., uptake, distribution, and degradation, were investigated in crops and monocotyledon weeds when applied at both pre-emergence and early post-emergence timings. The relationship of the biokinetic properties of cinmethylin and its efficacy and crop safety will be discussed.

Keywords Cinmethylin, mode of action, fatty acid thioesterases, lipid biosynthesis, uptake, distribution.

Agixa™ herbicide – a new treatment for selective weed control in drill-sown rice

Gregory S. Wells[1], Malcolm Taylor[2] and Colin Plater[3]
[1] PO Box 838, Sunbury, Victoria 3429, Australia
[2] 283 Torgannah Road, Cobram, Victoria 3644, Australia
[3] 12 Derwent Avenue, Wagga Wagga, New South Wales 2650, Australia
(wells1@dow.com)

Summary Agixa™ herbicide with Rinskor™ active (12 g L^{-1} Rinskor + 160 g L^{-1} cyhalofop) is a new treatment that will allow selective removal of grass and broadleaf weeds at three-leaf rice stage onwards. Rinskor active which is Group I mode of action, is one component of the formulation and as such, is a new mode of action for control of barnyard grass. Agixa herbicide is likely to be registered for Australian rice in the 2019/2020 season. It has a broad spectrum and is a strong new control option that has demonstrated value when used in sequence after the drill-sown rice foundation treatment.

Keywords Rice, drill-sown rice, barnyard grass, silvertop grass, weed control, Agixa, Rinskor active.

INTRODUCTION

The Australian rice industry has a major goal to improve water use efficiency, whilst increasing yield. By 2023 the industry aims to grow 1.5 T rice ML^{-1} water used, with present levels about 1 T ML^{-1}. In part, this will be achieved by the use of drill-sown rice, with water strategically applied and removed, to allow rice establishment as well as best weed control, whilst minimising water use.

Under these aerobic conditions, weeds like barnyard grass (*Echinochloa crus-galli* (L.) P.Beauv.) readily establish after planting, as rice emerges. Therefore early weed removal to minimise competition with young rice is critical for effective establishment and to maximise yield. Grass weeds like barnyard grass and silvertop grass (*Leptochloa fusca* (L.) Kunth ssp. *fusca*) readily establish and are aggressive competitors with young rice.

The Australian rice industry has established an effective foundation (at planting) treatment combination of paraquat + pendimethalin + clomazone, which is selective to rice and effective on many key weeds in drill-sown rice. It has been very successful due to multiple herbicide modes of action being used at planting time, to remove a significant portion of the weed population, through both knockdown and residual control. This is generally applied post-flush, pre-emergence to maximise effectiveness. However, the window of time is often very short from planting to emergence of rice. Another treatment was required to augment the foundation treatment and also enable effective weed control when paraquat was not able to be used at foundation timing, due to rice already being emerged.

There had been no new herbicides released for weed control in rice in Australia until 2016, when Rinskor was first introduced (Wells and Taylor 2016).

Ubeniq™ herbicide with Rinskor™ active is a new rice selective herbicide that suppresses dirty dora (*Cyperus difformis* L.) and controls broad leaf weeds and has short persistence in soil. It has been registered for weed control in water seeded rice.

This paper summarises new research trials to confirm rice selectivity and weed efficacy for registration of Agixa herbicide (12 g L^{-1} Rinskor™ + 160 g L^{-1} cyhalofop) for post-emergence weed control in drill-seeded rice in Australia.

MATERIALS AND METHODS

Trials were conducted in southern New South Wales and northern Victoria on rice research farms or commercial rice farms from 2012–2017 to determine the selectivity and weed control efficacy of combinations of cyhalofop and Rinskor active. Trials were designed as randomised complete block with three or four replications. Applications were made with gas powered small plot booms delivering 100 L ha^{-1} of spray solution. Data were summarised using Agricultural Research Manager Summary Across Trials software (Gylling 2015) and analysed using Analysis of Variance (ANOVA) at the 5% significance level.

Rice selectivity Eleven field trials were conducted with tank mixes of cyhalofop and Rinskor. Tank mixes were applied at 30 g ha^{-1} Rinskor plus 285 g ha^{-1} cyhalofop, either alone or mixed, to determine rice selectivity and weed efficacy. Crop oil concentrate was added to all treatments at 1–2% v v^{-1}.

A further seven trials were conducted with formulated product (Agixa herbicide), containing 12 g L^{-1} Rinskor plus 160 g L^{-1} cyhalofop and applied at a rate of 2 L ha^{-1} alone or in sequence after foundation treatment to determine either rice selectivity or weed efficacy. Crop oil concentrate was added to all treatments at 1-2% v v^{-1}.

Crop injury was measured by subjective visual assessment where 0 = no injury and 100 = complete crop loss. Crop effect was considered acceptable if injury was up to 20%.

Weed efficacy In all 18 field trials outlined above, where tank mixes of Rinskor and cyhalofop or Agixa (formulated product) were applied, weed control was also assessed. Weed control was assessed using subjective visual assessment where 0 = no control and 100 = complete control. Weed efficacy was considered acceptable if control was more than 95% at final assessment.

RESULTS AND DISCUSSION

Rice selectivity Tank mixes of Rinskor™ active plus cyhalofop at 30 g ha^{-1} plus 285 g ha^{-1} resulted in less than 15 % rice crop injury assessment across 11 trials (Table 1), which was commercially acceptable. Rice injury was seen as mild leaf curling, twisting or height reduction. There was no impact on yield.

Table 1. Mean rice injury (%) by Rinskor, cyhalofop or Rinskor plus cyhalofop tankmix.

Treatment (g a.i. ha^{-1})	Rice injury (%)
Untreated	0 c
Rinskor 30 g	11 a
cyhalofop 285 g	4 b
Rinskor 30 g + cyhalofop 285 g	14 a
LSD (p=0.05)	2.4

Agixa™ herbicide applied at 2 L ha^{-1} resulted in slight injury to rice, with increased injury observed when applied in sequence after foundation treatment, but it was considered commercially acceptable (Table 2).

Table 2. Mean rice injury (%) by Agixa herbicide at 2 L ha^{-1} and in sequence after foundation treatment. (F = foundation treatment at planting, P = post-emergence and fb = followed by).

Treatment (g a.i. or prod ha^{-1})	Timing	Rice injury (%)
Untreated		0 b
paraquat 200 g + pendimethalin 1496 g + clomazone 288 g	Foundation (F)	13 a
cyhalofop 320 g	F fb Post (P)	15 a
Agixa herbicide 2 L	F fb P	20 a
cyhalofop 320 g	P	0 b
Agixa herbicide 2L	P	2 b
LSD (p=0.05)		10.4

Weed efficacy Tank mixes of Rinskor active plus cyhalofop at 30 g ha^{-1} plus 285 g ha^{-1} resulted in excellent barnyard grass and silvertop grass control across 11 and three trials respectively (Table 3).

Table 3. Mean weed control (%) by Rinskor, cyhalofop or Rinskor plus cyhalofop tankmix.

Treatment (g a.i. ha^{-1})	Barnyard grass	Silvertop grass
Untreated	0 c	0 c
Rinskor 30 g	96 a	35 b
cyhalofop 285 g	71 b	92 a
Rinskor 30 g + cyhalofop 285 g	97 a	100 a
LSD (p=0.05)	2.5	14.4

Agixa herbicide applied at 2 L ha^{-1} resulted in excellent weed control (Table 4), which was stronger and broader in spectrum than individual components. Best control was achieved when it was applied in sequence after foundation treatment.

Table 4. Mean weed control (%) by Agixa herbicide at 2 L ha^{-1} and in sequence after foundation treatment. (F = foundation treatment at planting, P = post-emergence and fb = followed by).

Treatment (g a.i. or prod ha^{-1})	Timing	Barnyard grass	Silvertop grass
Untreated		1 d	0 c
paraquat 200 g + pendimethalin 1496 g + clomazone 288 g	Foundation (F)	80 bc	97 b
cyhalofop 320 g	F fb Post (P)	99 a	100 a
Agixa Herbicide 2 L	F fb P	98 a	100 a
cyhalofop 320 g	P	70 c	100 a
Agixa herbicide 2 L	P	88 ab	100 a
LSD (p=0.05)		11.1	0.9

CONCLUSIONS

Agixa herbicide premix of Rinskor active and cyhalofop is a new product that has acceptable rice selectivity, with strong weed control over a broad spectrum. It is a significant new tool for weed control in drill-seeded rice in Australia that will help the Australian rice industry move toward its new water use efficiency goal.

ACKNOWLEDGMENTS

The authors would like to acknowledge Malcolm Taylor, Agropraisals for conduct of many of the field sites and research trials.

REFERENCES

Gylling, S. (2015). Agricultural Research Manager, Summary Across Trials. http://gdmdata.com/Products/Summary Across Trials (ST). Gylling Data Management Inc, Brookings, South Dakota.

Wells, G. and Taylor, M. (2016). Rinskor™ active – a new herbicide for rice weed control in Australia. Proceedings of the 20th Australasian Weeds Conference, eds R. Randall, S. Lloyd and C. Borger, pp. 269-71. Weed Society of Western Australia, Perth.

Innovations for weed management – From new herbicide modes of action to novel technologies for weed management

Stefan Tresch
Bioscience Research, BASF SE, Germany
(stefan.tresch@basf.com)

Summary Chemical weed control is currently the method of choice for provision of efficient weed management in broadacre crops. For decades the introduction of novel small molecules with improved efficacy as well as ecological and economic properties has been the key for the delivery of systems that can be readily integrated into modern agricultural practices. One driver for the selection of new molecules is the rapid evolution of weed resistance to herbicides and therefore there is increasing need for herbicides with novel modes of action. Understanding the basis for mode of action therefore allows the development of useful properties associated with novel molecules and thus new applications.

Current status and progress in determination of herbicide mode of action to identify the relevant target(s) out of a massive number of potential target sites will be presented. Our findings suggest that a structured approach using large data sets, data analytics and automatisation has paved the way for straightforward determination of herbicide mode of action. Key technological trends will be reviewed and examples presented to describe how these technologies drive practice change in agriculture. Novel chemistry thus enables the development of new agronomic practices, based on informed decision making with less input, higher efficacy and the production of better products for a more sustainable agriculture.

Keywords Mode of action, weed management tools, data analytics, automatisation.

Topramezone (Frequency® Herbicide), a new flexible tool for weed control in winter cereals across Australia

Ian Francis, Marco Montagna and Russell Ison
BASF Australia Ltd, Level 12, 28 Freshwater Place, Southbank, Victoria 3006, Australia
(ian.francis@basf.com)

Summary Frequency® is a new herbicide for the Australian cereal market containing the active ingredient topramezone, an inhibitor of 4-hydroxyphenylpyruvate dioxygenase (HPPD). The efficacy and crop safety of topramezone was tested across the major cereal growing regions of Australia from 2013 to 2017 seasons. Topramezone, when applied post emergence in combination with bromoxynil, provides high level control of a wide range of broadleaf weeds, including species known to be resistant to other mode of action herbicides, such as wild radish (*Raphanus raphanistrum* L.), milk thistle (*Sonchus oleraceus* L.) and flaxleaf fleabane (*Conyza bonariensis* L.). Moreover, excellent weed control was obtained from the tank mix with other mode of actions herbicides: MCPA, picolinafen, diflufenican, and metsulfuron allowing high levels of use flexibility in managing varying weed spectrums and resistant populations. In addition to broadleaf weed control, topramezone has shown a significant level of reduction in wild oat (*Avena sterilis* L.) seed set introducing a new mode of action as a tool in managing this important grass weed of northern cropping areas. Due to the high level of control on a wide range of species, along with good crop safety and flexibility of tank mix partners, topramezone (Frequency® Herbicide) represents a significant new tool for weed control in the Australian cereal production industry.

Keywords Topramezone, cereals, broadleaf weed, wild oats, resistance.

A new mode of action (MOA) pre-emergence herbicide for the control of annual ryegrass (*Lolium rigidum* Gaud.) and other monocotyledon weeds in cereals in Australia

Ian Francis, Marco Montagna, Russell Ison and Gavin Heard
BASF Australia Ltd, Level 12, 28 Freshwater Place, Southbank, Victoria 3006, Australia
(ian.francis@basf.com)

Summary With many weeds resistant to common herbicide active ingredients (PSII-, ALS- and ACCase-inhibitors), farmers around the world are seeking new weed management options. In Australian winter cereals there has been a shift away from reliance on post-emergence graminicides to an increased emphasis on pre-emergence applications. Cinmethylin currently belongs to the cluster of herbicides with an unknown mode of action (HRAC Group Z). However, new research from BASF has identified this molecule to act *via* a totally new mode of action. Cinmethylin was tested as a pre-emergence herbicide in Australia during the 2014–2017 growing seasons. Its efficacy and crop safety was evaluated at different application timings: incorporated by sowing (IBS), post-sowing pre-emergence (PSPE) and early post-emergence (EPE). Field trials were conducted at various rates across the main cereal growing regions of Australia, targeting a number of monocotyledon weeds. Compared, primarily, to pyroxasulfone and S-metolachlor + prosulfocarb, cinmethylin demonstrated effective control of annual ryegrass (*Lolium rigidum* Gaud.) including resistant biotypes. Furthermore, good activity was displayed against other cool season grasses such as wild oat (*Avena* spp.), brome grass (*Bromus* spp.) and barley grass (*Hordeum* spp.). Different crops showed variable level of tolerance to cinmethylin, which was related to parameters such as soil application timing, conditions at sowing, sowing depth and rainfall patterns after sowing. Furthermore, cinmethylin was screened in the laboratory to determinate efficacy on annual ryegrass (*L. rigidum* Gaud.) resistant to other modes of action, showing high levels of control on all tested biotypes.

Keywords Cinmethylin, pre-emergence, cereals, annual ryegrass.

ForageMax™ herbicide – a new treatment for selective weed control in graze and grain canola

Gregory S. Wells[1] and Colin Plater[2]
[1] PO Box 838, Sunbury, Victoria 3429, Australia
[2] 12 Derwent Avenue, Wagga Wagga, New South Wales 2650, Australia
(wells1@dow.com)

Summary Graze and grain or dual purpose canola has gained popularity due to improved income versus growing canola for grain only. There are several benefits to this dual purpose approach: increased flexibility with sowing time to spread farm workload; reduced crop biomass for easier windrow and harvest; reduced lodging risk in high yield years; improved grass weed control; winter pasture spelling value; disease breaks; and additional source of income from stock earlier in season rather than just grain at harvest.

Historically, weed control treatments like imidazolinones, triazines or clopyralid have been used in canola for weed control. They can have persistent crop or soil residues, a narrow weed control spectrum, limited product compatibility and have extended grazing withholding periods.

ForageMax™ herbicide with Arylex™ active (100 g L^{-1} Arylex plus 50 g L^{-1} aminopyralid) is already registered for use in rape and turnips in Australia. It has a broad weed control spectrum, shorter soil persistence, improved product compatibility and a shorter grazing withholding period making it a potentially useful alternative for weed management in graze and grain canola. This paper outlines research to determine the selectivity of ForageMax herbicide when used in either spring or autumn sown canola, for dual purpose graze and grain.

Keywords Canola, graze and grain, weeds, ForageMax, Arylex active.

INTRODUCTION

Use of canola for grazing and grain production has gained popularity, due to improved profitability. Dove and Kirkegaard (2014) conducted a review of dual purpose wheat and canola, which showed there was little impact on yield for late maturing types when early sown crops were grazed prior to stem elongation.

Canola may be sown in either spring or autumn. Spring sown canola may be sown, grazed over summer and then stock removed to allow for regrowth, prior to harvest in the following spring. Paridaen and Kirkegaard (2015) showed that spring sown winter canola could be used as a biennial dual-purpose crop, to provide additional forage for summer and autumn grazing, before recovery to produce an oilseed crop.

Autumn sown canola is planted early (late summer), then grazed and stock removed for normal grain harvest. Sprague *et al.* (2015) reported research on canola varieties of winter, winter × spring and spring maturity in various parts of the high rainfall zone. The experiments confirmed the potential of dual-purpose canola across all regions of the high rain zone of Australia when suitable maturity types are sown, managed and grazed appropriately.

There are limited herbicide options for dual purpose canola, so farmers have often used imidazolinone tolerant types (Clearfield* varieties), to allow residual herbicide use during the winter. In some cases the weed spectrum or soil persistence of these treatments has limited their utility. Industry needed another herbicide that had a broad weed control spectrum and shorter soil persistence, with short grazing withholding periods.

This paper outlines new research to determine the selectivity of ForageMax herbicide for use in graze and grain (dual purpose) canola.

MATERIALS AND METHODS

Ten trials were conducted in southern New South Wales and Victoria on commercial farms from 2015–2017 to determine the selectivity of ForageMax. The trials were designed as randomised complete block with three or four replications. Herbicide treatments were applied with gas powered small plot booms delivering 100 L ha^{-1} of spray solution. Data were summarised using Agricultural Research Manager Summary Across Trials software (Gylling 2015) and analysed with analysis of variance at the 5% significance level.

Canola selectivity Ten selectivity trials using commonly grown canola cultivars were conducted in 2015, 2016 and 2017. ForageMax™ herbicide was applied at label rate (100 mL ha^{-1}) and twice label rate (200 mL ha^{-1}) at the 4–8 leaf canola stage, as per the existing label recommendation for forage brassica use. Crop oil concentrate was added at 1% v v^{-1} in all treatments. Crop injury was measured by visual assessment where 0 = no injury and 100 = complete crop loss.

Canola yield At two sites there was sufficient rain or irrigation and conditions allowed grain yield to be taken. Grain yields were taken at normal harvest time and expressed as T ha^{-1} to compare to weed-free untreated control (untreated plots were hand weeded).

RESULTS AND DISCUSSION

Canola selectivity ForageMax applied at 100 or 200 mL ha^{-1} resulted in average injury of 2–3 percent 0–60 days after application (days after application, DAA), 1–2 % injury 60–120 DAA and no injury 120–180 DAA (Table 1). This is commercially acceptable for dual purpose canola.

Table 1. Mean canola injury (%) by ForageMax at 0 to >120DAA.

ForageMax Treatment	Days after application (DAA)		
	0–60	60–120	>120
100 mL	2 a	1 a	0
200 mL	3 a	2 a	0
Weed-free untreated	0 a	0 b	0
LSD (p=0.05)	2.2	0.5	–
N	59–60	24	12

Canola yield ForageMax applied at 100 or 200 mL ha^{-1} resulted in similar yield to weed-free untreated in a trial in Victoria, but slightly lower yield in a trial in southern New South Wales (Table 2). At the site in Victoria, treatments were applied 330 days prior to harvest and the site also had significant rain and some irrigation after herbicide application. At the New South Wales site, treatments were applied 121 days prior to harvest and the crop had little to no rain during the critical flowering to early pod set period in late August to early September, as well as being grazed past the stem elongation stage. We believe both these factors contributed to yield reduction, due to lack of time and conditions for growth compensation after treatment.

Further work is needed to determine the effect of ForageMax use on grain yield in dual purpose canola.

Table 2. Mean canola yield (T ha^{-1}) compared to untreated control, after treatment with ForageMax.

ForageMax Treatment	Trial number	
	163012GW Vic	KA17-0565 NSW
100 mL	3.4 a	2 a
200 mL	3.9 a	1.9 b
Weed-free untreated	3.5 a	2.5 a
LSD (p=0.05)	0.42	0.52

CONCLUSIONS

ForageMax is currently registered for use in forage brassicas. It has significant potential for use in dual purpose canola. Treatment at the early post-emergence 4–8 leaf canola stage resulted in little canola injury across 10 trials. However, further work is needed to determine what the impact, if any, will be on canola yield from treatment at 4–8 leaf canola stage.

ACKNOWLEDGMENTS

The authors would like to acknowledge Dr John Kirkegaard, CSIRO for provision of papers on dual purpose canola in Australia.

REFERENCES

Dove, H. and Kirkegaard, J. (2014). Using dual-purpose crops in sheep-grazing systems. *Journal of the Science of Food and Agriculture* 94, 1276-83.

Gylling, S. (2015). Agricultural Research Manager, Summary Across Trials. http://gdmdata.com/Products/Summary Across Trials (ST). Gylling Data Management Inc, Brookings, South Dakota.

Paridaen, A. and Kirkegaard, J. (2015). Forage canola (*Brassica napus*): spring-sown winter canola for biennial dual purpose use in high-rainfall zone of southern Australia. *Crop and Pasture Science* 66, 275-86.

Sprague, S., Kirkegaard, J., Graham, J., Bell, L., Seymour, M. and Ryan, M. (2015). Forage and grain yield of diverse canola (*Brassica napus*) maturity types in the high rainfall zone of Australia. *Crop and Pasture Science* 66, 260-74.

* Clearfield is a registered trademark of BASF Corporation.

™ Trademark of The Dow Chemical Company ('Dow') or an affiliated company of Dow.

Encapsulated dry herbicides: A novel approach for control of trees

Ken C. Goulter[1], Victor J. Galea[2] and Peter Riikonen[1]
[1] Bioherbicides Australia Pty Ltd, The University of Queensland, Gatton Campus, Gatton, Queensland 4343, Australia
[2] School of Agriculture and Food Sciences, The University of Queensland, Gatton Campus, Gatton, Queensland 4343, Australia
(p.riikonen@bioherbicides.com.au)

Summary Bioherbicides Australia (BHA) is researching innovative and environmentally safe means of controlling tree weeds. Apart from our biocontrol product for parkinsonia (Di-Bak Parkinsonia™), the recent availability of dry formulations of several herbicides has provided an opportunity to apply them by insertion of a soluble capsule directly into tree stems. To facilitate delivery an easy-to-use applicator has been developed. The applicator is designed to drill a hole into the trunk, insert the herbicide capsule followed by a sealing plug. The sealing of the capsule ensures that moisture does not escape preventing oxidation of the immediate wound area and facilitating the dissolution of the capsule and herbicide allowing rapid distribution into the vascular stream.

Trials with a number of chemical formulations encapsulated in Size 0 hypromellose capsules have been conducted against WoNS such as prickly acacia and parkinsonia; other weed trees such as leucaena, calotrope, camphor laurel and privet, and plantation eucalypts such as *Eucalyptus saligna* Sm. and *E. dunnii* Maiden. Di-Bak G (glyphosate) capsules with ~350 mg (700 g kg^{-1} ai) have been shown to be highly efficacious against all species tested except for parkinsonia. Sampling of the point of insertion demonstrates that the capsule and chemical are quickly dissolved.

This application method will significantly reduce environmental and operator exposure to herbicides. It reduces waste as unused capsules can be stored for long periods and eliminates off-target placement. It is not subject to external environmental conditions and provides a targeted, readily calibrated herbicide application to the weed of interest.

Keywords Capsules, glyphosate, tree weeds, safety.

INTRODUCTION

Various strategies for herbicide-based control of undesirable trees exist. Foliar sprays where the herbicide is often dissolved in a water miscible solvent and then diluted into a water solution with the possible addition of adjuvants to aid solubility and adherence to and penetration of plant surfaces. The efficiency of spraying is affected by environmental factors such as wind, temperature and humidity. Basal bark application is performed at the base of target trees typically with a mixture of herbicide and diesel distillate to assist penetration through the bark. It is only recommended for plants less than 5 cm in diameter. For larger woody weeds, the cut stump method (Little and van der Berg 2006) is performed by felling the tree and painting or spraying the cut surface of the stump with a herbicide. Application of insufficient herbicide can allow reshooting or coppicing. Application of excess herbicide can lead to non-target damage or 'flashback' where neighbouring trees are also killed. It is thought that the herbicide travels between anastomosed roots of neighbouring trees (Ashton and Kelty 2017). This phenomenon is also seen with conventional stem injection methods where the herbicide is typically applied in either of two ways, 'frill and fill' (DiTomaso *et al.* 2013) or 'drill and fill' (McKenzie *et al.* 2010). With the former, an axe is used to cut into the stem to form a frill before the herbicide is applied to the cut. In drill and fill a hole is drilled that is then filled with herbicide. Other herbicides such as tebuthiuron can be applied as granules to the soil surface around target trees. This application requires sufficient rainfall to allow solubilisation and movement of the herbicide to tree roots, but excess rain can cause movement of the chemical away from the root zone.

Although suitable for weed control in many circumstances, the above described application methods are associated with various disadvantages. For example, the methods: may involve the handling of herbicides in toxic forms; maybe of limited effectiveness for certain weed types; maybe sensitive to environmental parameters; may involve imprecise or excessive herbicide application; once applied, may result in no control of subsequent distribution of the chemical in the environment; and/or may involve the use of relatively dangerous equipment.

Herbicides undergo constant reappraisal as to their safety to the human population and environment. Tebuthiuron which is currently applied on a broadacre basis in Australia has been banned in Europe due to concerns that this stable molecule will contaminate ground water. The widely used glyphosate was controversially relicensed in Europe late last year over concerns that is a group 1B carcinogen (presumed to have carcinogenic potential in humans based on animal studies).

With all chemicals there are issues in the use in agriculture and the subsequent residues in food and their safety to the environment and to humans. The goal therefore should be to minimise exposure to alleviate those concerns. This can be achieved by reducing dosage or improving application methods. We will present one approach to achieving these goals using encapsulated glyphosate (Di-Bak G). Other formulations under evaluation include Di-Bak H (hexazinone), Di-Bak M (metsulfuron methyl), Di-Bak I (imazapyr), Di-Bak Teb (tebuthiuron) and Di-Bak TyP (triclopyr and picloram).

MATERIALS AND METHODS

General Glyphosate (~330 mg 700 g kg^{-1} ai, Glymac Dri 700, Macspred Pty Ltd) was prepared in pharmaceutical grade capsules (HPMC Size 0 Natural TR.V700 colour cap and body Vcaps, Suzhou Capsugel®Ltd, China). Capsules were hand packed using a Capsugel Profiller apparatus. The BHA MkIII applicator was used to drill a hole into the stem, insert the herbicide capsule and seal with a polypropylene plug. Generally, one capsule was inserted per 10 cm stem circumference. Tree health was assessed on a 0–4 scale, 0 being dead and 4 being healthy. Observations were made for evidence of 'flashback' in nearby untreated trees. At later assessment times, a small sample of trees were cut down and cross sections of stems containing the sites of insertion were removed. These stem pieces were returned to the laboratory and split open longitudinally through the insertion point.

Case 1. Control of prickly acacia (*Vachellia nilotica* (L.) P.J.H. Hurter & Mabb) with capsules of dry glyphosate (Di-Bak G) The trial was established at Bilaban Station, St Lawrence, Queensland (GPS 22°20.913′S, 149°30.203′E) on 5 December 2016. At application (T0) trees were assessed as being healthy with no dead branches and significant foliage. The trees were mature and flowered regularly. Stem diameters at 30 cm above ground level varied from 4–20 cm and trees were up to four metres tall. Application treatments consisted of applying 1, 2, 3 or 4 glyphosate capsules equidistant around the tree trunks at ~30 cm above ground level of around 10 trees each. The first thorough assessment was done 9 MAT (months after treatment) on 15 September 2017. Stem circumferences at 30 cm above ground level, number of applications per tree, and foliage cover and percentage branch death were recorded. Assessment of foliage cover was confounded by an earlier natural infestation of the introduced biological control agent, the leaf-feeding geometrid looper caterpillar, *Chiasmia assimilis* (Warren) that led to significant defoliation from which the trees were recovering. Weather data was collected at the nearby (<5 km) St Lawrence Post Office.

Case 2. Use of Di-Bak G in the thinning of plantation grown *Eucalyptus saligna* Sm. (Sydney blue gum) The trial was established on Brooklyn Station, Upper Tooloom, near Urbenville on 25 May 2017. GPS co-ordinates of the first treated plot were 28°32.486′S, 152°24.782′E. The trees were about 15 years old and are planted in rows 4 m apart at 1.5 m spacing. Trees were 6–10 m tall. Desirable trees were marked and the other trees were treated. The trial consisted of four replicates with 15 trees per replicate. The number of capsules applied to each tree was dependent on tree circumference which was measured at chest height. The tree circumferences range was from 18–57 cm. Tree health was assessed at four months (25 August 2017) and seven months (20 December 2017). Samples were collected for observation of dissolution of capsule and herbicide. Rainfall data for the closest recording station (Urbenville) was taken from the Bureau of Meteorology (BOM) internet site. No recent temperature data records were available for this site.

Case 3. Dose rate effect of Di-Bak G on *Eucalyptus dunnii* Maiden (Dunn's white gum) In this experiment one, two or three capsules of glyphosate were inserted into *Eucalyptus dunnii* trees irrespective of their circumference. This trial did not include replications. The trial was established in a block adjacent to the trees of Case 2 on 26 May 2017. GPS co-ordinates were 28°32.488′S, 152°24.778′E. The trees were arranged and similar in age and size to the *E. saligna* trees described in Case 2. Desirable trees were marked and the other trees were treated.

The trial consisted of one replicate of 15 trees for each treatment. The number of capsules applied to each tree was independent of tree circumference which was measured at chest height. Tree health was assessed at four months (25 August 2017) and seven months (20 December 2017). Untreated trees were observed for signs of flashback. At the four month assessment, samples were collected for observation of dissolution of capsule and herbicide.

RESULTS

Case 1. The plant health rating scale was formulated as a linear model based on foliage cover and percentage of living branches. Given the defoliation caused by *C. assimilis* confounded the analysis it was decided to compare all treated trees to the mean health rating of the control untreated trees giving a proportional result between 0 (dead) to 1 (as healthy as control trees).

A relationship exists between stem circumference, number of capsules applied, and tree mortality, that is, application of 1 capsule seems sufficient to kill trees with stems up to 30 cm in circumference (n = 5) (Table 1). The same is true for 2 capsules (n = 3) whereas for trees 40–60 cm circumference at least 3 capsules should be applied (n = 5). It is also evident that treating prickly acacia trees with Di-Bak G capsules at any dose caused some deterioration in plant health.

Case 2. The range of stem circumferences of treated trees ranged from a minimum of 12 cm to a maximum of 70 cm. On average, a capsule was inserted every 13–14 cm of stem circumference.

Stem pieces that were cut and split four months after treatment revealed that the capsules had dissolved in all treatments and no traces of glyphosate were observed. At four months there were indications of deterioration in tree health as expressed by leaf drop and bark splitting.

There was over 90% mortality at four months. By the seven month assessment there had been good rainfall but no indication of reshooting or flashback was observed. The failure of full strength glyphosate to kill all treated trees may have been due to loss of the sealing plugs. If the plug, for whatever reason, is displaced soon after application it is likely that moisture loss will prevent solubilisation of the capsule and herbicide. More recent modifications to the applicator in the new design of BHA Applicator MkIV has reduced the probability of sealing plugs being lost.

Case 3. Applying 3 capsules of glyphosate had killed all treated trees by four months. At four months ~90% of trees treated with 2 capsules were dead and by seven months all had been killed. However, on average, the trees treated with 2 capsules had the largest girth of the treatments (Table 2). This might explain the delayed response. Use of a single capsule resulted in ~70% mortality at four months which increased to ~80% by seven months. The observation that a better kill was achieved with *E. dunnii* compared to *E. saligna* might be explained by the former having a thicker bark and softer wood which allowed more complete insertion of the plug. There was no sign of flashback even with the highest dose rate.

Table 2. 'Chest Height' measurements of tree circumferences and mean distance between capsules.

Treatment Di-Bak G	Stem circumference (cm) Min – (Average) – Max	Distance between capsules (cm)
1 Capsule	20 – (38.67) – 58	38.7
2 Capsules	26 – (45.60) – 59	22.8
3 Capsules	16 – (34.80) – 75	11.6

Table 1. Relationship of capsule number applied and tree circumference on prickly acacia tree health at nine months after treatment.

Stem circumference (cm)	# of capsules applied					
	1	2	3	4	5	6
0–20	0.0A (n = 2)B					
21–30	0.0 (n = 3)	0.0 (n = 3)				
31–40	0.1 (n = 2)	0.17 (n = 2)	0.0 (n = 2)			
41–50	0.35 (n = 6)	0.21 (n = 3)	0.01 (n = 2)	0.016 (n = 5)	0.0 (n = 2)	
51–60	0.58 (n = 4)	0.18 (n = 3)	0.01 (n = 1)	0.0 (n = 2)	–	0.01 (n = 3)
>61		0.42 (n = 1)		0.06 (n = 2)	0.18 (n = 1)	

Mean plant health of treated trees as a proportion of healthy untreated trees. 0.0 – plants dead to 1.0 plants as healthy as untreated trees.
Number of plants assessed in each cohort.

DISCUSSION

We have demonstrated that insertion of encapsulated dry glyphosate can be a potent control method for unwanted trees. This method delivers a minimum recommended fatal dose of chemical directly in the trees vascular system thereby utilising 100% of the active agent's potential. In comparison to other commonly used manual methods such as cut stump and basal bark spraying, only 20–30% of the chemical quantity is required with Di-Bak G to achieve the same result. Application of minimal dosages will reduce occurrence of 'flashback'. Also, no axes, hatchets, backpacks and tank mixes of liquid herbicide, chainsaws or diesel distillates were required. There was no waste, no need to dispose of unused chemicals and easy storage of capsules are all additional benefits. Application of a single capsule could kill prickly acacia trees with stem circumference of up to 30 cm. On larger stems insertion of a capsule every 10–15 cm was effective. Care needed to be taken to ensure that insertion was done below major branches. Evaluation of Di-Bak G and other herbicides on prickly acacia is currently underway in western Queensland. Di-Bak G was seen to be highly effective against the eucalypt species tested, and may find application as tool in plantation thinning.

The insertion of encapsulated Di-Bak G is highly targeted and does not expose the operator or the environment to accidental exposure to the herbicide. It should be broadly applicable to all trees sensitive to glyphosate. We are also developing a range of other herbicide capsules that maybe suitable to treat trees which are tolerant to glyphosate.

Although not presented in this paper, preliminary results to date show that Di-Bak G is active against leucaena, calotrope, camphor laurel, privet and ficus. This methodology could be seen as a ready replacement for cut stump application in plantations, weed tree control in sensitive areas such as waterways and National parks and in areas that are inaccessible to other equipment such as hillsides.

ACKNOWLEDGMENTS

We would like to thank the managers of Bilaban and Brooklyn stations for access to their trees and to Tanya Jobling (Landcare New South Wales) for coordinating and assisting with the eucalypt trials.

REFERENCES

Ashton, M.S. and Kelty, M.J. (2017). 'The practice of silviculture: applied forest ecology' (John Wiley and Sons, London).

DiTomaso, J.M., Kyser, G.B., Oneto, S.R., Wilson, R.G., Orloff, S.B., Anderson, L.W., Wright, S.D., Roncoroni, J.A., Miller, T.L., Prather, T.S., Ransom, C., Beck, K.G., Duncan C., Wilson K.A. and Mann, J.J. (2013). 'Weed control in natural areas in the western United States.' (Weed Research and Information Center, University of California).

Little, K.M. and van den Berg, G.J. (2006). First rotation *Eucalyptus macarthurii* cut stump control in KwaZulu-Natal, South Africa. *Southern African Forestry Journal* 207, 15-20.

McKenzie, J., Brazier D., Owen A., Vitelli J. and Mayer, B. (2010). Stem injection: a control technique often overlooked for exotic woody weeds. Proceedings of the 17th Australasian Weeds Conference, ed. S.M. Zydenbos, pp. 459-61. (New Zealand Plant Protection Society, Christchurch).

Stopping the invasion of *Persicaria chinensis* (Chinese knotweed) – a herbicide solution

Kim Hignell
Lake Macquarie City Council, Box 1906, Hunter Region Mail Centre, New South Wales 2310, Australia
(khignell@lakemac.nsw.gov.au)

Summary Invasive weeds impact the economy and society globally and contribute significantly to the ecological degradation to the environment both terrestrially and aquatically. Weed incursions have continually established themselves in Australia since European settlement and will continue to invade into the future. One new weed incursion which has been recently discovered invading two localised areas within NSW is *Persicaria chinensis* (L.) H.Gross, commonly known as Chinese knotweed. This weed has been assessed as a highly invasive species. With limited literature and research available on the control of this species, a field trial using registered herbicides was established to find effective herbicide options to control these incursions. The application of a metsulfuron-methyl was the most effective with 100% control at 47 days after treatment (DAT), followed by a mixture of metsulfuron-methyl and glyphosate with 100% control at 111 DAT. Repeated treatments of a foliar spray are required for two other herbicides used in the trial. Current research for an effective herbicide has been very limited, with this study identifying effective herbicide options for controlling Chinese knotweed in Australia.

Keywords Invasive weed, new incursion, herbicides, metsulfuron-methyl.

INTRODUCTION

This paper has two aims: to review what is known about the biology of Chinese knotweed; and, subsequently, to identify the most effective herbicide to treat and eradicate Chinese knotweed within riparian corridors of New South Wales.

CHINESE KNOTWEED (*Persicaria chinensis*)

Description *Persicaria chinensis* (L.) H.Gross, (syn. *Polygonum chinense*) known as Chinese knotweed, is a scandent, divaricately branched perennial plant with stems and branches that are reddish purple in colour, glabrous and spineless with prominent nodes, and have a zigzag appearance. The stipules form a sheath on the stem, are membranaceous and glabrous with no hairs at the margin (Stuart 2015, Sasidharan n.d). The leaves are simple, alternate, ovate- to oblong-shaped entire with an acuminate apex, truncate base, 50–100 mm long and 30–70 mm wide. The leaf margins are minutely crenulated; the midrib violet-red, the upper surface of leaf with a central darker green inverted 'V' shaped spot (eFloras 2011a and b, Sasidharan n.d.) The inflorescence is a compound corymb, arranged terminally or in terminal and auxiliary positions while the peduncles are glandular and hairy (eFloras 2011a and b). The flowers are campanulate, small, white or pinkish in colour, 2.0 to 3.5 mm bearing 5 tepals, and 8 stamens with purple tips (eFloras 2011a and b, EoL 2014). The fruits are globose in shape turning dark purple when ripe. The seeds are trigonous (triangular in cross-section), small 3 to 4 mm round and black (Stuart 2015, Sasidharan n.d) (Figure 1).

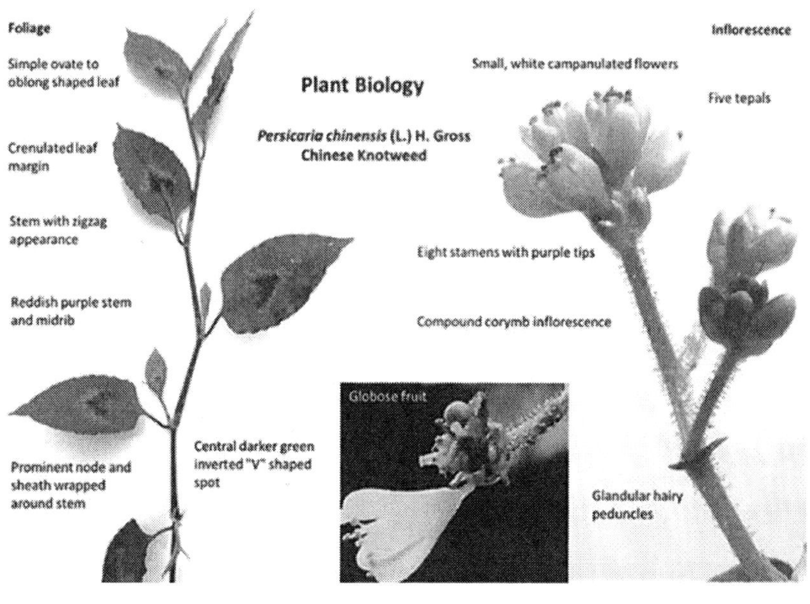

Figure 1. Plant biology close up of Chinese knotweed looking at the foliage, inflorescence and fruit. (Images K. Hignell).

Habit Chinese knotweed is a rhizomatous herbaceous perennial from the Polygonaceae family from South-east Asia (IPNI 2012) and is known as a high-risk invasive species (Wong *et al.* 2015). Chinese knotweed has been described as an upright, shrub-like perennial (Galloway and Lepper 2010), a twining perennial herb (Stuart 2015), a scrambling herb (Turner 1995) and a rhizomatous perennial that sometimes climbs (Wilson 2010). Chinese knotweed can reach a height of 1 to 1.5 m (Turner 1995, Wilson 2010, EoL 2014, Stuart 2015). It is closely related to other invasive *Persicaria* species such as princes feathers (*P. orientalis* (L.) Spach), Japanese knotweed (*P. capitate* (Buch.-Ham. ex D.Don) H.Gross), and mile-a-minute weed (*P. perfoliata* (L.) H. Gross) (Randall 2012).

This plant grows rapidly, forming thick canopies that have the ability to smother native plants (Galloway and Lepper 2010). Naturally growing from sea level to 3000 m (eFloras 2011b), Chinese knotweed can be found growing in open moist grounds, gardens, riparian zones, roadsides (Galloway and Lepper 2010, US Forest Service 2010, EoL 2014) and swamp margins (Wilson 2010, EoL 2014).

As a new incursion into New South Wales, particularly invading riparian zones (PlantNET 2016), research to identify the most effective control techniques to reduce its impact on the environment is required. There is limited information regarding effective control using herbicides (US Forest Service 2010).

MATERIALS AND METHODS

Trial site The study was conducted within the riparian corridor at Little Flaggy Creek, Kahibah Road, Kahibah, within the Lake Macquarie local government area on the Central Coast of New South Wales (NSW). Little Flaggy Creek flows from Charlestown to the west of the site to the east into Glenrock State Conservation Area then into the Tasman Sea. This site has been infested with Chinese knotweed since the discovery of the incursion in 2012 (PlantNET 2016).

The trial site has creek banks up to 2 m high with full sunlight and a minimal amount of dapple shading early morning on the eastern end of the site. The Chinese knotweed has been growing to 1–1.5 m high, covering the entire northern side of the bank. There were 25 plots established to allow 5 replicates of each treatment at the site. Each plot size was 1.5 m by 3 m with a 100% ground coverage of Chinese knotweed.

Application The applications of herbicides for this trial were performed under the APVMA permit, to allow the conduct of small scale trials with AGVET chemicals – PER 7250. Application of the herbicides was via a 10 L portable knapsack using a variable cone nozzle and a water rate of 3 L 4.5 m^{-2}. Four herbicide applications and one control were randomly selected in each of the 5 replicates. All the herbicide treatments were applied to actively growing Chinese knotweed before flowering stage.

Weather condition At the time of application on the 24 December 2015, starting at 8.00 am, weather conditions were a partly cloudy morning with the dry bulb temperature at 20°C, the humidity was at 87% and a light breeze from the south-east at five km h^{-1}. Light rain had fallen overnight but the moisture on vegetation had evaporated by application time. No rain fell within 24 hours after application.

Herbicides All herbicide applications were applied as a foliage spray and each treatment was applied according to recommended commercial standards as stated on the herbicide labels. Three individual herbicides and a mixture of two herbicides were used for this trial and were applied once only. Each herbicide was selected from a different 'mode of action' group to allow for the management of herbicide resistance.

The first herbicide was a glyphosate product, Eraze 360 Bi-aquatic™ herbicide, with an application rate of 100 mL 10 L^{-1}. Glyphosate is a broad-spectrum herbicide of choice in agriculture, urban parks, gardens, and waterways due to its low toxicity to the environment (Sánchez-Bayo *et. al.* 2010).

The second herbicide was a metsulfuron-methyl product, Metmac 600 WG™, with an application rate of 1 g 10 L^{-1}. Metsulfuron-methyl is a systemic selective herbicide targeting broadleaf and woody weeds and is very effective on weeds that include bulbs or tubers (Spencer 2012). There is an APVMA minor use permit PER14734 to use metsulfuron-methyl on alligator weed in waterways of NSW (APVMA 2015).

The third herbicide was a packaged mixture of amitrole and ammonium thiocyanate, labelled as Amitrole T™, with an application rate of 28 mL 10 L^{-1}. Amitrole and ammonium thiocyanate is another broad-spectrum herbicide routinely used for clearing weeds in irrigation channels (Sánchez-Bayo *et. al.* 2010).

The fourth herbicide treatment was the mix of glyphosate and metsulfuron–methyl. The application rate of the glyphosate was 100 mL 10 L^{-1} and metsulfuron-methyl at 1 g 10 L^{-1}.

Assessment Control assessments were undertaken in the centre 1 metre of area in each plot so that any influences from the adjacent plots were negated. A subjective visual rating of 0 to 10 (Table 1) was used to assess the control efficacy of each treatment. The scoring indicates a commercially accepted rate of

Table 1. The subjective visual rating score used control efficacy assessment.

Rate of Control	Visual appearance	% Control
0	Nil Control with no wilting or leaf colour change	0%
1	Wilting, Leaf drop, Leaf colour change	10%
2	Wilting, Leaf drop, Leaf colour change, Dieback and/or Regrowth	20%
3	Wilting, Leaf drop, Leaf colour change, Dieback and/or Regrowth	30%
4	Leaf drop, Leaf colour change, Dieback and/or Regrowth	40%
5	Leaf drop, Leaf colour change, Dieback and/or Regrowth	50%
6	Leaf colour change, Dieback and/or Regrowth	60%
7	Leaf colour change, Dieback and/or Regrowth	70%
8	Dieback and/or Regrowth	80%
9	Dieback and/or Regrowth	90%
10	Total Control with no regrowth	100%

control by considering the visual appearance of the plant material and the amount of dieback and/or regrowth each plot had without disturbance (Cook and Shoemark 2015). The site was assessed seven times throughout the trial, occurring at 14, 35, 47, 71, 91 and 125 Days After Treatment (DAT).

RESULTS

The results indicate that there is little difference in the control efficacy of Chinese knotweed between the three herbicide treatments, metsulfuron-methyl, glyphosate and the mix of these two herbicides (Figure 2). All three treatments significantly reduced the health and coverage Chinese knotweed in all plots at 35 DAT and continued to trend through to 125 DAT retaining a percentage control of 90% or above from 47 DAT.

The most effective herbicide of the trial was metsulfuron-methyl, with a control score of 10 from 47–125 DAT without regrowth. Metsulfuron-methyl was the slowest activating herbicide at 14 DAT, with only the visual appearance of leaf colour change observed. The mix treatment was second by achieving a control rate score of 10 at 111 DAT and this continued through to 125 DAT with no regrowth present. Glyphosate was the fastest activating herbicide at 14 DAT with wilting and leaf drop, then matched the rate

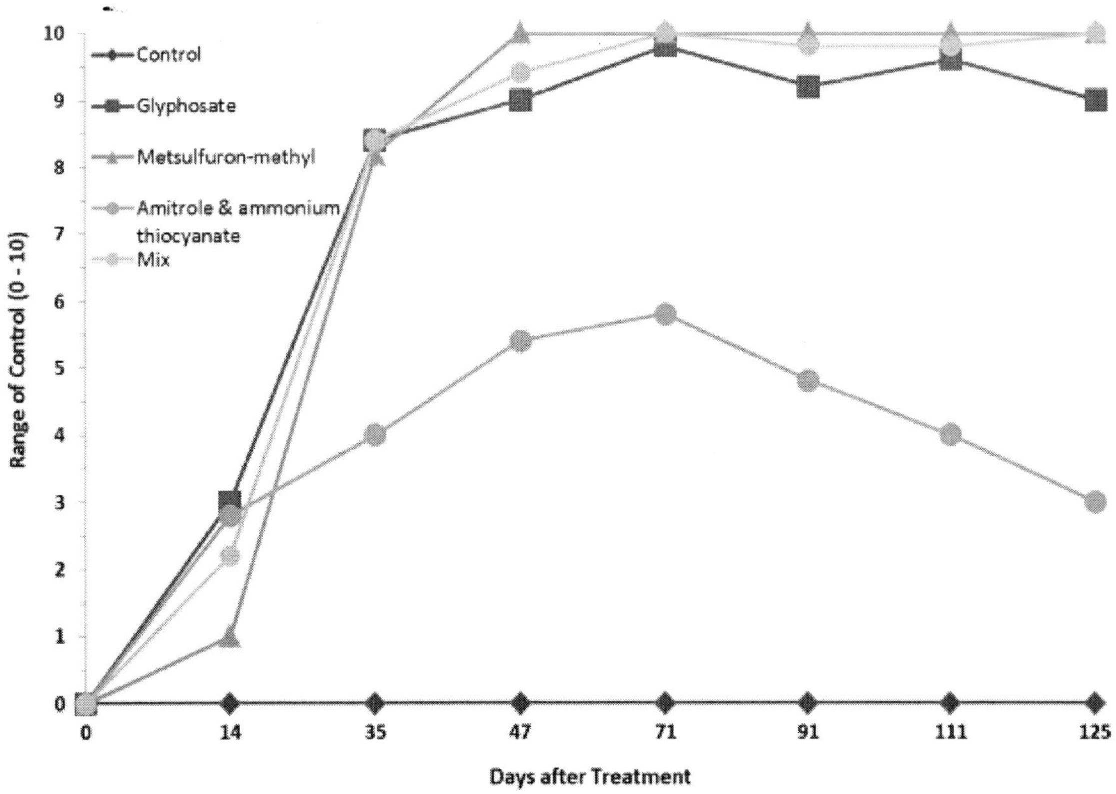

Figure 2. The average control results of each herbicide treatment on Chinese knotweed showing the variance of control from 0 to 125 DAT.

of control at 35 DAT with metsulfuron-methyl and the mix, but evidence of regrowth was present at 91 DAT and again at 125 DAT. The packaged mixture of amitrole and ammonium thiocyanate had poor control on Chinese knotweed. Its control efficacy peaked at 71 DAT with a score of 6, then declined steadily and regrowth became evident after 71 DAT.

At 35 DAT, metsulfuron-methyl, glyphosate and their mix had a control rate of 8, exhibiting a few living stems and green leaves. The packaged mixture of amitrole and ammonium thiocyanate was showing signs of approx. 40% of leaf drop and living stems plus the bleaching of foliage, with a control rate of 4.

At 91 DAT, the metsulfuron-methyl treatment had no living Chinese knotweed plants and had a control rating of 10. Both the mix and glyphosate exhibited a few live stems and a few discoloured leaves, while glyphosate had one branch of regrowth though both have a control rate of 9. The packaged mixture of amitrole and ammonium thiocyanate showed signs of leaf drop, bleaching of the leaves and tip regrowth with a control rate of 5.

DISCUSSION

No single herbicide can kill all weed species (Harrington 2000) but trialling various herbicides in a riparian zone, allows better choices to control Chinese knotweed in Australia. Consideration must be given to the accuracy and efficiency of an application of herbicide to ensure minimal export of the herbicide into the aquatic environment (Kent and Preston 2000). For this study, broad-spectrum herbicide registered for aquatic use or having an APVMA permit for minor use, directs the selection.

This study showed that metsulfuron-methyl delivered the highest rate of control at 47 DAT with 100% kill. That level of control continued over the length of the study to 125 DAT, with no emergence of other weeds within metsulfuron-methyl plots, which was probably due to the persistence of metsulfuron-methyl in the soil (Kent and Preston 2000).

Glyphosate is the most commonly used knockdown herbicide and can be applied over the full growing season and will translocate throughout the plant (Harrington 2000). This study has shown that although glyphosate was the quickest to show signs of control such as wilting and leaf drop at 14 DAT, and continued to increase in control, one application is insufficient to control Chinese knotweed due to the regrowth that occurred from 91 DAT.

Many herbicides have a narrow spectrum of weed control, so to increase control efficacy, a combination of herbicides applied to the target weed can retain longer residual, increase the herbicide effectiveness and delay the development of herbicide resistance (Damalas 2004). The mix, a knockdown herbicide glyphosate mixed with metsulfuron-methyl, a systemic known for its effectiveness on bulbs and tubers (Department of Primary Industries, Parks, Water and Environment, 2002), gave 100% control after 111 DAT. Therefore, the mix appears to be a desirable combination to control Chinese knotweed.

Another broad-spectrum herbicide treatment, amitrole and ammonium thiocyanate, is an efficacious product to control broadleaf weeds and suitable to use as an alternative to glyphosate (Sánchez-Bayo *et al.* 2010). Its systemic action, though slower than glyphosate at 14 DAT, showed what is commonly observed with a Group Q herbicide, such as the bleaching of leaves and cessation of growth at 35 DAT. This stunted growth habit allowed other weeds that germinated at the site to take over each amitrole and ammonium thiocyanate plot after 47 DAT. Amitrole and ammonium thiocyanate was not as effective as the other three herbicide treatments and is not recommended for the control of Chinese knotweed.

For similar plants of the Polygonaceae family, the United States of America (PCD 2016), New Zealand (Galloway and Lepper 2010), Indian subcontinent, Sri Lanka and West Java (Panda 2011), used a glyphosate based product to control *P. perfoliata* and *P. maculosa* Gray (syn. *Polygonum persicaria)*, (redshank) and Chinese knotweed. Research is limited with the use of metsulfuron-methyl on the Polygonaceae family, although in Norway, it has been found that *P. maculosa* first evolved resistance to Group B/2 tribenuron-methyl herbicides in 2009 and they may be cross-resistant to other Group B/2 herbicides (Heap 2018).

The consideration of alternating herbicides rather than using the same mode of action herbicides can reduce the risk of weed resistance (Rattray *et al.* 2006). Rattray *et al.* (2006) states as herbicide resistance develops, there is often a decision to increase herbicide rates for control, but avoiding this situation suggests that herbicides can be used more effectively for longer. Three different 'mode of action' groups (B, M and Q) were used in this study. With the result of effectiveness with both herbicides, metsulfuron-methyl and glyphosate on Chinese knotweed, there is the choice of using either group or the combination of both groups as the mix, to prevent the herbicide resistance especially with Group B.

The use of herbicides as a control is the most efficient way to treat Chinese knotweed due to the morphology, especially its underground rhizomes. This study has shown that the preferred control for Chinese knotweed is a foliar application of metsulfuron-methyl or its mixture with glyphosate. Weed managers may choose to select a broad-spectrum herbicide and use the chemical glyphosate; however, this herbicide will give a lower level of control and will require further treatments. The continual monitoring of sites after herbicide applications will be required to control any further regrowth. If one small plant were left to regrow, it would develop into a significant infestation within a short period.

Herbicides are an integral part of weed management and to maximise their efficacy, weed managers needs to have an understanding of their characteristics so they can minimise adverse impacts to the environment while achieving an effective control of the targeted weed. For Chinese knotweed, current research for effective herbicide options has been very limited. There is a need for continuing research and monitoring of this study, and future studies to gain long-term data on effectiveness of the herbicide application. This research has identified some effective herbicide options for the control of Chinese knotweed in Australia and may help in acquiring a minor use permit from the APVMA to treat Chinese knotweed in riparian corridors.

ACKNOWLEDGMENTS

The following individuals and organisations are deserving of my thanks for their assistance in seeing this project to fruition.

Mr. Tony Cook, Research Scientist with NSW Department of Primary Industries for his guidance with designing the trial and selection of herbicide to use on the project.

Mr. Brad Shultz, field assistant, for setting up the trial site, applying the herbicide treatments and data recording. What an almighty task it would have been for me without your help, especially on Christmas Eve.

Finally, Dr. Stephen Johnson, Weed Ecologist with NSW Department of Primary Industries, for assessing Chinese knotweed under the NSW Weed Risk Assessment and his guidance in completing the project.

REFERENCES

APVMA, Australian Pesticides and Veterinary Medicines Authority (2015). Chemicals and Products. http://apvma.gov.au/node/1061. (accessed 17 July2018).

Cook, T. and Shoemark, V. (2015). Is there another herbicide to replace flupropanate? – The repercussion of resistance in serrated tussock? NSW Department of Primary Industries, Tamworth.

Damalas, C.A. (2004). Review – herbicide tank mixtures: Common interactions. *International Journal of Agriculture and Biology* 6, 209-12.

Department of Primary Industries, Parks, Water and Environment, (2002). AGVET Chemicals Information Sheet. Metsulfuron-methyl. Biosecurity Tasmania, http://dpipwe.tas.gov.au/Documents/Information%20Sheet%20-%20Metsulfuron-methyl_Feb2014.pdf (accessed 17 July 2018).

eFloras, (2011a). *Polygonum chinense* Linnaeus. Flora of China. Missouri Botanical Garden, St. Louis, Missouri & Harvard University Herbaria, Cambridge, Massachusetts. http://www.efloras.org/florataxon.aspx?flora_id=2&taxon_id=200006718 (accessed 17 July 2018).

eFloras, (2011b). South China Botanical Garden Checklist. Missouri Botanical Garden, St. Louis, Missouri & Harvard University Herbaria, Cambridge, Massachusetts. http://www.efloras.org/florataxon.aspx?flora_id=610&taxon_id=200006718 (accessed 17 July 2018).

EOL, Encyclopedia of Life. (2014). *Persicaria chinensis*. http://www.eol.org (accessed 26 January 2018).

Galloway, D.J. and Lepper, V.E. (2010). *Persicaria chinensis* – a new alien Asian invader? Proceedings of the 17th Australasian Weeds Conference, ed. S.M. Zydenbos, pp. 174-5. (New Zealand Plant Protection Society, Christchurch).

Harrington, K.C. (2000). Tree crop and viticultural weed management systems. *In* 'Australian Weed Management Systems', ed. B.M. Sindel, pp. 373-92. (R.G. and F.J. Richardson, Melbourne, Australia).

Heap, I. (2018). The International Survey of Herbicide Resistant Weeds. www.weedscience.org. (accessed on 18 July 2018).

IPNI, The International Plant Names Index (2012). *Persicaria chinensis*. http://www.ipni.org (accessed 17 July 2018).

Kent, J.H. and Preston, C. (2000). Application and fate of herbicides in the environment. *In* 'Australian Weed Management Systems', ed. B.M. Sindel, pp. 227-50. (New Zealand Plant Protection Society, Christchurch).

Panda, H., (2011). 'The complete book on cultivation and manufacture of tea'. (Asia Pacific Business Press Inc, India).

PCD, Pierce Conservation District (2016). Knotweed control program – Control methods, http://www.piercecountycd.org/184/Control-Methods (accessed 26 January 2018).

PlantNET, (2016). *Persicaria chinensis*. The NSW Plant Information Network System, Royal Botanic Gardens and Domain Trust, Sydney. http://plantnet.rbgsyd.nsw.gov.au (accessed 18 July 2018).

Randall, R.P. (2012). 'A Global Compendium of Weeds'. (Department of Agriculture and Food Western Australia, Perth, Australia).

Rattray, D.J., Freebairn, D.M. and Gurner, N.C. (2006). Management principles for residual herbicides and water quality. (Grain Research and Development Corporation, Canberra).

Sánchez-Bayo, F., Hyne, R.V. and Desseille, K.L. (2010). An amperometric method for the detection of amitrole, glyphosate and its aminomethylphosphonic acid metabolite in environmental waters using passive samplers. *Analytica Chimica Acta* 675, 125-31.

Sasidharan, N. (n.d.). *Persicaria chinensis* (L.) Nakai. India Biodiversity Portal. http://indiabiodiversity.org/biodiv/species/show/251881 (accessed 17 January 2018).

Spencer, H.J. (2012). Metsulfuron-methyl impact on native vegetation in the Daintree lowlands, far-north Queensland, Australia. Proceeding of the 18th Australasian Weeds Conference, ed. V. Eldershaw, pp. 122-5. (Weed Society of Victoria, Melbourne).

Stuart Jr., G.U. (2015). Philippine medicinal plants. Daynon. *Polygonum chinense* L. http://www.stuartxchange.com/Daynon.html (accessed 18 July 2018).

Turner. I. (1995). A catalogue of vascular plants of Malaya. The Gardens' bulletin, Singapore. Government. p. 407. http://biodiversitylibrary.org/page/43661004 (accessed 18 July 2018).

US Forest Service (2010). Pacific Island Ecosystems at Risk (PIER): Plant threats to Pacific ecosystems. United States Forest Service, United States Department of Agriculture. http://www.hear.org/pier/index.html (accessed 18 July 2018).

Wilson, K.L. (2012). PlantNET (The NSW Plant Information Network System). Royal Botanic Gardens and Domain Trust, Sydney. http://plantnet.rbgsyd.nsw.gov.au (accessed 18 July 2018).

Wong, M.M., Lim, C.L. and Wilson, J.J. (2015). DNA barcoding implicates 23 species and four orders as potential pollinators of Chinese knotweed (*Persicaria chinensis*) in Peninsular Malaysia. *Bulletin of Entomological Research* 105, 515-20.

Drive-by shooting: Increasing weed treatment speeds using a skattergun

Nathan March[1] and Wayne Vogler[2]
[1] Queensland Department of Agriculture and Fisheries, PO Box 53, Cloncurry, Queensland 4824, Australia
[2] Queensland Department of Agriculture and Fisheries, Tropical Weeds Research Centre, PO Box 187, Charters Towers, Queensland 4820, Australia
(nathan.march@daf.qld.gov.au)

Summary The Epple 'skattergun', a relatively new weed control device, is a compressed air driven applicator for soil applied herbicides. The device is mounted on a suitable vehicle and allows the operator to 'shoot' measured doses of tebuthiuron pellets up to 20 metres away from the distribution point. The device was developed to increase the herbicide application speed for control of prickly acacia (*Vachellia nilotica* ssp. *indica* L. Willd. ex Del. (Mimosoideae)) and other weeds.

A field trial to quantify the increased tebuthiuron pellet application speeds by tractor and buggy-mounted 'skatterguns' versus hand application of pellets from quad bikes was conducted in the Julia Creek area of north-west Queensland in 2016–2017. The trial involved the treatment of 60 × one hectare plots of varying densities of prickly acacia on Mitchell grasslands.

The trial found that hand application from quad bikes was comparable to slightly faster than the 'skattergun' at densities less than 20 plants per hectare. However, at higher densities, a 'scattergun' provided significantly faster application regardless of whether it was used from a tractor or buggy. At 50, 150 and 300 plants per hectare, a buggy-based 'skattergun' operator could treat 81%, 111% and 133% more plants per hour than a quad bike-based operator for the same densities, respectively.

The trial confirmed that the 'skattergun' is an efficient tebuthiuron application device that reduces labour whilst enabling significantly more prickly acacia and other weeds to be treated within rangeland and savanna environments.

Keywords Prickly acacia, *Vachellia nilotica*, 'scattergun', tebuthiuron, Mitchell grasslands.

INTRODUCTION

Prickly acacia (*Vachellia nilotica* ssp. *indica* L. Willd. ex Del. (Mimosoideae)) is an exotic thorny tree that grows 4–5 m high and occasionally to 10 m. Native to the Indian subcontinent, it was introduced into Australia for shade and fodder provision, but is now a landscape level weed problem affecting large expanses of western Queensland and other areas of Northern Australia (Spies and March 2004). It is recognised as a Weed of National Significance in Australia.

To improve prickly acacia management, the Queensland Department of Agriculture and Fisheries has led various initiatives (March *et al.* 2017) during 2013–2018. A priority activity of these initiatives has been the field testing and refinement of new control tools, including those which increase the efficiency of current treatment methods.

While the authors have observed a significant increase in tebuthiuron (e.g. Graslan herbicide™) pellet application for prickly acacia control since the 1990s, this has primarily been through hand application for low and medium densities. The 'skattergun', developed in 2013–2014, became a focus of control tool refinement as it potentially enabled improvements in treatment speeds for prickly acacia infestations of different densities. The device is a compressed air driven applicator for soil applied residual herbicides. When mounted on a suitable vehicle, it allows the operator to 'shoot' measured doses of tebuthiuron pellets up to 20 metres away from the distribution point.

A trial was conducted aimed at assessing the efficiency of the 'skattergun' compared with current hand application methods and identify optimal situations for use.

MATERIALS AND METHODS

Site details The trial was undertaken at 'Wyaldra Station' (20°21′S, 141°51′E), a cattle property located approximately 60 km north of Julia Creek in north west Queensland. The trial site consisted of flat Mitchell grasslands dominated by *Astrebla* spp. with deep shrink swell grey clay soils.

Experimental design The trial was set up during September 2016 with additional plots for buggy-based 'skattergun' application set up in April 2017. Forty-eight randomised 250 × 40 m (1 ha) plots were established across a 6.6 km range of grassland to incorporate 4 treatments and 12 replicates of varying plant density. The plot width was restricted to enhance visibility of plants to be treated and reduce search effort as a factor of subsequent treatment times. All plots were marked with star pickets on each corner and midway along the two longer sides of the plot.

All plants in each plot were counted and recorded according to height categories of <0.5, 0.5–1, 1–2, 2–4 and >4 m. There were 4 replicates of 3 densities categories for each treatment. The density categories were low (<50 plants ha^{-1}), medium (50–150 plants ha^{-1}) and high density (>150 plants ha^{-1}). Densities of individual plots varied from 10 to 336 plants ha^{-1}.

The treatment chemical was tebuthiuron, with the Graslan™ herbicide pellets used for both 'scattergun' and hand application. Tebuthiuron at 200 g kg^{-1} of product is registered for prickly acacia at a treatment rate of 1.5 g m^{-2} calculated on the basis of the area of the canopy extending to 30 cm beyond the drip line. This means that more herbicide needed to be applied for larger trees.

The treatment types included a buggy-mounted 'skattergun', tractor-mounted 'skattergun', hand application from a single quad bike, hand application from two quad bikes working together and a control. Brett Epple, the developer of the 'skattergun' applied 'skattergun' treatments while an experienced weed control contractor was used for plots treated by hand application.

The time and amount of herbicide to treat each plot was recorded. The average amount of herbicide used per plant was calculated. Regression analysis was used to determine differences in treatment application speeds expressed as plants per hour and hectares per hour. Genstat 16th Edition (2013) was used for regression analysis.

Due to insufficient rainfall following treatment, further monitoring is required to assess mortality rates. This has not hindered the trial in achieving its principal aim of determining treatment efficiency differences and potential advantages or otherwise of the 'skattergun'.

RESULTS

Herbicide usage Average herbicide use $plant^{-1}$ was generally lower for 'skattergun-based' application with the greatest difference of 5.1 g $plant^{-1}$ observed between the tractor-based 'skattergun' and hand application from two quad bikes (Table 1).

A buggy-based 'skattergun' and hand application from a single quad bike are the best relative measures for comparison of treatments. Both use a single operator and the 'skattergun' is more likely to be used from a buggy than a tractor. The difference in herbicide usage between these two application methods was 0.8 g $plant^{-1}$ or 5.1%.

Treatment speed – plants per hour Regression analysis accounted for 87.4% of variance and determined that the treatment speeds between treatment methods were significantly different ($F<0.001$, $P<0.05$) (Figure 1). Treatment speeds expressed in terms of plants $hour^{-1}$ were similar for densities up to about 20 plants ha^{-1} (Figure 1). Hand application provided some efficiency advantages below this density although this only occurred for two plots (one with hand application from a single quad bike and one from two quad bikes).

By ~30 plants ha^{-1}, treatment speeds were consistently faster for both types of 'skattergun' treatment compared to hand application from quad bikes – even the two quad bike treatment (Figure 1). The 'skattergun' demonstrated significantly higher treatment speeds (plants treated $hour^{-1}$) above densities of ~30 plants ha^{-1}. At 50, 150 and 300 plants ha^{-1}, a buggy-based 'skattergun' operator could treat 81%, 111% and 133% more plants per hour than a quad bike-based operator for the same densities, respectively (Figure 1).

Treatment speed – area per hour Regression analysis accounted for 94.6% of variance and determined that the treatment speeds between a single quad bike and a buggy with a 'skattergun' were significantly different ($F<0.001$, $P<0.05$) (Figure 2). In particular, there was a significant treatment speed advantage for the buggy-mounted 'skattergun' relative to hand ap-

Table 1. Average herbicide usage per plant for trial treatments.

Treatment	Average herbicide usage (g $plant^{-1}$)	Percentage difference between 'Skattergun' – buggy and other treatments
'Skattergun' – tractor	14.8	– 6.3%
'Skattergun' – buggy	15.8	–
Hand application – 1 × quad bike	16.6	+ 5.1%
Hand application – 2 × quad bikes	19.9	+ 25.9%

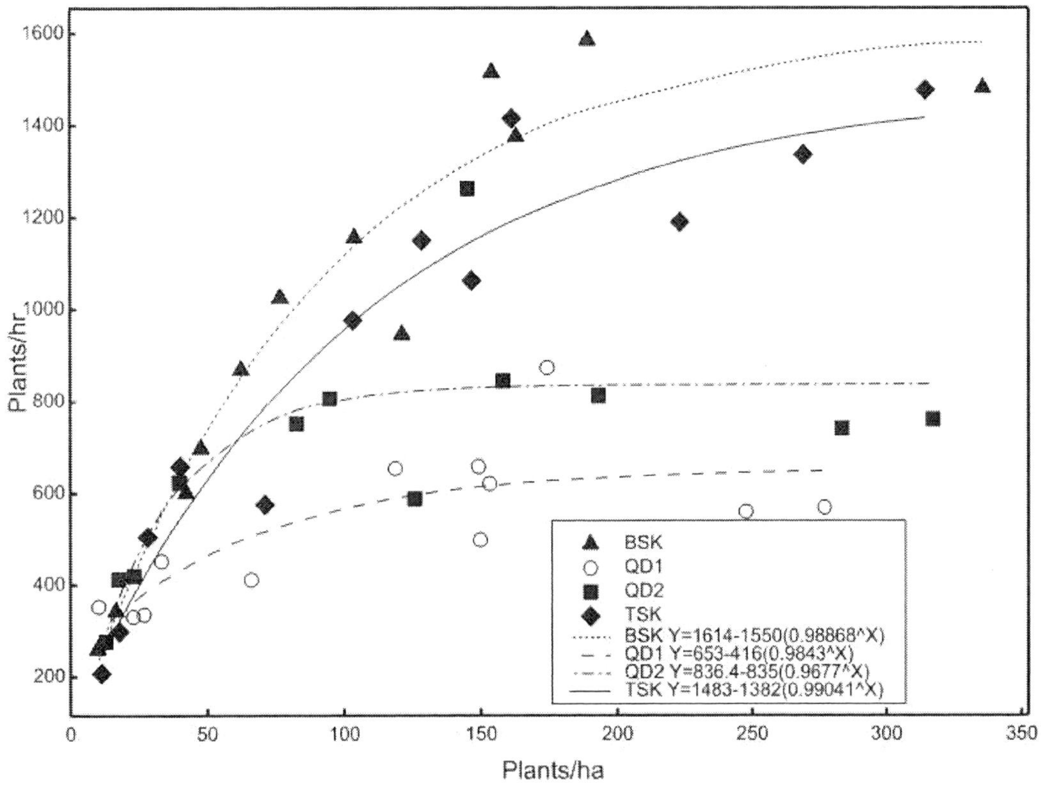

Figure 1. Comparative treatment speeds (plants per hour) of 'skattergun' versus hand applied treatments. (BSK – Buggy 'skattergun'; QD1 – One quad bike; QD2 – Two quad bikes; TSK – Tractor 'skattergun').

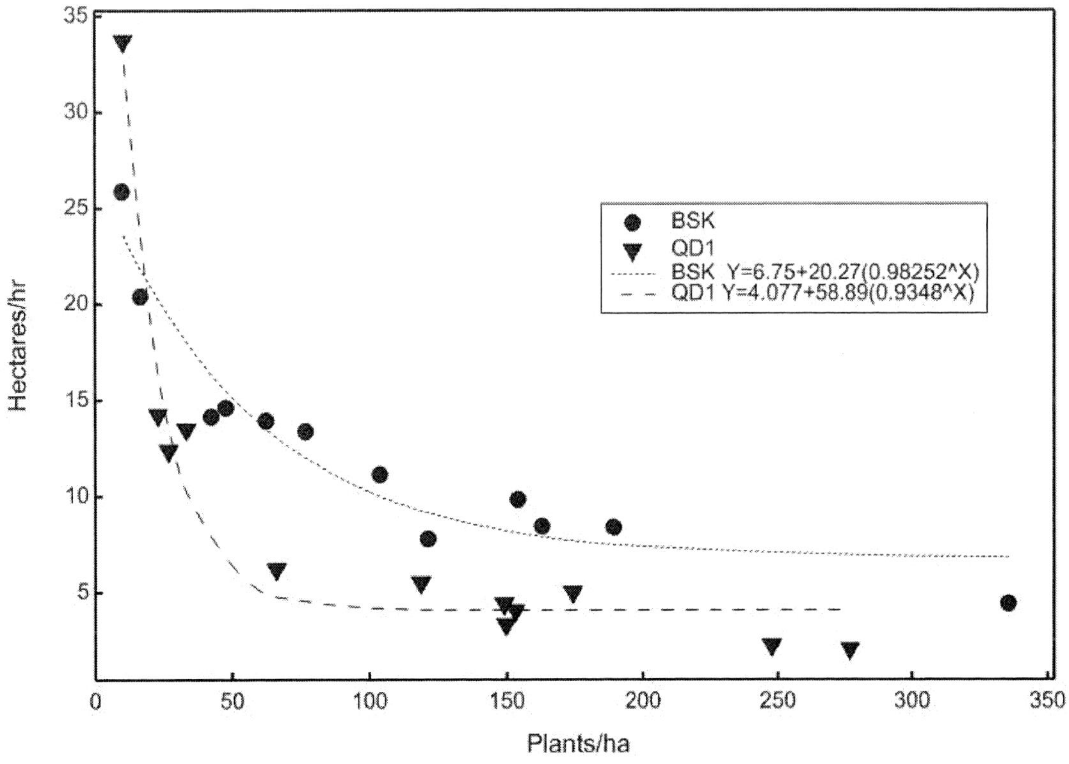

Figure 2. Comparative treatment speeds (hectare per hour) of the buggy-mounted 'skattergun' versus hand application from a single quad bike. (BSK – Buggy 'skattergun'; QD1 – One quad bike).

plication from a single quad bike with approximately twice the treatment rate observed for all plant densities above about 50 plants ha^{-1} (Figure 2).

DISCUSSION

Herbicide usage Herbicide usage for both 'skattergun' treatments was lower than hand applied treatments, though only markedly for hand application when using two quad bikes. Plant heights were relatively consistent for all treatments except those treated using two quad bikes. The plots for this treatment had a higher proportion of >4 m trees (4–6% higher) relative to other treatment plots requiring a higher rate of herbicide application per tree in accordance with the herbicide label. This, coupled with individual operator variance likely explains the higher usage of herbicide for the hand applied treatment using two quad bikes.

Treatment speed Based on observations during the trials, the 'skattergun' was able to achieve significant treatment speed advantages as the operator could dispense herbicide up to 20 m from the target plant. Hand application from quad bikes required operators to be closer to target plants (generally 0.5–3 m), necessitating greater distances to be travelled and thus requiring more time.

The treatment speed advantage of the 'skattergun' becomes pronounced when expressing the data as area treated time per unit. For example, at a density of 100 plants per hectare, a buggy-mounted 'skattergun' can treat ~40 ha more in an 8 hour day than a single quad bike which is about double the area that could be treated by the single quad bike.

Field observations have indicated the operational costs of using a buggy versus a quad bike are similar. The potential speed advantage of a quad bike, when traversing Mitchell grasslands, was only evident in the result for the 10 plants ha^{-1} plot.

It is interesting to note that the use of two quad bikes for hand application treatment did not double the treatment speed relative to using one quad bike. This result may be due to operators overlapping their search area either from over-checking their arbitrary common search boundary or the 40 m plot width was too narrow to allow the two quad bikes to operate efficiently.

Optimal situations for use Based on the outcomes of this and previous trials and case studies undertaken by Queensland Department of Agriculture and Fisheries (QDAF) through the War on Western Weeds initiative, optimal situations for use of the 'skattergun' have been identified.

The 'skattergun' is best used for prickly acacia control on flat to undulating plains at low to medium densities, from upwards of 20 plants ha^{-1}. Densities of 10–20 plants ha^{-1} could be treated by either 'skattergun' or hand application while sparse infestations under 10 plants ha^{-1} may be more efficiently treated by hand application from a quad bike.

Further analysis is required, but it is expected that treatment of high density infestations will be more efficient using broad-scale aerial application of tebuthiuron or perhaps double chain pulling with bulldozers. In either case, at high to very high densities, 'skattergun' use may be limited when air pressure cannot be maintained for effective pellet dispersal.

It is also noted the 'skattergun' treatment using tebuthiuron cannot be used in close proximity to native trees or watercourses due to herbicide registration restrictions associated with tebuthiuron based chemical products.

ACKNOWLEDGMENTS

The author wishes to thank: Brett Epple for the opportunity to field trial the 'skattergun'; Denise and Graham Abdy whose property hosted the trials; and, Southern Gulf NRM staff who assisted trial establishment. Funding for the trials was provided through the Queensland Government War on Western Weeds initiative and Australian Government-funded War on Northern Invasive Weeds project.

REFERENCES

March, N., Vogler, W. and Kunjithapatham, D. (2017). Advancing prickly acacia management through the War on Western Weeds initiative. Proceedings of the 14th Queensland Weeds Symposium, ed. T. Sydes pp. 164-9. (Weed Society of Queensland, Port Douglas).

Spies, P. and March, N.A. (2004). 'Prickly acacia national case studies manual' (Department of Natural Resources, Mines and Energy).

A magic bullet for the Northern Australian neem nightmare?

Jessica Miller[1,2] and Louise Beames[1]
[1] Environs Kimberley, PO Box 2281, Broome, Western Australia 6725, Australia
[2] The University of Queensland, School of Earth and Environmental Science, Brisbane, Queensland 4072, Australia
(jessica@environskimberley.org.au)

Summary Planted as a shade tree in many Kimberley communities, neem (*Azadirachta indica* A.Juss), is now a highly invasive weed, harming the natural environment, agricultural production and cultural sites across the region and northern Australia. In 2013, neem was ranked among the Kimberley's top environmental weeds, and voted the highest priority weed at the 2015 'Kimberley Weed Forum'. The Northern Territory lists neem as a Class B and Class C weed, yet it remains undeclared in Queensland and Western Australia. In 2017, Environs Kimberley developed the Interim Kimberley Neem Management Plan which, for the first time, collated Kimberley mapping data, presented case studies and made recommendations; including trialling new control technology and advocating for improved recognition of neem as a serious weed at a state and national level.

Currently, Environs Kimberley is working with several groups to run experimental control trials, and Nyikina Mangala Rangers to undertake control and mapping of neem at Oongkalkada, an important cultural site. Control poses a wide range of challenges for land managers, including low efficacy of some treatment methods, cost of equipment and materials, labour intensity and occupational, health and safety risks in remote areas. The control method being trialled uses dry herbicide in capsules developed by Bioherbicides Australia. Previously with one option of control, operators might spend a day sweating it out with chainsaws and loppers to treat a relatively small area of neem infestation. Use of the capsules could allow hundreds of trees to be treated per person per day at reduced cost and greatly improved human and environmental safety.

Keywords Neem, Northern Australia, weed control, herbicide technology.

INTRODUCTION

Neem (*Azadirachta indica* A.Juss) is a weed tree invading and dominating areas in tropical Australia, and impacting natural, cultural, agricultural and other values across the Kimberley, Northern Territory and Queensland. Neem is able to invade and quickly expand into intact natural areas, directly reducing the diversity and abundance of native species and interfering with natural ecosystem processes (Northern Territory Government Department of Environment and Natural Resources 2015). It is not compatible with agricultural crops owing to its aggressively invasive nature (Northern Territory Government Department of Environment and Natural Resources 2015). Neem is also encroaching on significant Aboriginal cultural sites, having a medium to high impact (Beames *et al.* 2017). Such weed problems can affect the health and well-being of people and country, as well as cultural economies including foods and medicines (Duff 2012, Duff and Weir 2013).

Neem is thought to have originated in India and Myanmar and has a rich cultural and medicinal history (Biosecurity Queensland 2016). It has insecticidal properties and strong links with Ayurvedic medicine (Kumar and Navaratnam 2013). It is used in a range of products including for health and cosmetics, cleaning and agriculture. These attributes are behind unsuccessful attempts to grow commercial quantities in northern Australia, with plantations established from 1965–1988. The species was also heavily promoted to communities, townships and cattle stations in the Kimberley as a shade tree, and these individuals are now acting as source populations, allowing the species to be spread into important ecosystems. Seeds germinate in abundance around parent trees, displacing natives. Birds eat the fruit and spread seeds widely, especially under resting or roosting trees. Large eucalypts and boab trees in the Kimberley are now being suffocated by neems, which kill the parent tree and reduce the germination of new stands (Biosecurity Queensland 2016, Reynolds *et al.* 2018). In the East Kimberley, there are large infestations that land managers and government are increasingly unable to control.

Neem was voted as the top priority weed by attendees at the inaugural 2015 Kimberley Weeds Forum. It had previously been identified by the then Western Australian Department of Environment and Conservation as a top environmental weed in a 2013 Weed Prioritisation Process.

The Northern Territory Government listed neem as a Class B and Class C weed under the *Weeds Management Act* in 2015. This means that neem may no longer be brought into the Northern Territory, and land managers are required to control the growth and spread of neem. Queensland conducted an Invasive Plant Risk Assessment on neem in 2008, which was subsequently revised in 2016 (Biosecurity Queensland 2016). They noted neem has become naturalised and is spreading into native habitat, and that, while there is a lack of data to enable robust decision-making, the high seed production could result in substantial growth of populations in the future. To date, the Western Australian Government has taken no investigative or legislative action regarding neem. A handful of groups and individuals continue to advocate for a range of actions to address the significant threat posed by neem to the Kimberley environment and communities.

INTERIM NEEM MANAGEMENT PLAN

The Interim Neem Management Plan (Beames *et al.* 2017) benchmarks current knowledge and understanding of neem biology and ecology, along with the history of its spread and impacts in the Kimberley. For the first time, the plan collated distribution points from multiple operators, mostly non-government organisations and committed individuals, to create a Kimberley neem distribution map. Case studies were used to share knowledge, observations and experiences of neem management to inform Kimberley project priorities and enhance control practices. The plan includes results from a qualitative stakeholder survey conducted by Environs Kimberley in 2017. The results of the survey demonstrate the significant concerns about neem among stakeholders as well as capturing information about efficacy of current control methods.

To best apply the information compiled and forge a positive future where neem invasion can be prevented and contained in the Kimberley, the plan presented 11 recommendations. A key recommendation was to conduct applied research on herbicide capsule technology, a new method described within the plan, to determine its capacity to assist in neem eradication.

NEW METHODS IN WOODY WEED CONTROL

Recent technological and methodological developments are enabling Kimberley people to trial large-scale control of neem and other woody weeds, improving the efficacy, efficiency and cost effectiveness of programs, while reducing human and environmental risks.

For example, Nyamba Buru Yawuru Land and Sea Team have adapted older methods of treating young woody weeds, by using long nozzle containers to apply a thin line of liquid herbicide around sapling stems (up to 5 cm diameter at breast height (DBH)), in effect ring-barking without the need to paint or spray herbicide. This reduces effort and time involved and improves occupational health and safety for operators.

The second method utilises technology developed by Bioherbicides Australia and promises an even greater gain in occupational health and safety, and an efficient use of human and financial resources. Preliminary results show the method can be applied to neem trees from 5 cm DBH upwards and involves drilling a hole in the trunk and inserting a capsule that contains dry herbicide, followed by a wooden plug. The auto-applicator allows insertion of up to 30 capsules and plugs before requiring refilling. Results will be made publicly available and disseminated to stakeholders.

Should trials of the capsule method prove to be effective, treatment of neem will still require an integrated approach utilising a variety of methods to control it at various life stages, undertake follow-up treatment and manage seed bank and spread. This may include foliar spraying and hand-pulling of seedlings across different seasons, burning, revegetation and other changes to land management such as grazing regimes and the management of feral animal populations. There is also a need for more research aimed at determining the best treatment programs for different land management scenarios and environments across northern Australia.

Weeds on private lands are at risk of low management input as landowners have less access to weed management resources, training and time to undertake control. These new methods present exciting community engagement and participation opportunities, as the reduced costs, risk and greater efficacy will mean that private landholders can more easily undertake their own control activities.

PROGRESS ON RECOMMENDATIONS

Environs Kimberley has begun to respond to the recommendations in the Interim Neem Management Plan, by leading collaborative actions to control neem. We outline progress in the Kimberley against each recommendation below. A more thorough evaluation should be conducted by regional experts and operators at a second Kimberley weed forum, scheduled for October 2018, 12 months after the date of publication of the first neem plan.

1. **Undertake comprehensive distribution and density mapping of neem, particularly in unsurveyed areas of the Kimberley region** This is underway. Preliminary discussions have been held with Aboriginal ranger groups and other stakeholders in the Kimberley to identify neem spatial data not currently accessible through databases. A

more comprehensive mapping program is needed and requires input and funding from the state.

2. **Work with stakeholders to identify the current and potential ecological, cultural, agricultural and other values threatened by neem invasion** Funding for a research project should be sought to investigate this. Environs Kimberley has commenced preliminary discussions with potential research partners.

3. **Conduct applied research with on-ground weed control projects and universities/academics to test new herbicide capsule technology and its capacity to eradicate neem** This is underway. Environs Kimberley is leading field trials in the west Kimberley to compare costs, risks and success rates of some of the current and emerging control methods. More comprehensive research is needed to answer questions such as: optimal application rates for dry herbicide capsules; impacts of herbicide capsules on the environment and people; and strategies for regional control.

4. **In line with both the Northern Territory and Queensland, the state of Western Australia to conduct a weed risk assessment/invasive plant assessment for neem** State cooperation is being sought.

5. **Further develop documentation and advocate for the listing of Neem under the *Biosecurity and Agriculture Management Act 2007*, in line with the listing of this weed under corresponding legislation in the neighbouring Northern Territory** This is underway. Preliminary work commenced by the Kimberley Rangelands Biosecurity Association on a submission to Western Australian Department of Primary Industries and Regional Development (DPIRD) requesting that neem be considered for listing as a declared pest. Environs Kimberley and several other groups have provided letters of support for this process.

6. **Develop a pan-northern taskforce to share neem information, develop documentation and advocate for national recognition of the current and potential impact to northern Australian systems i.e. WoNS** Environs Kimberley is maintaining current relationships with stakeholders in order to help progress this in the future.

7. **Address gaps in the Western Australian 'off label' permit to allow effective herbicide treatments to be applied** Results of current trials coordinated by Environs Kimberley will be submitted to DPIRD and discussion held regarding permit structure.

8. **Develop a focused public education and awareness communication strategy to ensure the public and the nursery industry understand the implications of neem establishment and spread in the Kimberley, are prepared to undertake measures required to manage and control neem, and are engaged in communities. Early identification and control of new outbreaks** Environs Kimberley is in the preliminary phase of awareness raising, through media and community events is building on work carried out previously and by current partners. There is a need for a comprehensive awareness campaign targeting communities across the Kimberley. Initial feedback suggests interest and support from current and potential partners.

9. **Work with the four Kimberley shires to introduce measures that will prevent new introductions of neem; require neem to be removed from any new developments and support landowners to remove and control it** Environs Kimberley is maintaining current relationships with stakeholders in order to help progress this in the future.

10. **Investigate the feasibility of formally gazetting neem as a weed under the local laws of the four Kimberley shires and assess what, if any, influence this may have on the state to secure greater recognition of neem as a weed** No progress.

11. **Work with the five Kimberley towns, Aboriginal communities and pastoral stations to remove and replace neems that were once planted as shade trees** Environs Kimberley has initiated discussions with some groups. There are a range of factors influencing uptake, including propagation and availability of local provenance replacement species and an absence of funding for materials and extension work.

COLLABORATION AND LEADERSHIP

Environs Kimberley has been working with Aboriginal ranger groups and their communities, pastoralists, environmental community groups and scientists to coordinate and support weed management projects since 2007. The isolated location and challenges around access to resources makes collaborating and engaging with a variety of partners an integral part of doing business and working on Country in the Kimberley. A number of groups and individuals are leading this collaborative approach and key guidelines are emerging to improve working relationships and outcomes (e.g. Lincoln *et al.* 2017). Earlier work by Duff (2012) showed that processes and priorities of weed control need to have greater consideration for the vast native title lands and Traditional Owners throughout the Kimberley and that certain mechanisms for engagement, support and participation are better than others.

While developing the neem plan the authors received information from weed operators and land managers about trends in weed management resourcing, particularly in relation to the staff and funding allocated to combatting weed spread across northern Australia (L. Beames, pers. comm.). Funding for weed control in Australia's tropics continues to be inadequate, with little long-term, consistent program funding. This plays a large role in dismantling effective working relationships and control programs. Neem control is dually beset by a lack of legislative prioritisation and targeted funding.

CONCLUSION

Several groups have been working alongside Environs Kimberley to implement the recommendations outlined in the Interim Neem Management Plan, both explicitly and implicitly through management plans (e.g. Healthy Country Plans). The organisation intends to secure further funding and resources to progress recommendations and identify and reach targets. The authors aim to seek further collaboration with research institutions and government departments to improve understanding of the challenges and threats posed by neem and enlist greater support and leadership across northern Australia for better protection of natural and cultural assets.

ACKNOWLEDGMENTS

We are grateful for the support of Dick Pasfield; John Szymanski (West Kimberley Rubber Vine Eradication Project – Willare); Department of Primary Industries and Regional Development; Yawuru Country Managers from Nyamba Buru Yawuru Land and Sea Team; Department of Biodiversity Conservation and Attractions Yawuru Rangers and colleagues; Donna and Jeffrey Foy; Bioherbicides Australia; and Environs Kimberley staff and supporters. This project was funded by the Western Australian Government State Natural Resource Management Program.

REFERENCES

Beames, L., Pickard, I. and Hamaguchi, N. (2017). 'Interim Neem (*Azadirachta indica*) Management Plan for the Kimberley, WA'. (Environs Kimberley, Broome).

Biosecurity Queensland (2016). Invasive plant risk assessment: Neem tree. *Azadirachta indica*. State of Queensland.

Duff, N. (2012). Managing Weeds on Native Title Lands: Workshop Report Broome WA. 26–27 October 2011. Native Title Report. AIATSIS, Canberra.

Duff, N. and Weir, J.K. (2013). Weeds and Native Title: Law and assumption. Publication No. 13/078. Australian Government Department of Rural Industries Research and Development Corporation, Canberra.

Kumar, V.S. and Navaratnam, V. (2013). Neem (*Azadirachta indica*): prehistory to contemporary medicinal uses to humankind. *Asian Pacific Journal of Tropical Biomedicine* 3, 505-14.

Lincoln, G., Austin, B.J., Dobbs, R.J., Mathews, D., Oades, D., Wiggan, A., Bayley, S., Edgar, J., King, T., George, K., Mansfield, J., Melbourne, J., Vigilante, T., Balangarra-, Bardi Jawi-, Dambimangari-, Karajarri-, Nyul-, Wunambal Gaambera- and Yawuru Traditional Owners (2017). Collaborative science in Kimberley Saltwater Country – A guide for researchers V17.03. Prepared by the Kimberley Land Council for the Kimberley Indigenous Saltwater Science Project (KISSP), Western Australian Marine Science Institute (WAMSI). Broome Western Australia.

Northern Territory Government Department of Environment and Natural Resources (2015). Neem (*Azadirachta indica*): NT weed risk assessment technical report. Northern Territory Government, Darwin.

Reynolds, S., Beames, L., Willing, T. and Parker, C. (2018). Distribution, ecology and cultural importance of Gunurru or Cable Beach Ghost Gum *Corymbia paractia* in the Broome area, Western Australia. Environs Kimberley, Broome.

Herbicidal control of environmental weeds from sub-Antarctic Macquarie Island

Brian M. Sindel[1], Waqas Zahid[1], Paul E. Kristiansen[1], Susan C. Wilson[1], Justine D. Shaw[2] and Laura K. Williams[1]

[1] Department of Agronomy and Soil Science, School of Environmental and Rural Science, University of New England, Armidale, New South Wales 2351, Australia

[2] Centre for Biodiversity and Conservation Science, School of Biological Sciences, The University of Queensland, St Lucia, Queensland 4072, Australia

(bsindel@une.edu.au)

Summary Macquarie Island in the sub-Antarctic is an inhospitable place for weeds due to its remote location and severe climate. Only three species of non-native environmental weeds are well established in this World Heritage Area – *Poa annua* (winter grass), *Stellaria media* (chickweed) and *Cerastium fontanum* (mouse ear chickweed) – where rabbits, rats and mice were recently eradicated. Attention has now turned to understanding the ecology of the non-native plants and potential for their control. The efficacy and selectivity of several herbicides was tested on the target weed *S. media,* the other two dominant weeds and several co-occurring native grasses and herbs under simulated cold sub-Antarctic climatic conditions (between 5 and 10°C) *ex situ* at the University of New England's cold climate research facility, Armidale, NSW. The most effective post-emergent herbicides tested for *S. media* control were glyphosate and linuron, both of which showed reduced damage to several of the native species, particularly at lower application rates. The most effective pre-emergent control of *S. media* was achieved with simazine, although linuron also significantly reduced weed emergence and growth. Effective and selective herbicidal control of *S. media* appears achievable under sub-Antarctic conditions. Field testing on Macquarie Island is now required to confirm these findings.

Keywords Environmental weeds, eradication, *Stellaria media*, *Cerastium fontanum*, *Poa annua*.

More than a Chore: The pleasures of weed management in high amenity rural landscapes

Nicholas Gill
School of Geography and Sustainable Communities | Australian Centre for Cultural Environmental Research, Faculty of Social Sciences, University of Wollongong, New South Wales 2522, Australia

Summary Weed management by rural landowners is conventionally seen as a chore, as task that must be undertaken for reasons that include legal obligations, conforming to social norms, keeping neighbours happy, or meeting higher order property management goals. At best it may be a means to an end. Previous research has framed weed management as a burden; a factor in selling land, an unexpected obstacle to land use goals among newer landowners, an extra cost for graziers. Moreover, divergent weed management effort can represent a source of neighbourly friction.

An alternative is to consider weed management in terms of the processes of learning, struggle, care, and transition that owning and managing land embodies. This framing of weeds will be considered in terms of environmental weed management by amenity or lifestyle-oriented rural land owners. Interviews and property walks with lifestylers in southern New South Wales suggest that management of weeds such as Lantana offers certain pleasures or benefits. Weeding is the vehicle by which they explore, learn about, and (sometimes literally) uncover their land. This process embodies and cements relations of care and potentially the knowledge and experience for managing landscapes that have been formed by almost two centuries of settlement, clearing, farming, effective abandonment, (often) weedy forest regrowth, and lifestyle 'resettlement'. More than just a chore or a means to higher land use goals, weed management is itself an activity that embodies and generates a range of positive, constructive, and normatively desirable outcomes.

Buffel grass management in Indigenous communities

Troy A.J. Bowman
PIRSA – Rural Solutions SA, PO Box 469, Murray Bridge, South Australia 5253, Australia
(Troy.bowman@sa.gov.au)

Summary Buffel grass has been recognised as one of the greatest threats to biodiversity, in South Australia's arid and semi-arid rangelands, (Williams and Barauch 2000). It has the capacity to transform ecosystems through habitat loss, competition with native plants and alteration of natural fire regimes. Buffel grass is increasingly impacting on the culture and health and safety of Indigenous communities and new approaches to the management of this devastating weed are being employed.

Advances gained from South Australian research has resulted in the application of new control options, increasing the efficiency and effectiveness of herbicide application. The emergence of new technologies has resulted in the use of drones and user-friendly platforms for the mapping of buffel grass infestations in Indigenous communities.

These new and innovative ways of effectively managing buffel grass are being employed to improve the condition of country and manage the threats posed to the environment and culture in Australia's rangelands.

A focus on building the capacity of Indigenous communities has seen a changing face of Indigenous managed lands. 'Healthy country planning' is being used to develop achievable management objectives for a range of cultural and environmental issue such as weed management.

Keywords Great Victoria Desert, natural resources, Indigenous communities, herbicides, technology.

INTRODUCTION

Buffel grass (*Cenchrus ciliaris* L.) is a deep-rooted, perennial grass native to eastern Africa, Saudi Arabia, Afghanistan, Pakistan and India. Buffel grass was widely planted between the 1960s and 1980s across vast expanses of central Australia as both a pasture grass and dust suppressant in remote communities. It has since spread across large areas of Northern Territory (NT), Queensland, South Australia (SA) and Western Australia (WA) threatening the very things that make Australia unique, our Indigenous culture and world-renowned biodiversity.

Considered one of the world's worst invaders, buffel grass is listed as a key threatening process to Australia's biodiversity and is considered the most significant invasive grass of the Australian arid heartlands with the capacity to alter entire ecosystems in a relatively short time frame (Williams and Baruch 2000).

The ability of buffel grass to change the fire regimes of entire landscapes could be the most significant impact. Being more flammable than native grasses and invading areas at high density, buffel grass produces high-intensity fires more frequently, damaging ecosystems never previously exposed to such extreme temperatures. Increased frequency of fires results in an inability of plants such as *Triodia* (spinifex) to grow, mature and set seed due to their evolution under natural fire regimes. High frequency and intense fire regimes kills many woodland trees and shifts woodlands towards a degraded grassland ecosystem dominated by buffel grass and without the biodiversity that once thrived in these areas. People in the remote outback also are at risk from more intense frequent fires threatening communities. Buffel grass is threatening specific species of high conservation and cultural importance, such as the endemic Everard garland lily (*Calostemma abdicatum* P.J. Lang) and EPBC (*Environment Protection and Biodiversity Conservation Act 1999*) listed Black-footed Rock Wallaby and Malleefowl. Impacts in Indigenous communities include a loss of bush foods and bush medicines, reduced availability of hunting opportunities through species decline, reduced ability to track prey and infestation of rock holes and sites of cultural significance as depicted in this short film *Storm on the Horizon* (https://vimeo.com/206163898)

The 'Healthy country planning' (HCP) process has a strong emphasis on traditional owner engagement. An inaugural Southern Desert Ranger Forum was held in the Great Victoria Desert (GVD) in 2017 as part of the 'Buffel Free GVD' project, providing an opportunity for Indigenous rangers from SA, NT and WA to learn from the experiences of researchers, fellow rangers and traditional owners. This presentation outlines research findings and provides an overview of buffel grass management in Indigenous communities.

MATERIALS AND METHODS

A three-year project titled 'Buffel Grass Control in Arid Rangelands' commenced in 2013 with a focus on developing best practice buffel grass control options through extensive herbicide trials. Trials were undertaken in three locations throughout the state including the Anangu Pitjantjatjara Yankunytjatjara (APY) Lands (a remote Indigenous community in North-Western South Australia). Further trials have recently been undertaken in the Anangu Pitjantjatjara Yankunytjatjara Lands to refine the application rate and assess the effectiveness of aerial application of a granular form of flupropanate, enabling efficient treatment of buffel grass in remote, difficult to access terrain without a requirement for large quantities of water (often a limiting factor in the arid rangelands of central Australia).

Since 2007, numerous Indigenous ranger groups have been established, providing opportunities for rangers to participate in both formal and informal training opportunities as well as conducting a variety of works on country to protect both the cultural and environmental assets of the region.

Extensive HCP is being undertaken in Indigenous communities throughout the GVD in an effort to identify threats to the health of Indigenous culture and the environment in which it is intrinsically interconnected. Healthy Country Plans are developed in close consultation with communities in an effort to build ownership over land management issues such as fire management, rock-hole management, threatened species management and predator control. Buffel grass management has been rated the highest priority in many of the HCPs with aspirational goals identified such as eradicating buffel grass from the region within a 10-year period. Supported by Ten Deserts org, Rangelands NRM, Indigenous Desert Alliance, Alinytjara Wilurara Natural Resource Management (AW NRM) and others, project planning is underway for a collaborative cross-border Buffel Free GVD project.

RESULTS

A key outcome from this project includes the results of trials identifying a grass selective residual herbicide Taskforce® (RTH and RTL) effective in controlling mature buffel grass tussocks and suppressing new germinations for approximately 18 months (Figure 1). Use of this herbicide enables treatment all year round as opposed to the limited window when buffel grass is actively growing and dramatically reduces the amount of follow-up control required as detailed in Bowman and Prider (2016). For further information regarding the results of the herbicide trials can be found elsewhere (Bowman and Prider 2016).

Trials undertaken by a University of Adelaide student (Tschirner 2016) led to the discovery of an organic herbicide (Bioweed™) proven to be effective in destroying both aerial and surface seed in addition to the temperature and duration of heat exposure required to destroy buffel grass seed (Figure 2).

DISCUSSION

The Buffel Free GVD project will provide opportunities for Indigenous land managers and ranger groups from across the GVD (both WA and SA) to prioritise and implement buffel grass control strategies together, not only through on-ground control efforts but also through planning and ranger exchanges to build capacity within all groups. Indigenous land managers will be able to communicate with others about the negative impacts of buffel grass and the long-term goals of a

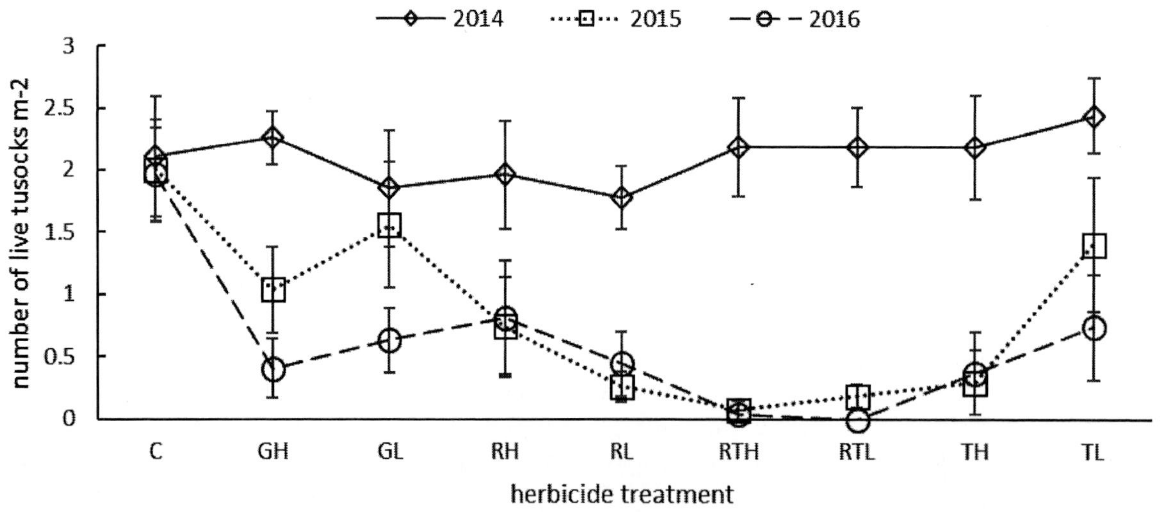

Figure 1. Results of the buffel grass herbicide trials.

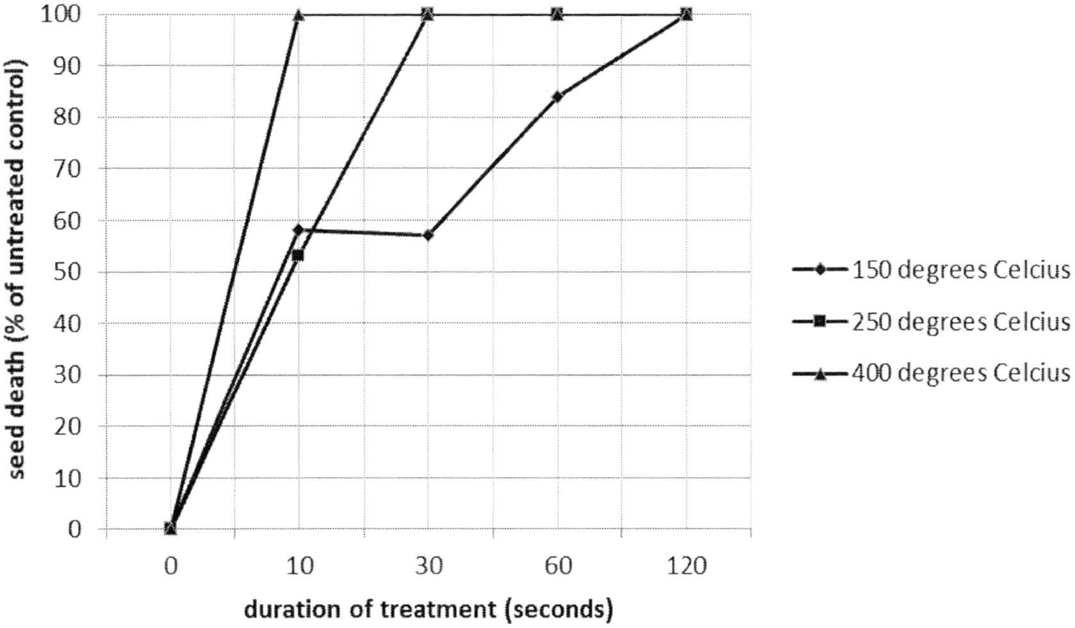

Figure 2. Temperature and duration of heat exposure required to destroy buffel grass seed.

Buffel Free GVD to ensure community buy-in and long-term stakeholder support across the region. The development of a 'GVD Bioregional Buffel Grass Management Plan' will incorporate existing plans, consult stakeholders and develop an integrated plan with maps to highlight priority areas including ecological hotspots, cultural sites and introduction pathways for treatment both at an individual organisational level and suitable for strategic responses. The development of the Plan will incorporate digital and cloud based data collection and monitoring.

Indigenous communities in the GVD are embracing new control tools and are now moving toward an integrated approach using liquid and granular flupropanate in combination with the organic product and burning to destroy aerial and surface seed.

Indigenous communities throughout SA and the GVD are increasingly adopting new tools and technologies for managing buffel grass. Communities have undertaken extensive mapping and monitoring of infestations through the use of tablets and mobile phones and in the case of Spinifex Land Management, recently undertaken aerial mapping of the Tjuntjuntjara community through the use of a drone.

Spinifex rangers have not only proactively tackled buffel grass infestations, they have also been instrumental in delivery of training to other Indigenous communities in the region. This peer-to-peer training methodology has proven hugely successful and has been praised by those involved. Spinifex Land Management undertook a total of 2078 km of roads and tracks in 2016, representing 79% of known tracks (up from 22% in 2015).

In addition to adoption of new control tools and technologies, Indigenous communities are increasingly involved in opportunities to share learnings and experiences to improve the condition of country at a landscape scale. The inaugural Southern Deserts Ranger Forum was held in Ilkurlka in June 2017 with the 2018 forum held in the WA Goldfields. The forums were a resounding success and all of the Indigenous rangers in attendance strongly supported the idea of such a forum being an annual event.

The events have attracted more than 150 people from across Western Australia, South Australia and the Northern Territory, 102 of whom were Indigenous rangers. Collectively the ranger groups in attendance manage an area of 1.2 million square km across Western Australia and South Australia.

In addition to oral and practical sessions on buffel grass management, the forum was an opportunity for ranger groups to discuss camel management, rock hole and threatened species management, firearm safety, the importance of monitoring and data management and predator control.

AW NRM Board are the authority responsible for planning, coordinating and facilitating buffel grass management in Indigenous communities of North-Western South Australia. The region is committed to eradication of all infestations in the Maralinga Tjarutja (MT) Lands and protection of key sites of cultural and environmental significance in the APY Lands.

The region will provide ongoing support to build the capacity of Indigenous communities to manage buffel grass through the provision of both formal and informal training opportunities and the purchase of equipment to enable best practice surveillance and control.

CONCLUSION

Buffel grass has been recognised as one of the greatest threats to biodiversity in South Australia's rangelands. Co-design of HCP and formal and informal training opportunities are a focus on Aboriginal managed lands. Indigenous ranger forums provide a valuable capacity building opportunity for Indigenous land managers and improving buffel grass management throughout South Australia, Western Australia and the Northern Territory. Improved best practice management of buffel grass is being used throughout Aboriginal managed lands with a tri-state buffel grass management project commencing throughout the GVD in 2018.

REFERENCES

Bowman, T. and Prider, J. (2016). Buffel grass herbicide trails. http://www.pir.sa.gov.au/__data/assets/pdf_file/0004/288652/Buffel_Grass_Herbicide_Trial_Report.pdf (accessed 1 August 2018).

Tschirner, A (2016). The effects of temperature extremes and two herbicides on the germination of buffel grass (*Cenchrus ciliaris* L.) and implications for its management. Honours Thesis, University of South Australia.

Williams, D.G. and Baruch, Z. (2000). African grass invasion in the Americas: ecosystem, consequences and the role of ecophysiology. *Biological Invasions* 2, 123-40.

Restoring coastal ecosystems for the Great Barrier Reef and the RAMSAR wetlands of Bowling Green Bay

Scott Fry
15–20 Wills Street, Townsville City, Queensland 4810, Australia
(scott.fry@nqdrytropics.com.au)

Summary The Lower Burdekin floodplain supports the nationally important cane industry. Changes in the landscape for cane irrigation purposes have led to a rather unique problem, too much water. Now relatively high groundwater levels and perennial surface flows (previously ephemeral) have led to a loss of seasonality in the coastal wetlands flowing into the Great Barrier Reef. Too much water also provided ideal conditions for the proliferation of invasive species resulting in significant loss of ecological service and function in these internationally recognised Ramsar Wetlands of Bowling Green Bay. Previous efforts to manage these threats to the wetlands relied on traditional means of mechanical and chemical control. While this effort was in itself successful, involving farmers and institutional land managers, it ignored the underlying cause, the overwhelming problem of too much water. Our previous research had measured floristic composition (as an indicator of weediness), and freshwater/salinity and wetting/drying regimes for 88 wetlands across three land zone types. This research supported the hypothesis that wetland condition was dependant on these underlying features of the landscape. Our current project tested this hypothesis by strategically mimicking the natural wetting/drying regime to manage weeds and to ultimately restore ecological function to these priority coastal wetlands. To do this wetlands were dried out by improving efficiencies in the water delivery systems to reduce excess water and improved irrigation practices to reduce runoff. Under these more natural conditions, large areas of invasive aquatic weed infestations were controlled resulting in immediate responses of preferential aquatic species to return.

Keywords Great Barrier Reef (GBR) Ramsar, Bowling Green Bay, seasonality, land zones, coastal ecosystems, invasive species, wetting/drying regimes, aquatic weed management.

Managing bitou and boneseed on Long Reef Headland

Jillian Macintyre
Northern Beaches Council, 725 Pittwater Road, Dee Why, New South Wales 2099, Australia
(Jillian.macintyre@northernbeaches.nsw.gov.au)

Summary Long Reef Headland is located on Sydney's Northern Beaches approximately 15 km from the Sydney CBD between the suburbs of Dee Why and Collaroy. The area of Long Reef is approximately 70 hectares and known in the land register as Griffith Park. Long Reef's natural area contains a distinctive coastal headland and cliffs, beaches and rock platforms, wetlands, coastal heath and threatened Themeda grasslands all of which combine to form an area of unique landscape. Bitou bush and Boneseed invasion has been a major threat to biodiversity on Northern Beaches headlands since its introduction in the 50s and 60s. The invasion and establishment of bitou bush in the area was recognised as a key threatening process and Long Reef Headland was listed as a priority 1 Bitou Threat Abatement Plan (TAP) site. As part of a Sydney wide Bitou and Boneseed management plan, a control program commenced at Long Reef Headland in 2004.

This presentation will show how strategic and coordinated weed control management can achieve long term results. It will demonstrate by what means Council has been able to oversee community interaction of sensitive ecological areas in a more sustainable manner.

Keywords Weed management, bitou, boneseed, Northern Beaches.

Weed management is not quite 'Bush Regeneration' – An opinion

Peter Harper[1] and Nimal Chandrasena[2]
[1] Bettersafe Pest and Weed Management, 33 Lagonda Drive, Ingleburn, New South Wales 2565, Australia
[2] GHD Pty Ltd, 20 Smith Street, Parramatta, New South Wales 2650, Australia
(peter@bettersafe.com.au)

Summary As long-term practitioners of managing weeds in varied circumstances, we notice the increasing trend towards bush-regeneration as a broad approach to weed management, particularly in New South Wales (NSW). Recent observations in the Sydney basin and other parts of NSW show that bush-regeneration has not achieved the desired weed control outcomes for major weed species that have plagued eastern Australia for decades. These include semi-terrestrial species – alligator weed (*Alternanthera philoxeroides* (Mart.) Griseb), primrose willow (*Ludwigia peruviana* (L.) Hara), long-leaf willow primrose (*Ludwigia longifolia* (DC.) Hara), and full-terrestrial species such as blackberry (*Rubus fruiticosus* L. species aggregate), Chilean needle grass (*Nassella neesiana* (Trin. & Rupr.) Barkworth), Coolatai grass (*Hyparrhenia hirta* (L.) Stapf) and green cestrum (*Cestrum parqui* L'Hér.). Added to this are infestations of: interstate invaders such as golden wreath wattle (*Acacia saligna* (Labill.) H.L.Wendl.); fully aquatic invaders, such as egeria (*Egeria densa* (L.) Planch); and terrestrial invaders expanding their range southwards, such as groundsel bush (*Baccharis halimifolia* L.). Following an initial survey in 2009, we now observe these weeds greatly expanded invading new areas. We attribute inadequate local control primarily to the change in approach adopted by some agencies: i.e. a greater portion of funding used for bush-regeneration, instead of, as previously, weed management. We appreciate the noble goal of bush-regeneration, that is to assist regeneration of native bushlands from a diminished quality to a better condition. However, we lament the techniques used have limited value in controlling the large-scale infestations of highly invasive weeds. Indeed, the adoption of bush-regeneration as the primary paradigm for funding appears to be self-serving for some stakeholders. We suggest that well-coordinated, direct control, deployed on a regional-, catchment- and local-site-scale, is the only way to contain highly invasive weeds. We contend, that unless there is a change in weed management investment, these well-known invaders will soon become so entrenched that they will become widespread and likely to be given lower priority for management.

Keywords Bush-regeneration, willow primrose, alligator weed, egeria.

INTRODUCTION

Weeds are generally pioneering, opportunistic species, which are adapted to rapidly colonise, establish, and then thrive in disturbed environments, largely associated with human disturbances and habitation. Humans are primarily responsible for moving plants from one location to another, and across continents. As global trade and human habitations expand, more and more opportunities are created for this special group of plants (weeds) to thrive. These plants are not at fault. Plants we call 'weeds' are supremely well adapted to occupy vacant niches created by disturbances caused by humans, or by natural causes.

That weeds cause significant problems (negative impacts), goes without saying. Their management requires a holistic and integrated approach, incorporating various control methods, at both spatial (catchment to infested patch) and temporal scales (short, medium, or long-term control). Although the principles underlying the above approaches are well-established, often, weed management fails not because of lack of control options, but because control is reactive, poorly executed, and also poorly resourced. In this paper, we aim to highlight some aspects of the impact of the recent changes that have occurred within New South Wales (NSW), and how they are affecting weed management in the Sydney basin and elsewhere.

MATERIALS AND METHODS

Review of legislative changes To understand how jurisdictional responsibilities have changed since 1993, we undertook a brief review of legislative changes, particularly, as noted in Montaya (2012), NRC (2014a,b), Grain Growers (2014), and Local Land Services (LLS 2017, 2018). We also reviewed the frameworks around weed management in NSW (NSW DPI 2013, 2017, 2018a,b), the extent of on-ground knowledge of existing weed problems (often, the missing component in weed management), and the expectations of various responsible parties, apparent in these reports. Further, we obtained input from a wide network of land managers in agencies and local government clients.

Weed survey We conducted an extensive mapping of alligator weed infestations in the Sydney and Hunter regions, considered 'core areas' for this weed in 2007–2008, (Chandrasena 2008). This data were subsequently used in prioritising areas for alligator weed control in the Sydney basin (Sydney Weeds Committees 2010). During this extensive geographical survey, we were also able to opportunistically obtain information on the extent of infestations of many other weeds. Approximately 8–9 years later, during 2015–2017, we visually sampled many the same sites, including rivers, creek lines and semi-terrestrial sites, to assess the status of alligator weed infestations, as well as other species, which had been historically known to occur at these sites. We focused on some of the State's most problematic weeds, and compared the current levels of infestations, including new infestations, with historically known and documented infestation levels.

RESULTS

Legislative changes In NSW, the responsibilities of managing problematic weeds primarily reside with State and Local governments, and with various agencies that own or manage large tracts of land (such as roads and rail authorities, Sydney Water, Water NSW, and the National Parks and Wildlife Service). Most weeds were declared in 1907–1931 *'for the purposes of primary production'* (Johnson 2013) and in the areas of NSW that we surveyed, weeds used to be the domain of the former NSW Department of Agriculture, now the NSW Department of Primary Industries with various Catchment Management Authorities (CMAs).

The NSW Department of Primary Industries (NSW DPI 2018a) is currently the lead agency for invasive species policy in NSW. It also takes a lead role in managing new terrestrial and aquatic invasive species incursions and in managing established aquatic pests. Its other responsibilities include the administering of key legislation (the *Biosecurity Act 2015*) and leading programs, such as the NSW Weeds Action Program 2015–2020 (NSW DPI 2018a).

Local Land Services (LLS), a newly-formed arm of the NSW government, established in 2013 under the *Local Land Services Act 2013*, now has major responsibilities for coordinating the management of weeds (LLS 2017), previously performed by regional local government groups. LLS take a supervisory and advisory role for most of the on-ground actions implemented at a local level. The functions of the LLS' are to be *'the interface between landholders and Government across invasive species management'*. LLS' also provide operational assistance during invasive species incursions and surveillance, play a key role capacity building and technical advice, and act as the lead regional organisation for implementing the NSW Weeds Action Program 2015–2020 (NSW DPI 2017).

In the past three decades, practitioners have been confused by not just various classes and categories of weeds, but also about required action necessary against individual weeds, some of which has changed over time. Anecdotally, on-ground evidence suggests that weed management in the Sydney basin, and in NSW more broadly, has changed significantly. We suggest that this is because allocated funding for weed management has dramatically declined, or has been aligned with other 'asset management' priorities by many stakeholders. Historically, there has not been much emphasis on monitoring the effectiveness of any weed management program. Reporting was mostly based on areas sprayed and not on outcomes. Today, there are tighter objective-based outcomes to be reported. However, results from our surveys find that there is much variation in the way many stakeholders perceive the reporting requirements, much to the detriment of previously gained successes over many individual weed species.

For at least two decades, weeds in NSW were managed under the *Noxious Weeds Act 1993*. The *Biosecurity Act 2015* has now replaced the *Noxious Weeds Act* and many other separate pieces of 'biosecurity' legislation, some of which dealt with plant diseases and pest animals. A key objective of the *Biosecurity Act 2015* is to streamline and simplify existing procedures and provide greater flexibility in managing biosecurity risks posed by diseases, weeds and pest animals. Overall, the aim is to identify ways to reduce the negative impact of weeds on the economy, environment and/or community. A 'State Weeds Committee' is established to oversee implementation of weed management. Eleven LLS-based Regional Weed Committees are expected to facilitate community participation in promoting a landscape-scale approach to managing prioritised weeds. Weed management is implied as everyone's 'General Biosecurity Duty' (NSW DPI 2013) and a collaborative effort (NSW DPI 2018a). Although not explicitly stated, we assume that strict adherence to the Act will result in enforcement. Regional Strategic Weed Management Plans (RSWMPs) are the new mechanisms for articulating community and stakeholder expectations, and working together to identify, minimise, respond to and manage 'high-risk' weeds (LLS 2017, 2018). High risk weeds in NSW have been identified through the Weed Risk Management System (Johnson 2009, NSW DPI 2018b) and the lists are available from the LLS websites (LLS 2018).

Existing local control authorities have responsibilities for on-ground weed management within their jurisdictions. This includes high-risk weeds and those managed for 'asset protection' purposes. Those weeds to be controlled via a 'Biosecurity Duty' (as defined in the Act) are not always clear, particularly, if the latter group of species includes those that are not high-risk species. For an integrated and systematic weed management approach across a local control area ALL stakeholders need to be corralled, without any stragglers!

To implement the weed management across the State broadly, and the RSWMPs more specifically, the *NSW Weeds Action Program* [WAP] *2015–20* (NSW DPI 2017) has been developed, which *'follows from the successful implementation of the NSW Weeds Action Program 2010–2015'*. The WAP is an initiative to reduce the impact of weeds through implementation of the *NSW Biosecurity Strategy 2013–2021* (NSW DPI 2013) and the *NSW Invasive Species Plan 2016–23*. Our day-to-day experience is that the claim of success in implementing the WAP 2010–2015 is debatable, as evident in the proliferation of infestations of major weeds across the Sydney basin (see Table 1), and particularly because there has been no independent audit. Our observations concur with the view strongly expressed by the Natural Resources Commission (NRC) that limited stakeholder accountability and absence of a consistent audit of weed management programs in NSW is a major limitation for success on the ground (NRC 2014a).

Our weed and land management experiences, working with many clients and other contractors, is that there is quite a deal of confusion about the roles and expectations associated with the *Biosecurity Act 2015*. Some of the current workforce is also poorly trained in weed identification and control methods, as responsibilities have shifted, and most local government funding has been shifted towards Bushcare and Landcare initiatives, which make little impact on existing infestations of aggressive weeds.

Ground-truthing by the authors has revealed that during the past 6–8 years, many weeds that should have received some control treatment, have not been treated at all; or have only been treated sporadically, or, inadequately. We recognise a crisis looming regarding the overall weed management, at least in the Sydney basin, particularly in the Hawkesbury-Nepean River, Georges River, South Creek and areas affected by developments in western Sydney. We observe that the overall result of so many legislative and policy changes, particularly over the past 25 years, and changes in approaches with these, is that many of the State's worst weeds have expanded their territory (Table 1). In some cases, there is evidence that these expansions have not gone unnoticed and that responsible individuals and agencies are managing these weeds, but that they receive inadequate attention and funding, primarily because the existing system does not require obligatory action, compliance and reporting.

Weed survey Results of the initial and most recent weed surveys are summarised in Table 1. Relevant observations and comments are summarised below:

Alligator weed There has been a vast expansion of alligator weed in the past 10 years, into new areas within 'core areas' of the infestation (some areas of Sydney and the Hunter). Historical infestations remain deeply entrenched at most sites, despite the availability of metsulfuron for control. We fear that the current failure to control this invader will most likely leave the State with no option but to accept that the weed has become too widespread and given lower priority for management in much the same way, as has been done with other widespread species, such as lantana (*Lantana camara* L.) and egeria (*Egeria densa* (L.) Planch).

Long-term practitioners have witnessed, how unsuccessful the management of alligator weed in the State has been. Despite spending millions of dollars over the past 25 years on biological control and other research, lessons learnt through research, which has been amply documented (and reviewed in Chandrasena 2008), are yet to be adopted. Fifteen years have passed since we contributed to the CRC alligator weed management guide (CRC 2003) on successful control with targeted programs. Despite this, in most areas of Sydney, alligator weed thrives, because on-ground implementation of effective control methods is poor. A much more hands-on coordination approach is required for alligator weed management, instead of leaving it to stakeholders alone.

Willow primroses To our knowledge, infestations of *Ludwigia peruviana* and *L. longifolia* have not been properly mapped or controlled in the Sydney basin. Many of the wetlands in the Sydney basin and the Hunter region continue to be at great risk of being invaded by these two species. Since the late-1980s, these species have been well researched, and reasonably well controlled (e.g. Chandrasena *et al.* 2002). Our recent experience is that agencies and their contractors have little appreciation of the risk these species pose, nor how to control them. One stark example is the continued use of the non-selective Roundup Biactive® to control primrose willows (which is ineffective), in contrast to the broad-leaf herbicide 2,4-D, to which

Table 1. Comparative weed survey results from the Sydney and Hunter regions of New South Wales. Surveys were conducted in 2009, and then again in the period 2015–2017. LGAs refer to Local Government Areas and Hwy to highways.

Weed species	Botanical name	Comparative weed survey observations
Alligator weed	*Alternanthera philoxeroides* (Mart.) Griesb.	Vastly expanded its territory since 2009: deeply entrenched, historical infestations in all of Botany Wetlands' ponds, Sydney Airport's Engine Ponds, Rouse Hill – Second Ponds Creek, Smalls Creek, and other smaller tributaries; the upper Nepean River, and the Hawkesbury River; new and spreading infestations within LGAs – Hills Shire, Penrith, Bankstown, Blacktown, Botany and Rockdale, Camden, Campbelltown, Liverpool, Sutherland; South Creek, Anzac Creek, Lane Cove River, Macquarie University, entrenched in low lying creeks and lagoons away from major tributaries; new infestations in Erskine Park; and Orchard Hills.
Willow primrose	*Ludwigia peruviana* (L.) Hara	Vastly expanded its territory since 2009: new regrowth in Botany Wetlands, Engine Ponds, Second Ponds Creek, Smalls Creek, and many locations along Nepean River and Hawkesbury River (Sackville Ferry) etc. Infested creeks include Darling Mills Creek, Toongabbie Creek, Parramatta River; other sites – Bankstown Airport, Camden, Campbelltown, Liverpool, Glenfield Creek, Georges River, and Warwick Farm Sewage Treatment Plant wetland.
Long-leaf willow primrose	*Ludwigia longifolia* (DC.) Hara	Significant infestations in Penrith Lakes and basins: Darling Mills Creek, Toongabbie Creek, Parramatta River, Kenthurst (Hills Shire), Penrith; and Port Macquarie (north of the Hunter region).
Blackberry	*Rubus fruiticosus* L. spp. aggregate	Vastly expanded territory since 2009 last survey: major infestations in western Sydney, Wollondilly, Wingecarribee, Camden, Campbelltown, Liverpool, Fairfield, and Penrith LGAs.
Green cestrum	*Cestrum parqui* L'Hér.	Vastly expanded territory since 2009 last survey: major infestations in Parramatta, Liverpool and Fairfield LGAs.
Chilean needle grass	*Nassella neesiana* (Trin. & Rupr.) Barkworth	Major infestations in Campbelltown, Liverpool, and Fairfield LGAs.
Coolatai grass	*Hyparrhenia hirta* (L.) Stapf.	Major infestations in Sutherland and Wollondilly Shire LGAs, Picton Road, Cumberland Hwy, along the M4, M7 and other major roads.
Golden wreathe wattle	*Acacia saligna* (Labill.) H.L. Wendl.	Major infestations along major arterial roads – Hume Hwy, from Picton to Liverpool, Picton Road, Pacific Hwy to Wollongong, etc.
Egeria	*Egeria densa* (L.) Planch	Vastly expanded territory since 2009: major infestations in Fairfield, Liverpool, Hills Shire, and Parramatta LGAs – Toongabbie and Pendle Hill Creeks, Georges River, Nepean River, Kangaroo River; new infestations in many creeks in the Shoalhaven areas (south of Sydney region).
Groundsel bush	*Baccharis halimifolia* L.	Significant infestations along the Pacific Hwy to Coffs Harbour; from Taree to the Queensland Border; Isolated infestations in the Sydney basin (i.e. Parramatta, Liverpool LGAs, and at Menangle).

both species are highly susceptible. Inadequate sharing of information on effective weed management solutions, and poor training of the workforce are two reasons for this failure.

DISCUSSION

In this paper, we express our sincere concern that all is not rosy in weed management in NSW, despite the efforts of responsible authorities, over several recent decades. The absence of proper audits and reporting, over decades, has led to complacency and a lack of accountability for tracking whether outcomes are being achieved with public funds spent on weed management. Many stakeholders and agencies appear to be rather 'out-of-touch' with on-ground realities and are indifferent to certain weeds and their management. However, instead of being pessimistic, we offer some practical suggestions, with the hope that these may be seriously considered by relevant authorities.

We agree with the view that weed management is a *shared responsibility* between the public and private including all government landowners. All parties are equally responsible for stopping the spread of weeds, and the proposed collaborative approach, if coordinated well, is likely to result in success. Conversely, failure by any single party regarding on-ground implementation will lead to an increased risk of weeds becoming a greater problem than they have to be.

We share the concerns expressed by Arcioni (2003), that is that the *Noxious Weeds Act 1993* was too broad in its scope, and that its expectations were not being fully met due to the prevailing policies, the interplay with other regimes, international politics, and practical issues, such as insufficient funding. We contend that after this analysis, the situation has changed very little, except that weed management, across the board has become even more diffuse and ineffective 'on the ground' in the ensuing period.

Inadequate priority and legislative positioning How major changes in approaches affect outcomes is best illustrated again using the example of alligator weed. Before 2003, alligator weed was declared as a '**W1**' noxious weed throughout NSW under the *Noxious Weeds Act 1993*. As such, its presence needed to be notified to the Local Control Authority (usually the Local Government), and mandatory actions had to be taken to '*fully and continuously suppress and destroy*' the weed. Subsequent Act and declaration amendments resulted in the declaration of alligator weed as a '**Class 3**' noxious weed within 'core' infestation areas in NSW. For all intents and purposes the mandatory management actions did not change; i.e. the plant was to be '*fully and continuously suppressed and destroyed*', where it was feasible.

That said, we question if the declaration amendment reflected the NSW DPI's view that infestations of alligator weed had reached unmanageable levels. We suggest that there was a lack of confidence in the likelihood of containment, and where relevant, eradication of alligator weed. However, in so far as managing a potential that was called a '*National threat*', the apparent demotion of alligator weed to a **Class 3** weed in 2005 was not in line with the invader's capacity to spread further, adversely impacting waterways, and potentially, Australia's national economy. The demotion led to several State agencies, water utilities and LGAs assigning much *less* priority to alligator weed in their asset management plans. The disbanding of the NSW DPI aquatic weed management task force (NSW DPI 2001), which existed until about 2006–2007, also contributed to an inadequate focus on the '*national threat*' posed by alligator weed, which has continued unabated to this date. Before this time, Australia had spent several million dollars funding projects, through the now extinct CRC for Weed Management, looking for suitable biocontrol agents for alligator weed, with no notable success. For example in the field, we occasionally observe that the flea-beetle *Agasicles hygrophila* Selman & Vogt, introduced in 1977, is still the only agent actively doing some damage, albeit sporadically, to alligator weed. Given that no other agent has thus far been successfully field-deployed, despite more than 25 years of alligator weed bio-control research, the public would be excused for doubting whether funds have been correctly spent!

Separate bush regeneration from weed management We attribute the inadequate local control of many major weeds primarily to the change in approach adopted by some agencies, which have allocated greater portions of annual funding for bush-regeneration, instead of straight forward weed management. We appreciate the noble goal of bush-regeneration, i.e. to assist regeneration of native bushlands from a diminished quality to a better condition, possibly, to a quality that might have existed before weed invasion. However, bush-regeneration methods have limited value in arresting the large-scale establishment of highly invasive weeds. Indeed, we contend that the primary motive for adoption of bush-regeneration as the primary paradigm for weed management funding appears only to be self-perpetuating.

We are aware of a growing realisation among some land managers that bush-regeneration contracts do not adequately address weeds within their assets. As to why this is the case, we offer the following, relevant observations:

- actions on-the-ground to achieve bush-regeneration goals, even on principle, are very localised. These cannot be easily replicated across sub-catchments or broader regions. The scales of operation are often so small that of the most major weed infestations do not get adequately treated;
- the tools available, i.e. small-scale herbicide spot sprays, cut-and-paint applications, etc. leave large infestations behind, as bush-regenerators are overwhelmed by the scale of weed control required at most sites; and
- the self-perpetuating nature of the work, apparent only with some stakeholder agencies and contractors. Under bush regeneration contracts, some major weed infestations may go unreported for years, because the primary focus is on facilitation and slow recovery, and rehabilitation of targeted sites; yet, it creates more bush regeneration contracts!

Multiple stakeholder cooperation Models of multiple stakeholder engagement and cooperative approaches are well established in natural resource management. However, leadership is required to garner goodwill, invoke effective partnerships and obtain commitment from multiple stakeholders to reduce the national threats posed by weeds. Having the legislative requirements to support this cooperation is a must. The key to obtaining effective cooperation is obligatory compliance reporting, and penalties for non-compliance.

Inadequate resourcing Most weed managers agree that the current expansion of weeds in NSW is attributable to severe under-resourcing, both of funding, and human resources, committed to a sustained and coordinated management effort across the State. Even a cursory check would reveal that, in submitting applications for weed control funding in a given area, the tactic used by local managers is to '*not ask for much*' for the fear of not getting any! Inadequate funding means insufficient effort to contain major weed infestations in many areas. Inadequate accountability with regard to tracking how weed funding has been used, against outcomes, continues largely unaddressed.

Our on-ground experiences reveal that local managers tend to use various defences for inadequate action. These include: a lack of clarity on boundaries; diffused responsibilities; and the insufficient priority given to specific weeds. We fear that this unsatisfactory State-wide problem will continue if the new LLS Regional Weed Committees, set up under the new *Biosecurity Act 2015* lack the legislative authority required to oversee the management efforts across jurisdictions. The authority established must be for independent verification and compliance reporting, particularly on the effectiveness and adequacy of weed control programs. To achieve this, it would be prudent to set up a special Weed Management Task force in NSW to manage all high-profile species that pose a '*national threat*'.

CONCLUSIONS

Our paper is not an exhaustive study of why weed management has largely been ineffective across the Sydney basin or other parts of NSW, although we sense such a review is timely. We argue that all stakeholders approach weed management within their jurisdictions, purely and simply, to achieve a minimal level of compliance only, and not because they have '*to do the right thing*'. Asset protection and compliance are strong drivers, which motivate stakeholders to allocate some level of funding to weed management.

A primary failure in weed management in the last two decades, during which so many legislative changes have occurred, is the inadequacy of compliance reporting. Further, the lack of will to enforce weed legislation is a political decision, but the price society pays is rampant growth of many species, and wastage of resources. If the approach evident in the most recent Greater Sydney Strategic Weed Management Plan (LLS 2017) is to be successful, establishing an accurate baseline of key weeds and their abundance would be a first step. In conclusion, we re-iterate that a review of on-ground weed management, actual actions, funding and performances of ALL stakeholders will reveal inadequacies, a need that has already been highlighted elsewhere (NRC 2014a,b).

REFERENCES

Arcioni, E. (2003). Out damned weeds! Weed management in Australia – Keeping them at bay. *Australasian Journal of Natural Resources Law and Policy* 8, 2003, 75-122.

Chandrasena, N. (2008). Assessment of risk of spread for strategic management of the core alligator weed infestations in Australia – 'Taking stock'. Australian Government Department of Agriculture, Fisheries and Forestry, Canberra.

Chandrasena N., Pinto L. and Sim R. (2002). Reclaiming Botany wetlands, Sydney through integrated management of *Ludwigia peruviana* and other weeds. Proceedings of the 13th Australian Weeds Conference, eds H. Spafford Jacob, J. Dodd and J.H. Moore, pp. 134-7. (Plant Protection Society of Western Australia, Perth).

CRC, Cooperative Research Centre [for Australian Weed Management] (2003). Weeds of National Significance. Alligator weed – *Alternanthera philoxeroides*. Weed Management Guide.

Grain Growers (2014). The economic cost of weeds in NSW. A Grain Growers research report, May 2014.

Johnson, S.B. (2009). The New South Wales Weed Risk Management System. NSW Department of Primary Industries, Orange. 127 p.

Johnson, S.B. (2013). Don't let the truth get in the way of a good story. The declaration of weeds that affect the environment started in 1907 in New South Wales. *Plant Protection Quarterly*, 28, 77-9.

LLS, Local Land Services [New South Wales] (2017). Greater Sydney Regional Strategic Weed Management Plan 2017–2022. Greater Sydney Local Land Services. 80 p.

LLS, Local Land Services (2018). NSW Weed Reforms https://www.lls.nsw.gov.au/biosecurity/weed-control (accessed 21 April 2018).

Montoya, D. (2012). Noxious Weeds. Briefing Paper No. 02/2012, NSW Parliamentary Library Research Service, Sydney.

NRC, Natural Resources Commission [of NSW] (2014a). Issues Paper: Review of Weed Management in NSW. Document: D13/4149.

NRC Natural Resources Commission [of NSW] (2014b) Review of weed management in NSW. Document: D13/6066.

NSW DPI, New South Wales Department of Primary Industries (2001). Strategic plan for the management of alligator weed in the Hawkesbury-Nepean catchment. Hawkesbury-Nepean Aquatic Weeds Task Force, NSW Department of Primary Industries, Sydney.

NSW DPI New South Wales Department of Primary Industries (2013). NSW Biosecurity Strategy, 2013–2021.

NSW DPI, New South Wales Department of Primary Industries (2017). NSW Weeds Action Program, 2015–2020 Guidelines.

NSW DPI, New South Wales Department of Primary Industries (2018a). NSW Invasive Species Plan 2018–2021.

NSW DPI, New South Wales Department of Primary Industries (2018b). NSW Weed Risk Management System. https://www.dpi.nsw.gov.au/biosecurity/weeds/strategy/nsw-weed-risk-management-system (accessed 21 April 2018).

Sydney Weeds Committees (2010). Priorities for the control of Alligator Weed in the Sydney Region. Unpublished report.

Effects of soil nitrogen reduction on competition between *Phragmites australis* and *Melaleuca ericifolia*

Md N. Uddin[1,2], Randall W. Robinson[2,3] and Takashi Asaeda[1]

[1] Department of Environmental Science and Technology, Saitama University, 255 Shimo-Okubo, Sakura-Ku, Saitama, Japan-338-8570

[2] Department of Ecology and Environmental Management, College of Engineering and Science, Victoria University, St Albans Campus, Melbourne, Victoria 8001, Australia

[3] Institute for Sustainable Industries and Liveable Cities, Victoria University, PO Box 14428, Melbourne, Victoria 8001, Australia

(nazim.uddin@vu.edu.au)

Summary Nutrient enrichment, particularly nitrogen, is an important determinant of plant community productivity, diversity and invasibility in wetlands. It is thought to contribute to increasing colonisation and dominance of *Phragmites australis*, especially during restoration. Manipulating soil resource availability with understanding of plant-soil microbe interactions may explain the success of some invasive plant species and thereby, such type of management tools may be required for ecological restoration. We examined competition between *Phragmites australis* and *Melaleuca ericifolia* in greenhouse with range of AC treatments followed by cutting of *Phragmites* shoots in nutrient enriched soil. Additionally, we evaluated the effect of AC on plant free microcosms in laboratory to isolate direct effects from indirect effects on soil microbial community function. Overall, the objective was to test whether lowering nitrogen availability might be an effective strategy for suppressing *Phragmites* invasion in wetland. Activated carbon significantly decreased *Phragmites* growth by increasing *Melaleuca* biomass from 41% to 68% and thereby, reduced suppressive effect significantly on *Melaleuca* at higher concentration, together with repeated cutting of *Phragmites* shoots. AC also caused significant changes to microbial community functions, namely microbial respiration, N mineralisation, nitrification, and dehydrogenase enzyme activity that could potentially explain changes in plant growth. The overall effects on plant growth, however, may be partially microbially mediated, which was demonstrated through soil microbial functions. Results support the idea that reducing community vulnerability to invasion through manipulations of resource availability with understanding of plant-soil microbe interactions may provide a promising approach to invasive species management and ecology restoration.

Keywords *Phragmites australis*, nutrient enrichment, resource manipulations, soil microbes, restoration.

The invasive *Parthenium hysterophorus* L. has limited impact on soil chemistry and enzyme activities but influences above and below ground biodiversity

Olusegun O. Osunkoya[1], Boyang Shi[1], Kerri Moore[2] and Kunjithapatham Dhileepan[1]
[1] Invasive Plant and Animal Science Unit, Biosecurity Sciences, Department of Agriculture and Fisheries, Eco-sciences Precinct, Brisbane, Queensland 4102, Australia
[2] CSIRO, Eco-sciences Precinct, Brisbane, Queensland 4102, Australia
(Olusegun.osunkoya@daf.qld.gov.au)

Summary The invasive *Parthenium hysterophorus* L., a Weed of National Significance, is pervasive in agriculture, conservation and disturbed lands of central Queensland. In recent years, it has spread to the eastern and southern parts of the State and has the potential for incursion to the neighbouring States of New South Wales and Northern Territory, especially in view of climate change. However, very little work (most done overseas with inconsistent findings) has been reported on the weed's impact on soil processes and native biodiversity. The work reported herein (involving soil sampling across multiple sites in central Queensland in parthenium weed infested and non-infested habitats) showed that due to the weed's annual growth habit, a null effect was detected for soil chemistry (both micro- and macro-nutrients) and enzyme activities (β-glucosidase, fluorescein diacetate (FDA) hydrolysis and total microbial nitrogen and carbon). In contrast, significant negative impacts of parthenium weed infestation were observed for both below-ground (soil seed bank) and above-ground composition and diversity of co-occurring plant species. The results are discussed in terms of sampling methodology, adaptive pasture management, the weed's growth habit and its 'perceived' position on Queensland weed list for risk assessment (based on impact, abundance and spread) and management.

Keywords Australia, biological invasion, soil chemistry, soil enzymes activity, biodiversity, weed impact.

INTRODUCTION

Parthenium hysterophorus L (henceforth parthenium weed) is an annual herbaceous plant with its origin in the Gulf of Mexico and/or central south America (Dhileepan and McFadyen 2012). This weed now has a pantropical distribution and is found in more than 80 countries (Shi *et al.* 2015). It is a major weed in Australia especially in Queensland. Parthenium weed has substantial negative impacts on many natural and managed ecosystems, including open grassland, road-sides, and agricultural lands as well as health costs to grazing animals and humans.

Several studies have reported on the pervasive negative impacts of plant invasions on above ground communities (Pyšek *et al.* 2012, Mollot *et al.* 2017). In contrast, few studies have examined the impact of plant invasions on below-ground processes, including seed bank diversity, soil chemistry and nutrient cycling, and soil enzyme activities (Osunkoya and Perrett 2011, Giora and Pyšek 2016, Osunkoya *et al.* 2017). For parthenium weed, the findings on its ecological impact have been inconsistent. We had earlier reported that its invasion across varying land–use type (n = 3) and sites (n = 12) had minimal impact on soil physico-chemical properties (Osunkoya *et al.* 2017). We then hypothesised that perhaps other ecosystem properties, such as above-ground biota, may be more impacted by parthenium weed invasion. Additionally, the null effect was attributed, amongst other factors, to a large environmental variation (idiosyncrasies of site and land-use factors) and low sample size (n = 3–6 per treatment) used in the previous study. Consequently, a larger sample size per site was advocated to increase the power of the test and thus limit committing type II error (i.e. declaring a significant result non-significant when in actual fact there is a difference). In this study, in view of logistical constraints, we limited the number of sites surveyed (from n = 12 to n = 4) and increased substantially the sample size (n = 15 per treatment per site) to assess parthenium weed impact on above and below ground processes. Our ecological markers were below-ground (soil seedbank, soil chemistry and enzyme activities) and above-ground (plant species identity and assemblages) processes.

MATERIALS AND METHODS

The four field survey and soil collection sites are in central Queensland, Australia. All these sites have been heavily invaded by parthenium weed for at least 10 years. These sites are representative of grazing grassland (Gracemere and Hutton creek), cropping

(Morebridge) and riparian (Moolayamber Creek) lands typicality infested by parthenium weed in the region (see Osunkoya *et al.* 2017). The data for above-ground (standing vegetation) and below ground (seed bank and soil chemistry) assessments were collected from paired parthenium weed infested and parthenium weed free patches (5–10 m apart) using a standardised procedure (Osunkoya *et al.* 2017). Each site has 15 replicates of each vegetation (or soil) patch type sampled, except for Hutton Creek site which consisted of only five replicates. For assessment of above-ground species plant composition, we used 0.25 m × 0.25 m quadrats. Soils from both infested and non-infested patches were collected using a 10 cm diameter × 7 cm depth soil corer. Each replicated soil sample was made up of five randomly selected cores in a 1 m^2 area that were subsequently combined into a composite sample. Soil seedbank was evaluated in germination trays under adequate moisture and temperature (17°C to 32°C) for nine months at Ecosciences Precinct glasshouse in Brisbane, (Queensland, QLD) Australia. Germinated seedlings of parthenium weed and other plant species were identified and counted monthly. Soil moisture and a series of chemical traits (pH, electrical conductivity, total carbon, total nitrogen, microbial biomass carbon and microbial carbon nitrogen) and enzyme activities (fluorescein diacetate [FDA] hydrolysis and β-glucosidase – two major enzymes involve in the first phase of plant cell tissue decomposition) were assayed (see details of analytical procedures in Osunkoya *et al.* 2017).

We normalised that data and used common parametric analyses (ANOVA and generalised linear modelling) and multivariate ordination techniques to detect differences in mean performance of infested and uninfested (control) patches. For species assemblages we converted the abundance data to presence-absence data, and used non-metric multidimensional scaling ordination (NMDS) involving Bray-Curtis dissimilarity matrix to explore general trends in the data sets. For soil chemistry and soil enzyme data, values were log transformed prior to NMDS ordination. All ordinations were performed via PRIMER version 7.0.

RESULTS

A higher number of seedlings of parthenium weed germinated from the soil of infested compared with those from non-infested patches (Figure. 1). Also, the greater abundance of the parthenium weed in the soils of the infested patches had a detrimental effect on the seed bank composition of other species (both native and non-native species, Figure 1).

Above ground plant species diversity varied across land use type ($F_{2, 194}$ = 98.58, P<0.001) or site ($F_{3, 192}$ = 66.55, P<0.001). Standing plant density (no. $0.0625m^{-2}$) of parthenium weed was significantly (P<0.0001) higher in the infestation plots (14.49 ± 1.27) compared to the parthenium weed free areas (3.90 ± 1.20). In all and across the four sites surveyed, 33 plant species were identified from the above ground vegetation, with greater species number (Figure 2) and significant difference in species composition in uninfested (control) compared to infested vegetation patches (Figure 3). The depressing effect of the parthenium weed infestation on species diversity was of the order cropping > riparian ≥ grazing and there were no significant differences in soil chemistry and enzyme activities both within and across sites and land use type (Figure 4).

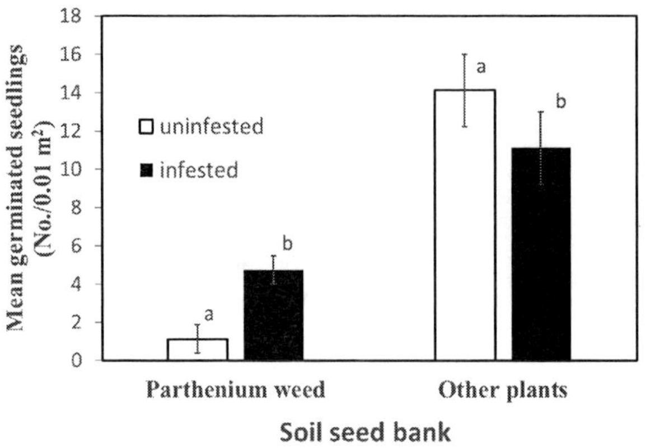

Figure 1. Mean (±SE) seed bank population in parthenium weed infested and uninfested vegetation patches. Data have been pooled across the four sites surveyed. Within paired vegetation patches, bars with different letters are significantly different at P≤0.05.

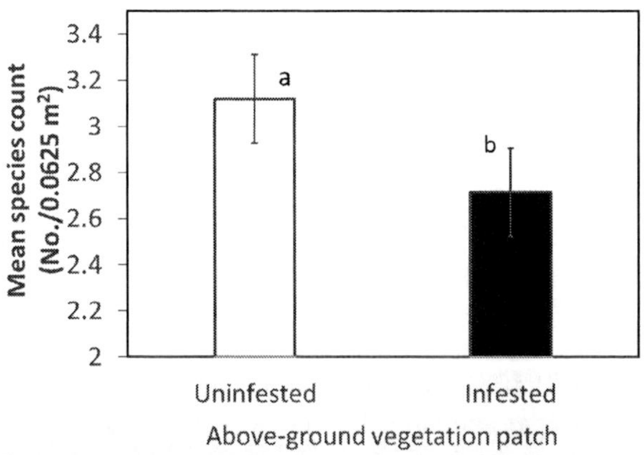

Figure 2. Mean (±SE) species richness in parthenium weed infested and uninfested vegetation patches. Data pooled across the four sites surveyed. Bars with different letters are different at P≤0.05.

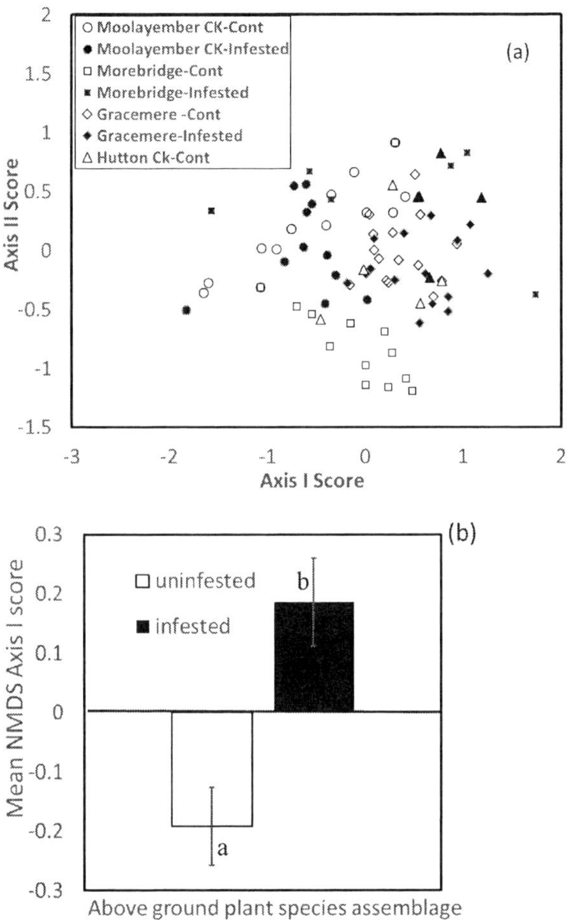

Figure 3a,b. NMDS ordination on the first two principal axes of above-ground vegetation patches based on presence-absence species data. Each data point represents a parthenium infested (closed symbol) or uninfested (open symbol) vegetation patch (2D stress = 0.07) (Figure 3a). A significant difference (P<0.02) exists between the two vegetation patches in species composition, especially along Axis I scores and this difference is summarised (±SE) in Figure 3b.

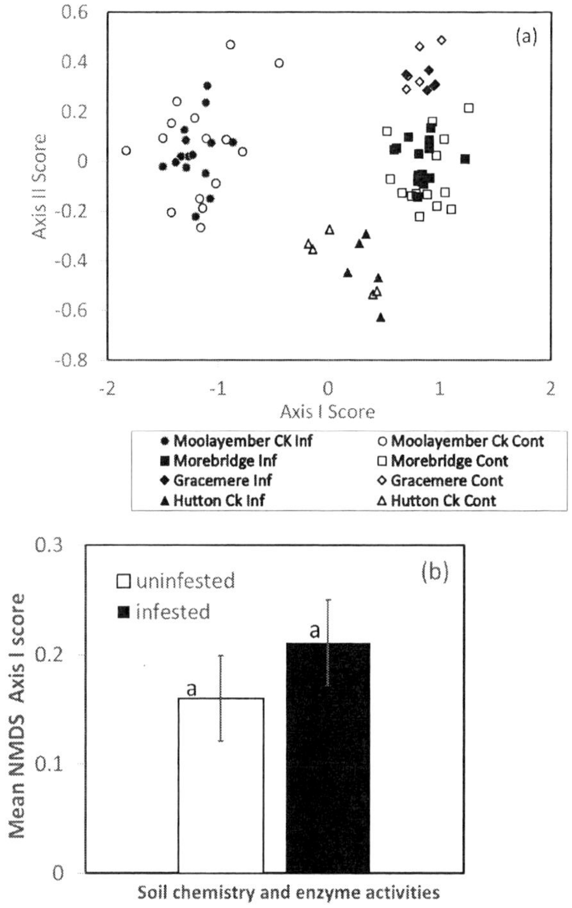

Figure 4a,b. NMDS ordination on the first two principal axes of parthenium infested (closed symbol) and uninfested (open symbol) soil patches across the four sites surveyed in Central QLD, Australia based on chemical traits and enzyme activities (2D stress = 0.03) (Figure 4a). No significant differences were detected in soil chemical traits and enzyme activities both within and across sites/land use type between the two vegetation patches as indicated by the mean (±SE) score for Axis I (Figure 4b).

DISCUSSION

It is interesting to note the presence of seeds of parthenium weed in the soil seed bank from some of the so called 'non-infested' (control) patches (the seeds are minor in abundance and detected in ~35% of weed-free soil patches surveyed). This suggests that parthenium weed-free habitat patches that we sampled may have, in the immediate past seasons, been infested by the weed but unknown to us due to its the annual life form. This could mask any soil differences we expected to observe between infested and non-infested patches, especially if a legacy effect of parthenium weed invasion persists under such ephemeral conditions (Gioria and Pyšek 2016). The finding also reinforces the belief that a 'good' cultural/pasture management with minimal tillage may help to suppress germination of buried parthenium seeds, hereby limiting its competitive impact on native plants and/or crop productivity.

Impact studies of invasive species are becoming increasingly available, and will contribute immensely towards a better estimation of pest risk assessment and management. Parthenium weed has always been ranked highly in terms of invasiveness and management needs in QLD Australia (see Osunkoya *et al.* 2018). While this assertion maybe correct, we have shown in previous work (Osunkoya *et al.* 2017), and in this study, that parthenium weeds overall (ecological) effect may be context- and trait-of-interest-specific.

The detrimental effect on above-ground native and non-native species and below-ground soil seed bank assemblage is real (Figure 1), but its influence as driver of soil processes (e.g. nutrient cycling) is minimal.

Sample size was substantially increased in this work compared to that of our previous study (Osunkoya *et al.* 2017; from n = 3–6 to n = 15), and we could still only detect minimal influence of parthenium weed on soil processes (i.e., a change was only detected in the soil seed bank assemblage). This is in contrast to many invasive plant species that we have worked on, including *Lantana camara* L. and cat's claw creeper vine (*Dolichandra unguis-cati* (L.) L.G.Lohmann) (Osunkoya and Perrett 2011, Perrett *et al.* 2012). Parthenium weeds annual growth habit with minimal ephemeral litter input may be a contributing factor to its limited influence on soil processes. In a review of the global assessment of invasive plant impacts of more than 150 species, Pyšek *et al.* (2012) reported that short-statured plants (<4.8 m), including many annuals like parthenium weed (but excluding grasses) were least likely to exert significant impacts on outcomes related to species diversity and soil attributes. Additionally, rather than invoking a change in major nutrients *per se*, the mode of action of parthenium weed on soil processes could be indirect via the release of allelopathic phytotoxic exudates (e.g. phenols) from its roots and/or residue decomposition. These phytotoxins are known to interact with soil chemistry, can suppress seed germination and growth of native and other weedy species, and can potentially lead to the establishment of monocultures of parthenium weed (Belz *et al.* 2007, Shi *et al.* 2015). Phytotoxins can also impact negatively on soil food-web structures of nematodes, insects, and mammals (Van der Putten *et al.* 2013). Assuming the phytotoxicity hypothesis is real, there is a need for microcosm and controlled laboratory trials to explore further and explicitly the impact of the parthenium weed on soil processes, including possible feedback effects (Van der Putten *et al.* 2007).

ACKNOWLEDGMENTS

We thank the Queensland Government, through the Land Protection Trust Fund, for travel support that greatly assisted in field work and accommodation during the course of the study. The assistance of Christine Perrett in the seed bank germination trials is appreciated.

REFERENCES

Belz, R.G., Reinhardt, C.F., Foxcroft, L.C. and Hurle, K. (2007). Residue allelopathy in *Parthenium hysterophorus* L.—Does parthenin play a leading role? *Crop Protection* 26, 237-45.

Dhileepan, K. and McFadyen, R.E. (2012). *Parthenium hysterophorus* L. – parthenium. *In* 'Biological control of weeds in Australia' eds M. Julien, R.E. McFadyen and J. Cullen, pp. 448-62. (CSIRO Publishing, Melbourne)

Gioria, M. and Pyšek, P. (2016). The legacy of plant invasions: changes in the soil seed bank of invaded plant communities. *BioScience* 66, 40-53.

Mollot, G., Pantel, J., and Romanuk, T. (2017). The Effects of invasive species on the decline in species richness: A global meta-analysis. *Advances in Ecological Research* 56, 61-83.

Osunkoya, O.O. and Perrett, C. (2011) *Lantana camara* L. (Verbenaceae) invasion effects on soil physicochemical properties. *Biology and Fertility of Soils* 47, 349-55.

Osunkoya, O.O., Akinsanmi, O.A., Lim, L.S., Perrett, C., Callander, J., and Dhileepan, K. (2017). *Parthenium hysterophorus* L. (Asteraceae) invasion had limited impact on major soil nutrients and enzyme activity: Is the null effect real or reflects data insensitivity? *Plant and Soil* 420, 177-94.

Osunkoya, O.O., Perrett, C., Moore, K., Nicol, S., Froese, J., Gooding, K. and Campbell, S. (2018). An inventory of Queensland prioritised invasive plant species for management and research. Proceedings of the 21st Australasian Weeds Conference, 9–13 September, Sydney, New South Wales, eds S. Johnson, L. Weston, H. Wu and B. Auld, pp. 392-96. (Weed Society of NSW).

Perrett, C., Osunkoya, O.O. and Clark, C. (2012) Cat's claw creeper vine, *Macfadyena unguis-cati* (Bignoniaceae), invasion impacts: Comparative leaf nutrient content and effects on soil physicochemical properties. *Australian Journal of Botany* 60, 539-48.

Pyšek, P., Jarošík, V., Hulme, P.E., Pergl, J., Hejda M., Schaffner, U. and Vilà, M. (2012). A global assessment of invasive plant impacts on resident species, communities and ecosystems: The interaction of impact measures, invading species' traits and environment. *Global Change Biology* 18, 1725-37.

Shi, B., Aslam, Z. and Adkins, S. (2015). The invasive potential of parthenium weed. A role for allelopathy. *In* 'New Developments in Allelopathy Research', ed. J.E. Price, pp. 135-73. (Nova Science Publishers, United States of America).

van der Putten, W.H., Bardgett, R.D., Bever, J.D., Bezemer, T.M., Casper, B.B., Fukami, T., Kardol, P., Klironomos, J.N., Kulmatiski, A., Schweitzer, J.A., Suding, J.A., van de Voorde, T.F.J. and Wardle J.A. (2013). Plant-soil feedbacks: the past, the present and future challenges. *Journal of Ecology* 101, 265-76.

Weeds in Natural Systems (WINS): Weighing up the importance of different invasion mechanisms

Kristine French
Centre for Sustainable Ecosystem Solutions, School of Biological Sciences, University of Wollongong, Wollongong, New South Wales 2522, Australia
(kris@uow.edu.au)

Summary There is some confusion about how weeds 'win' in natural systems and replace native species and a range of mechanisms are often vaguely assumed. Do seedling plants kill native adults through some allelopathic mechanism? Or is seedling competition for nutrients or light the most important mechanism? This talk investigates a number of examples of research that investigate the mechanism behind invasion for a range of species to highlight what we know and what we really don't know.

This talk will precede a symposium on weeds in natural systems. The actual mechanisms of why weeds establish in an environment are often assumed and only occasionally tested. Yet this understanding may well provide important information that leads to appropriate management options. A plethora of hypotheses (e.g. enemy release, allelopathy, resource competition) have been put forward and some evidence has been collected globally for some weeds. In this symposium, we will develop a conceptual framework and gather the evidence together for Australasian weeds to assess what we currently know about the different ways that species 'win' space in natural landscapes.

Keywords Natural systems, competition, invasion mechanisms.

Global learnings for research impact in weed management in Australia

Rachel Melland
Spiegare Pty Ltd, Kambah, Australian Capital Territory 2902, Australia
(Rachel.Melland@spiegare.com.au)

Summary The Australian government is increasingly seeking evidence of a pathway to impact, and demonstration of impact, for research, as a measure of its return on invested taxpayer dollars. The government is seeking evidence of positive impact arising from research efforts, on society, the economy and the environment. This follows international efforts by countries such as the United Kingdom (UK), Canada and the Netherlands to do the same. In 2014, the UK Research Excellence Framework (REF) received 6679 case studies from UK Universities, demonstrating research impact on 'the economy, society, culture, public policy or services, health, the environment or quality of life, beyond academia', across individual and multiple disciplines.

Avenues of research impact can include results of collaborations with industry and community stakeholders. Demonstrating the impact of weed research, including management programs, is likely to become increasingly important in Australia. As such, active engagement in the development, or use of designated impact metrics from research idea to results adoption will become necessarily important, as will the development of case studies. This paper highlights some key international learnings for describing and/or measuring research impact as they relate to weed management.

Keywords Research impact, weed management, collaboration, industry, university, community, case studies, metrics, multidisciplinary.

INTRODUCTION

Sustainable weed management will be critical to global food security (Paini 2016, Westwood 2018) and natural ecosystem conservation (Pyšek *et al.* 2012) into the future. To be able to describe the value proposition of our weed research and management work to the public, as well as to policy makers and funders, the 'weeds community' needs to clearly articulate the positive and vital impacts we have on national and global production systems and the environment.

Increasingly, Australian research granting and investment bodies are seeking evidence of research impact (ARC 2018). This can either be retrospective, as is the case for the current ARC system, or as a forward forecast that sets out the business case for the research in competitive funding calls. This paper encompasses ideas for multiple uses of describing the positive real-world impact of weed research beyond academia.

There are many reasons for, and ways to assess, the complex mechanisms enabling the successful movement of knowledge from research to application and the delivery of tangible positive impact. For example, in utilising one such mechanism, (the Research Contributions Framework, Morton 2015), Morton (2017) and Morton and Casey (2017) found four lessons that could help maximise impact from future research. These would be equally applicable to weed research: '1. plan an impact strategy that addresses complexity from the start, and identifies key monitoring criteria, as well as risks and assumptions; 2. include stakeholder mapping to identify key actors who will be essential to taking recommendations forward, and allow time to build partnerships with them; 3. keep communication lines open from the start and throughout any project with partners and wider stakeholders; and 4. support and recognise key staff who act as knowledge-brokers.' This current paper investigates the ways authors of the 6679 Case Studies submitted to the 2014 United Kingdom (UK) Research Evaluation Framework (UK REF 2014) demonstrated positive impact on the environment, economy, society or technology from studies either directly about, or similar to, weed management research.

MATERIALS AND METHODS

The case studies in UK REF (2014) were searched employing the search terms 'weeds' and 'invasive alien species' in April 2018. Case studies identified under these search terms were evaluated and categorised into the pre-defined impact types of environmental, economic, societal or technological impact. Within these groups an analysis of actions leading to positive impact of the research in the real world was undertaken and is described.

RESULTS

A search of the UK REF (2014) database using the search term 'weeds' revealed 10 case studies, seven of which were directly relevant to weeds. Searches under the term 'invasive alien species' showed 617 case studies, of which case studies under the following 'Units of Assessment' (UoA) are most relevant to weeds: Biological Sciences, 104; Agriculture, Veterinary and Food Sciences, 42; and Earth Systems and Environmental Sciences, 54. Within each UoA, it is possible to search for 'Impact Type' and the Impact Types evaluated as having most relevance to weeds are listed below for each of the three UoAs searched (Table 1).

Learnings and ideas for demonstrating the positive impact of research into weeds have been taken from both the case studies directly and also other similar case studies in the search results (Table 1).

DISCUSSION

The UK REF (2014) case study authors described and measured the positive impact of their research on society, the environment, and on both via improved technological solutions, at scales ranging from local to global. Advice to all levels of government and key non-government bodies improving policy, planning, strategy and action planning likewise were described at levels from influencing conservation plans covering target habitats, to advice imbedded in national and international policies such as those of the United Nations. These actions have led to widespread improved conservation status and farming practices. A wide range of media and public engagement activities has also been shown to increase public understanding and interaction in the scientific/management research area leading again to improved conservation and land use practices. Research advice was found to be included into policy and planning as an impact type of all three of the following: social; environmental; and technological areas. The impact activities learned from the case studies have been illustrated here as they would relate to weed management.

Social Impacts Research that resulted in changes in international and national public understanding of the research area and increased public actions towards the goals of the research area. Demonstrating social impact can include a measurement/estimate of the number of people exposed to various forms of media. Media included film production such as on nature channels, public lectures, images on stamps and focus images for parks signage. Media use examples described evidence of public discussion and debate through twitter, blogs, sharing of, and comments on, articles. Public awareness was also enhanced through numbers of books written for non-specialist audiences, and translated into multiple languages. Encouraging 'citizen science', using interactive websites, was an avenue for measuring public use and increased understanding, especially if the site invited public involvement in a project, for example, to count or monitor species spread or reduction.

Educational activities at primary and secondary levels illustrated research that had been used as building blocks for educational resources, with free lesson plans and linked multimedia resources. These case studies showed the resources had been used by large percentage of schools and had reached a large number of pupils in a given timeframe. The studies highlighted when the education resources were used internationally. At the tertiary education level, authors described impact in the context of inspiring young people to study their research area.

Research was shown to have had a broad impact once it had been incorporated into key public policies, planning, strategies and action plans.

Technological (often Economic) Impacts Some examples included commercialisation of a new method or diagnostic or treatment, and often the establishment of a spin out company. Technological impacts included improved manufacturing processes, and reduction in manufacturing wastage, leading to reduction in management costs in particular or improved conservation over multiple land uses. For technological equipment, the case studies described the number of units sold,

Table 1. 'Impact Types' relevant to weeds found in a search under the terms 'Invasive Alien Species' in three 'Units of Assessment' (UoA) on the UK REF (2014) Case Studies searchable database in April 2018, and for some, the percentage directly related to weeds.

Impact types	Biological Sciences	Agriculture, Veterinary and Food Sciences	Earth Systems and Environmental Sciences
Environmental	55(22%)	12(0%)	43(26%)
Economic	3	1	3
Social	4	5	3
Technological, also often Economic	30(0%)	18(0%)	3

including if the new method of making them/program/algorithm had been taken up by other major/global companies, which increased its usefulness and made it a mainstream tool. Research was classified as having high impact when it was used globally or across a range of crops/weeds/land management practices. Widespread or global use of knowledge and patented techniques, was therefore claimed by authors as having impacts on food security, sustainable crop production and farming profitability.

Predictive technology for growers was described, for example to sense disease/weeds and plan planting or treatments, as having positive impacts on decreasing both crop loss and use of herbicides. Along with these useful impacts, the author of this current paper would suggest also, technology that differentiates crops, native animals, insects, or whole of landscape sensitivity to weed control disturbance leading to the positive impact of improved timing and type of control on farms, and improved conservation. Biopesticides research was shown to have led to commercialised products that help farmers (and potentially others) overcome pest resistance. Whether similar avenues exist for the production and use of bioherbicides remains to be seen.

A technological impact was also the existence of a 'legacy' of young scientists trained by industry. Close collaborations between researchers, government bodies and scientific societies meant that research has been used by policy makers within that particular sector. Research was shown to contribute to weed control guidelines, regulations and policy planning in the UK, as well as abroad.

Environmental Impacts Use of research by government and non-governmental public bodies to produce targeted management plans or prescriptions was an indicator of environmental impact. Research also provided supportive evidence for management advice and advocacy work, and to inform government legislation and policy that led to incorporation into high level planning. This included scientific data on negative effects on biodiversity and economic costs resulting in increased general understanding of why weeds are problem species and need to be considered in a policy, planning strategy and legislation context. Robust scientific evidence was described as being used to underpin eradication and control programs, especially in complex land use systems. Similarly, research predictions of increased weed threats under climate change were incorporated into policy and planning documentation of various levels of government and peak environment bodies, especially in areas under threat from weeds. Evidence of environmental impact was also presented as improved control techniques that reduce negative environmental effects of existing techniques.

Case studies measured the extent of a natural ecosystem that had been restored utilising outputs from a particular body of research and plans developed using the research. They showed that researchers had improved conservation by providing expert advice and guidance on survey methodologies, taxonomy and basic biology, particularly if this advice had been widely adopted, regionally, nationally or globally. An important avenue of impact was where research influenced the 'management options' put forward under government schemes and grants by public and community bodies applying for funds to undertake biodiversity and weed projects. Overseas research investigating degradation caused by multiple land uses delivered impact by informing sustainable development plans across industry and the environment, and had increased use of 'Fairtrade' or equivalent certifications. Research was also shown to have underpinned the international agreement, the 'ICUN Red List of Threatened Species'.

Research to improve protocols for growth and use of biocontrol agents led to practice change at institutions/industries that produce/sell biocontrol agents. New genetic diagnostic tools for pest plant identification were shown, that led to improved plant tracking. Case studies described the assessment of risk to native or other species resulting in restrictions on imports of species known to be invasive and development of licensing arrangements to regulate the use of invasive species if for commercial purposes, and as a mechanism to impose penalties for misuse.

Research utilised in the above ways therefore led to best practice management for the target species and for whole landscapes. These kinds of activities by weeds researchers can lead to weed management that conserves species and ecosystems. This type of impact was described as a paradigm shift in weed prevention and management methods across the landscape, across land uses and between multiple partners.

Gaps Economic analysis methods used to assess degradation caused by weeds and amelioration in different systems were not directly described, although some case studies stated they had made a large contribution to the public and governmental understanding of the economic impact of invasive species.

CONCLUSION

The Australian and global 'weeds community', including researchers, industry members, policy makers and community land managers or practitioners need to fully describe and measure, where possible, our efforts to improve weed biosecurity and management in Australia. As always, there is finite funding available from governments, industry and even individuals who volunteer their time, and this needs to be spent on the weed activities that result in the most positive outcomes for land conservation, farming practices, the amenities we all use, and the underlying environmental systems we all rely upon.

We need to embed the measurement and description of the positive impact of our weed research and management into our work. This will enable us to describe the value proposition of our work to funders, policy makers, planners and to the community in general. Our impact statements and examples will inevitably be used by funders for media commentary, and by others in social media posts. As such, impact measures need to be described with the intended audience and dissemination method in mind. This could build to create an improved cultural acceptance of the importance of weed biosecurity and management, which in turn will help those who are making strong weed management efforts in their industries, and in their policy and planning work. The increased literacy of our weed research community in defining our positive impacts gives us further license to operate. Therefore, if planning your research impact narrative becomes a core component of undertaking weed research work, it will give our community the mechanism to do more of the vital weed research that is needed to protect Australia's and the worlds ecosystems, farming systems and overall environments.

ACKNOWLEDGMENTS

Thank you to Cameron Begley, of Spiegare Pty Ltd for his continuing interest in the area of demonstrating the impact of research. Thank you also to all those who pursue a goal common to many of us, which is that our combined research and other work has meaning and use in the real world, as well as to those involved in basic research, who enable so much of what we know to be known.

REFERENCES

ARC, Australian Research Council http://www.arc.gov.au/ (accessed 1 August 2018).

Morton, S. (2015). Progressing research impact assessment: A 'contributions' approach, *Research Evaluation* 24, 405-19.

Morton, S. (2017). Bridging the gap between research and policy: recommendations for social science research in low/middle-income countries. http://blogs.lse.ac.uk/impactofsocialsciences/2017/11/06/bridging-the-gap-between-research-and-policy-recommendations-for-social-science-research-in-lowmiddle-income-countries/ (accessed 1 August 2018).

Morton, S. and Casey, T. (2017). Changing National policy on violence affecting children: An impact assessment of UNICEF and partners' multi-country study on the drivers of violence affecting children in Peru. University of Edinburgh. https://www.era.lib.ed.ac.uk/bitstream/handle/1842/22063/UNICEF%20impact%20assesment%20Peru.pdf?sequence=1&isAllowed=y (ACCESSED 1 August 2018).

Paini, D.R. Sheppard, A.W., Cook, D.C., De Barro, P.J., Worner, S.P. and Thomas, M.B. (2016) Global threat to agriculture from invasive species. *PNAS* 113, 7575-9.

Pyšek, P., Jarošík, V., Hulme, P.E., Pergl, J., Hejda, M., Schaffner, U. and Vilà, M. (2012). A global assessment of invasive plant impacts on resident species, communities and ecosystems: The interaction of impact measures, invading species' traits and environment. *Global Change Biology*. 18, 1725-37.

UK REF (2014). Search REF Impact case studies. http://impact.ref.ac.uk/CaseStudies/ (accessed 1 August 2018).

Westwood, J.H., Charudattan, R., Duke, S.O., Fennimore, S.A., Marrone, P., Slaughter, D.C., Swanton, C. and Zollinger, R. (2018). Weed management in 2050: Perspectives on the future of weed science. *Weed Science* 66, 275-85.

Emergence patterns and herbicide control of prickly lettuce (*Lactuca serriola* L.)

Hanwen Wu, Adam Shephard and Michael Hopwood
Wagga Wagga Agricultural Institute, Department of Primary Industries, Wagga Wagga 2650, New South Wales, Australia
(hanwen.wu@dpi.nsw.gov.au)

Summary Prickly lettuce (*Lactuca serriola* L.), an annual or biennial plant, is becoming an increasing problem in southern New South Wales; mostly in cereals and legume pastures. This study showed that eight prickly lettuce populations differed in their final cumulative emergence in a seedling tray trial, ranging from 39 to 341 plants m^{-2}, however, they had similar emergence patterns, with 69% emergence in late autumn and early winter, 27% in later winter, 2.4% in spring in the first year after burial and only 1.6% emergence between autumn and winter in the second year. Prickly lettuce in wheat also followed similar emergence patterns, but it had higher emergence (7.5%) between autumn and winter in the second year. In addition, the total emergence in wheat was only one tenth of that in the seedling trays (without competition). Glyphosate, 2,4-D amine + glyphosate, metsulfuron-methyl + glyphosate, glufosinate ammonium and fluroxypyr controlled mature prickly lettuce plants by 90–95%. Paraquat consistently achieved 100% control either used alone or as a 'double knock' treatment.

Keywords Herbicides, weeds, weed biology, weed management.

INTRODUCTION

Prickly lettuce (*Lactuca serriola* L.) is of Eurasian origin and is adapted to a summer-dry Mediterranean climate (Weaver and Downs 2003). It is distributed globally: in Australia; Europe; North Africa; North America; and West Asia (Prince *et al.* 1978). It has spread northwards since the beginning of the 19th century (D'Andrea *et al.* 2009).

In Australia, it was first recorded in 1899 in the Upper Hunter, New South Wales (NSW) and is now widely distributed across a number of states, particularly in NSW, Victoria and South Australia (ALA 2018). It is a common weed of fallows, roadsides, waste lands and gardens, and it has recently become an increasing problem in cereals and lucerne pastures in southern NSW.

Lactuca serriola can grow up to 2 m and is highly competitive with crops and pastures. If left uncontrolled, it competes for soil moisture and nutrients in summer. *Lactuca serriola* causes soybean yield losses by 10% at a density as low as 0.2–1.2 plants m^{-2} and by up to 80% at densities more than 50 plants m^{-2}. However, it did not cause significant yield loss in winter wheat (Weaver *et al.* 2006). Similarly, no yield losses of cereals or grain legumes due to prickly lettuce were reported in Australia, but grain quality and harvesting efficiency were severely compromised (Amor 1986a). Flowering buds are cut together with grain during harvest, resulting in grain contamination and reductions in value. In addition, cut stems release a milky sap which can clog the harvesting machinery and increase the moisture content of the grain (Amor 1986a).

Lactuca serriola can quickly build up its population due to its prolific production of wind-dispersed seeds that have little or no innate dormancy (Marks and Prince 1982). Seed production per plant ranged from 2200 to 67,000 in soybeans, and up to 87,000 in a non-crop area (Weaver *et al.* 2006). In Australia, Amor (1986b) estimated that prickly lettuce plants in crop stubble produced 48,000 seeds per plant while it produced 900 seeds per plant when in competition with a wheat crop.

The light-weight achene with a pappus is easily dispersed by wind and through surface run-off. The tall stem also facilitates dispersal (Weaver and Downs 2003). Optimum temperatures for germination are 12–24°C, with little or no germination at constant temperatures below 8°C (Marks and Prince 1982). Temperatures 26–35°C induced secondary dormancy (Marks and Prince 1982). However, Mikulka and Chodová (2003) reported that 95% germination was achieved at 30°C.

Lactuca serriola emerges predominantly in autumn, followed by limited emergence in spring (Amor 1986a, Marks and Prince 1982, Weaver *et al.* 2006). It forms a rosette of leaves after emergence and develops a strong taproot. Plants are difficult to control with herbicides once they start to elongate (Weaver and Downs 2003). Mechanical control is ineffective as the plant regrows after cutting or harvesting and progresses to set seeds (Amor 1986a, Mikulka and Chodová 2003).

The weed has evolved resistance to ALS inhibitor herbicides in both Australia and the United States of

America (USA), to synthetic auxins in USA in 2007, and most recently to EPSP synthase inhibitor in Australia in 2015 (Heap 2018).

Field emergence information of *L. serriola* is scarce in southern NSW. Effective herbicide control options are limited for mature plants. This study was conducted to compare emergence patterns of *L. serriola* with and without a growing crop, in the field, and to evaluate a range of herbicide options of distinct modes of action on mature *L. serriola* plants.

MATERIAL AND METHODS

Prickly lettuce field emergence Seeds of four populations of prickly lettuce were each collected from low and high rainfall areas in southern NSW in January 2016. On 1 February 2016, 150 seeds of each of the eight populations were placed in seedling emergence trays (34 cm × 28 cm) filled with a mixture of coco peat potting mix and garden soil (1:1). Seeds of each population were evenly spread across the tray surface and then lightly covered by the peat/soil mix. The trays were maintained in the field under natural conditions in Wagga Wagga.

Two populations (one each from the high and the low rainfall zones) were used to determine the emergence pattern in wheat. Glyphosate was used as a pre-sowing knockdown. Wheat cv. Corack was sown at 60 kg ha^{-1} on 14 June 2016 and on 8 June 2017, respectively, with a 6-row cone seeder at 23 cm row spacing using coulter knife points and press wheels. A basal fertiliser DAP was applied at 135 kg ha^{-1} each year at sowing. Prickly lettuce seeds (1000 seeds m^{-2}) were sown in plots (3 × 1.3 m^2) on 15 April 2016. Any surviving or late emerging plants of prickly lettuce were counted and manually removed to prevent any new seed replenishment.

Field emergence of prickly lettuce was recorded at monthly intervals from April 2016 to December 2018 in both trials.

Herbicide options Two trials were established after wheat harvest in December 2015, one at Lake Cowal and the other at Temora, NSW. Average prickly lettuce density was 9.5 plants m^{-2} at Lake Cowal and 10 plants m^{-2} at Temora. The plot size was 2 × 16 m. A total of 17 and 19 treatments were imposed, respectively, at Lake Cowal and Temora sites, each including an untreated control. After the initial application, each plot was equally sub-divided with one half of the plot receiving an additional application of paraquat as a 'double knock' treatment.

Herbicides were applied using a hand-held boom sprayer, calibrated to deliver 100 L ha^{-1} at 2 bar pressure. The first application of herbicides was applied at the Lake Cowal and Temora on 10 and 15 December, respectively, and the 'double knock' paraquat application (600 g a.i. ha^{-1}) at Lake Cowal and Temora on 16 and 21 December 2015, respectively. At the time of herbicide applications, prickly lettuce plants were at the elongation/re-branching stage after being cut during crop harvest and not under moisture stress.

A visual rating (% of control) in the single knock treatments was conducted on 8 January 2016. The number of surviving prickly lettuce plants was also recorded in a 1 × 6 m strip in each plot on 2 February 2016.

Design and statistical analysis The experiments were designed as a randomised complete block with four replications in the two emergence trials. A split plot design with three replications was used for in the two herbicide trials. The main plot was the absence and presence of the 'double knock' paraquat treatment and the subplot was the single knock treatments. An ANOVA was conducted using Genstat.

RESULTS AND DISCUSSION

Emergence patterns of prickly lettuce in the field Prickly lettuce populations differed in field emergence patterns and there were three distinct groups (Figure 1).

The first group had two populations (BARM 01 and WAGG 12), and had the highest cumulative emergence between April and August (323–340 plants m^{-2}) in the first year and no further emergence in the second year. The second group had five populations, which had the major emergences between April and October (170–259 plants m^{-2}), followed by limited emergences between June and September in the second year. The final group had one population (L-COW 06P) and had the lowest emergence, which occurred between April and June (39 plants m^{-2}) in the first year, followed by little emergence afterwards. It is not clear why this population had low cumulative emergence as this population had a high germination (97.8%) tested soon after collection (data not shown). The average of the eight populations, about 69% emergence occurred in late autumn and early winter, 27% in late winter and 2.4% in spring in the first year after burial, with only 1.6% emergence between autumn and winter in the second year.

The two prickly lettuce populations from different rainfall zones also differed in the emergence patterns in the wheat crop, with higher emergence between April and August 2016 in the high rainfall population COOT 07 when compared to the low rainfall population BARM 01 (Figure 2). Both populations also had limited emergence between May and September in

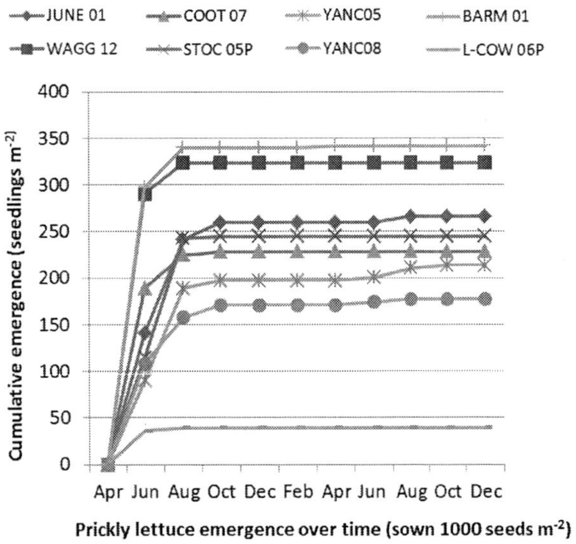

Figure 1. Emergence of prickly lettuce in seedling trays without crop competition under field conditions from April 2016 to December 2017. The number of seeds sown onto each tray was expressed as seeds m^{-2}.

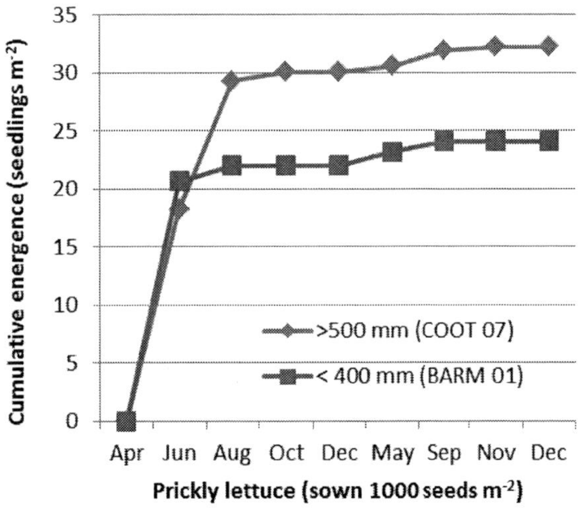

Figure 2. Emergence of prickly lettuce in a wheat crop grown under field conditions from April 2016 to December 2017 (the populations COOT07 and BARM 01 were collected from high and low rainfall zones, respectively).

the second year after sowing. On average, there was 69% emergence in late autumn and early winter, 22% in later winter and 1.4% in spring in the first year after burial, with 7.5% emergence between autumn and winter in the second year. The final cumulative emergence at 18 December 2017 was 32 and 24 plants m^{-2} for the high and low rainfall populations, respectively, while the final cumulative emergence for the two populations (COOT07 and BARM 01) in the above seeding tray trial in the absence of crop competition was 228 and 341 plants m^{-2}, respectively. These results indicate that the emergence of prickly lettuce in a standing crop is only one tenth of that in a bare ground.

The emergence patterns with and without a competing crop in this study support the findings of Prince *et al.* (1978) who reported that prickly lettuce emerged predominantly in autumn followed by limited emergence throughout winter, that a second emergence cohort occurred in spring and that no emergence occurred in summer due to the dry soil conditions. Prickly lettuce management should therefore focus on the major emergence cohorts in autumn prior to elongation stage to achieve effective control. A fallow control soon after harvest is also necessary to control the spring emergence cohort.

Effects of herbicides on prickly lettuce in summer fallows At Lake Cowal site, a single knock application of herbicides differed significantly in controlling prickly lettuce (Table 1). Five treatments, glyphosate, 2,4-D amine + glyphosate, metsulfuron-methyl + glyphosate, glufosinate ammonium and fluroxypyr achieved good control with more than 90% mortality, while the remaining 11 treatments only controlled 30–88% of prickly lettuce. The follow-up 'double knock' treatment with paraquat at 600 g a.i. ha^{-1} provided 100% control on prickly lettuce, even in the untreated plots which did not have the first knock of herbicide applications (data not shown).

The control of the single knock herbicide application at Temora was generally less effective than at Lake Cowal. Only four treatments had >80% control efficacy including 2,4-D amine + glyphosate, a prepacked mixture of 2,4-D amine and picloram + glyphosate, glufosinate ammonium and a prepacked mixture of amitrole (250 g L^{-1}) and ammonium thiocyanate (220 g L^{-1}). The other 14 treatments resulted in poor control of prickly lettuce (7–72%, Table 2). The follow-up 'double knock' treatment with paraquat also provided near 100% control on prickly lettuce, with less than 0.1 surviving plant m^{-2}.

In general, no single treatments, except paraquat achieved 100% control of mature prickly lettuce plant after crop harvest. Many plants, even though severely damaged, managed to survive and re-branch. The paraquat application was the only outstanding treatment, achieving complete control of the mature prickly lettuce plants.

ACKNOWLEDGMENTS

The authors gratefully acknowledge the financial support of Grains Research and Development Corporation for funding this research.

Table 1. Herbicide control efficacy on mature prickly lettuce plants at Lake Cowal.

Treatment	Rate (mL or g ha^{-1})	Adjuvant	Single_knock[A]	
			Visual rating[A] (%)	Density (plants m^{-2})[B]
2,4-D amine (700 g L^{-1})	1150 mL	Liase at 2%	30.0	3.7
2,4-D amine + glyphosate (540 g L^{-1})	515 mL + 1300 mL	LI700 at 0.3%	93.3	0.6
Glyphosate	1300 mL	LI700 at 0.3%	90.0	2.0
Fluroxypyr (333 g L^{-1})	600 mL	Uptake 0.5%	95.0	0.8
Fluroxypyr + glyphosate	600 mL + 1300 mL		88.3	2.4
Metsulfuron-methyl (600 g kg^{-1}) + glyphosate	7 g + 1300 mL		95.0	1.6
Oxyfluorfen (240 g L^{-1}) + glyphosate	75 mL + 1300 mL	BS1000 at 1%	86.7	2.1
Dicamba (500 g L^{-1}) + glyphosate	240 mL + 1300 mL		85.0	4.5
Glufosinate ammonium (200 g L^{-1})	4000 mL		91.7	1.8
Mixture of amitrole (250 g L^{-1}) and ammonium thiocyanate (220 g L^{-1})	5600 mL	LI700 at 0.3%	85.0	1.6
Mixture of 2,4-D amine (300 g L^{-1}) and picloram (75 g L^{-1})	700 mL		53.3	3.9
Mixture of MCPA (250 g L^{-1}) and diflufenican (25 g L^{-1})	1000 mL		81.7	3.7
Carfentrazone-ethyl (240 g L^{-1}) + MCPA amine (750 g L^{-1})	100 mL + 333 mL		76.7	2.5
Clopyralid (300 g L^{-1}) + LVE MCPA (570 g L^{-1})	150 mL +1000 mL		63.3	2.4
LVE MCPA	1000 mL		40.0	4.3
Mixture of aminopyralid (10 g L^{-1}) and fluroxypyr (140 g L^{-1}) + LVE MCPA	750 mL +1000 mL		86.7	3.5
Control			0.0	6.2
$l.s.d_{0.05}$			17.93	1.89

REFERENCES

ALA, Atlas of Living Australia (2018). http://www.ala.org.au (accessed 10 April 2018).

Amor, R.L. (1986a). Incidence and growth of prickly lettuce (*Lactuca serriola* L.) in dryland crops in the Victorian Wimmera. *Plant Protection Quarterly* 1, 148-51.

Amor, R.L. (1986b). Chemical control of prickly lettuce (*Lactuca serriola*) in wheat and chickpeas in the Victorian Wimmera. *Plant Protection Quarterly* 1, 103-5.

D'Andrea, L., Broennimann, O., Kozlowski, G., Guisan, A., Morin, X., Keller-Senften, J. and Felber, F. (2009). Climate change, anthropogenic disturbance and the northward range expansion of *Lactuca serriola* (Asteraceae). *Journal of Biogeography* 36, 1573-87.

Heap, I. (2018). The international survey of herbicide resistant weeds (www.weedscience.com) (accessed 12 April 2018).

Marks, M.K. and Prince, S.D. (1982). Seed physiology and seasonal emergence of wild lettuce *Lactuca serriola*. *Oikos* 38, 242-9.

Mikulka, J. and Chodová, D. (2003). Germination and emergence of prickly lettuce (*Lactuca serriola* L.) and its susceptibility to selected herbicides. *Plant, Soil and Environment* 49, 89-94.

Prince, S.D., Marks, M.K. and Carter, R.N. (1978). Induction of flowering in wild lettuce (*Lactuca serriola* L.). *New Phytologist* 81, 265-77.

Weaver, S.E. and Downs, M.P. (2003). The biology of Canadian weeds, 122: *Lactuca serriola* L. *Canadian Journal of Plant Science* 83, 619-28.

Weaver, S., Cluney, K., Downs, M. and Page, E. (2006). Prickly lettuce (*Lactuca serriola*) interference and seed production in soybeans and winter wheat, *Weed Science* 54, 496-503.

Table 2. Herbicide control efficacy on mature prickly lettuce plants at Temora.

Treatment	Rate (mL or g ha^{-1})	Adjuvant	Visual rating[A] (%)	Density (plants m^{-2})[B]	
				Single_ knock[A]	Double_ knock[A]
2,4-D amine (700 g L^{-1})	1150 mL	Liase at 2%	51.7	4.6	0.0
2,4-D amine + glyphosate (540 g L^{-1})	515 mL + 1300 mL	LI700 at 0.3%	80.0	1.2	0.1
Glyphosate	1300 mL	LI700 at 0.3%	78.3	4.2	0.0
Metsulfuron-methyl (600 g kg^{-1}) + glyphosate	7 g + 1300 mL		28.3	5.5	0.0
Oxyfluorfen (240 g L^{-1}) + glyphosate	75 mL + 1300 mL	BS1000 at 1%	60.0	8.7	0.0
Dicamba (500 g L^{-1}) + glyphosate	240 mL + 1300 mL		71.7	2.6	0.0
Fluroxypyr (333 g L^{-1})	600 mL	Uptake at 0.5%	53.3	6.8	0.1
Fluroxypyr + glyphosate	600 mL + 1300 mL		63.3	5.7	0.1
Mixture of 2,4-D amine (300 g L^{-1}) and picloram (75 g L^{-1}) + glyphosate	700 mL + 1300 mL		81.7	2.6	0.1
Triclopyr (600 g L^{-1}) + glyphosate	700 mL + 1300 mL	Uptake 0.5%	53.3	5.8	0.1
Saflufenacil (700 g kg^{-1}) + glyphosate	34 g + 1300 mL	Bonza at 1%	53.3	5.0	0.1
Glufosinate ammonium (200 g L^{-1})	4000 mL		83.3	1.3	0.1
Mixture of amitrole (250 g L^{-1}) and ammonium thiocyanate (220 g L^{-1})	5600 mL	LI 700 at 0.3%	81.7	4.9	0.1
Mixture of 2,4-D amine (300 g L^{-1}) and picloram (75 g L^{-1})	700 mL		30.0	6.1	0.1
Carfentrazone-ethyl (240 g L^{-1}) + MCPA amine (750 g L^{-1})	100 mL + 333mL		6.7	5.4	0.1
Clopyralid (300 g L^{-1}) + LVE MCPA (570 g L^{-1})	150 mL +1000 mL	Uptake at 0.5%	30.0	6.3	0.0
LVE MCPA	1000 mL		28.3	5.7	0.0
Mixture of aminopyralid (10 g L^{-1}) and fluroxypyr (140 g L^{-1}) + LVE MCPA	750 mL +1000 mL		58.3	5.8	0.0
Control			0.0	7.4	0.0
l.s.d$_{0.05}$			21.6	3.9	0.3

[A] Single knock treatments applied on 15 December 2015 and the 'Double knock' paraquat on 21 December 2015, followed by a visual rating in the single-know treatments conducted on 08 January 2016.
[B] The surviving prickly lettuce plants were recorded on 02 February 2016.

The effect of integrating defoliation with herbicides on barley grass survival in a legume pasture

Jane E. Kelly[1,2], William Brown[1,2], John Broster[1,2], Paul A. Weston[1,2], Wayne Robinson[3], Jane Quinn[1,2] and Leslie A. Weston[1,2]
[1] Charles Sturt University, Locked Bag 588, Wagga Wagga, New South Wales 2678, Australia
[2] Graham Centre for Agricultural Innovation, Charles Sturt University, Locked Bag 588, Wagga Wagga, New South Wales 2678, Australia
[3] QCU, Research Office, Charles Sturt University, Locked Bag 588, Wagga Wagga, New South Wales 2678, Australia
(janekelly@csu.edu.au)

Summary Barley grass (*Hordeum* spp.) is an annual weed of southern Australian mixed farming regions causing significant crop revenue loss and carcass damage in grazing sheep. Increased reliance on herbicides for management of barley grass has led to the development of herbicide resistance across southern Australia. Integrated weed management tactics have recently been used to manage barley grass populations, and have proven effective in reducing reliance on herbicides for management. Field studies performed in 2016 and 2017 examined the use of post-emergent herbicides and strategic defoliation through mowing on a barley grass infestation in a mixed legume pasture in Wagga Wagga, NSW. This paper presents data from the first year of the study. Significant differences between herbicide-only treatments revealed propaquizafop to be 99% effective in reducing barley grass survival and seed production in contrast to paraquat, which was generally ineffective. A significant interaction between mowing and date also influenced barley grass biomass, and complex interactions between herbicide, mowing and date influenced barley grass height. Barley grass control, as a result of propaquizafop application, was accompanied by an increase in the biomass of clover and other weeds including *Vulpia* spp., also a known carcass contaminant.

Keywords *Hordeum*, *Vulpia*, mowing, carcass contamination, integrated weed management.

INTRODUCTION

Historically, both pre- and post-emergent herbicides have been used extensively for management of barley grass (*Hordeum* spp.), a problematic weed of Australian crops and pastures, which is also associated with significant carcass damage in sheep across the Australian mixed farming zone (Kelly *et al.* 2018). Herbicide resistance has recently been noted in several barley grass populations in Western Australia and South Australia (Owen *et al.* 2012, Shergill *et al.* 2015), suggesting that IWM strategies for management are likely to be particularly useful in regions where herbicide resistance is more prevalent. Past reports have suggested that defoliation by mowing was an effective alternative tactic to herbicides in controlling annual pasture weeds, with success relying on the timing of mowing and the species targeted (Bowcher 2002). When mowing is timed to coincide with the reproductive phase of the target annual weed, a reduction in seed rain and significant shifts in botanical composition have been shown to occur (Bowcher 2002). Reductions in barley grass fecundity were also noted when plants were defoliated to a 5 cm height (El-Shatnawi *et al.* 1999) at the boot stage of growth (Zadok *et al.* 1974). Furthermore, mowing barley grass plants as the inflorescences turned colour resulted in smaller seeds (Smith 1968b), likely reducing seed viability. The combined efficacy of the integration of herbicide usage and mowing in controlling barley grass populations has not yet been investigated.

This study thus investigated the effect of integrating defoliation by mowing with the application of two commonly used graminicides, propaquizafop (Group A) and paraquat (Group L), to reduce barley grass survival within a heavily-infested standing legume crop in southern New South Wales (NSW). The integration of both tactics were compared to their use in isolation after 12 months.

MATERIALS AND METHODS

The experiment was conducted in Wagga Wagga NSW at Charles Sturt University's Ashmont field research site (35.0578°S, 147.3544°E, elevation 215 m). Soil was a red silty loam sodosol. The paddock was previously sown in 2012 to lucerne and a mixture of other legumes. A yearly application of superphosphate was applied from 2012 to 2016.

Plots measured 6 × 4 m with 2 m spacing between plots. Treatments consisted of an untreated control, two herbicide treatments, two mowing treatments, and various combinations of herbicide and mowing.

Propaquizafop at 100 g a.i. L^{-1} (as Shogun®), was applied at a rate of 250 mL ha^{-1} in autumn during tillering. Alcohol alkoxylate at 1000 g a.i. L^{-1} (as Cropwett 1000®) was added at a rate of 250 mL ha^{-1} as a wetting agent. Paraquat at 250 g a.i. L^{-1} (as Sinmosa 250®) was applied at a rate of 1.6 L ha^{-1} during rapid growth of barley grass in late winter. The single mowing treatment slashed the plots to an average height of 7.5 cm when 50% of the barley grass plants across each plot reached boot stage of growth (Zadoks 49) (Zadoks *et al.* 1974). The second mowing treatment was a repeated defoliation, when 50% of the barley grass regrowth across the plot reached boot stage of growth. If no barley grass regrowth was observed, the second mowing was not performed. Mown material was not removed. Each of the treatments was randomised in a strip plot design for ease of management and replicated five times.

Data collection was performed after all treatments were imposed by comparison with the untreated control, and all values were assessed on a square metre basis. Due to heavy rainfall events in winter and spring in 2016, treatments were applied when weather conditions allowed for optimal application.

Biomass of main species (barley grass, *Medicago sativa* L. (lucerne), mixed clover species, *Vulpia* spp. and other weed species) was collected by cutting to ground level, drying and separating herbage fractions within two random quadrats per plot (25 × 25 cm). The number of barley grass inflorescences m^{-2} was also obtained from biomass samples at plant maturity. Seed number per inflorescence was calculated by averaging the number of seeds contained within 20 random inflorescences per plot at plant maturity, where possible. Fecundity was determined by multiplying inflorescence number m^{-2} in each plot by the mean seed number per inflorescence. Seed mass was calculated as the mass in grams of 100 randomly collected mature seeds taken from each plot at plant maturity.

Plant density was determined by counting the number of barley grass plants within a 10 × 10 cm area representative of the species composition within each of two random 25 × 25 cm quadrats per plot. Barley grass plant height was assessed fortnightly, and measured from ground level to the tip of the inflorescence, at three equally spaced and random points along a diagonal transect and averaged across each plot. The experiment was concluded at the time of seed fall in November 2016.

Statistical analysis Data were subjected to a factorial ANOVA for a strip plot experimental design using IBM SPSS software, version 20 (SPSS 2011). Assumptions of normality and homoscedasticity were investigated using residual plots and when violated, the data were transformed using square root or log transformations as necessary to meet the assumptions. A comparison of means was performed using the least significant difference (LSD) test (P<0.05). Pooled standard errors are presented.

RESULTS

Herbicide effects Herbicide treatment resulted in significant differences in barley grass (F = 40.272, df = 2, P<0.001), clover (F = 9.177, df = 2, P<0.001) and weed biomass (F = 64.493, df = 2, P<0.001), as well as barley grass inflorescence number (F = 16.155, df = 2, P<0.05), plant density (F = 62.155, df = 2, P<0.001) and fecundity (F = 15.04, df = 2, P<0.05). No significant treatment effects or interactions were observed for lucerne biomass or barley grass seed weight (P>0.05). Propaquizafop in particular significantly reduced barley grass biomass, inflorescence number, fecundity and plant density by at least 99% compared to both the untreated control and the paraquat treatment (Figure 1). No significant differences were observed between the control and paraquat application for barley grass establishment and fecundity. Conversely, after barley grass was completely controlled by the application of propaquizafop, the biomass of clover was increased by 79.4% and other weeds by 128%. In contrast, the application of paraquat significantly reduced the biomass of clover and other weeds by 65% and 36.5%, respectively, in relation to the control (Figure 1). The eradication of barley grass plants by propaquizafop resulted in a lack of production of barley grass seed and no seed rain in plots in late 2016. No significant differences in barley grass fecundity were observed between the control (33602.5 ± 4856.2 seeds m^{-2}) and paraquat-treated plots (31474.2 ± 4856.2 seeds m^{-2}).

Mowing*date effects Barley grass biomass was significantly reduced (F = 4.843, df = 2, P<0.05) in the one-mow treatment during October (160.8 ± 38.3 g m^{-2}) and also in the repeat-mow treatment during November (120.3 ± 38.3 g m^{-2}) when compared to the zero-mow treatment in November (349.2 ± 38.3 g m^{-2} (Figure 2).

DISCUSSION

A single application of propaquizafop was most effective in reducing the biomass, density and reproductive capacity of barley grass compared to all other treatments one year following application (Figure 1). This was not unexpected, given the reported efficacy of other Group A herbicides in managing barley grass

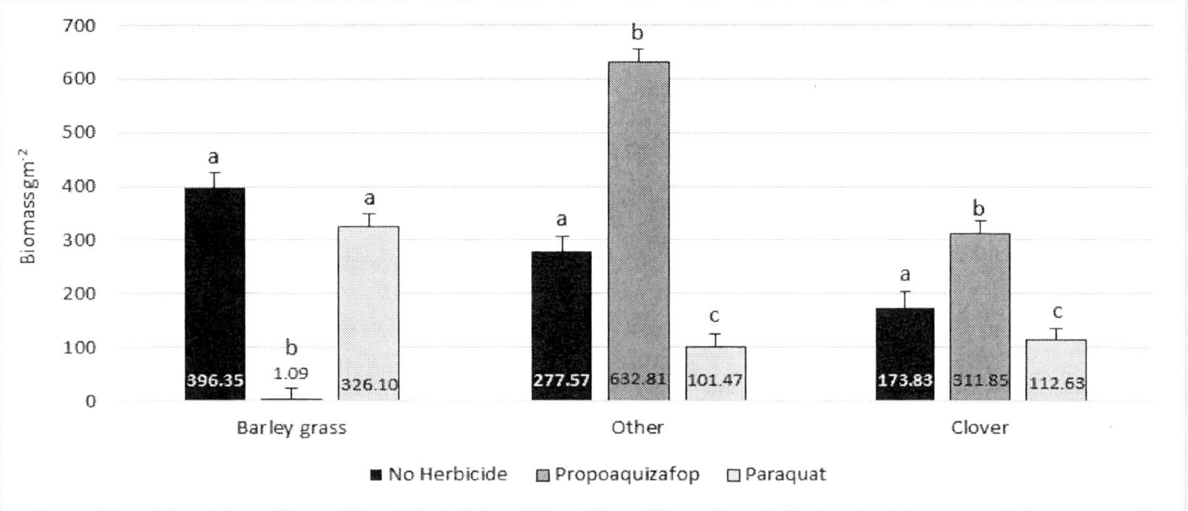

Figure 1. Mean barley grass, other weed and clover biomass (g m^{-2}) following herbicide application during 2016. Separate analyses were conducted across treatments within each species group. Bars with the same letter are not significantly different (P<0.05). Numbers within bars signify treatment means.

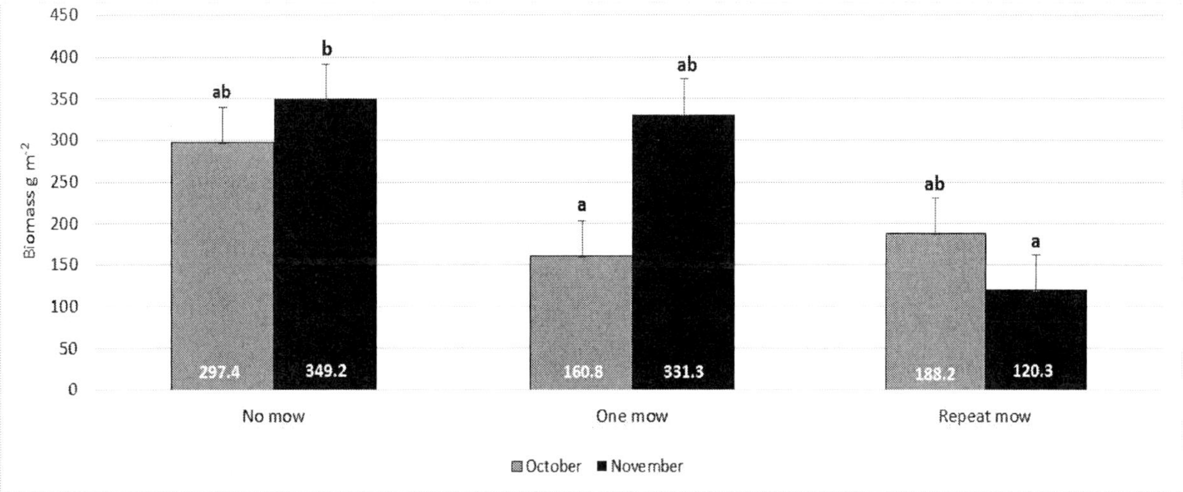

Figure 2. Mean barley grass biomass (g m^{-2}) as influenced by mowing during the months of October and November 2016. Bars with different letters are significantly different (P<0.05). Numbers within and above bars signify treatment means.

populations (Stephenson and Mitchell 1993). Further, the increase in clover and other accompanying weeds such as *Vulpia* spp. and capeweed was consistent with previous findings, where the ingress of capeweed was observed to accompany an increase in clover following applications of propyzamide (Thorn and Perry 1987). It is likely that the rapid and virtual eradication of barley grass in the current study provided spatial gaps in existing vegetation, allowing the introgression of legume and weed species. Weed species proliferation was likely aided by the availability of soil nitrogen and phosphorus resulting from the long-standing presence of legumes and the regular history of annual superphosphate applications, conditions known to create an environment well suited for encouraging proliferation of these other weeds (Rossiter 1964). Furthermore, subsequent to barley grass removal, the conditions present in the paddock may have also stimulated a competition response in the legume component of the crop.

Less efficacious control of barley grass by paraquat (Figure 1) was consistent with results of Stephenson and Mitchell (1993), who reported similar results in clover pastures after 12 months. Paraquat is a common Group L herbicide used in the control of grass weeds in crops (Owen *et al.* 2012). As a contact herbicide, it requiries sufficient leaf area for high efficacy (Fuerst and Vaughn, 1990). It is likely that the timing of late spring paraquat applications may have resulted in reduced herbicide efficacy due to shading from more mature plants, thereby preventing adequate contact of herbicide with barley grass foliage. The late paraquat application also reduced clover biomass in comparison to the control and propaquizafop treatments, a result also consistent with previous studies (Barrett *et al.* 1973).

Despite the lack of efficacy of mowing in reducing barley grass biomass during October, mowing was effective in November, and repeat mowing treatments resulted in significantly lower barley grass biomass than the control. As mowing results were inconsistent over time, it is possible that mowing at an even later barley grass maturity stage would have been be more effective, particularly when rainfall events were limited (Butterfield and Malmström 2009). Additional experimentation performed under more typical rainfall conditions will be useful in evaluating the use of mowing in conjunction with herbicide application for reduction in barley grass fecundity.

CONCLUSIONS AND FUTURE RESEARCH

Propaquizafop treatment was more effective in controlling barley grass 12 months following application than various combinations of herbicide and mowing. Despite the virtual eradication of barley grass by propaquizafop, the pasture was infested by later emerging weeds such as *Vulpia* spp. (Collins *et al.* 2013). Data obtained from additional years of experimentation will be useful in determining the success of propaquizafop and other herbicides in sustaining pasture composition and may offer further insight into the value of integrating a timely mowing with herbicide application for barley grass control over the longer term

ACKNOWLEDGMENTS

The authors acknowledge the funding support of Meat and Livestock Australia, the AW Howard Memorial Trust, the Grains Research and Development Corporation and the Graham Centre for Agricultural Innovation (Charles Sturt University and Department of Primary Industries). Authors also gratefully acknowledge the assistance of Allison Chambers and Graeme Heath in the collection of data and application of treatments.

REFERENCES

Barrett, D., Arnold, G. and Campbell, N. (1973). Effects of time and rate of paraquat application on yield and botanical composition of annual pastures containing subterranean clover in a Mediterranean climate. *Australian Journal of Experimental Agriculture* 13, 556-62.

Bowcher, A.J. (2002). Competition between temperate perennial pasture species and annual weeds: the effect of pasture management on population dynamics and resource use. (PhD), Charles Sturt University, Australia.

Butterfield, H. and Malmström, C. (2009). The effects of phenology on indirect measures of aboveground biomass in annual grasses. *International Journal of Remote Sensing* 30, 3133-146.

Collins, A., Fitzsummons, P., Behrendt, K., Camac, M., McCormick, L., Latta, R., John, C., Sheehan, M., Prance, T., Humphries, A., Hooke, T., McEachern, S., Jolly, S., Dickson, H., Shands, C., James, R. and Evans, N. (2013). Winning against seeds: Management tools for your sheep enterprise (1st ed.). Australia: Meat and Livestock Australia Ltd.

El-Shatnawi, M.J., Ghosheh, H.Z., Shannag, H.K and Ereifej, K.I. (1999). Defoliation time and intensity of wall barley in the Mediterranean rangeland. *Journal of Range Management* 52, 258-62.

Fuerst, E.P. and Vaughn, K.C. (1990). Mechanisms of paraquat resistance. *Weed Technology* 4, 150-6.

Kelly, J.E., Quinn, J.C., Loukopoulos, P., Broster, J.C., Behrendt, K. and Weston, L.A. (2018) Seed contamination in sheep: new investigations into an old problem. *Animal Production Science.* https://doi.org/10.1071/AN17771

Owen, M., Goggin, D. and Powles, S. (2012). Identification of resistance to either paraquat or ALS-inhibiting herbicides in two Western Australian *Hordeum leporinum* biotypes. *Pest Management Science* 68, 757-63.

Rossiter, R. (1964). The effect of phosphate supply on the growth and botanical composition of annual type pasture. *Crop and Pasture Science* 15, 61-76.

Shergill, L.S., Malone, J., Boutsalis, P., Preston, C., and Gill, G. (2015). Target-site point mutations conferring resistance to ACCase-Inhibiting herbicides in smooth barley (*Hordeum glaucum*) and hare barley (*Hordeum leporinum*). *Weed Science* 63, 408-15.

Smith, D. (1968a). The growth of barley grass (*Hordeum leporinum*) in annual pasture. 2. Growth comparisons with Wimmera ryegrass (*Lolium rigidum*). *Australian Journal of Experimental Agriculture* 8, 484-90.

Smith, D. (1968b). The growth of barley grass (*Hordeum leporinum*) in annual pasture. 4. The effect of some management practices on barley grass content. *Australian Journal of Experimental Agriculture* 8, 706-11.

Stephenson, D. and Mitchell, G. (1993). Barley grass control with herbicides in subterranean clover pasture. 1. Effect on pasture in the year of spraying. *Animal Production Science* 33, 737-41.

SPSS (2011) IBM Corp. Released 2011. IBM SPSS Statistics for Windows, Version 20.0. Armonk, NY: IBM Corp.

Thorn, C. and Perry, M. (1987). Effect of chemical removal of grasses from pasture leys on pasture and sheep production. *Animal Production Science* 27, 349-57.

Zadoks, J., Chang, T. and Konzak, C. (1974). A decimal code for the growth stages of cereals. *Weed Research* 14, 415-21.

Performance of INTERCOM for predicting rice-barnyard grass interference in dry seeded rice systems

Tahir Hussain Awan[1] and Bhagirath Singh Chauhan[2]
[1] Research Officer/Weed Scientist, Weed Science, Agronomy Department, Rice Research Institute, Kala Shah Kaku, Lahore 54000, Punjab, Pakistan
[2] The Centre for Plant Science, Queensland Alliance for Agriculture and Food Innovation (QAAFI), The University of Queensland, Gatton 4343, Queensland, Australia
(tahirawanrri@gmail.com)

Summary Crop yields on-farm rarely approach their production potential, partly as a result of growth reduction due to weed interference. Cost-effective weed management requires accurate estimates of yield and the potential yield loss resulting from weed infestations. However, crop yield and the effects of weeds are highly variable across weed density and their time of emergence. This may be accomplished through early crop vigour or delaying the weed emergence. An ecophysiological model of rice-barnyard grass competition (INTERCOM) may be useful for predicting the effects of weed density and their time of emergence on crop and weed growth and competitive ability. To evaluate the ability of the model to predict rice growth and yield, the effect of barnyard grass interference on rice yield loss was evaluated using two season data sets collected at the IRRI experimentation field. Predicted and observed monoculture leaf-area-index, leaf, stem, aboveground, and panicle biomass of both rice and barnyard grass were in close agreement. For inter-competition between species, height is very important, and the model captured the height very well. The normalised deviation for all values was near zero, which means the model calibration efficiency is good. Model-simulated results were very much similar to our field-observed results. Predicted and observed weed-free rice yields were ranged from 6.42–7.47 t ha^{-1}. The simulated results depicted that rice panicle yield was affected by the increasing weed density at early emergence. As weed emergence was delayed, there were no observed effects of different densities as the weed was usually unable to survive and did not cause much reduction in panicle yield. Percentage yield reduction decreased with the decrease in weed density and the delay in their emergence. Barnyard grass at 70 plants m^{-2}, emerging 2 days after rice emergence (DARE), reduced the grain yield by 65–70%. When the weed emergence was delayed to 45 DARE, the reduction in grain yield was only 2–5%. This reduction in yield was not so much different from the reduction caused by the weed density of 10 plants m^{-2} (0.7–3.2%) having the same emergence time as with rice. The model predicted that barnyard grass emergence with any densities at 45 DARE had negligible effect on rice growth and yield. The model suggests that the use of competitive rice cultivars or delaying the weed emergence may reduce the need for chemical weed control.

Keywords Economic threshold, integrated weed management, weed ecology, IPM, weed-crop interference.

Effect of high temperature and soil moisture on seed germination of four summer weeds

Sudheesh Manalil and Bhagirath S. Chauhan
The Centre for Crop Science, Queensland Alliance for Agriculture and Food Innovation (QAAFI),
The University of Queensland, Gatton, Queensland 4343, Australia
(s.manalil@uq.edu.au)

Summary Heat waves are increasingly common in Australia and may affect the germination of weeds. The effect of elevated temperature on bladder ketmia, sesbania pea, button grass and windmill grass was assessed in a factorial experiment (with three replications) with four levels of temperature (ambient, 45, 55 and 65°C), two levels of soil moisture (dry and 50% field capacity) and two burial depths (surface and 2 cm). Thirty seeds of two populations of each weed species from high and medium rainfall regions of Queensland were used in this study. Soil was filled in trays and seeds enclosed in nylon bags were placed at two soil depths. One set of trays was moistened to 50% of field capacity. Seeds were incubated for 4 hours in a hot air oven set at the required temperature and germination was assessed in the laboratory. Thermochron loggers were used to monitor the soil and oven temperature. Germination of button grass and wind mill grass was unaltered by the treatments; however, a significant increase in germination of sesbania pea compared to the ambient temperature condition (25–28% germination) was observed at surface layers incubated at 45°C and for all the treatments maintained at 55 and 65°C. Maximum germination (67%) was observed for the moist soil at 2 cm maintained at 65°C. Bladder ketmia germination was improved only for the moist soil at 2 cm incubated at 65°C. The studies indicate that heatwaves may alter the germination and emergence of weeds.

Keywords Biology, climate change, heat waves.

Specificity of LAMP for genetic diagnostics of Chilean needle grass and serrated tussock

David Gopurenko, Aisuo Wang and Hanwen Wu
Department of Primary Industries, PMB, Wagga Wagga, New South Wales 2650, Australia
Graham Centre for Agricultural Innovation (an alliance between NSW Department of Primary Industries and Charles Sturt University), Locked Bag 588, Wagga Wagga, New South Wales 2678, Australia
(david.gopurenko@dpi.nsw.gov.au)

Summary Chilean needle grass (*Nassella neesiana* [Trin. & Rupr.] Barkworth) and serrated tussock (*Nassella trichotoma* [Nees] Hack. ex Arechav.) are related Weeds of National Significance (WoNS) undergoing rapid range expansion in Australia. They are difficult to distinguish from native tussock grasses and emergences of these weeds often remain undetected until after they have established as difficult to control populations. Genetic diagnostics using loop-mediated isothermal amplification (LAMP) may provide a simple cost-efficient means for rapid identification of these weeds, under laboratory and field conditions. Provision of LAMP diagnostics would assist in minimising the time-lag to initial recognition of tussock weed outbreaks in new areas, and subsequently improve the time to response for eradication of the weeds where they are recently emergent.

Here we report laboratory development and tests of seven LAMP assays incorporating novel primer suites matched to species-specific DNA barcode sequences of Chilean needle grass and serrated tussock, and three additional suites designed for general LAMP validation. LAMP assays conducted using a real-time PCR platform tested positive for replicates of the target weeds at several of the primer suites and in less than 20 min. Three of the suites are particularly promising (89–100% repeatability and specificity) and following optimisation to further minimise false positive or false negative test outcomes they are likely to be reliable for field use. As promising as these results are, the real challenge will be next-stage development of a simple but effective mobile LAMP platform for use in the field by officers engaged in weed surveillance efforts.

Keywords Genetic assay, *Nassella neesiana*, *Nassella trichotoma*, DNA barcodes, loop-mediated isothermal amplification.

INTRODUCTION

Nassella neesiana and *N. trichotoma* (commonly known as Chilean needle grass (CNG) and serrated tussock (ST) respectively) are introduced Weeds of National Significance (WoNS) in Australia, requiring co-ordinated national approaches to control their spread and reduce their serious impacts on natural and agricultural ecosystems in the country (McLaren *et al.* 2002, CRC 2003a,b, Iaconis 2004). Both weeds are currently established in substantial portions of temperate south-eastern Australia, and if uncontrolled are likely to expand into broader areas across the country due to their high dispersal potential coupled with their high seed set and soil seedbank longevity.

Early detection and management to eradicate novel emergences of these two weeds, limiting their ability to establish into new locations, is preferable to the costly alternative of their perpetual control once they have established. Such early intervention is however reliant on fast and accurate identification of these two WoNS across all stages in their development. Descriptive morphology keys used for identification of CNG and ST require examination of subtle features, several of which are only apparent during later flowering stages. The issue is compounded by the close resemblance of these two species to a broad variety of tussock grasses native to Australia which overlap the distributions of the weeds. Subsequently field officers involved in weed surveillance may not recognise the two weeds until after they have increased their seed set and have established as difficult to control populations.

The potential development of in-field genetic diagnostics has received much focus in pathogen research, where uses of rapid genetic diagnostics have proven to be essential in the containment of emergent pathogens in host populations (Parida *et al.* 2008). Similar approaches could be adopted for diagnostics of weeds (Gopurenko *et al.* 2017). Loop-mediated isothermal amplification (LAMP, Notomi *et al.* 2000) in particular, has been developed extensively for use in pathogen field diagnostics, because of its increased sensitivity, specificity and portability over conventional PCR-based approaches (Parida *et al.* 2008). The feasibility of using LAMP diagnostics for in-field identification of weeds is appealing for multiple reasons, but mainly because of its potential to be based on growing DNA barcode sequence libraries useful for genetic based weed identifications, and by its capacity to be compartmentalised as a modular and simple to use device (Gopurenko *et al.* 2017).

We believe a laboratory-based LAMP platform could be developed for targeted diagnostics of CNG and ST. Further, we propose the simplicity of the LAMP method allows it to be potentially developed as a mobile assay platform suitable for targeted diagnostics by non-specialist field officers. The primary laboratory development and proofing stages required for this proposed LAMP platform are reported below.

MATERIALS AND METHODS

Whole mature plants of serrated tussock (ST, n = 40) and Chilean needle grass (CNG, n = 40) specimens were sampled from Canberra and three sites in New South Wales (Bungendore, Cooma and Goulburn). Specimens were catalogued and retained as vouchers at Wagga Wagga Agricultural Institute (NSW DPI). An additional 56 specimens of four other species of *Nassella* (*N. charruana* (Arechav.) Barkworth, *N. hyalina* (Nees) Barkworth, *N. leucotricha* (Trin. & Rupr.) Pohl and *N. tenuissima* (Trin.) Barkworth) and ten native Poaceae (*Austrostipa aristiglumis* (F.Muell.) S.W.L.Jacobs & J.Everet, *A. densiflora* (Hughes) S.W.L.Jacobs & J.Everett, *A. variabilis* (Hughes) S.W.L.Jacobs & J.Everett, three varieties of *Poa labillardieri* Steud., and four varieties of *P. sieberiana* Spreng.) similar in appearance to CNG and or ST were sampled as leaf cuttings, preserved in >90% ethanol from sites in east Australia, and included for LAMP testing (Table 1). Leaf tissue (<1 mg) was digested overnight at 55°C in 280 μL of DXT tissue digest reagent (QIAGEN, Doncaster, Australia) with 1% added proteinase K (Sigma – Aldrich). Genomic DNA was isolated from digestions using a Corbett Research 1820 X-tractor Gene robot and associated DX buffers (QIAGEN) as reported in Gopurenko *et al.* (2013).

Species identities of voucher specimens were tested using DNA barcoding (Hebert *et al.* 2003). DNA barcode loci screened included two linked nuclear ribosomal DNA regions, (the external transcribed spacer [*ETS*], and the internal transcribed spacers [*ITS*] bounded by 18S and 28S genes). Both loci have utility for DNA barcode identifications of the two target *Nassella* species (Wang *et al.* 2017), and are potentially useful for LAMP assay development. Specimens were DNA barcoded at the *maturase* K (*matK*) chloroplast DNA gene to test for possible introgression among sympatric specimens of the two *Nassella* species. DNA barcoding methods and analysis used here follow that reported in Gopurenko *et al.* (2013), and specifically for grasses in Wang *et al.* (2017).

Seven independent primer suites specific for LAMP amplification of the two target *Nassella* species, were identified among nuclear encoded ribosomal *ETS*, *ITS* and chloroplast *matK* sequences reported earlier for a variety of invasive and native Stipeae grasses in Australia (Wang *et al.* 2017). As recommended for LAMP applications (Tomlinson *et al.* 2010), each suite included two external (F3 & B3), two internal (FIP & BIP) and two loop (LF & LB) primers. Primer suites were newly designed from DNA barcode sequences using PrimerExplorer V5 (https://primerexplorer.jp/e/) and synthesised at Sigma-Aldrich (Sydney, Australia). Suites were tested in LAMP assays for their specificity and repeatability among specimen replicates. Three additional primer suites were assayed for use as positive controls of Poaceae grass DNA, including suites for two mitochondrial cytochrome c oxidase subunit I (*COI*) targets (one novel, the other reported by Tomlinson *et al.* (2010) as LAMP control for various plants), and a chloroplast DNA ribulose-1,5-bisphosphate carboxylase oxygenase (*rbcL*) target in native Poacea (Wang *et al.* 2015). In total, ten LAMP primer suites were independently screened among replicates.

LAMP reactions (25 μL total) contained 5 μL DNA (or nuclease free water for negative controls), 1 μL (8 units μL^{-1}) of *Bst* DNA polymerase (M0538L, New England Biolabs), 40 mM betaine (Sigma), 40 mM Tris–HCl (pH 8.8), 20 mM KCl, 10 mM $(NH_4)_2SO_4$, 12 mM $MgSO_4$, 4% Triton X-100, 8% DMSO, 400 μM each dNTP, 5 μL of a primer suite, and 1 μL of SYBR Green I (1:32000, Invitrogen, Karlsruhe, Germany). Primer suites consisted of 5.0 μM of each F3/B3 external primers, 15.0 μM of each FIP/BIP internal primer, and 10.0 μM of each loop F/B primer. Real-time isothermal LAMP assays of 48 reactions at a time (includes eight negative controls) were run using a Rotor-Gene Q (QIAGEN) real-time PCR cycler set to run at a constant 65°C cycle for 30 min. with SYBR Green emission detection for intervals of 20 seconds every minute. Reactions were terminated at 80°C for 5 min. Fluorescent emission recordings were quantified and graphically represented using Rotor-Gene Q Series Software 1.7 (Build 94). For each primer suite, replicates of target CNG and ST weeds, other *Nassella* and native grass species were qualitatively scored at LAMP assays as positive/negative for a defined presence of LAMP amplification products. We defined a positive outcome at a replicate if its peak fluorescence intensity was >20% of that observed at the time during the cycle of the 1st instance of LAMP fluorescence saturation (indicative of high-yield LAMP product) among replicates in an assay run.

RESULTS

DNA barcodes of the sampled WoNS were correctly matched to existing sequence accessions of CNG and ST confirming their species identity. The DNA barcodes were reciprocally monophyletic for CNG and ST at each of three loci (data not shown) and minimum sequence differences between the two species varied as 3.31%, 2.77%, and 0.11% at *ETS*, *ITS* and *matK* respectively.

Saturation level LAMP assay products were detected by a real-time PCR platform after 15 20 min. of isothermal temperature cycling at 65°C. Contamination of LAMP master mix identified as presence of saturation level products in negative controls was observed at <5% of runs, and in all these instances, replicate scores were discarded and runs repeated using fresh reagents.

Specificity and repeatability of LAMP assays was highly variable among the ten primer suites (Table 1). Positive control (+) assays using primer suites 08 and 10 failed across all replicates; in contrast + assays using primer suite 09 targeted to mitochondrial *COI* (Tomlinson *et al.* 2010) were successful at detection of 83% of target CNG and ST replicates, but only 32% of other grass replicates.

Of the seven LAMP assays specific for the two target *Nassella* spp., the highest repeatability and specificity was observed using primer suite 05 which exclusively detected >97% of tested target CNG. Suite 02 detected 100% of target CNG, but also 15% of other *Nassella* spp. replicates. Suite 03 exclusively detected >89% of target ST. Other suites had mixed propensity for partial detection across the two target WoNS (Suite 01) or across other tested grasses (suites 06 and 07). Suite 04 failed overall.

DISCUSSION

LAMP specificity testing of seven primer suites designed from DNA barcode sequences specific for Chilean needle grass (CNG) and serrated tussock (ST) demonstrated high accuracy and repeatability of two suites (05 and 02) for laboratory based diagnostic detection of CNG, and one suite (03) for ST. LAMP products generated using these three primer suites amplified to saturation levels of detection within 15–20 min. using a real-time PCR platform. In contrast, non-target specimens either failed to amplify using these suites or amplified at fewer replicates, at much later assay stages (>25 min.) and at sub-saturation levels of detection. Chimeric formation among primers during isothermal amplification may be inducing some of these observed false positive LAMP results. Attaining greater rates of repeatability and specificity (i.e. near 100%) of these LAMP assays for the target *Nassella* species is critical to the reliability of the proposed LAMP platform for in-field uses, where misdiagnosis would lead to an erroneous control response and potentially result in emergence of a new infestation and or loss of local native flora. Optimisation of LAMP reagent composition (Tomlinson *et al.* 2010) used with these primer suites and or primer modification will improve both specificity of primers to the targets

Table 1. Specificity and repeatability of LAMP primer suites (01–07) designed from loci (*ETS*, *ITS*, *matK*) for detection of target weeds Chilean needle grass (CNG, n = 40) or serrated tussock (ST n = 40), and primer suites (08–10) from loci (*COI* and *rbcL*) for use as positive (+) controls in Poaceae detection. Extra grasses tested include four other *Nassella* spp. (NS; n = 20) and ten native Poaceae (PO; n = 36) of similar appearance to the target weeds. All suites are novel except (09) Tomlinson *et al.* 2010.

	Target		n detected / n tested			
Suite	loci	grass	CNG	ST	NS	PO
01	*ETS*	ST	10/40	30/38	0/20	0/36
02	*ETS*	CNG	40/40	0/40	6/20	0/36
03	*ETS*	ST	0/40	34/38	0/20	0/36
04	*ITS*	ST	0/10	0/10	0/5	0/9
05	*ITS*	CNG	39/40	0/38	0/20	0/36
06	*matK*	ST	39/40	31/38	3/20	1/36
07	*matK*	CNG	28/40	28/38	6/20	5/36
08	*COI*	+	0/10	0/10	0/5	0/9
09	*COI*	+	35/40	30/38	4/20	14/36
10	*rbcL*	+	0/10	0/10	0/5	0/9

(minimising false positives) and repeatability of the assay among target replicates (minimising false negatives). Unexpected partial success in detection of both target weeds (to the exclusion of all other assayed grasses) using suite 01 may allow for the possibility of a single LAMP assay useful for diagnostics of either weed, however modifications of the constituent primers in the suite are required to improve detection success (currently 51% detection across the two target *Nassella* species)

Integral to our future proposed use of LAMP for in-field detection of the *Nassella* WoNS is the need for a secondary LAMP test, ancillary to the primary assay for use as a positive control of Poaceae grasses in general. This secondary test would allow accurate interpretation of negative test outcomes observed in the primary assay as resulting either from absence of the target organism (true negative) or due to corruption of the LAMP reaction itself (false negative). Of the three suites of primers tested for this purpose as a general LAMP positive control, only one (suite 09) showed some measure of success across replicates of the assayed taxa. The suite was originally designed as a control in detection of Rhododendrons but shown to have broader specificity across a variety of plants (Tomlinson *et al.* 2010). The suite provided high levels of positive repeatability (83%) for detection of the two target weeds but less repeatability among the remaining grasses (32%). Modifications to constituent primers in this suite to better anneal to Poaceae will likely improve its reliability as a positive diagnostic of that group of grasses.

Ultimately, preparation of all these diagnostic LAMP assays for in-field identification of CNG and ST will require: 1) identification of a rapid and crude DNA release method that is non-inhibitive to LAMP reactivity; and 2) development of a simple low-cost isothermal platform incorporating inexpensive LAMP product visualisation capacity. As discussed earlier (Gopurenko *et al.* 2017) existing methods for crude DNA extraction, isothermal heating and visual detection of LAMP products using inexpensive lateral flow devices have been developed for in-field diagnostics of pathogens in a range of host tissues. Similar approaches may be modified and applicable here for diagnostics of CNG and ST and will be the focus of the next stage of our research.

ACKNOWLEDGMENTS

Funding for this project (CT-17) was provided through the Agricultural Competitiveness White Paper (Commonwealth of Australia) Control Tools and Technologies for Established Pest Animals and Weeds Programme, administered by the Department of Agriculture and Water Resources. We gratefully thank Brett Jones (Snowy Monaro Regional Council), Neville Plumb (Queanbeyan-Palerang Regional Council), Malcolm Ross (Goulburn Mulwaree Council), Steve Taylor and Jenny Conolly (Australian Capital Territory Parks and Conservation Service) for invaluable assistances in field sampling of weeds examined here. We acknowledge edits by Gretchen Kay Foster and anonymous reviewers to improve our manuscript.

REFERENCES

CRC for Australian Weed Management (2003a). Chilean needle grass (*Nassella neesiana*) weed management guide. Weeds of National Significance. Department of the Environment and Heritage and CRC for Australian Weed Management.

CRC for Australian Weed Management (2003b). Serrated tussock (*Nassella trichotoma*) weed management guide. Weeds of National Significance. Department of the Environment and Heritage and CRC for Australian Weed Management.

Gopurenko, D., Fletcher, M.J., Löcker, H. and Mitchell, A. (2013). Morphological and DNA barcode species identifications of leafhoppers, planthoppers and treehoppers (Hemiptera: Auchenorrhyncha) at Barrow Island. *In* 'Records of the Western Australian Museum, Supplement 83: The Terrestrial Invertebrate Fauna of Barrow Island', eds N.R. Gunawardene, J.D. Majer, C.K. Taylor and M.S. Harvey, pp. 253-85. (Western Australian Museum Publications, Perth).

Gopurenko, D., Wang, A. and Wu, H. (2017). Developing a portable DNA-based platform, for rapid on-site identification of Chilean needle grass and serrated tussock. Proceedings of the 19th NSW Biennial Weeds Conference, pp. 145-51. (Armidale).

Hebert, P.D.N., Cywinska, A., Ball, S.L. and DeWaard, J.R. (2003). Biological identifications through DNA barcodes. *Proceedings of the Royal Society B, Biological Science* 270, 313-21.

Iaconis, L. (2004). How is the national Chilean needle grass program progressing? Proceedings of the 14th Australian Weeds Conference, eds. B.M. Sindel and S.B. Johnson, pp. 636-9. (Weed Society of New South Wales, Sydney).

McLaren, D.A., Coventry, R. and Thorp, J. (2002). Stipoid grasses as Weeds of National Significance. *Plant Protection Quarterly* 7, 119-24.

Notomi, T., Okayama, H., Masubuchi, H., Yonekawa, T., Watanabe, K., Amino, N. and Hase, T. (2000). Loop-mediated isothermal amplification of DNA. *Nucleic Acids Research* 28, e63.

Parida, M., Sannarangaiah, S., Dash, P.K., Rao, P.V.L. and Morita, K. (2008). Loop mediated isothermal amplification (LAMP): a new generation of innovative gene amplification technique; perspectives in clinical diagnosis of infectious diseases. *Reviews in Medical Virology* 18, 407-21.

Tomlinson, J.A., Dickinson, M.J. and Boonham, N. (2010). Rapid detection of *Phytophthora ramorum* and *P. kernoviae* by two-minute DNA extraction followed by isothermal amplification and amplicon detection by generic lateral flow device. *Phytopathology* 100, 143-9.

Wang, A., Gopurenko, D., Wu, H., Stanton, R. and Lepschi, B.J. (2015). Identification of invasive grass species via DNA barcoding. Proceedings of the 18th NSW Weeds Conference, pp. 68-81. (Cooma).

Wang, A., Gopurenko, D., Wu, H. and Lepschi, B. (2017). Evaluation of six candidate DNA barcode loci for identification of five important invasive grasses in eastern Australia. *PLoS ONE* 12, e0175338.

Molecular identification of *Solanum elaeagnifolium* in Australia using DNA barcoding, a solution for better management

Xiaocheng Zhu[1,2], David Gopurenko[1,2], Laurence A.R. Haegi[3] and Hanwen Wu[1,2]
[1] Department of Primary Industries, DPI Biosecurity and Food Safety, Wagga Wagga, New South Wales 2650, Australia
[2] Graham Centre for Agricultural Innovation (Charles Sturt University and NSW Department of Primary Industries), Charles Sturt University, Wagga Wagga, New South Wales 2678, Australia
[3] State Herbarium of South Australia, GPO Box 1047, Adelaide, South Australia 5001, Australia
(xiaocheng.zhu@dpi.nsw.gov.au)

Summary The problematic weed silverleaf nightshade (*Solanum elaeagnifolium* Cav.) was introduced to Australia during the early 20th century from North America and has since emerged as a Weed of National Significance. This species is almost impossible to eradicate after establishment, with its management in areas of new emergence therefore much aided by prompt identification and early action. Identification of the weed is often confounded because of the weed's remarkable similarity (especially pre-floral stages) to a variety of native *Solanum* species. DNA barcoding for improved identification may therefore improve its management. Sequences of the nuclear *waxy* gene and two chloroplast genes (*matK* and *trnL-trnF*) were compared among *S. elaeagnifolium* (N = 12) and 10 native species (*S. brownii*, *S. centrale*, *S. coactiliferum*, *S. esuriale*, *S. jucundum*, *S. lasiophyllum*, *S. lithophyllum*, *S. petrophilum*, *S. quadriloculatum* and *S. sturtianum*) with which the weed is commonly confused. A further five species including *S. campanulatum*, *S. chenopodinum*, *S. cleistogamum*, *S. nummularium* and *S. succosum*, were also analysed to provide a wide cross-section of Australian native species. Clear and well supported genetic separations of all species were evident at each of the three gene regions. The maximum observed genetic distance in *S. elaeagnifolium* was 0.41%, and the minimum distance to its nearest genetic neighbour species was 1.12%. Our study provides a reliable genetic tool for *S. elaeagnifolium* identification irrespective of developmental stage, assisting efficient management aimed at reducing the risk of further spread of this weed to new areas.

Keywords Granule-bound starch synthase 1, silverleaf nightshade, Australian native *Solanum*, WoNS.

INTRODUCTION

Silverleaf nightshade (*Solanum elaeagnifolium* Cav.) is a deep-rooted, summer-growing perennial. It is listed as Weed of National Significance, infesting over 350,000 ha in Australia (Stanton *et al.* 2009). It competes for soil water and nutrients and causes up to 77% yield loss in cereal (Stanton *et al.* 2009). Due to the deep root system and its high re-generation ability, this species is almost impossible to eradicate after establishment (Stanton *et al.* 2009).

Prompt identification and early action is required for effective management especially in the areas of new incursions. However, although *S. elaeagnifolium* originated in North America, it is remarkably similar to several Australian native species especially at pre-floral stage. Like the vast majority of Australian native *Solanum*, silverleaf nightshade belongs to the *Leptostemonum* or 'spiny solanum' clade in *Solanum*. Distinguishing these native species from *S. elaeagnifolium* is currently based on subtle morphological features requiring specialised knowledge. Field identifications are therefore often unreliable (Zhu *et al.* 2013). Misidentification of *S. elaeagnifolium* as *S. esuriale* Lindl. in South Australia (1918) contributed to the delay of co-ordinated control until 1958 (Hosking *et al.* 2000). Similarly, native *Solanum* species are also commonly misidentified as *S. elaeagnifolium*, resulting in erroneous species distribution records, and unnecessary control of native biodiversity. Our previous study demonstrated how use of microsatellite (SSR) markers could differentiate *S. elaeagnifolium* but was restricted to only one native species, *S. esuriale* (Zhu *et al.* 2011).

DNA barcoding is a sequence-based method widely used to assist alpha-taxonomic identifications (Gopurenko *et al.* 2013). Phylogenetic relationships among *Solanum* species have been reported using various nuclear and chloroplast genes (Aubriot *et al.* 2016, Levin *et al.* 2006, Zhang *et al.* 2013), and such genes are potentially also useful for DNA barcoding purposes here. However, these studies generally used single replicates of examined species and some did not include both *S. elaeagnifolium* and similar Australian native species. Limited replicate sampling can underestimate intraspecific genetic diversity and lead to erroneous DNA barcode identifications (Levin *et al.* 2005, Zhang *et al.* 2013). This study aims to provide a reliable tool to distinguish *S. elaeagnifolium* from 15 Australian native *Solanum* species irrespective of growth stage.

MATERIALS AND METHODS

Solanum elaeagnifolium (N = 12) and vouchered samples of 15 Australian native spiny *Solanum* species were obtained from recent field collections, and specimens available at the State Herbarium of South Australia, the National Herbarium of New South Wales and the Queensland Herbarium. These included ten species morphologically similar to *S. elaeagnifolium* (*S. brownie* Dunal., *S. central* J.M.Black, *S. coactiliferum* J.M.Black, *S. esuriale* Lindl., *S. jucundum* A.R.Bean, *S. lasiophyllum* Poir., *S. lithophilum* F.Muell., *S. petrophilum* F.Muell., *S. quadriloculatum* F.Muell. and *S. sturtianum* A.R.Bean (Table 1) and a further five species from diverse groups within the 'spiny solanums' (*S. campanulatum* R.Br., *S. chenopodinum* F.Muell., *S. cleistogamum* Symon, *S. nummularium* S.Moore and *S. succosum* A.R.Bean & Albr.) for comparative purposes. Replicates were included for all species except *S. centrale*.

DNA extraction, PCR and sequencing were performed as described (Gopurenko *et al.* 2013), but modified for plant DNA barcoding (Zhu *et al.* 2017). Chloroplast DNA genes (*matK* and *trnL-trnF*) were PCR amplified using primer pairs 390F/1326R (Cuenoud *et al.* 2002), and ucp_c/ucp_f (Taberlet *et al.* 1991), respectively. The nuclear *waxy* gene (exons 2–7) was targeted using primers 181F & 1171R designed by Walsh and Hoot (2001) as reported in Levin *et al.* (2005). To improve amplification of *waxy* we modified primer 181F as ("waxyF") 5' CGGG-TAATGACAATATCCCC 3'. Gene sequences were concatenated using FABOX (Villesen 2007). Inter- and intra- specific pairwise genetic distances (unweighted) were calculated using MEGA7 (Kumar *et al.* 2016). Optimal nucleotide-substitution models for six unlinked partitions of the sequence alignment (*matK* codons 1, 2 and 3; *trnL-trnF*; *waxy* introns; *waxy* exons) were identified using Jmodeltest 2.0 (Darriba *et al.* 2012) and implemented in a Bayesian phylogenetic analysis run for 80 million generations using MrBayes 3.2.4 (Ronquist and Huelsenbeck 2003).

RESULTS AND DISCUSSION

High quality sequences were obtained for all samples at all three gene regions. Alignments were truncated to 771, 1042 and 950 bp for *matK*, *trnL-trnF* and *waxy*, respectively. There were 168 parsimoniously informative sites and 66 singletons. Most species have a very low within-species diversity, ranging from 0 to 0.29 %, with exception of *S. elaeagnifolium*, *S. nummularium* and *S. jucundum* where within species diversity was higher at 0.41, 0.44 and 0.45 %, respectively.

Table 1. Species compared in this study. N: number of samples sequenced; D_{ma}(%): maximum % intraspecific genetic distance; D_{mi} (%): minimum % interspecific genetic distance between *S. elaeagnifolium* and other species.

Species	N	D_{ma} (%)	D_{mi} (%)
S. brownii	3	0.18	1.28
S. campanulatum	2	0.18	1.35
S. centrale	1	NA	1.40
S. chenopodinum	2	0.18	1.61
S. cleistogamum	4	0.25	1.12
S. coactiliferum	2	0.07	1.23
S. elaeagnifolium	12	0.41	NA
S. esuriale	4	0.11	1.55
S. jucundum	4	0.45	1.38
S. lasiophyllum	2	0.00	1.42
S. lithophyllum	2	0.04	1.23
S. nummularium	2	0.44	1.53
S. petrophilum	2	0.04	1.43
S. quadriloculatum	4	0.15	1.39
S. sturtianum	2	0.11	1.38
S. succosum	2	0.29	1.19

Although remarkably similar in morphological traits, *S. elaeagnifolium* is genetically very different from all tested Australian native species (Table 1). The maximum genetic distance observed in *S. elaeagnifolium* (0.41%, Table 1) was less than half the distance (1.12%) observed to *S. cleistogamum*, its nearest neighbour (Table 1, Figure 1). The large genetic distance between *S. elaeagnifolium* and Australian native *Solanum* is consistent with a previous study which highlighted limited allelic overlap between *S. elaeagnifolium* and *S. esuriale* using SSR markers (Zhu *et al.* 2011). Four, four and nine alleles specific to *S. elaeagnifolium* were identified at *matK*, *trnL-trnF* and *waxy* genes, respectively. These genetic markers could potentially be used in development of a Loop mediated isothermal amplification (LAMP) assay for targeted molecular diagnostics and rapid field identification of the species (Gopurenko *et al.* 2017).

The phylogenetic tree clearly separated all species at high posterior probabilities (Figure 1). *Solanum elaeagnifolium* formed a distinct clade from other species. This is consistent with previous molecular

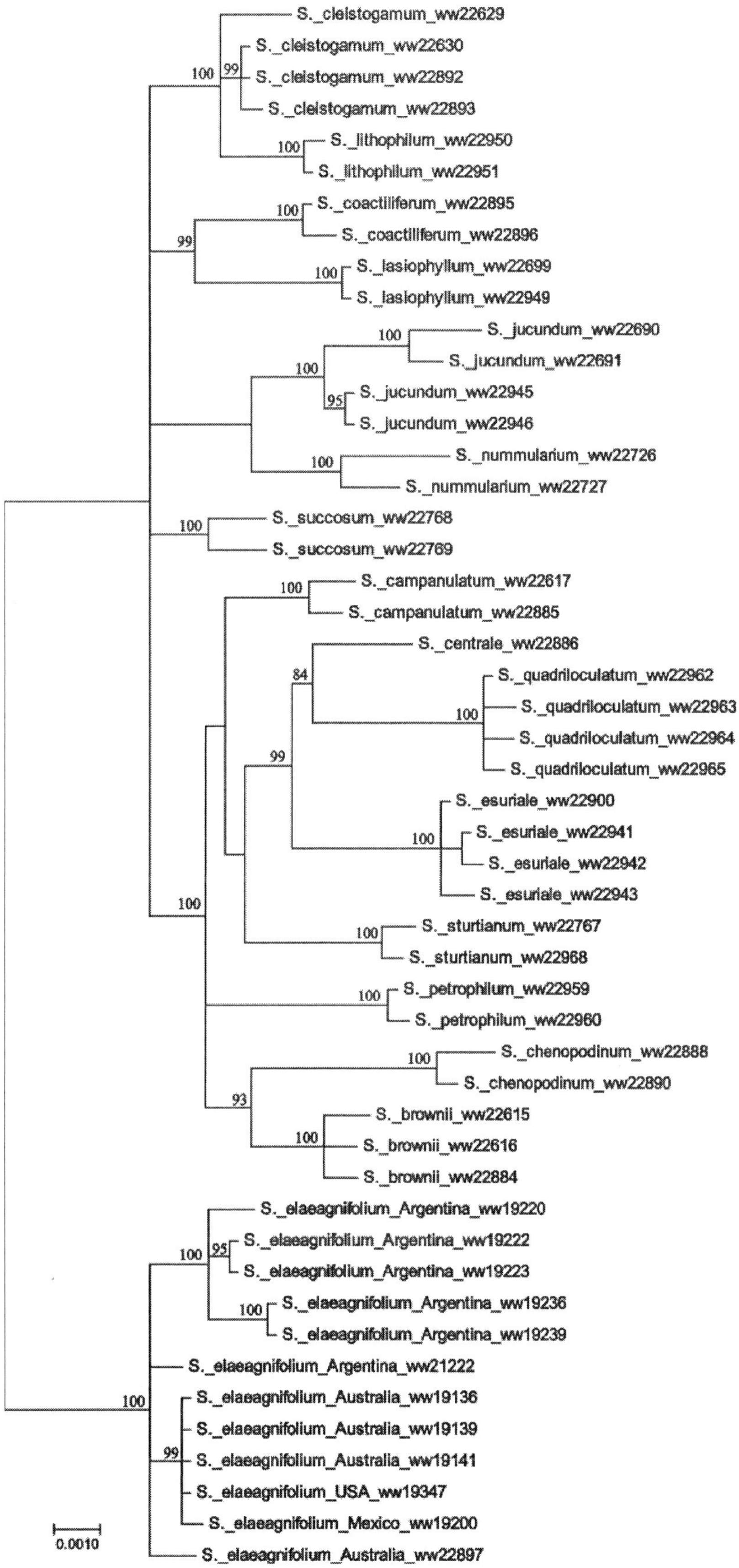

Figure 1. Bayesian phylogenetic relationships among *S. elaeagnifolium* and 15 Australian native *Solanum* species. Clade posterior probability support >80% indicated at nodes.

phylogenetic studies of 'spiny solanums' (Levin *et al.* 2006, Vorontsova *et al.* 2013) and extends this outcome to a wider range of Australian species. A more comprehensive molecular phylogenetic study currently underway by the authors is expected to throw more light on the relationships among Australian native 'spiny solanums' and *S. elaeagnifolium.*

In conclusion, DNA barcoding using three genes clearly distinguished *S. elaeagnifolium* from 10 morphologically very similar native *Solanum* species and five other closely related *Solanum* species. All three genes have alleles specific for *S. elaeagnifolium;* these could potentially be used in molecular genetic diagnostics for identification of the species. This in turn could improve management of SLN especially at the early stage of the invasion and avoid the unnecessary control of morphologically similar native species.

ACKNOWLEDGMENTS

We acknowledge the Biosecurity SA, PIRSA and Meat and Livestock Australia for funding support. Dr Franco Chiarini (Argentina), Dr Miguel Mellado (Mexico), Edwin Morris (United States of America), Dr John Heap and organisations including the State Herbarium of South Australia, National Herbarium of New South Wales and the Queensland Herbarium are thanked for access to specimens used in analysis.

REFERENCES

Aubriot, X., Singh, P. and Knapp, S. (2016). Tropical Asian species show that the Old World clade of 'Spiny Solanums' (*Solanum* subgenus *Leptostemonum* pro parte: Solanaceae) is not monophyletic. *Botanical Journal of the Linnean Society* 181, 199-223.

Cuenoud, P., Savolainen, V., Chatrou, L.W., Powell, M., Grayer, R.J. and Chase, M.W. (2002). Molecular phylogenetics of Caryophyllales based on nuclear 18S rDNA and plastid rbcL, atpB, and matK DNA sequences. *American Journal of Botany* 89, 132-44.

Darriba, D., Taboada, G.L., Doallo, R. and Posada, D. (2012). jModelTest 2: more models, new heuristics and parallel computing. *Nature Methods* 9, 772-72.

Gopurenko, D., Fletcher, M.J., Löcker, H. and Mitchell, A. (2013). Morphological and DNA barcode species identifications of leafhoppers, planthoppers and treehoppers (Hemiptera: Auchenorrhyncha) at Barrow Island. *In* 'Records of the Western Australian Museum, supplement 83: the terrestrial invertebrate fauna of Barrow Island', eds N.R. Gunawardene, J.D. Majer, C.K. Taylor and M.S. Harvey, pp. 253-85. (Western Australian Museum Publications).

Gopurenko, D., Wang, A. and Wu, H. (2017). Developing a portable DNA-based platform, for rapid on-site identification of Chilean needle grass and serrated tussock. Proceedings of the 19th NSW Biennial Weeds Conference, pp. 145-51. (Weed Science Society of NSW, Armidale)

Hosking, J.R., Sainty, G.R. and Jacobs, S.W.L. (2000). Certainty and uncertaintly in plant identification. *The New Mexico Botanist* 15, 1-4.

Kumar, S., Stecher, G. and Tamura, K. (2016). MEGA7: Molecular Evolutionary Genetics Analysis Version 7.0 for Bigger Datasets. *Molecular Biology and Evolution* 33, 1870-74.

Levin, R.A., Nicole, R.M. and Bohs, L. (2006). Phylogenetic relationships among the 'Spiny Solanums' (*Solanum* Subgenus *Leptostemonum*, Solanaceae). *American Journal of Botany* 93, 157-69.

Levin, R.A., Watson, K. and Bohs, L. (2005). A four-gene study of evolutionary relationships in *Solanum* section *Acanthophora. American Journal of Botany* 92, 603-12.

Ronquist, F. and Huelsenbeck, J. (2003). MrBayes 3: Bayesian phylogenetic inference under mixed models. *Bioinformatics* 19, 1572-4.

Stanton, R., Heap, J.W., Carter, R.J. and Wu, H. (2009). *Solanum elaeagnifolium. In* 'The biology of Australian weeds', volume 3, ed. F.D. Panetta, pp. 274-93. (R.G. and F.J. Richardson, Melbourne).

Taberlet, P., Gielly, L., Pautou, G. and Bouvet, J. (1991). Universal primers for amplification of three non-coding regions of chloroplast DNA. *Plant Molecular Biology* 17, 1105-9.

Villesen, P. (2007). FaBox: an online toolbox for fasta sequences. *Molecular Ecology Notes* 7, 965-68.

Vorontsova, M.S., Stern, S., Bohs, L. and Knapp, S. (2013). African spiny *Solanum* (subgenus *Leptostemonum*, Solanaceae): a thorny phylogenetic tangle. *Botanical Journal of the Linnean Society* 173, 176-93.

Walsh, B.M. and Hoot, S.B. (2001). Phylogenetic relationships of capsicum (Solanaceae) Using DNA sequences from two noncoding regions: the chloroplast atpB-rbcL spacer region and nuclear waxy introns. *International Journal of Plant Sciences* 162, 1409-18.

Zhang, W., Fan, X., Zhu, S., Zhao, H. and Fu, L. (2013). Species-specific identification from incomplete sampling: applying DNA barcodes to monitoring invasive *Solanum* plants. *Plos One* 8, e55927.

Zhu, X., Weston, P.A., Skoneczny, D., Gopurenko, D., Meyer, L., Lepschi, B.J., Callaway, R.M., Gurr, G.M. and Weston, L.A. (2017). Ecology and genetics affect relative invasion success of two

Echium species in southern Australia. *Scientific Reports* 7, srep42792.

Zhu, X.C., Burrows, G., Wu, H., Raman, H., Stanton, R. and Lemerle, D. (2011). Identification of silverleaf nightshade using microsatellite markers and microstructure. Proceedings of the 23rd Asian-Pacific Weed Science Society Conference, pp. 604-9. (Asian-Pacific Weed Science Society, Cairns).

Zhu, X.C., Wu, H.W., Stanton, R., Burrows, G.E., Lemerle, D. and Raman, H. (2013). Morphological variation of *Solanum elaeagnifolium* in south-eastern Australia. *Weed Research* 53, 337-43.

Identification of eight *Panicum* species in Riverina region of NSW using DNA sequence analysis

Yuchi Chen[1,2], Xiaocheng Zhu[2], David E. Albrecht[3], Panayiotis Loukopoulos[1,2], Jane C. Quinn[1,2] and Leslie A. Weston[2,4]

[1] School of Animal and Veterinary Sciences, Charles Sturt University, Wagga Wagga, New South Wales 2650, Australia

[2] Graham Centre for Agricultural Innovation; Charles Sturt University and NSW Department of Primary Industries, Wagga Wagga, New South Wales 2650, Australia

[3] Australian National Herbarium, Centre for Australian National Biodiversity Research, GPO Box 1700, Canberra, Australian Capital Territory 2601, Australia

[4] School of Agriculture and Wine Science, Charles Sturt University, Wagga Wagga, New South Wales 2650, Australia

(yucchen@csu.edu.au)

Summary Australia has over 30 *Panicum* spp. (panic grass) including several non-native species that cause crop and pasture loss due to competition. To develop appropriate management strategies for each species, it is critical to correctly identify panic grass species encountered. Currently, panic grass identification relies on microscopic examination of the inflorescence and spikelets, an approach that is only useful for flowering specimens and requires significant taxonomic expertise. To overcome this limitation, we applied both morphological and molecular techniques for identification of *Panicum* spp. in the Riverina region of New South Wales. We identified three molecular markers: one nuclear gene region (ITS) and two chloroplast gene regions (*mat*K and *trn*L intron-*trn*F) capable of differentiating eight *Panicum* spp. Concatenation of sequences from ITS, *mat*K and *trn*L intron-*trn*F gene regions provided clear separation of eight species collected regionally and identified a maximum intraspecific distance of 0.22% and minimum interspecific distance of 0.33%. Based on comparison with verified voucher specimens, *P. hillmanii* Chase was prevalent and constituted 78.9% of all samples collected and identified. DNA barcoding represents an accurate and potentially cost effective tool for distinguishing *Panicum* spp. at the species level regardless of growth stage. Molecular markers may also be useful for accurate demographic analysis of *Panicum* grass invasion in Australia and abroad.

Keywords *Panicum* spp., *Panicum hillmanii* Chase, DNA barcoding, *matK*, *trnL* intron-*trnF*, ITS, phylogenetic tree.

INTRODUCTION

The genus *Panicum* is one of the largest genera in the Poaceae, and is globally distributed (Aliscioni *et al.* 2003). Depending on the taxonomic concepts followed, up to 500 species are recognised (Byng 2014). Major photosynthetic types (C3, C4, and C3/C4 intermediate) reported in the Poaceae are represented in the genus (Zuloaga *et al.* 2010).

Panicum ssp. are widely distributed in Australia (Figure 1) with 24 indigenous and 9 non-native species currently recognised (CHAH 2005 onwards). They are currently ranked as one of the most economically important weeds of summer fallows in Australia (Llewellyn *et al.* 2016). Their toxic profile in livestock is suggested to relate to the presence of saponins or sapogenins (Miles *et al.* 1992). However, variable steroidal saponin profiles may exist in different *Panicum* spp., and could be associated with their differential ability to cause hepatogenous or secondary photosensitisation in livestock (Quinn *et al.* 2014).

Panicum is a taxonomically challenging group due to the wide range of variation in both morphological and physiological features (Zuloaga *et al.* 2010). Subtle morphological differences required for species differentiation have not been well documented (Zuloaga *et al.* 2010). Existing identification keys to *Panicum* spp. in Australia are based primarily on reproductive features and therefore are only effective for plants in their reproductive phase (Coissac *et al.* 2016). Furthermore, keys for identification can be difficult to use with confidence and even experienced taxonomists find *Panicum* identification problematic at the species level. Therefore, establishment of robust genotyping methods are desirable for unequivocal plant species identification. Identification of *Panicum* at an individual plant level may also be critical for development of targeted management strategies for invasive and potentially toxic *Panicum* spp. (Quinn *et al.* 2014).

DNA barcoding, a simplified DNA-based species identification approach (Hebert *et al.* 2003), has been

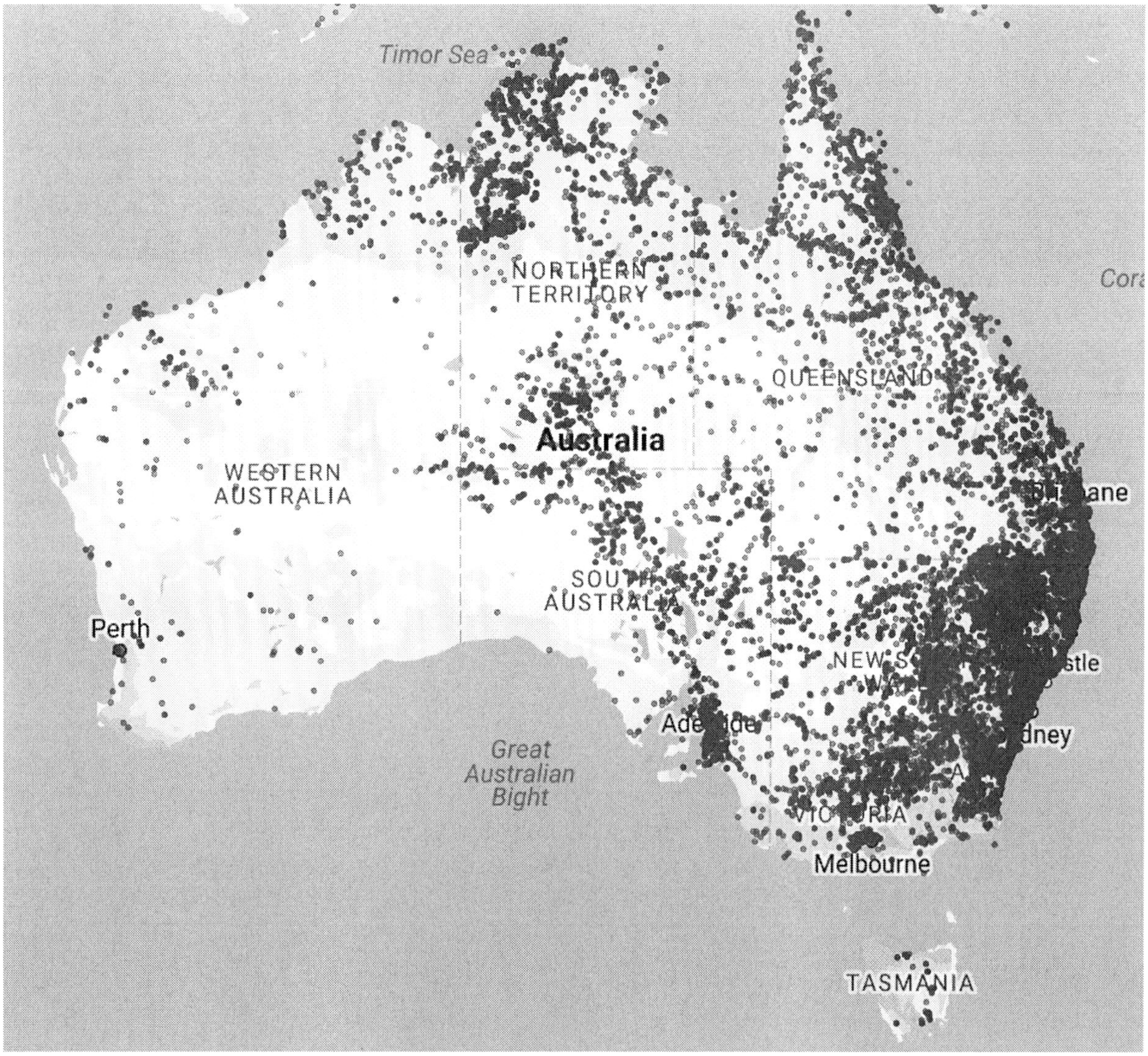

Figure 1. Distribution of *Panicum* spp. in Australia based on herbarium specimens represented in major Australian herbaria. Collection data extracted from The Australasian Virtual Herbarium.

utilised as an important and complementary method to traditional morphological identification (Zhu *et al.* 2014). DNA barcoding uses standardised DNA regions ('barcodes') to identify different species (CBOL Plant Working Group 2009). However, the optimal suite of barcoding loci for *Panicum* spp. identification in Australia has not yet been established. Following on from Hollingsworth *et al.* 2011, one nuclear locus (ITS) and two chloroplast loci (*mat*K and *trn*L intron-*trn*F) were evaluated in this study for their ability to differentiate eight *Panicum* spp. occurring in the Riverina region of New South Wales.

MATERIALS AND METHODS

In early 2017, fresh leaf samples and accompanying herbarium voucher specimens were collected from 90 *Panicum* plants dispersed across multiple sites within a 200 km radius range of Wagga Wagga, NSW (Figure 2). Voucher specimens were lodged at the Australian National Herbarium (ANH). Fresh leaf samples were stored in silica gel prior to extraction.

To supplement the field-collected materials, an additional 19 dried leaf samples, representing seven *Panicum* spp., were sampled from specimens within the ANH collection. A total of 36 samples, representing eight species found in Riverina, were selected for further analysis in this study (Table 1).

DNA extraction, PCR and sequencing were performed as described (Zhu et al. 2014) using primer pairs ITS5A-ITSR, 390F-1326R and ucp_c-ucp_f for ITS, matK and trnL intron-trnF regions, respectively. Geneious version 11.0.5 (Kearse et al. 2012) was used to read, edit and align the sequences. Sequence alignments were analysed using MEGA7.0.26 (Kumar et al. 2016) to calculate intra- and interspecific distances with the Kimura 2-parameter (K2P) model. Sequences of all loci for each Panicum specimen were concatenated and phylogenetic relationships among species were inferred by MrBayers 3.2.6 (Huelsenbeck and Ronquist 2001) using default settings with GTR+R substitution model as suggested by JModelTest 2.1.10 (Darriba *et al.* 2012).

Table 1. *Panicum* specimens used for DNA extraction and sequencing.

Species identification	Fresh leaf	Dried leaf (ANH)
P. capillare L.	1	2
P. decompositum R.Br.	3	5
P. effusum R.Br.	4	1
P. gilvum Launert	5	0
P. hillmanii Chase	4	2
P. laevinode Lindl.	0	3
P. miliaceum L.	0	3
P. queenslandicum Domin	0	3

RESULTS AND DISSCUSION

The majority of field-collected specimens were identified as *P. hillmanii* (78.9%, 71/90) by comparison with verified voucher specimens at the ANH (Figure 2). Other species identified included *P. gilvum* (10.0%, 9/90), *P. decompositum* (5.6%, 5/90), *P. effusum* (4.4%, 4/90) and *P. capillare* (1.1%, 1/90). The predominance of *P. hillmanii*, a non-native species that was previously rarely reported in the Riverina region, suggests the species has spread considerably in recent years. Therefore, further investigation in regards to its invasion ecology should be conducted for the development of a more holistic management strategy.

Successful PCR amplification and sequencing was achieved in three selected gene loci (ITS, *mat*K and *trn*L intron-*trn*F) for all specimens. Alignments were then truncated to 641, 730, and 750 bps for ITS, *mat*K and *trn*L intron-*trn*F, respectively. In addition, concatenated data over all three gene loci were evaluated for phylogenetic analysis. Intraspecific and interspecific genetic distances were calculated for the eight species identified (Table 2).

The efficacy of DNA barcoding depends on the extent of intraspecific and interspecific divergence in a selected locus or combined loci, and the ideal situation

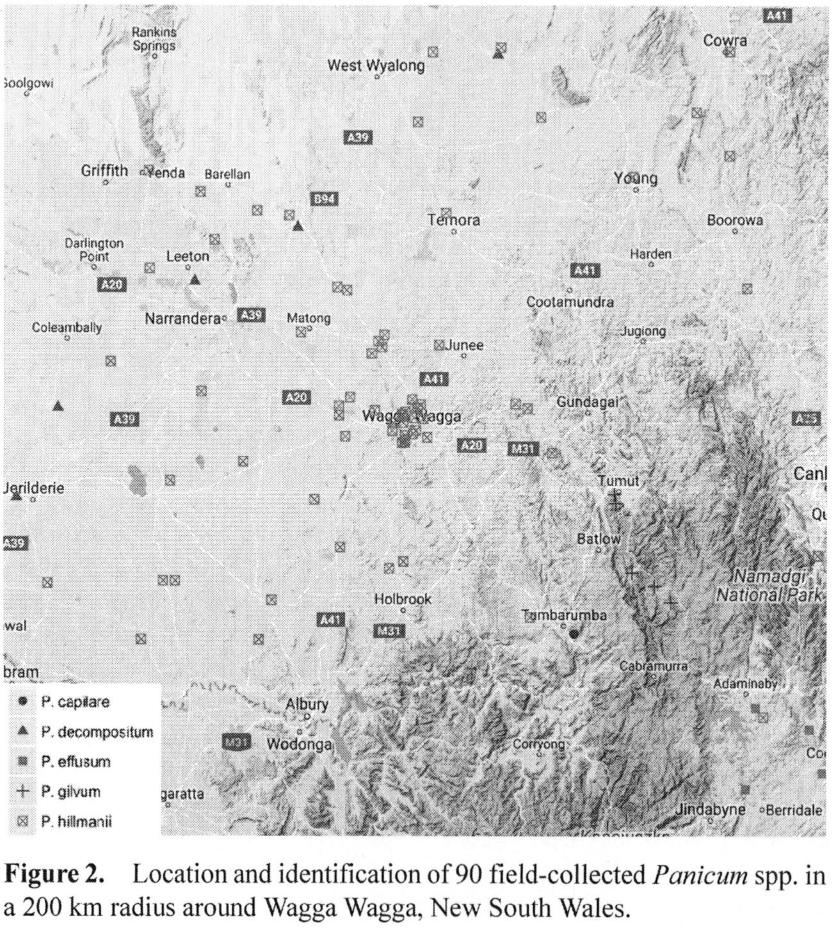

Figure 2. Location and identification of 90 field-collected *Panicum* spp. in a 200 km radius around Wagga Wagga, New South Wales.

Table 2. Intraspecific and interspecific K2P distances (%) of eight *Panicum* spp. (MinInterD: Minimum interspecific distance; MaxID: Maximum intraspecific distance; UAA: *trn*L intron-*trn*F; PC: *P. capillare*; PD: *P. decompositum*; PE: *P. effusum*; PG: *P. gilvum*; PH: *P. hillmanii* Chase; PL: *P. laevinode* Lindl; PM: *P. miliaceum*; PQ: *P. queenslandicum* Domin).

	ITS	*mat*K	UAA	ITS-*mat*K-UAA
MinInterD %	0.71	0.07	0	0.33
MaxID in PC%	0.33	0.09	0	0.22
MaxID in PD%	0.03	0.06	0	0.03
MaxID in PE%	0.34	0	0	0.15
MaxID in PG%	0	0	0	0
MaxID in PH%	0	0	0	0
MaxID in PL%	0.07	0.09	0	0.06
MaxID in PM%	0	0.09	0	0.03
MaxID in PQ%	0.14	0.14	0.27	0.16

is a lack of overlap between these two distance values (Aliabadian *et al.* 2009). For *Panicum* spp., ITS showed the highest minimum interspecific distance (0.71%), which was over 2-fold higher than the highest maximum intraspecific distance (*P. effusum*, 0.34%). This result confirmed that the ITS locus was used as an efficient barcode for separation of the eight *Panicum* spp. in this study. Conversely, the overlap between intraspecific and interspecific values of either the *mat*K or the *trn*L intron-*trn*F regions suggests that use of these barcodes alone may prove problematic for differentiation of some *Panicum* spp.

In order to gain strong interspecies discriminatory power, the three loci were concatenated for phylogenetic analysis. The phylogenetic tree showed clear differentiation between the eight *Panicum* spp. evaluated (Figure 3). Investigation of the efficacy of this approach with other *Panicum* spp. will be undertaken in future studies using these loci.

When considering the phylogenetic tree generated, the primary division of interest with respect to the relationship between indigenous and non-native species is presented (Figure 3). Not surprisingly, all four indigenous species (*P. decompositum, P. effusum, P. laevinode* and *P. queenslandicum*), clustered together on clade 2, while the four non-native species (*P. capillare, P. gilvum, P. hillmanii* and *P. miliaceum*) clustered together on clade 1. This deviation suggests that indigenous Australian *Panicum* spp. maintain a unique genetic identity despite the opportunity for possible introgression or hybridisation events experienced during the naturalisation process by their non-native relatives (Hovick and Whitney 2014). However, a broader and geographically defined set of global *Panicum* spp. specimens are required to further explore this observation.

In conclusion, this study indicates that a genetic approach using three concatenated loci provided an accurate and potentially cost effective tool for distinguishing *Panicum* spp. at the species level, regardless of growth stage. Using this method, we confirmed that the majority of *Panicum* specimens collected across the Riverina in 2017 were *P. hillmanii*. This was somewhat unexpected since very few (6) herbarium collections had been made in NSW prior to our study, indicating that the species has been overlooked or its range has expanded relatively recently. Although voucher specimens keyed well to *P. hillmanii* based on morphology, further molecular analysis of samples from American herbaria will be undertaken to confirm whether the name is being correctly applied in Australia.

The genetic approach utilised in this study could have application in: 1) comprehensive surveillance studies of *Panicum* spp.; and 2) the development of more useful diagnostic tools, including loop-mediated isothermal amplification (LAMP), for rapid identification of *Panicum* spp. by stakeholders. Results may also provide improved options for invasive *Panicum* spp. control in Australian pastures and summer fallows.

ACKNOWLEDGMENTS

The authors acknowledge financial support from Meat and Livestock Australia Project B WEE 0146, the Graham Centre for Agricultural Innovation and CSU School of Animal and Veterinary Science, and assistance with sample collection by G. Health, R. Powell, J. Moore and S. Gurusinghe.

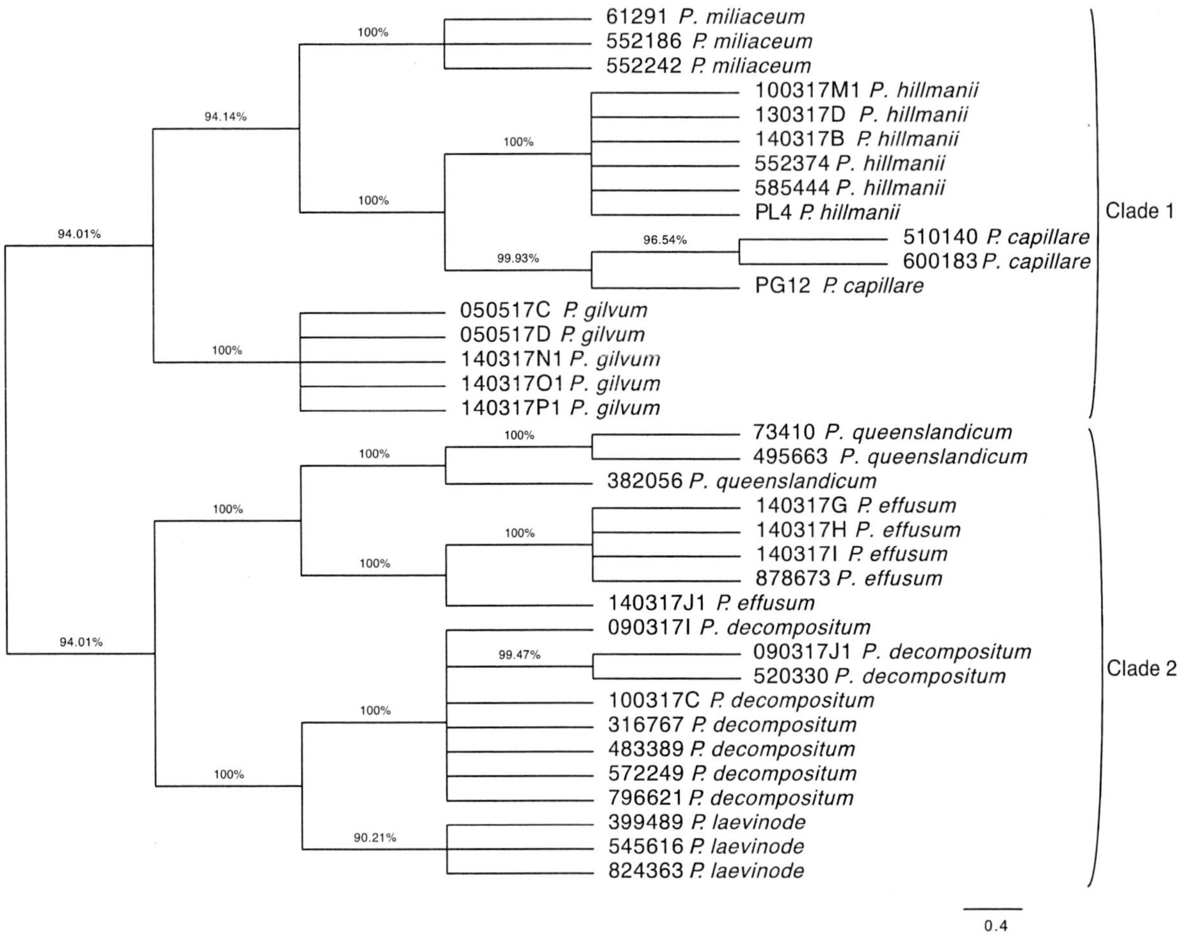

Figure 3. Bayesian phylogenetic relationships among eight *Panicum* spp inferred from the concatenation of three gene sequences. Clade posterior probability support >90% indicated at nodes.

REFERENCES

Aliabadian, M., Kaboli, M., Nijman, V. and Vences, M. (2009). Molecular identification of birds: performance of distance-based DNA barcoding in three genes to delimit parapatric species. *PLoS One* 4, e4119.

Aliscioni, S.S., Giussani, L.M., Zuloaga, F.O. and Kellogg, E.A. (2003). A molecular phylogeny of *Panicum* (Poaceae: Paniceae): tests of monophyly and phylogenetic placement within the Panicoideae. *American Journal of Botany* 90, 796-821.

Byng, J.W. (2014). The flowering plants handbook: a practical guide to families and genera of the world (Plant Gateway, Hertford, United Kingdom).

CBOL Plant Working Group (2009). A DNA barcode for land plants. *Proceedings of the National Academy of Sciences of the United States of America* 106, 12794-97.

CCHAH, Council of Heads of Australasian Herbaria (2005 onwards) Australian Plant Census. https://biodiversity.org.au/nsl/services/APC (accessed 10 April 2018).

Coissac, E., Hollingsworth, P.M., Lavergne, S. and Taberlet, P. (2016). From barcodes to genomes: extending the concept of DNA barcoding. *Molecular Ecology* 25, 1423-28.

Darriba, D., Taboada, G.L., Doallo, R. and Posada, D. (2012) jModelTest 2: more models, new heuristics and parallel computing. *Nature Methods* 9, 772.

Hebert, P.D.N., Cywinska, A., Ball, S.L. and deWaard, J.R. (2003). Biological identifications through DNA barcodes. *Proceedings of the Royal Society B: Biological Sciences* 270, 313-21.

Hollingsworth, P.M., Graham, S.W. and Little, D.P. (2011). Choosing and using a plant DNA barcode. *PLoS One* 6, e19254.

Hovick, S.M. and Whitney, K.D. (2014). Hybridisation is associated with increased fecundity and size in invasive taxa: meta-analytic support for the hybridisation-invasion hypothesis. *Ecology Letters* 17, 1464-77.

Huelsenbeck, J.P. and Ronquist, F. (2001). MR-BAYES: Bayesian inference of phylogenetic trees. *Bioinformatics (Oxford, England)* 17, 754-5.

Kearse, M., Moir, R., Wilson, A., Stones-Havas, S., Cheung, M.,Sturrock, S., Buxton, S., Cooper, A., Markowitz, S., Duran, C., Thierer, T., Ashton, B., Meintjes, P. and Drummond, A. (2012). Geneious Basic: an integrated and extendable desktop software platform for the organization and analysis of sequence data. *Bioinformatics (Oxford, England)* 28, 1647-9.

Kumar, S., Stecher, G. and Tamura, K. (2016). MEGA7: Molecular Evolutionary Genetics Analysis Version 7.0 for Bigger Datasets. *Molecular Biology and Evolution* 33, 1870-4.

Llewellyn, R.S., Ronning D., Ouzman J., Walker S., Mayfield A. and Clarke M. (2016). Impact of weeds on Australian grain production: the cost of weeds to Australian grain growers and the adoption of weed management and tillage practices Report for GRDC. CSIRO, Australia. https://grdc.com.au/__data/assets/pdf_file/0027/75843/grdc_weeds_review_r8.pdf.pdf (accessed 17 April 2018).

Miles, C.O., Wilkins, A.L., Munday, S.C., Holland, P.T., Smith, B.L., Lancaster, M.J. and Enabling, P.P. (1992). Identification of the calcium salt of epismilagenin β-D-Glucuronide in the bile crystals of sheep affected by *Panicum diehotomiflorum* and *Panicum schinzi*i toxicoses. *Journal of Agricultural and Food Chemistry* 40, 1606-9.

Quinn, J.C., Kessell, A. and Weston, L.A. (2014). Secondary plant products causing photosensitization in grazing herbivores: Their structure, activity and regulation. *International Journal of Molecular Sciences* 15, 1441-65.

Zhu, X., Meyer, L., Gopurenko, D. and Weston, P.A. (2014). Selection of DNA barcoding regions for identification and genetic analysis of two *Echium* invaders in Australia: *E. plantagineum* and *E. vulgare*. Proceedings of the 19th Australasian Weeds Conference, ed. M. Baker, pp. 396-400. (Weed Society of Tasmania, Hobart)

Zuloaga, F.O., Scataglini, M.A. and Morrone, O. (2010). A phylogenetic evaluation of *Panicum* sects. Agrostoidea, Megista, Prionitia and Tenera (Panicoideae, Poaceae): Two new genera, Stephostachys and Sorengia. *Taxon* 59, 1535-46.

Screening of gene regions for genetic diversity in global parthenium weed (*Parthenium hysterophorus* L.) populations

Ali A. Bajwa[1,2], Xiaocheng Zhu[3], Bhagirath S. Chauhan[2], Steve W. Adkins[1,2] and Leslie A. Weston[3]
[1] School of Agriculture and Food Sciences, The University of Queensland, Gatton, Queensland 4343, Australia
[2] The Centre for Crop Science, Queensland Alliance for Agriculture and Food Innovation, The University of Queensland, Gatton, Queensland 4343, Australia
[3] Graham Centre for Agricultural Innovation (Charles Sturt University and New South Wales Department of Primary Industries), Charles Sturt University, Wagga Wagga, New South Wales 2678, Australia
(a.bajwa@uq.edu.au)

Summary Parthenium weed (*Parthenium hysterophorus* L.) is an invasive species in more than 40 countries and a Weed of National Significance in Australia. It causes substantial losses to crop and pasture production and health problems in humans and livestock. The high invasive potential of this species is often attributed to its unique morphological features, competitive ability, and allelopathic potential; however, the level of genetic diversity in this species and how diversity contributes to its success is not currently clear. This study aimed to identify informative gene regions and sequences for the study of parthenium weed population genetics using samples collected from its introduced range in Australia (including two distinct biotypes) and the native ranges in Costa Rica, Mexico and the United States of America. Five gene regions were investigated including one nuclear region (ITS) and four chloroplast regions (*trn*H-*psb*A, *trn*L intron, *trn*L-*trn*F spacer and *matK*). Limited genetic diversity was found both in Australian and native populations, and our findings suggest that these gene regions are potentially not optimal for evaluation of population genetic diversity investigation in parthenium weed due to lack of polymorphism or that such genetic diversity is extremely limited. Our findings agree with previous reports of limited genetic variation observed in Australian, Chinese and Pakistan parthenium weed populations. Therefore, for future studies, we suggest the use of next generation sequencing approach such as targeted sequencing, RADseq (Restriction site Associated DNA Sequencing) or MassARRAY® to develop informative genetic markers for this species in an attempt to evaluate population diversity.

Keywords Parthenium weed, invasive alien species, DNA sequencing.

INTRODUCTION

Parthenium weed is a highly invasive species, but the actual mechanism of its invasion is not well understood (Bajwa *et al.* 2016). Several factors could potentially contribute to its spread including aggressive growth habit, high seed production, absence of natural enemies in introduced range, and abiotic stress tolerance ability (Bajwa *et al.* 2016, Nguyen *et al.* 2017). However, only a few studies have been conducted to: 1) evaluate the role of genetic diversity in parthenium weed invasion; and 2) to screen the gene regions or markers suitable to study the genetic diversity in this species.

In Australia, parthenium weed has two distinct biotypes, Toogoolawah and Clermont, both named from collection sites across Australia, but introduced on separate occasions from the United States of America (USA). The first introduction (Toogoolawah biotype) has remained uniquely non-invasive. However, the second introduction (Clermont biotype) of parthenium weed has spread across a large area in Queensland (Adkins and Shabbir 2014). The Clermont biotype is not only more invasive but also has higher germination and growth rates, reproductive output and competitive ability, as compared to the Toogoolawah biotype (Hanif 2014, Bajwa *et al.* 2017, 2018). Hanif (2014) reported that the Toogoolawah and Clermont biotypes were introduced independently in Australia, possibly from the northern and southern Texas races, respectively. However, limited genetic diversity was observed between both biotypes. Earlier studies have evaluated random amplified polymorphic DNA (*RAPD*), inter-simple sequence repeats (*ISSR*), and microsatellite (*SSR*) markers to assess genetic diversity in parthenium weed and reported very limited genetic diversity among various populations (Graham and Lange 1998, Tang *et al.* 2009, Qian *et al.* 2012, Jabeen *et al.* 2015).

Studying genetic lineages can help in identifying the point of origin of a specific invasive species, which might be useful for its management, particularly when considering biological control. In addition, a high level of genetic diversity is generally considered as key mechanism associated with successful invasion (Prentis *et al.* 2008). Therefore, it is important to evaluate genetic diversity among different populations of parthenium weed, and to do that, the screening of both chloroplast and nuclear gene regions is a crucial preliminary step. This study therefore aimed to select reliable chloroplast and nuclear gene regions for study of the population genetics of parthenium weed using both native and Australian populations.

MATERIALS AND METHODS

Parthenium weed leaf tissue was sampled from the field both in Australia and its native range in Costa Rica, Mexico and the USA. The sampled locations in each native country covered 2–5 geographically distant locations (Table 1). The Australian samples represented two local biotypes, Clermont and Toogoolawah. Five individuals were sampled in each population. Leaf tissue was stored in silica gel until DNA extraction.

The DNA extraction, PCR, sequencing, alignment and sequence analysis were performed as described by Zhu *et al.* (2017). Based on previous plant-based studies, five gene sequences were tested for their suitability as a potential candidate for population genetics, including one nuclear region (ITS) and four chloroplastic regions (*matK*, *trn*H-*psb*A spacer, *trn*L intron and *trn*L-*trn*F spacer).

RESULTS AND DISCUSSION

PCR and sequencing were nearly 100% successful for all gene regions. All sequences obtained from this study were deposited in the GenBank with accession numbers from MH017879–MH17987. As previously reported, very limited genetic diversity was found among samples, even within the native populations. Only one haplotype was found using the *trn*L-*trn*F spacer, and two haplotypes were found using each of *trn*L intron, *trn*H-*psb*A spacer and *matK* regions, while three were observed using ITS. There was no correlation between haplotypes and biotypes, or between haplotypes and geographical distribution.

This is the first study that evaluated multiple native populations covering vast geographical distances. The low levels of diversity apparent across the native and Australian populations may indicate the examined loci are insufficiently polymorphic to track the population genetics and invasion history of this species. In contrast, Tang *et al.* (2009) reported that the morphologically similar parthenium weed biotypes in China had contrasting genetic makeup and, thus, were attributed to independent introductions.

The results of this study are in agreement with those of previous studies of the species which also found very limited genetic diversity using genetic analysis (Hanif 2014, Jabeen *et al.* 2015). Jabeen *et al.* (2015) studied the genetic structure of 11 populations (95 individuals) from across Pakistan including the two populations (10 individuals) from Australia using *ISSR* fingerprinting. It was found that 18% genetic diversity existed among the populations, while 82% existed within the populations. Genetic diversity was highest among the populations from Pakistan; however, gene flow was limited. An earlier study using the ITS markers reported the Australian biotypes to be more closely related to Mexican biotypes, but significantly different from Indian biotypes of parthenium weed (Graham and Lange 1998).

In conclusion, the gene regions tested in this study showed limited genetic diversity in parthenium weed populations from native and Australian ranges. Future research aiming to study the population genetics of this species may potentially benefit from the use of polymorphic markers including *SSR* or *AFLP*. Next Generation Sequencing (NGS) may also be applied to sequence the chloroplast genome of parthenium weed to search gene regions exhibiting greater polymorphism. In addition, for large population genetic studies, NGS approaches including RADseq or MassARRAY® could be considered to try to obtain highly variable SNPs in the parthenium weed genome.

ACKNOWLEDGMENTS

This manuscript is part of the PhD research project of Ali Ahsan Bajwa who is thankful to the Australian Government and The University of Queensland (UQ), Australia for the provision of a Research Training Program Scholarship and UQ Centennial Scholarship, respectively. This study was performed at the Charles Sturt University (CSU) National Life Sciences Research Hub facility in Wagga Wagga, New South Wales with support provided by UQ, the Graham Centre for Agricultural Innovation at CSU and Meat & Livestock Australia (MLA). The authors are thankful to Dr. Fernando Ramirez Munoz, Mr. Eder Ramos Hernández and Dr. Muthu Bagavathiannan for providing the leaf tissue samples of parthenium weed from Costa Rica, Mexico and the United States of America, respectively.

Table 1. Parthenium weed samples, their locations and the GenBank accession number of each gene region.

Sample ID	Source country	Latitude (N)	Longitude (E)	GenBank accession number				
				ITS	*matK*	*trn*L intron	*trn*L-*trn*F	*trn*H-*psb*A
PH0001	Costa Rica	9.960	−84.112	MH017936	MH017961	MH017921	MH017894	MH017879
PH0002	Costa Rica	9.960	−84.112	MH017937	MH017962	MH017922	MH017895	MH017880
PH0013	Costa Rica	10.460	−84.292		MH017963	MH017923	MH017896	MH017881
PH0014	Costa Rica	10.460	−84.292	MH017938	MH017964	MH017924	MH017897	MH017882
PH0015	Costa Rica	10.460	−84.292	MH017939	MH017965	MH017925	MH017898	MH017883
PH0016	Australia	−22.876	147.775	MH017940	MH017976		MH017899	
PH0017	Australia	−22.876	147.775	MH017941	MH017977		MH017900	
PH0018	Australia	−22.876	147.775	MH017942	MH017978		MH017901	
PH0019	Australia	−22.876	147.775	MH017943	MH017979		MH017902	
PH0020	Australia	−22.876	147.775	MH017944	MH017980		MH017903	
PH0022	Australia	−22.876	147.775	MH017945	MH017981		MH017904	
PH0027	Australia	−27.139	152.377	MH017946	MH017982		MH017905	
PH0028	Australia	−27.139	152.377	MH017947	MH017983		MH017906	
PH0029	Australia	−27.139	152.377	MH017948	MH017984		MH017907	
PH0030	Australia	−27.139	152.377	MH017949	MH017985		MH017908	
PH0031	Australia	−27.139	152.377	MH017950	MH017986		MH017909	
PH0035	Australia	−27.139	152.377	MH017951	MH017987		MH017910	
PH0161	Mexico	18.006	−93.301	MH017952	MH017966	MH017926	MH017911	MH017884
PH0162	Mexico	18.006	−93.301	MH017953	MH017967	MH017927	MH017912	MH017885
PH0168	Mexico	18.017	−93.315		MH017968	MH017928	MH017913	MH017886
PH0180	Mexico	17.972	−93.381	MH017954	MH017969	MH017929	MH017914	MH017887
PH0186	Mexico	17.855	−93.394	MH017955	MH017970	MH017930	MH017915	MH017888
PH0333	USA	28.180	−97.767	MH017956	MH017971	MH017931	MH017916	MH017889
PH0348	USA	27.838	−97.720	MH017957	MH017972	MH017932	MH017917	MH017890
PH0373	USA	28.550	−96.645	MH017958	MH017973	MH017933	MH017918	MH017891
PH0383	USA	29.313	−96.275	MH017959	MH017974	MH017934	MH017919	MH017892
PH0388	USA	30.538	−96.421	MH017960	MH017975	MH017935	MH017920	MH017893

REFERENCES

Adkins, S. and Shabbir, A. (2014). Biology, ecology and management of the invasive parthenium weed (*Parthenium hysterophorus* L.). *Pest Management Science* 70, 1023-9.

Bajwa, A.A., Chauhan, B.S. and Adkins, S. (2017). Morphological, physiological and biochemical responses of two Australian biotypes of *Parthenium hysterophorus* to different soil moisture regimes. *Environmental Science and Pollution Research* 24, 16186-94.

Bajwa, A.A., Chauhan, B.S. and Adkins, S.W. (2018). Germination ecology of two Australian biotypes of ragweed parthenium (*Parthenium hysterophorus*) relates to their invasiveness. *Weed Science* 66, 62-70.

Bajwa, A.A., Chauhan, B.S., Farooq, M., Shabbir, A. and Adkins, S.W. (2016). What do we really know about alien plant invasion? A review of the invasion mechanism of one of the world's worst weeds. *Planta* 244, 39-57.

Graham, G.C. and Lange, C.L. (1998). Genetic analysis of relationship of Parthenium occurrences in Australia and indications of its origins. Research report. Cooperative Research Centre for Tropical Pest Management, The University of Queensland.

Hanif, Z. (2014). Characterization of genetic diversity and invasive potential of parthenium weed in Australia. PhD Thesis, School of Agriculture and Food Sciences, The University of Queensland, Australia.

Jabeen, R., Prentis, P., Anjum, T. and Adkins, S.W. (2015) Genetic structure of invasive weed *Parthenium hysterophorus* in Australia and Pakistan. *International Journal of Agriculture and Biology* 17, 327-33.

Nguyen, T., Bajwa, A.A., Navie, S., O'Donnell, C., and Adkins, S. (2017). Parthenium weed (*Parthenium hysterophorus* L.) and climate change: The effect of CO_2 concentration, temperature, and water deficit on growth and reproduction of two biotypes. *Environmental Science and Pollution Research* 24, 10727-39.

Prentis, P.J., Wilson, J.R.U., Dormontt, E.E., Richardson, D.M. and Lowe, A.J. (2008). Adaptive evolution in invasive species. *Trends in Plant Science* 13, 288-94.

Qian, M., Li, G., Tang, S., Qu, X., Lin, Y., Yin, J., van Klinken, R.D. and Geng, Y. (2012). Development of microsatellites for the invasive weed parthenium weed (Asteraceae). *American Journal of Botany* 99, 277-9.

Tang, S.Q., Wei, F., Zeng, L.Y., Li, X.K., Tang, S.C., Zhong, Y., Geng, Y. (2009). Multiple introductions are responsible for the disjunct distributions of invasive *Parthenium hysterophorus* in China: evidence from nuclear and chloroplast DNA. *Weed Research* 49, 373-80.

Zhu, X., Weston, P.A., Skoneczny, D., Gopurenko, D., Meyer, L., Lepschi, B.L., Callaway, R.M. Gurr, G.M. and Weston, L.A. (2017). Ecology and genetics affect relative invasion success of two *Echium* species in southern Australia. *Scientific Reports* 7, 42792.

A conversation starter: Standardising terminology and definitions for weeds status in Australia – What do you think?

Chris Brodie[1], Peter Lang[1] and Michelle Waycott[1,2]
[1] State Herbarium of South Australia, Adelaide Botanic Gardens, Hackney Road, Adelaide, South Australia 5001, Australia
[2] The University of Adelaide, School of Biological Sciences, Adelaide, South Australia 5001, Australia

Summary A 'weed' is classified by the State Herbarium of South Australia (State Herbarium) as a plant that has originally been introduced to an area by humans deliberately or accidentally, has spread and established itself by self-propagation where it is not wanted. Scientific verification and 'official' recognition of existing and new naturalised or 'weed' taxa in South Australia is achieved through the lodgement of voucher specimens and validation by taxonomic botanists in the State Herbarium. For taxa not previously recorded for South Australia, an entry is then made in the Census of South Australian Vascular Plants, Algae and Fungi (http://flora.sa.gov.au/census.shtml). Other states in Australia vary in how they recognise new weed taxa 'officially'.

A major effort in recent years has been made by the State Herbarium and its associates in collecting weeds to document the occurrence and spread of new and existing weed species. Many new weed incursions, such as, *Cardiospermum grandiflorum* (Balloon vine), *Cupressus goveniana* (Californian cypress) and *Eucalyptus* species and relatives have been recorded through this process. However, some taxa do not fit the criteria for a clearly established weed. In these cases, the State Herbarium assigns them the only other weed status currently available on our census system, 'questionably naturalised'. This status has been applied where there is insufficient evidence to be confident of establishment status, or where establishment is only in its early stages, or for casual occurrences that do not persist, or when established plants are highly restricted and present in low numbers at only one or a few locations. Having only two categories does not adequately inform on the true 'weed status' of these taxa in the landscape.

All states and territories have varying terminology and definitions for the status for introduced plants including those called weeds. These categories in South Australia need reviewing, expanding and updating. There would be clear benefits in having common methodology, terminology and definitions applied uniformly across Australia. Standard terminology and definitions are proposed and discussed using case examples.

The challenges of herbicide resistance in non-agricultural weed management systems

Andrew Storrie
AGRONOMO, PO Box 5944, Albany, Western Australia 6332, Australia
Executive Officer, Australian Glyphosate Sustainability Working Group
(andrew@agronomo.com.au)

Summary The development of weeds with resistance to glyphosate has paralleled the increasing reliance on this herbicide. Glyphosate is the world's most widely used herbicide with global use exceeding 850,000 tonnes in 2015. Western agricultural systems are increasingly reliant on glyphosate due to the expanding cultivation of glyphosate-resistant crops and the widespread use of glyphosate fallows for reduced or no-tillage farming systems. These systems have environmental benefits of reduced soil erosion and fuel consumption, with production benefits of lower labour and capital costs, and higher and more reliable crop yield through improved stored soil water.

While most cases of glyphosate resistance have been confirmed in agricultural environments, the number of glyphosate-resistant weed populations being found in non-agricultural land use is also increasing.

The world's first case was found in 1996 in annual ryegrass (*Lolium rigidum* Gaudin.) in no-till small grains farming in Australia. Currently there are 38 species that have developed resistance to glyphosate, of which Australia has 17.

The increasing incidence of glyphosate resistance is concurrently increasing the use and pressure on other herbicides with different modes of action such as paraquat, glufosinate and grass selective herbicides.

Most Australian broadacre farmers have been exposed to the herbicide resistance management message for over 30 years. However, a 2011–2012 national study of glyphosate use practices in the non-agricultural sector found little awareness of the issues. Current evidence suggests little has changed in the last 6 years except for more cases of glyphosate resistance. Key targets for many weed management decision-makers are based on cost and aversion risk to staff and the public.

This paper discusses the current direction of the development of herbicide resistance and suggests appropriate awareness and extension programs for non-agricultural users of herbicides.

Keywords Glyphosate, herbicide-tolerant, paraquat, amitrole, glufosinate, non-crop, lineal reserve, roadside, railway, irrigation channel.

INTRODUCTION

Glyphosate is the world's most widely used herbicide with annual use estimated at over 826,000 tonnes in 2014 (Anon. 2018). A large proportion of this usage will be with the increase in herbicide-tolerant crops. It has been estimated that in 2016 there were 186 million hectares of genetically modified crops grown in the world (James 2017). Glyphosate is also widely used for weed control in other sectors including native vegetation rehabilitation, residential yards, parks and gardens, industrial premises, roadsides, rail lines and forestry. Its popularity is due to: low cost; the large number of species it controls; low odour; minimal soil activity; and low mammalian toxicity. The Australian Pesticides & Veterinary Medicines Authority (APVMA) has determined that glyphosate poses little risk to human health (Anon. 2017).

The wide usage of glyphosate in Australia is demonstrated by registration of 584 products containing glyphosate with nearly 40,000 registered uses (Infopest 2018).

Heavy reliance on any herbicide leads to the selection weed populations that can be both resistant or tolerant to that herbicide mode of action (Storrie 2014). Outside of agricultural uses, roadsides are one of the highest risk areas for selecting herbicide-resistant or tolerant weeds due to the repeated use of a herbicide with few non-herbicidal control strategies being used.

This paper updates the current world and Australian development of glyphosate-resistant weed populations and discusses alternatives to current use patterns with particular emphasis on roadsides. It also explores the possibility of applying an awareness strategy used in weed biosecurity on intractable roadside weeds.

CURRENT GLOBAL STATE OF GLYPHOSATE RESISTANCE

There are now 41 weed species confirmed to have resistant populations globally in 29 countries (Heap 2018). This has doubled since 2011.

The concerning and growing complication is that at least one quarter of these populations are also resistant to one or more other herbicide modes of action. Australia holds the multiple resistant record so far with 5 modes of action in populations of annual ryegrass (*Lolium rigidum* Gaudin) and winter grass (*Poa annua* L.).

Resistance to multiple modes of action pose huge increases in weed control difficulty. This is called multiple resistance and it is a result of using a number of different herbicide modes of action without managing the seed set of spray survivors. Unfortunately, rotating herbicide modes of action continues to be one of the first recommendations in many herbicide-resistance management strategies.

In the last 7 years some countries have had large increases in the number resistant species while other countries have had no or little change (Table 1). Australia's more than tripling of glyphosate-resistant species largely comes from the widespread adoption of no-till crops with a concurrent decline in cultivation for weed control. This is compounded by the extensive use of fallow that is kept weed-free by the repeated use of glyphosate (Storrie 2014). The Americas have had the next largest increase (excepting Chile) largely due to the widespread adoption of glyphosate-resistant crops. European countries do not grow significant areas of herbicide-tolerant crops.

Other knockdowns There is a problem with the increasing resistance to other knockdown (non-selective) herbicides such as paraquat (Group L), glufosinate (Group N) and amitrole (Group Q). To date Australia has 10 species with Group L resistance. Globally three species have glufosinate resistant populations and all but one have multiple resistance to 2, 3 or 4 modes of action. Six species have resistance to amitrole, including annual ryegrass in Australia. Three of these species are also resistant to 2 or 3 modes of action.

Table 1. The number of species with populations resistant to glyphosate in 2011 and 2018 for a range of countries. (Heap 2018).

Country	Number of species	
	2011	2018
Australia	5	17
USA	11	17
Canada	1	6
Brazil	5	8
Argentina	3	11
Chile	1	1
Spain	5	5
France	2	2
China	2	2
Israel	2	2
South Africa	3	3

Glufosinate has been widely used in fruit trees and vines following registration in 1995. Glufosinate is now also being used in broadacre cropping as a fallow herbicide and as a post-emergent in glufosinate-resistant LibertyLink™ crops including cotton, maize, soybeans, sugar beet and canola. There are over 24 million hectares of LibertyLink™ crops being grown globally (James 2017).

GLYPHOSATE RESISTANCE IN AUSTRALIA

There are currently 17 species with populations confirmed resistant to glyphosate including 10 grass species and 7 broadleaf species of which 6 are winter-growing weed species and 11 are either non-seasonal or summer-growing species (Preston 2018). A detailed list can be found elsewhere (AGSWG 2018).

Many of these populations have been selected in broadacre cropping or related areas such as around buildings and along irrigation channels. Annual ryegrass, flaxleaf fleabane (*Conyza bonariensis* (L.) Cronquist), tall fleabane (*Conyza sumatrensis* (Retz.) E. Walker), windmill grass (*Chloris truncata* R.Br.) and feathertop Rhodes grass (*Chloris virgata* Sw.) are however now significant roadside weeds.

THE PROBLEM WITH ROADSIDES

Australia has nearly 800,000 km of roads (Anon. 1987) that are at risk of developing weeds with resistance to glyphosate. Road safety and infrastructure maintenance are the key drivers for weed control. Roadsides must have clear lines-of-site around and up to posts and signs and the allowable distance will vary with the allowed speed limit. Often there is a 30 cm maximum height for roadside vegetation. Road edges or shoulders are also managed to keep them clear of vegetation to minimise movement of water under the 'seal' or into the road-base to minimise pot-holing and increased maintenance costs (Storrie *et al.* 2012). The two main methods of vegetation management are slashing and spraying.

Slashing is relatively slow and often requires several staff to alert motorists to the tractor-slasher ahead. In wet years and higher rainfall areas such as the coast it can be difficult to keep vegetation at the allowable height. The design of many slashers also makes it impossible to prevent weed seed spread without significant down-time for cleaning. Also, most slashers are not able to get close to roadside furniture. Some managers deal with this by spraying around posts and roadside furniture.

Spraying the road shoulder is now the major form of vegetation management outside town boundaries.

The width of the sprayed area varies from 1–5 m (Storrie *et al.* 2012). Glyphosate is the main herbicide used. Tank-mixing with other herbicide modes of action is often used to broaden the range of weeds controlled. The use of tank-mix partners, if any, will vary greatly depending on the management authority and state legislation.

The number of glyphosate applications in a year is usually determined by use situation and rainfall. Drier areas normally have 1–2 glyphosate applications per year, whereas wetter areas range from 1–5 applications. In most environments spring is the key spray time, while summer spraying is dependent on summer rainfall (Storrie *et al.* 2012).

Market research has found most non-agricultural land managers are ill prepared to deal with glyphosate resistance and found that many councils would only use glyphosate for simplicity, safety and cost (Storrie *et al.* 2012). Field staff and contractors wanted training, but this was not mirrored by management.

A review of weed management plans by authorities involved with roadside management shows no mention of the threat of herbicide resistance. Roadside weed management focuses on managing remnant vegetation, and declared and Weeds of National Significance.

A 2011–2012 project on non-agricultural glyphosate use in Australia included a physical survey that targeted four weeds – annual ryegrass, fleabanes, windmill grass, and awnless barnyard grass (*Echinochloa colona* (L.) Link) (Malone *et al.* 2012). Testing showed that half the annual ryegrass and fleabane samples were resistant to glyphosate as were a smaller number of windmill grass populations. The majority of resistant samples came from roadsides.

Glyphosate-tolerant weeds are also becoming an increasing roadside problem. Glyphosate-tolerant weeds include *Hyparrhenia* species, African lovegrass (*Eragrostis curvula* (Schrad.) Nees), windmill grass, feathertop Rhodes grass, flaxleaf and tall fleabanes, crownbeard (*Verbesina encelioides* (Cav.) A.Gray subsp. *encelioides*), stinkwort (*Dittrichia graveolens* (L.) Greuter) and dove weed (*Croton setiger* Hook.).

Failure to manage for glyphosate resistance on roadsides is well demonstrated in South Australia (SA). In the early 2000s the SA government decided that only glyphosate would be used on roadsides for environmental work, and for health and safety reasons. In 2011 the Department of Planning, Transport & Infrastructure were alerted to widespread infestations of glyphosate-resistant annual ryegrass on many SA roads. This was raised by irate farmer organisations concerned about the spread of glyphosate-resistant weeds from roadsides into their paddocks. The SA government then had to quickly develop a state-wide management plan to arrest the problem. Resistant infestations on roadsides are now mapped and receive additional control tactics as well including a tank-mix of paraquat plus amitrole following glyphosate.

How to manage roadsides There are two basic options to manage difficult roadside weeds. First is the reactionary approach where the decision makers are forced into playing 'catch-up', often implementing systems that are expensive and less than ideal from a cost and worker safety perspective. The second and preferable option is proactive implementation of relatively simple monitoring and management systems to identify and deal with hard-to-control weeds.

As discussed there is ample evidence across the nation of increasing incidence of both glyphosate-resistant and glyphosate-tolerant weeds on roadsides, however despite this, little proactive management takes place, due to either decision makers for remaining unaware of the problem, or they consider it a low priority (Congreve *et al.* 2012).

For change to occur contractors, field staff and managers must understand the threats and cascading effects resulting from hard-to-control weeds. Many authorities have already started mapping 'significant' roadside vegetation so it is only a small step to map glyphosate-resistant and tolerant weeds as they are the ones still growing after treatment.

Resistance management costs will need to be included in contract costs with the contractor not bearing the additional cost.

To initiate change, messages need to be framed in positive terms to which managers can relate. Messaging to get managers 'on-board' might include:

cheaper overall weed management – monitoring and mapping of problem weeds allows site-specific programs to be used, rather that treat all roadsides with more expensive strategies;

better relationships with ratepayers / Councillors. Spread of glyphosate resistant or tolerant weeds into farmers' paddocks creates unnecessary conflict; and

well trained and involved staff will be happier and work more effectively.

A great start to the management of these weeds would be a modification of the 'Red guide post program' (Bosse 2017). This program aimed to increase awareness of declared weeds on roadsides and was supported by a well planned extension campaign that included a wide range of pathways targeting all road users and managers.

Using a similar strategy for resistant and tolerant roadside weeds could easily improve adoption of better management, reduce the risk of increasing roadside vegetation management costs and reduce the spread of these weeds.

ACKNOWLEDGMENTS

The author would like to thank the RIRDC for having the foresight to fund Project PRJ-006914 – *Management of glyphosate resistance in non-agricultural areas.*

REFERENCES

AGSWG, Australian Glyphosate Sustainability Working Group, http://www.glyphosateresistance.org.au/register_summary.html (accessed 1 August 2018).

Anon. (1987). Assessment of the Australian road system: 1987. Bureau of Transport Economics. https://bitre.gov.au/publications/1987/files/report_061.pdf (accessed 18 April 2018).

Anon. (2017). Final regulatory position: Consideration of the evidence for a formal reconsideration of glyphosate. Australian Pesticides and Veterinary Medicines Authority, Canberra, Australian Capital Territory.

Anon. (2018). Glyphosate use worldwide from 1994 to 2014. 'Statista' https://www.statista.com/statistics/567250/glyphosate-use-worldwide/ (accessed 9 July 2018).

Bosse, P. (2017) Red Guide Posts – a weed awareness tool! (Riverina Regional Weed Committee: Albury, Australia). http://www.riverinaweeds.org.au/rgp/ (accessed 20 March 2018)

Congreve, M., Storrie, A.M., Cameron, A.J. (2012). Glyphosate resistance in non-agricultural areas – New South Wales, Victoria, South Australia and WA – Market survey report – RIRDC PRJ-006914. Rural Industries Research & Development Corporation, Canberra, Australian Capital Territory.

Heap, I. (2018). The International Survey of Herbicide Resistant Weeds. http://www.weedscience.org (accessed April 17, 2018).

Infopest (2018). www.infopest.com.au (accessed 1 August 2018).

James, C. (2017). Global status of commercialized biotech/GM crops in 2017: Biotech crop adoption surges as economic benefits accumulate in 22 years. International Service for the Acquisition of Agri-biotech Applications (ISAAA), Ithaca, New York.

Malone, J., Cook, T., Wu, H., Hashem, A., Boutsalis, P. and Preston, C. (2012). Glyphosate resistance in non-cropping areas of Australia. *In* 'Giving a RATS' Issue 1 – Autumn 2012 p. 7. http://www.agronomo.com.au/giving-a-rats/ (accessed 17 April 2018).

Preston, C. (2018). The Australian Glyphosate Sustainability Working Group. http://www.glyphosateresistance.org.au/ (accessed 17 April 2018),

Storrie. A.M., Cameron. A.J. and Congreve. M. (2012) Glyphosate resistance in roads and rail – how prepared are we? *In* 'Giving a RATS' – Issue 2 – Winter 2012 p. 4-7. http://www.agronomo.com.au/giving-a-rats/ (accessed 17 April 2018).

Storrie, A.M. (ed.) (2014). 'Integrated weed management manual for Australian Cropping Systems.' 2nd Edition. (Grains Research and Development Corporation, Canberra, Australia).

Using detection dogs to assist weed eradication in conservation areas

Hillary Cherry
Pests and Weeds Unit, National Parks and Wildlife Service, NSW Office of Environment and Heritage,
PO Box 1967 Hurstville, New South Wales 1481, Australia
(hillary.cherry@environment.nsw.gov.au)

Summary Eradication is a cost-effective management objective at the early stages of weed invasions. However, timely interventions that allow successful eradications can be challenging and resource intensive. Characteristics of the target species and the invaded environment can impact on the ability to delimit the infestation or to detect every plant within the infestation, both of which are critical for eradication. In New South Wales, Australia, we are attempting eradication of mouse ear (*Hieracium pilosella* L.) and Orange hawkweed (*Hieracium aurantiacum* L.). Invaded plant communities include tussock grasslands and heathlands, where hawkweed plants can be difficult to detect among dense vegetation. Additionally, small non-flowering hawkweed rosettes can be difficult for humans to distinguish visually. Thus, traditional human surveillance requires extensive time and resource allocation. To improve detection speed and accuracy, we are using scent detection dogs to complement human surveillance. Over a two-year period, two spaniels were trained to successfully detect hawkweeds in field environments. One dog has also been trained to detect alligator weed (*Alternanthera philoxeroides*), and has successfully identified plant fragments in aquatic and streamside vegetation. This presentation outlines the use of these 'weed eradication detector dogs' and discusses: i) dog training procedures and field trials, including challenges related to biosecurity and presenting the correct target scent to dogs, ii) factors contributing to dog detection ability, including weather, terrain, and handler skills and ability, iii) the unexpected benefits of extensive media interest, and iv) the costs, benefits and challenges of implementing detector dog surveillance to improve weed eradication efforts.

Keywords Weed detection dog, eradication, hawkweed, alligator weed.

Evolved resistance to glyphosate in barley grass (*Hordeum* spp.) in South Australia

Patricia Adu-Yeboah, Jenna Malone, Benjamin Fleet, Samuel Kleeman, Gurjeet Gill and Christopher Preston
School of Agriculture, Food and Wine, University of Adelaide, PMB 1, Glen Osmond,
South Australia 5064, Australia
(patricialowery.adu-yeboah@adelaide.edu.au)

Summary The barley grasses (*Hordeum* spp.) are problematic weeds in South Australian agricultural systems. Recently barley grass suspected of resistance to glyphosate was reported from a fence line on the Yorke Peninsula in South Australia and a vineyard in the Barossa Valley in situations where glyphosate had been used extensively to control weeds. Preliminary dose response experiment conducted on samples collected showed 4.8-fold resistance greater than the susceptible populations. The barley grasses have already evolved resistance to paraquat, the Group A herbicides and the Group B herbicides in South Australia and the evolution of glyphosate resistance further limits control options for this species.

Keywords Glyphosate, resistance, barley grass, *Hordeum* spp.

Longevity of blady grass (*Imperata cylindrica*) seeds

Husham A. Alqaderi[1], Brian M. Sindel[1], Paul E. Kristiansen[1] and Ralph D.B. Whalley[2]
[1] Department of Agronomy and Soil Science
[2] Department of Botany
School of Environmental and Rural Science, University of New England, Armidale, New South Wales 2351, Australia
(halqader@myune.edu.au)

Summary Blady grass (*Imperata cylindrica* (L.) P.Beauv.) is considered one of the 10 worst weeds globally. It has the ability to reproduce by seeds and by rhizomes, making management more difficult than if reproduction was through one method alone. Once established, it is a strong competitor for resources such as nutrients, water and light. The objective of this research was to determine the longevity of blady grass seeds in order to improve weed management. Seeds were purchased from a commercial seed supplier in May 2017, having been collected from an area near Coonabarabran in Western New South Wales (NSW) in 2015 and subsequently stored by the seed company at 10°C and 50% relative humidity. After purchase, seeds were stored at room temperature (~12–22°C) at the University of New England in Armidale, NSW. Germination testing on 16 June 2017 indicated a germination percentage of 89%. Seeds were then subsequently germinated under the same conditions on 7 November 2017 and the germination percentage had declined to 37%. This rapid decrease in germination rate during the nearly five months of storage at room temperature indicates that blady grass seed may be very short lived under field conditions in temperate areas of Australia as has been found elsewhere in the world and that sexual reproduction may play a less significant role in dispersal and establishment than plant rhizomes.

Keywords Rhizomes, germination, weed management.

INTRODUCTION

Blady grass (*Imperata cylindrica* (L.) P.Beauv.) is a problematic perennial weed which causes serious risks for cropping systems and uncultivated areas in more than 73 countries (MacDonald 2004, Mahdi 2012, Estrada and Flory 2015) across the tropics and sub-tropics and on all continents except Antarctica (MacDonald 2004). In many places it threatens both biodiversity and ecosystem structure (MacDonald 2004, Daneshgar and Jose 2009). Blady grass occurs in all Australian states (Richardson *et al.* 2011), and grows in a wide range of soils, particularly in poor soils with low pH (e.g. 4.7), low fertility and low organic matter (Wang 2008, Richardson *et al.* 2011). Blady grass reproduces sexually by seeds and vegetatively by a complex network of rhizomes that makes management difficult (Shilling 1997, Chikoye *et al.* 2002, MacDonald 2004, Wang 2008, Omezine and Skiri-Harzalla 2009). Individual plants of blady grass can produce new rhizomes with a yearly length ranging from 1.5–12 m, which is a significant factor for long term survival and for dispersal over short and long distances. The latter may occur due to movement of contaminated soil, equipment and machinery from a blady grass patch to a new location (Tominaga 2003, Omezine and Skiri-Harzalla 2009).

Rhizomes start development in blady grass between the third and fourth leaf stage and have the ability to conserve water, nutrients and resist breakage, disruption, drought and heat from fire (Holm *et al.* 1977, Tominaga 2003). Rhizomes are long, white, tough and scaly with prominent nodes and short internodes and have sharp apical ends that are able to penetrate the roots of neighbouring plants (Senkosi 2014, Zaccaro 2016). A single plant can produce over 3000 seeds, which are small, light and attached to a plume of silky white hairs that help them to travel by wind, animals and equipment over long distances (MacDonald 2004, Zaccaro 2016). Viable seeds are produced only by cross pollination and are reported to lose their viability in less than 12 months (Chikoye *et al.* 2002, Zaccaro 2016). Seeds have germination rates as high as 95% after harvest which means that they have no dormancy period (Chikoye *et al.* 2002). Blady grass invasions generally start when the seeds reach a new site and, after establishment, blady grass spreads asexually by rhizome development (Tominaga 2003). Flowering is highly variable depending on the region and environment and occurs in response to stress or human disturbance such as desiccation, burning, overgrazing, repeated slashing, mowing, frost or the addition of nitrogen (Chikoye *et al.* 2002, MacDonald 2004). There is limited information in both Australia and Iraq about the longevity of seeds of blady grass under different environmental conditions and so the aim of this study was to investigate the longevity under laboratory storage conditions.

MATERIALS AND METHODS

Two germination experiments were carried out in a controlled temperature room (25°C) under constant light (average light intensity was 36.6 μmol) at the University of New England in Armidale, New South Wales (NSW), Australia (latitude 30.515 S and longitude 151.665 E).

On each occasion 25 seeds were placed in each of four Petri-dishes lined with filter-paper and kept moist with rainwater. The seeds were purchased from the Nindethana Seed Service in Albany, Western Australia 6331, Australia and all seeds had been collected from one location near Coonabarabran in Western NSW in 2015.

The two germination tests were started on the 16 of June and 7 of November 2017 and lasted for 18 and 15 days respectively. During the intervening five months, the seeds were stored at room temperature (~12–22°C) in a laboratory.

Data analysis The experiments had a Completely Randomized Design. Non-linear regression was used to analyse cumulative germination for *I. cylindrica* over time. A Michaelis-Menten function was used,

$$y = (a \times x) / (b + x)$$

where y = cumulative germination, x = time (days), a = upper asymptote and b = time at the mid-point (a/2) (Thornley and Johnson 1990). Confidence bands were used to indicate the variability at the widths of one and two standard errors.

RESULTS

Most seeds of blady grass germinated in the first 4 days from the initiation of the first germination test (Figure 1). The germination percentage reached 29% on the first day of the test.

In the second test, there was no germination of seeds recorded in the first three days and most seeds of blady grass germinated from the 3rd to the 6th days (Figure 1).

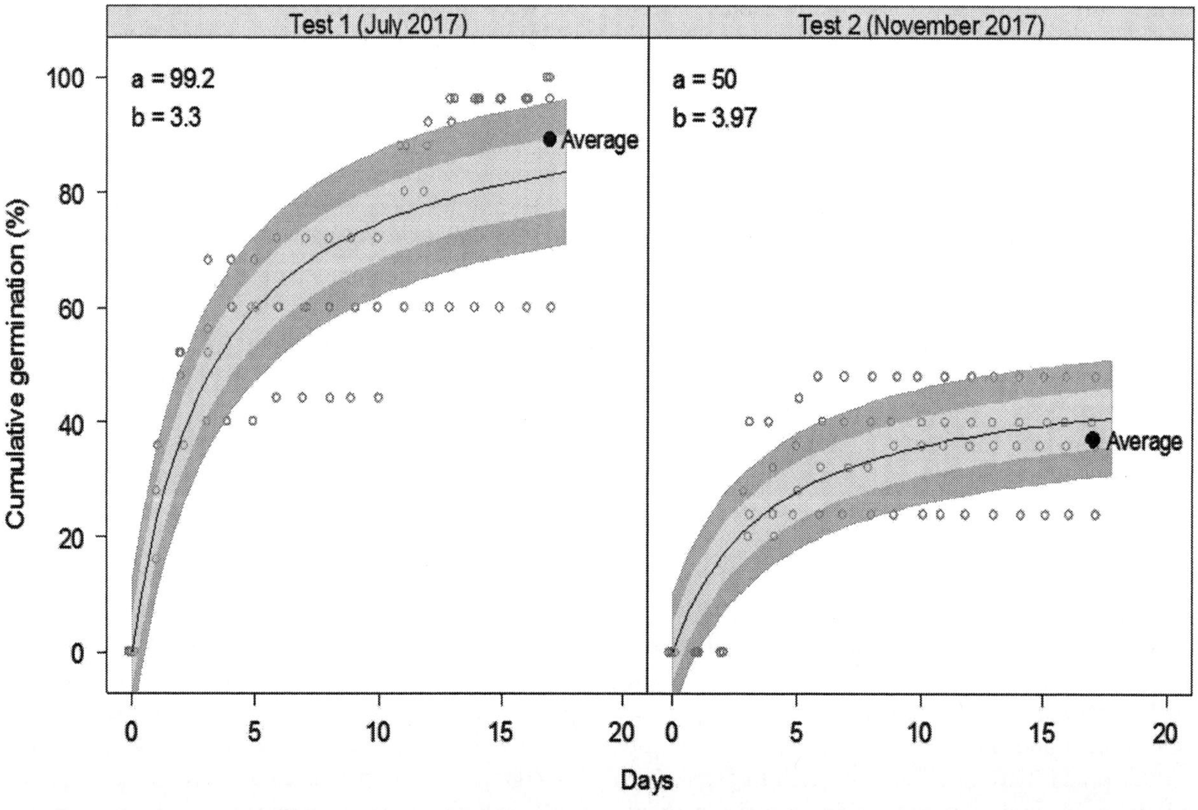

Figure 1. Cumulative germination of *Imperata cylindrica* (blady grass) over time for two test runs. Note: circles = raw data, solid line = fitted non-linear regression (Michaelis-Menten function), grey bands = 1 and 2 standard errors, a = upper asymptote, and b = mid-point in time.

The maximum cumulative germination was significantly different for the two tests (P=0.003), with averages of 89% and 37% for the first and second tests respectively (Figure 1). Non-linear regression confirmed that the maximum potential level of germination was significantly lower in the second test (a = 53% compared with 99% in the first test) and also indicated that the speed of germination was lower in the second test (b = 3.97 days compared with 3.30 days in the first test).

DISCUSSION

The seeds of blady grass initially had a high level of viability a few days after receipt from the supplier, germinating readily from the first day and reaching 89% germination after 16 days. However, after five months of storage at room temperature seed germination was down to 37%, having lost 52% viability. Shilling (1997) found similar results, and also that fresh seeds had no dormancy mechanism. Dickens and Moore (1974) also found that germination percentage decreased when seeds were stored at room temperatures. When the seeds were stored in cold conditions their viable period was extended. The seed company had stored their seed stocks in a cold room at 10°C and 50% relative humidity.

Dickens and Moore (1974) found that blady grass seed germinated at temperatures ranging from 25–35°C and that the optimal temperature was 25–30°C (Wilcut *et al.* 1988). Florescent light was used during the germination period in both germination tests and light had been previously shown to increase both the total germination and speed of germination (Dickens and Moore 1974). The rainwater used to water the Petri dishes in both germination tests had a pH of 6.0 and was considered suitable for germination and growth of blady grass (Bryson *et al.* 2010). Previous research also showed that seeds germinated and survived better in moist conditions than when seeds were overtopped with water (King and Grace 2000a, 2000b).

Despite their small size, light weight and ability to disperse by wind over long distances, seeds appear to have less of a role in the dispersal and invasion process than rhizomes (Estrada *et al.* 2016). Our results indicate that blady grass seeds are likely to be short lived under field conditions in temperate Australia.

ACKNOWLEDGMENTS

I would like to acknowledge the Ministry of Higher Education and Scientific Research in Iraq for their financial support and sponsorship of my PhD study of which this work forms a small part.

REFERENCES

Bryson, C.T., Krutz, L.J., Ervin, G.N., Reddy, K.N. and Byrd Jr, J.D. (2010). Ecotype variability and edaphic characteristics for cogongrass (*Imperata cylindrica*) populations in Mississippi. *Invasive Plant Science and Management* 3, 199-207.

Chikoye, D., Manyong, V., Carsky, R., Ekeleme, F., Gbehounou, G. and Ahanchede, A. (2002). Response of speargrass (*Imperata cylindrica*) to cover crops integrated with handweeding and chemical control in maize and cassava. *Crop Protection* 21, 145-56.

Daneshgar, P. and Jose, S. (2009). Role of species identity in plant invasions: experimental test using *Imperata cylindrica. Biological Invasions* 11, 1431-40.

Dickens, R. and Moore, G. (1974). Effects of light, temperature, KNO_3, and storage on germination of cogongrass. *Agronomy Journal* 66, 187-8.

Estrada, J. A. and Flory, S. L. (2015). Cogongrass (*Imperata cylindrica*) invasions in the US: mechanisms, impacts, and threats to biodiversity. *Global Ecology and Conservation* 3, 1-10.

Estrada, J.A., Wilson, C.H., NeSmith, J.E. and Flory, S. L. (2016). Propagule quality mediates invasive plant establishment. *Biological Invasions* 18, 2325-32.

Holm, L.G., Plucknett, D.L., Pancho, J.V. and Herberger, J.P. (1977). 'The world's worst weeds'. (University Press, Honolulu, Hawaii).

King, S.E. and Grace, J.B. (2000a). The effects of gap size and disturbance type on invasion of wet pine savanna by cogongrass, *Imperata cylindrica* (Poaceae). *American Journal of Botany* 87, 1279-86.

King, S.E. and Grace, J.B. (2000b). The effects of soil flooding on the establishment of cogongrass (*Imperata cylindrica*), a nonindigenous invader of the southeastern United States. *Wetlands* 20, 300-6.

MacDonald, G.E. (2004). Cogongrass (*Imperata cylindrica*) biology, ecology, and management. *Critical Reviews in Plant Sciences* 23, 367-80.

Mahdi, A.S. (2012). Competition between two wild plant species (*Imperata cylindrica* and *Cynodon dactylon*) in semi-natural pasture. *Diyala Agricultural Sciences Journal* 4, 17-27.

Omezine, A. and Skiri-Harzalla, F. (2009). Regenerative capacity of speargrass *Imperata cylindrica* (L.) P.Beauv. *Pakistan Journal of Weed Science Research* 15, 53-69.

Richardson, F.J., Richardson, R.G. and Shepherd, R.C.H. (2011). 'Weeds of the south-east: An identification guide for Australia' (2nd ed.). (Meredith, Victoria).

Senkosi, K. (2014). Soil fertility determinants of spear grass abundance along a cultivation chronosequence in Buikwe District, Uganda. (Doctoral dissertation, Makerere University.)

Shilling, D.G. (1997). Ecology, physiology, and management of cogongrass (*Imperata cylindrica*). Retrieved from Research report prepared for Florida Institute of Phosphate Research. University of Florida.

Thornley, J.H. and Johnson, I.R. (1990). 'Plant and crop modelling'. (Clarendon, Oxford).

Tominaga, T. (2003). Growth of seedlings and plants from rhizome pieces of cogongrass (*Imperata cylindrica* (L.) Beauv.). *Weed Biology and Management* 3, 193-5.

Wang, J. (2008). Mechanisms of cogongrass (*Imperata cylindrica* (L.) Beauv.) competition, low light survival, and rhizome dormancy. (Doctoral dissertation, University of Florida).

Wilcut, J.W., Truelove, B., Davis, D.E., and Williams, J.C. (1988). Temperature factors limiting the spread of cogongrass (*Imperata cylindrica*) and torpedograss *Panicum repens*). *Weed Science* 36, 49-55.

Zaccaro, M.L.M. (2016). Cogongrass (*Imperata cylindrica* (L.) Beauv.) control using chemical treatment with cover cropping systems. (Doctoral dissertation, Mississippi State University).

Efficacy of some selective herbicides against broad leaved weeds of wheat crop grown under moisture deficit conditions of Pakistan

Muhammad Asad[1], Muhammad Mudassar[2], Zahid Mahmood[3] and Muhammad Rasheed[1]
[1] Department of Agronomy, PMAS Arid Agriculture University Rawalpindi, Pakistan
[2] Department of Environmental Economics, Pakistan Institute of Developmental Economics, Islamabad, Pakistan
[3] Wheat Wide Crosses Program, National Agricultural Research Centre, Islamabad, Pakistan

Summary Herbicides are mostly used to control weeds in irrigated areas, but neglected in rainfed areas due to uncertain and erratic rainfall patterns. A field experiment was conducted to investigate the efficacy of some herbicides for the control of broad leaved weeds in wheat crop during winter 2014–2015. Four herbicides, i.e., Ally Max (metsulfuron methyl + Tribenuron methyl), Lihua (chlorsulfuron), Wheat Star (chlorpyrifos + carfentrazone ethyl + tribenuron methyl) and Buctril Super 60% EC (bromoxynil + MCPA) were included in the study. Three different doses of each herbicide were used during the experiments. A non-treated control (weedy check) and hand weeded plots were also included in the study. All tested doses of the herbicides provided good control of broad leaved weeds by significantly decreasing their density. The results indicated that the hand weeding, with minimum weed counts (14 m^{-2}) and dry weight (24.5 gm^{-2}) resulted in the highest weed control efficiency (77.85%) and mortality (81.77%). Among herbicides, Wheat Star with recommended dose observed the lowest weed counts (17 m^{-2}), dry weight (16.5 g m^{-2}) along with maximum weed control efficiency (74.25%) and mortality (63.37%). It was followed by Buctril Super with recommended dose resulting in weed counts (20.2 m^{-2}), dry weight (25 g m^{-2}) and weed index (−0.68%). Based on these results hand weeding can be recommended for small landholders having sufficient available family labour, while Wheat Star and Buctril Super at the recommended dose could be the best weed control option in areas where wheat fields are largely infested by broad leaved weeds.

Keywords Efficacy, broad leaved weeds, herbicides, moisture deficiency.

Effect of parthenium weed on maize yield at different competition durations in Ethiopia

Ali A. Bajwa[1,2], Tamado Tana[3], Bhagirath S. Chauhan[2] and Steve W. Adkins[1,2]
[1] School of Agriculture and Food Sciences, The University of Queensland, Gatton, Queensland 4343, Australia
[2] The Centre for Crop Science, Queensland Alliance for Agriculture and Food Innovation, The University of Queensland, Gatton, Queensland 4343, Australia
[3] Haramaya University, Dire Dawa, Ethiopia
(a.bajwa@uq.edu.au)

Summary Parthenium weed (*Parthenium hysterophorus* L.) is an invasive weed species in more than 40 countries, including Australia. It infests several major crops and causes substantial yield losses. It is important to determine its critical competition duration in different crops to devise a suitable management strategy. A field study was conducted to evaluate the effect of parthenium weed on maize at different weed-crop competition durations in Ethiopia. Six treatments applied included: season-long weed-free; weed-free after 2, 4, 6 and 8 weeks of crop emergence; and season-long weedy. Parthenium weed biomass increased with increasing competition duration. Parthenium weed negatively affected the growth, yield and yield related traits of maize at all competition durations; however, more reductions in all these parameters were observed at longer parthenium weed competition durations. The highest and lowest values for leaf area index, ears per plant, thousand grain weight, grain yield and biological yield were observed in season-long weed-free and season-long weedy treatments, respectively. Parthenium weed competition for 2, 4, 6 and 8 weeks after crop emergence caused 3, 6, 8 and 15% losses in maize grain yield, respectively, when compared with the season-long weed-free treatment. It is important to note that the yield reductions in plots which were kept weed-free after 8 weeks were similar as in those kept weedy throughout the season. The results suggest that management strategies should be devised to control parthenium weed in maize during the critical competition period of 4–8 weeks after crop emergence.

Keywords *Parthenium hysterophorus*, weed-crop interference, *Zea mays*, weed management, yield losses.

INTRODUCTION

Parthenium weed is an invasive alien species which infests non-cropped areas, pastures, diverse cropping systems and forest ecosystems in more than 40 countries around the world (Adkins and Shabbir 2014, Bajwa *et al.* 2016). It is also a Weed of National Significance in Australia and a Class-2 declared pest in Queensland (Adkins and Shabbir 2014). Parthenium weed causes substantial losses to pasture and crop productivity, livestock production and native biodiversity (Adkins and Shabbir 2014). It is also notorious as a source of allergies, respiratory problems and other health complications in humans and animals.

Parthenium weed infests many field crops in countless countries of Asia and Africa (Tamado *et al.* 2002, Safdar *et al.* 2016). It causes severe yield losses in major food crops and therefore, poses serious threat to regional and global food security. It interferes with crop production directly through competition and through allelopathic interactions (Bajwa *et al.* 2016). Parthenium weed infests a large area in Ethiopia where it has become a major problem for sustainable crop production. For instance, sorghum (*Sorghum bicolor* L. Moench) is one of the most important grain crops in dry low-land areas of Ethiopia and suffers significantly from parthenium weed infestation (Tamado *et al.* 2002). In eastern Ethiopia, grain yield losses in sorghum were up to 97% at 100 parthenium weed plants m^{-2} (Tamado *et al.* 2002). Even at the density of three plants m^{-2}, parthenium weed caused up to 69% yield reductions. The critical period for parthenium weed management to avoid the substantial yield loss was up to 60 days after crop emergence.

Parthenium weed infestation in maize (*Zea mays* L.) crop is also very troublesome in different regions. Maize is an important grain crop in Ethiopia along with sorghum and both suffer from heavy parthenium weed infestations. However, information on the parthenium weed critical competition period in the Ethiopian maize system is lacking. Such information is very important to understand the dynamics of parthenium weed competition with maize and to devise management strategies. This field study was conducted to evaluate the impact of parthenium weed on maize growth and yield at different competition durations in Ethiopia.

MATERIALS AND METHODS

A field study was conducted at Haramaya, eastern Ethiopia (latitude of 9°26′ N, longitude of 42°03′ E and an altitude of 1980 m above sea level). The soil of the experimental site was a well-drained deep alluvial with a sub-soil stratified with loam texture. The total average annual rainfall was 827 mm, and the average temperature was 16.8°C. The experiment was laid out in a randomised complete block design with three replications per treatment. Maize cultivar Melkassa-2 was planted at 75 × 25 cm spacing. Phosphorus (P) fertiliser in the form of triple super phosphate at 46 kg P_2O_5 ha^{-1}, and half of the nitrogen (N) fertiliser in the form of urea (87 kg N ha^{-1}) were applied uniformly in all plots by band application method at the time of planting and the remaining half of N-fertiliser was applied at the knee height stage (50 cm) of seedlings.

Seeds of parthenium weed were collected from local, naturally growing populations. About 10 g parthenium weed seed was uniformly broadcasted over each experimental plot just after sowing the crop to ensure a dense population of parthenium weed; except in the season-long weed-free treatment. Six different parthenium weed competition treatments were used:

- season-long weed-free;
- weed-free 2 weeks after crop emergence;
- weed-free 4 weeks after crop emergence;
- weed-free 6 weeks after crop emergence;
- weed-free 8 weeks after crop emergence; and
- season-long weedy.

The season-long weedy and weed-free treatments were established as controls. A uniform parthenium weed density of 15 plants m^{-2} was maintained in all treatments except the weed free treatment. All the other weeds were removed by hand pulling in all the treatments throughout the crop season. For parthenium weed biomass, plants were harvested from two different locations from within the crop (each 1 m^2) and from different treatments just before starting the parthenium weed control in the respective treatments. Biomass was recorded after oven drying at 70°C for 3 days. Leaves were collected from all treatments to measure the leaf area index 10 weeks after crop emergence. The number of ears per plant was recorded before harvesting from 20 plants in each treatment. The 100-grain weight, biological and grain yield were recorded after harvesting the crop. All the data were subjected to analysis of variance using Statistix-8.1 software. Treatment means were separated by the least significant difference (LSD) test at P<0.05. The graph was prepared in Microsoft excel.

RESULTS

Parthenium weed competition duration had significant impact on parthenium weed biomass and maize growth and yield (P<0.05).

The highest parthenium weed biomass was observed in the season-long weedy treatment which was statistically similar to the treatment which was kept free from parthenium weed 8 weeks after crop emergence (Figure 1). The lowest parthenium weed biomass was obtained from plots where parthenium weed grew for only 2 weeks. Parthenium weed competition duration significantly (P<0.05) affected the leaf area index, number of ears per plant, 100-grain weight and biological yield of maize (Table 1). The highest leaf area index, number of ears per plant, 100-grain weight and biological yield were observed in the season-long weed free treatment. The lowest values for all these parameters were recorded for the season-long weedy treatment. Parthenium weed competition for 2–8 weeks caused the reductions of 4–29%, 5–16%, 4–18% and 2–16% in leaf area index, number of ears per plant, 100-index and biological yield, respectively, as compared with the season-long weed-free treatment.

Grain yield of the maize crop was also significantly reduced with increasing duration of parthenium weed competition (Table 2). The highest grain yield was obtained in the season-long weed-free treatment (9.4 t ha^{-1}) followed by the treatment which was kept weed-free 2 weeks after crop emergence (9.1 t ha^{-1}).

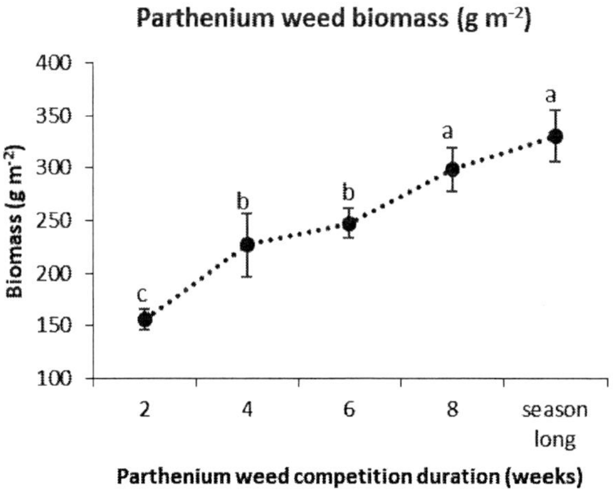

Figure 1. Effect of competition duration on parthenium weed biomass when growing in a maize crop. Error bars represent ± standard errors of the mean. Means sharing the same letter, do not differ significantly according to the least significant difference (LSD) test at P<0.05.

The lowest grain yield was recorded in the season-long weedy treatment. The grain yield in the 4 and 6-weeks competition treatments was statistically similar (Table 2). The season-long weedy and 8-weeks competition treatments provided the lowest grain yields (7.9 and 8.0 t ha^{-1}, respectively) which were statistically similar (Table 2). Overall, different competition durations caused 2.5–15.5% grain yield loss compared to the season-long weed-free treatment.

Table 1. Effect of parthenium weed competition duration on different growth and yield attributes of maize.

	Leaf area index	No. of ears per plant	100-grain weight (g)	Crop biomass (t ha^{-1})
T1[A]	5.1 a[B]	1.37 a	45.2 a	23.5 a
T2	4.9 ab	1.30 ab	43.2 b	23.1 a
T3	4.6 abc	1.29 ab	40.0 c	21.9 b
T4	4.4 bc	1.24 bc	39.0 c	20.9 c
T5	4.3 c	1.24 bc	38.6 cd	20.1 cd
T6	3.6 d	1.15 c	37.0 d	19.8 d
LSD at P<0.05	0.49	0.09	1.68	0.96

[A] T1 = Season-long weed-free, T2 = Weed-free after 2 weeks, T3 = Weed-free after 4 weeks, T4 = Weed-free after 6 weeks, T5 = Weed-free after 8 weeks, T6 = Season-long weedy

[B] Means sharing the same letter, in a column, do not differ significantly according to the least significant difference (LSD) test at P<0.05.

Table 2. Effect of parthenium weed competition duration on maize grain yield.

	Grain yield (t ha^{-1})	Yield loss (%)[C]
T1[A]	9.4 a[B]	–
T2	9.1 ab	2.5
T3	8.8 bc	5.7
T4	8.7 c	7.5
T5	8.0 d	14.5
T6	7.9 d	15.5
LSD at P<0.05	0.32	

[A] T1 = Season-long weed-free, T2 = Weed-free after 2 weeks, T3 = Weed-free after 4 weeks, T4 = Weed-free after 6 weeks, T5 = Weed-free after 8 weeks, T6 = Season-long weedy.

[B] Means sharing the same letter, in a column, do not differ significantly according to the least significant difference (LSD) test at P<0.05.

[C] Yield loss compared season-long weed-free treatment.

DISCUSSION

Weed-crop competition balance determines the crop growth and yield. Parthenium weed competition for longer periods negatively affected maize growth and yield. This may be mainly due to enhanced competition for water, nutrients and space (Bajwa et *al.* 2016, Safdar *et al.* 2016). The fast-growing habit and efficient resource acquisition of parthenium weed might have given it a competitive advantage over the crop (Bajwa *et al.* 2016). It has been reported that parthenium weed can germinate under a wide range of environmental conditions and it can grow very quickly by establishing a long tap root, extensive branching, large number of well-expanded leaves and a stiff, erect stem (Bajwa *et al.* 2016, 2017, 2018). These attributes are helpful for rapid canopy development, space acquisition and overall plant growth. In addition, parthenium weed uptakes water and nutrients from the rhizosphere very efficiently as compared with most crop plants. For instance, Safdar *et al.* (2016) reported that N, P and K uptake by parthenium weed increased with increasing competition period in the maize crop. Parthenium weed growth increased with the passage of time and it had higher biomass production in the treatments where it grew for the longer durations. This is possibly due to the continuous supply of moisture through irrigations and nutrients by fertiliser application (Safdar *et al.* 2016).

Parthenium weed competes strongly with neighbouring plants by not only resource competition but also by releasing allelopathic chemicals in its rhizosphere and aerial surroundings. Several earlier publications support the view that parthenium weed can suppress the growth of other plants through allelopathy (Singh *et al.* 2003, Belgeri and Adkins 2015). The plant can release potent allelochemicals like parthenin, and various phenolic acids and flavonoids (Bajwa *et al.* 2016) which are phytotoxic to growth of certain plants. So, the allelopathic potential of parthenium weed might also play an important role in the interference with maize growth.

The higher biomass production of parthenium weed at longer competition durations (4–8 weeks) caused a decline in growth and yield-related traits of maize, which ultimately led to grain yield losses. Lower yield losses in the 2-week competition treatment might be due to less competition of parthenium weed with crop growth. The critical competition period is between 4 and 8 weeks where the crop suffered heavy yield losses. Safdar *et al.* (2016) reported 21–44% yield losses in maize due to 5–8 weeks of parthenium weed competition. Earlier, Tamado *et al.* (2002) reported huge yield losses (40 to 95%) caused by parthenium weed competing with sorghum crop for 2–9 weeks. Yield losses in the 8-week competition treatment were equal to the unweeded control which indicates that any weed control after 8 weeks of competition may not provide economic benefits. Our results also show significant yield declines at prolonged parthenium weed competition durations; however, the magnitude of yield loss is comparatively less. This may be due to good rainfall during the crop growth season which benefited crop growth.

In conclusion, parthenium weed can cause a substantial yield loss in maize crops grown in Ethiopia. Growth and yield losses increased with increasing parthenium weed competition duration. The critical competition period was between 4–8 weeks. Any management strategy should target parthenium weed in maize during this period to enhance crop yield and profitability. Given the parthenium weed infestation in north and central Queensland cropping systems, this recommendation is also valid for maize growers of those areas. Similar studies should be carried out in Australia with focus on other major crops.

ACKNOWLEDGMENTS

This manuscript is part of the PhD research project of Ali Ahsan Bajwa who is thankful to the Australian Government and The University of Queensland, Australia for the provision of a Research Training Program Scholarship and University of Queensland Centennial Scholarship, respectively.

REFERENCES

Adkins, S. and Shabbir, A. (2014). Biology, ecology and management of the invasive parthenium weed (*Parthenium hysterophorus* L.). *Pest Management Science* 70, 1023-9.

Bajwa, A.A., Chauhan, B.S. and Adkins, S. (2017). Morphological, physiological and biochemical responses of two Australian biotypes of *Parthenium hysterophorus* to different soil moisture regimes. *Environmental Science and Pollution Research* 24, 16186-94.

Bajwa, A.A., Chauhan, B.S. and Adkins, S.W. (2018). Germination ecology of two Australian biotypes of ragweed parthenium (*Parthenium hysterophorus*) relates to their invasiveness. *Weed Science* 66, 62-70.

Bajwa, A.A., Chauhan, B.S., Farooq, M., Shabbir, A. and Adkins, S.W. (2016). What do we really know about alien plant invasion? A review of the invasion mechanism of one of the world's worst weeds. *Planta* 244, 39-57.

Belgeri, A. and Adkins, S.W. (2015). Allelopathic potential of invasive parthenium weed (*Parthenium hysterophorus* L.) seedlings on grassland species in Australia. *Allelopathy Journal* 36, 1-14.

Safdar, M.E., Tanveer, A., Khaliq, A. and Maqbool, R. (2016). Critical competition period of parthenium weed (*Parthenium hysterophorus* L.) in maize. *Crop Protection* 80, 101-7.

Singh, H.P., Batish, D.R., Pandher, J.K. and Kohli, R.K. (2003). Assessment of allelopathic properties of *Parthenium hysterophorus* residues. *Agriculture, Ecosystems and Environment* 95, 537-41.

Tamado, T., Ohlander, L. and Milberg, P. (2002). Interference by the weed *Parthenium hysterophorus* L. with grain sorghum: influence of weed density and duration of competition. *International Journal of Pest Management* 48, 183-8.

The effects of parthenium weed density on yield attributes and yield of maize in Ethiopia

Ali A. Bajwa[1,2], Tamado Tana[3], Bhagirath S. Chauhan[2] and Steve W. Adkins[1,2]
[1] School of Agriculture and Food Sciences, The University of Queensland, Gatton, Queensland 4343, Australia
[2] The Centre for Crop Science, Queensland Alliance for Agriculture and Food Innovation, The University of Queensland, Gatton, Queensland 4343, Australia
[3] Haramaya University, Dire Dawa, Ethiopia
(a.bajwa@uq.edu.au)

Summary Parthenium weed is an invasive alien species in numerous countries around the world, including Australia. It has become a major weed species in tropical and sub-tropical cropping systems, in which it causes significant crop yield losses. However, information about economic thresholds of parthenium weed in different crops is currently lacking which is imperative for sustainable management of this species. The present study was conducted to determine the effects of different densities of parthenium weed (0, 5, 10, 15, and 20 plants m^{-2}) on growth and yield of maize in Ethiopia. A strong linear relationship was observed between density and biomass of parthenium weed. Parthenium weed biomass also had strong negative correlations with yield and yield related traits of maize. Density of parthenium weed had a significant negative effect on growth and yield of maize. The highest density of parthenium weed (20 plants m^{-2}) caused the highest reductions in maize leaf area index (15%), ears per plant (21%), thousand grain weight (12%), biological yield (15%) and grain yield (18%) as compared with the plots free from parthenium weed. However, parthenium weed densities of 5–15 plants m^{-2} only caused 8–10% losses in the grain yield of maize. Therefore, it is suggested to reduce the density of this weed species to below 15 plants m^{-2} to avoid substantial yield losses.

Keywords *Parthenium hysterophorus*, economic threshold, *Zea mays*, weed management, yield losses.

INTRODUCTION

Parthenium weed (*Parthenium hysterophorus* L.) is one of the most invasive species in the world. It is not only a major problem for native biodiversity and human and animal health but also an increasing threat to crop production in over 40 countries (Adkins and Shabbir 2014, Bajwa *et al.* 2016). It is also a Weed of National Significance in Australia where it causes huge losses to pasture and crop productivity and livestock production (Adkins and Shabbir 2014). Parthenium weed infestations in cropping systems have increased dramatically in many countries in Asia and Africa (Tamado *et al.* 2002, Bajwa *et al.* 2016). Ethiopia is one such country where this noxious weed species is seriously affecting crop yields (Tamado *et al.* 2002). It infests staple food crops such as grain sorghum (*Sorghum bicolor* L. Moench) and maize (*Zea mays* L.) in this part of the world (Tamado *et al.* 2002).

Parthenium weed competes strongly with these crops and causes significant reductions in crop yield. In Pakistan, Safdar *et al.* (2015) reported up to 50% grain yield losses in maize by different densities of parthenium weed under irrigated conditions. Another study reported that parthenium weed, at the densities of 2 to 16 plants m^{-2}, caused up to 25% loss in fresh fodder yield of sorghum crop (Asif *et al.* 2017). However, competition dynamics of parthenium weed in maize crops in Ethiopia are not well-studied. It is crucial to determine the economic thresholds of parthenium weed in different crops to devise suitable and timely management strategies. This study was conducted to assess the negative effect of different densities of parthenium weed on growth and yield of a maize crop in Ethiopia.

MATERIALS AND METHODS

A field study was carried out at Haramaya, eastern Ethiopia (latitude of 9°26′N, longitude of 42°03′E and an altitude of 1980 m above sea level). The soil of the experimental site was a loam. The experiment was laid out in a randomised complete block design with three replications per treatment. Maize cultivar Melkassa-2 was planted at 75 × 25 cm spacing. Nitrogen (N) and phosphorus (P) fertilisers were applied at the rate of 87 and 46 kg ha^{-1}. The crop was grown under rainfed conditions (average annual rainfall in this region of Ethiopia is 827 mm). Seeds of parthenium weed were collected from local, naturally growing populations and uniformly broadcast at each experimental plot (ca. 10 g per plot) by hand just after sowing the crop except the control treatment (0 plants m^{-2}). Five different densities of parthenium weed (0, 5, 10, 15 and 20 plants m^{-2}) were established, then maintained throughout the duration of the study. More than required parthenium weed seedlings were removed manually as they appeared in the different treatments. Moreover, all other weeds were also removed by hand pulling in all the treatments throughout the crop season.

Aboveground plant biomass of parthenium weed was recorded from two different locations (each 1 m^2) in each experimental plot before harvesting the crop. Biomass was recorded after oven drying at 70°C for 3 days. Fresh leaves of maize were collected in the different treatments to measure the leaf area and to determine the leaf area index 10 weeks after crop emergence. The number of ears per plant was recorded before harvesting from 20 plants in each treatment. The 100-grain weight, biological and grain yield were recorded after harvesting the crop. All the data sets were subjected to an analysis of variance using Statistix-8.1 software. Treatment means were separated by the least significant difference (LSD) test at $P<0.05$. The data on parthenium weed biomass and maize grain yield were plotted by regression analysis in Microsoft Excel.

RESULTS AND DISCUSSION

Different densities of parthenium weed significantly affected its own biomass ($P<0.05$), increasing linearly with increasing densities (Figure 1). Parthenium weed densities also had a significant impact on growth, yield-related traits and yield of maize crop ($P<0.05$). The highest values for leaf area index, ears per plant, 100-grain weight and biological yield were observed in the plots where no parthenium weed was present (Table 1). The highest density of parthenium weed (20 plants m^{-2}) caused the highest reductions in leaf area index (15%), number of ears per plant (21%), 100-grain weight (12%) and biological yield (15%) as compared with the parthenium weed-free control.

The grain yield of maize was also significantly affected by different densities of parthenium weed (Figure 2). A strong negative relationship was observed between parthenium weed density and maize grain yield ($R^2 = 0.88$). Parthenium weed densities of 5, 10, 15 and 20 plants m^{-2} caused 7.6, 8.5, 9.7 and 17.7% loss in maize grain yield, respectively, when compared to the treatment that had no parthenium weed plants present.

The results of this study show that parthenium weed negatively affects maize growth and yield. However, the negative impact is more pronounced at weed densities higher than 15 plants m^{-2}. These higher weed densities also produced more biomass per unit area, which caused crop growth and yield reductions.

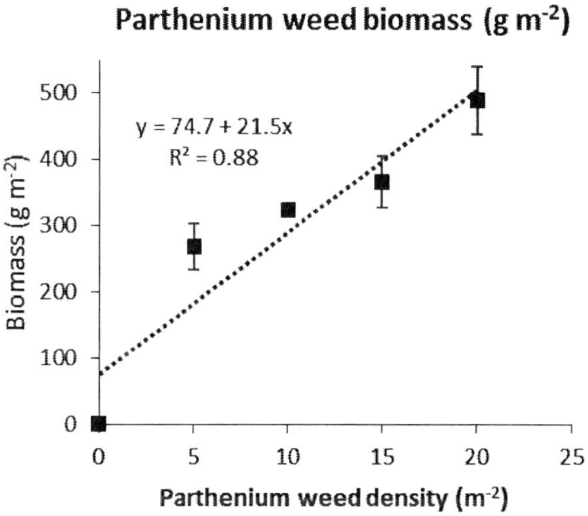

Figure 1. Effect of different densities of parthenium weed on its dry biomass when growing in a maize crop. Error bars represent ± standard errors of the mean.

Table 1. Effect of different densities of parthenium weed on growth and yield attributes of maize.

Weed density (m^{-2})	Leaf area index	No. of ears per plant	100-grain weight (g)	Biological yield (t ha^{-1})
0	4.6 a[A]	1.4 a	45.0 a	22.0 a
5	4.5 ab	1.3 b	43.5 ab	20.4 b
10	4.3 bc	1.3 b	42.6 bc	19.9 b
15	4.0 cd	1.2 c	41.3 cd	19.6 bc
20	3.9 d	1.1 d	39.8 d	18.6 c
$LSD_{0.05}$	0.10	0.03	1.80	0.53

[A] Means sharing the same letter, in a column, do not differ significantly according to the least significant difference (LSD) test at $P<0.05$.

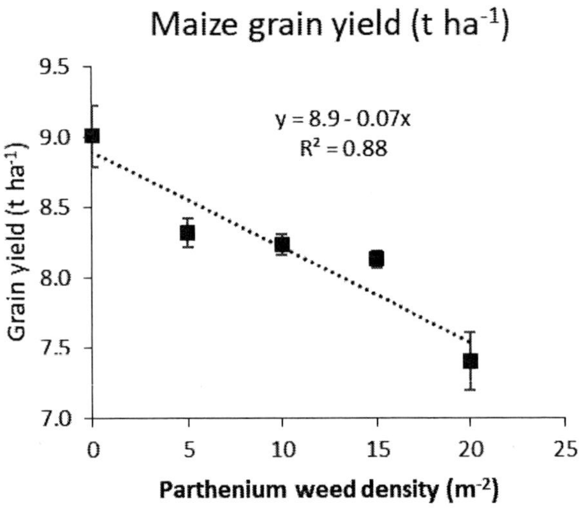

Figure 2. Effect of different densities of parthenium weed on grain yield of maize crop. Error bars represent ± standard errors of the mean.

The negative effects can be attributed to strong interference effects (competitive and allelopathic) provided by parthenium weed plants. It has been reported that parthenium weed efficiently acquires the essential resources such as moisture, nutrients and light, while growing alongside other plant species (Safdar *et al.* 2015, 2016). It has a tall stature, fast-growing habit, and the ability to tolerate harsh climatic conditions, which makes it highly competitive (Bajwa *et al.* 2016, 2018). Earlier studies have also shown similar negative effects of parthenium weed on maize and sorghum yields (Safdar *et al.* 2015, 2016, Asif *et al.* 2017).

Safdar *et al.* (2015) reported that parthenium weed attained more biomass at its higher densities and accumulated more N, P and potassium than maize plants. Parthenium weed interferes with crops not only through competition but it also suppresses crop growth through allelopathy (Singh *et al.* 2003, Bajwa *et al.* 2016). The plant can release potent allelochemicals like parthenin, and various phenolic acids and flavonoids which are phytotoxic to growth of certain plants. So, the allelopathic potential of parthenium weed might also play an important role in the interference with maize growth.

In conclusion, different densities of parthenium weed can cause a substantial yield loss in maize crops grown in Ethiopia. Growth and yield losses increased with increasing parthenium weed densities. However, the weed density above 15 plants m^{-2} caused a significantly higher yield loss. Therefore, this weed species must be kept below this threshold to avoid substantial yield loss.

ACKNOWLEDGMENTS

This manuscript is part of the PhD research project of Ali Ahsan Bajwa who is thankful to the Australian Government and The University of Queensland, Australia for the provision of a Research Training Program Scholarship and University of Queensland Centennial Scholarship, respectively.

REFERENCES

Adkins, S. and Shabbir, A. (2014). Biology, ecology and management of the invasive parthenium weed (*Parthenium hysterophorus* L.). *Pest Management Science* 70, 1023-9.

Asif, M., Ayub, M., Tanveer, A. and Akhtar, J. (2017). Estimating yield losses and economic threshold level of *Parthenium hysterophorus* in forage sorghum. *Planta Daninha* 35, e017164158.

Bajwa, A.A., Chauhan, B.S. and Adkins, S.W. (2018). Germination ecology of two Australian biotypes of ragweed parthenium (*Parthenium hysterophorus*) relates to their invasiveness. *Weed Science* 66, 62-70.

Bajwa, A.A., Chauhan, B.S., Farooq, M., Shabbir, A. and Adkins, S.W. (2016). What do we really know about alien plant invasion? A review of the invasion mechanism of one of the world's worst weeds. *Planta* 244, 39-57.

Safdar, M.E., Tanveer, A., Khaliq, A. and Maqbool, R. (2016). Critical competition period of parthenium weed (*Parthenium hysterophorus* L.) in maize. *Crop Protection* 80, 101-7.

Safdar, M.E., Tanveer, A., Khaliq, A. and Riaz, M.A. (2015). Yield losses in maize (*Zea mays*) infested with parthenium weed (*Parthenium hysterophorus* L.). *Crop Protection* 70, 77-82.

Singh, H.P., Batish, D.R., Pandher, J.K. and Kohli, R.K. (2003). Assessment of allelopathic properties of *Parthenium hysterophorus* residues. *Agriculture, Ecosystems and Environment* 95, 537-41.

Tamado, T., Ohlander, L. and Milberg, P. (2002). Interference by the weed *Parthenium hysterophorus* L. with grain sorghum: influence of weed density and duration of competition. *International Journal of Pest Management* 48, 183-8.

Emergence of great brome grass and barley grass

Catherine P.D. Borger and Abul Hashem
Department of Primary Industries and Regional Development, PO Box 483, Northam, Western Australia 6401, Australia
(catherine.borger@dpird.wa.gov.au)

Summary Barley grass and great brome grass seed were harvested from crop and adjoining roadside populations during December 2015 in three high rainfall and three low rainfall locations in the Western Australian wheatbelt (12 populations per species). Fifty seeds of each population were planted in trays and seedlings were counted and removed for a year. Most barley and great brome grass populations reached 50% emergence on 2 or 3 May 2016. However, two barley grass populations from cropping fields reached 50% emergence two weeks or four weeks later than the roadside populations (50% emergence of the crop populations on 19 May 2016 and 31 May 2016 compared to 2 May 2016 for the roadside populations). Only one great brome grass crop population had significantly delayed emergence compared to the roadside population, and emergence was delayed by less than a week (50% emergence of the roadside and crop population on 2 May and 8 May 2016). Rainfall in the seed collection zone had no impact on seed emergence time. Management programs for these weeds will need to be altered where delayed emergence allows seedlings to avoid the non-selective herbicides prior to seeding.

Keywords *Bromus diandrus*, *Hordeum leporinum*, dormancy, germination.

INTRODUCTION

Brome grass (*Bromus* sp.) is the most notable new weed in Australia, ranking nationally as the fourth worst weed in terms of area covered (1,414,297 ha), yield loss (91,392 t) and revenue loss ($22.5m) per year (Llewellyn *et al.* 2016). Barley grass (*Hordeum* sp.) is a less severe issue, but nationally ranks as the ninth worst weed in terms of area covered (244,558 ha) and fourteenth in terms of revenue loss ($1.7m) per year (Llewellyn *et al.* 2016). These weeds are difficult to control due to limited selective herbicide options, especially in cereal crops (Storrie 2014). Therefore, the development of an effective integrated weed management program relies on accurate knowledge of the biology and ecology of these species.

Research in South Australia (SA) has indicated that the dormancy characteristics of both species have changed. Gill and Fleet (2012) noted delayed emergence of smooth barley grass (*H. murinum* L. ssp. *glaucum* (Steud.) Tzvelev) collected from cropping pastures compared to populations collected from roadside or pasture. Likewise, Kleemann and Gill (2013) noted delayed emergence of great brome grass (*B. diandrus* Roth.) collected from cropped fields compared to roadsides. This research aims to test the hypothesis that great brome grass and barley grass (*H. leporinum* Link) in Western Australia (WA) will have developed delayed emergence in cropping fields compared to roadside populations.

MATERIALS AND METHODS

Seeds of barley grass and great brome grass populations were collected during December 2015. Seed was taken from three populations in cropped fields and matching populations on the adjacent roadside, in both high and low rainfall zones of the WA wheatbelt (12 populations per species, Table 1). Seeds were taken from at least 20 plants at each site, and the GPS coordinates and land use were recorded. Seed was kept in a storage room without air-conditioning, at the Department of Primary Industries and Regional Development, Northam, to keep the seed safe while partially mimicking the temperature fluctuations seed would experience outdoors. Trays measuring 30 × 30 × 10 cm were filled with sand to within 2 cm of the top. Within 2–3 weeks of collection, samples of 50 seeds of each population were placed in trays, replicated four times. Trays were placed in the field on 18/12/2015 and covered with netting to prevent seeds being introduced and to protect trays from animals. Seedlings were counted and removed each week for a year.

Analysis Seedling emergence was calculated as a percent of total emergence. The two great brome grass populations with emergence under 10% were removed from the analysis. Simple linear fit was used to estimate the time taken for each population to reach 50% emergence in year one. Standard errors were calculated from the mean of four replications (VSN International 2017).

Table 1. The species, nearest town to the collection site, rainfall zone and latitude/longitude of the roadside and crop collection site for each seed sample.

Species	Town	Rainfall	Roadside	Crop
Barley grass	Beacon	Low	−30.664652, 117.895068	−30.663937, 117.897033
	Wongan Hills	Low	−30.828428, 116.726661	−30.833077, 116.725389
	Merredin	Low	−31.483294, 118.221582	−31.483791, 118.221792
	Grass Valley	High	−31.629109, 116.814790	−31.629414, 116.814880
	Frankland	High	−34.353914, 117.107018	−34.353118, 117.109640
	York	High	−31.911945, 116.808146	−31.912359, 116.809094
Great brome grass	Meckering	Low	−31.623780, 117.052531	−31.623476, 117.052399
	Kellerberrin	Low	−31.634072, 117.744590	−31.634194, 117.743625
	Doodlakine	Low	−31.636054, 117.924943	−31.634575, 117.925650
	York	High	−31.911945, 116.808146	−31.912359, 116.809094
	Grass Valley	High	−31.591162, 116.829930	−31.592261, 116.830340
	Toodyay	High	−31.522843, 116.573438	−31.523573, 116.574209

RESULTS

Total barley grass emergence ranged from 13–72%. Great brome grass emergence ranged from 0.5–81%.

Most barley grass populations reached 50% emergence in the first week of May (on 2 or 3 May 2016) (Figure 2). There was high rainfall in March and April 2016, but average maximum and minimum temperature in these months was high (Figure 1). Barley grass seed collected from Beacon WA reached 50% emergence significantly later than other populations, on 6 May 2016 for the crop population and 8 May 2016 for the roadside population. Seed from Merredin WA had the crop seed reach 50% emergence a month after

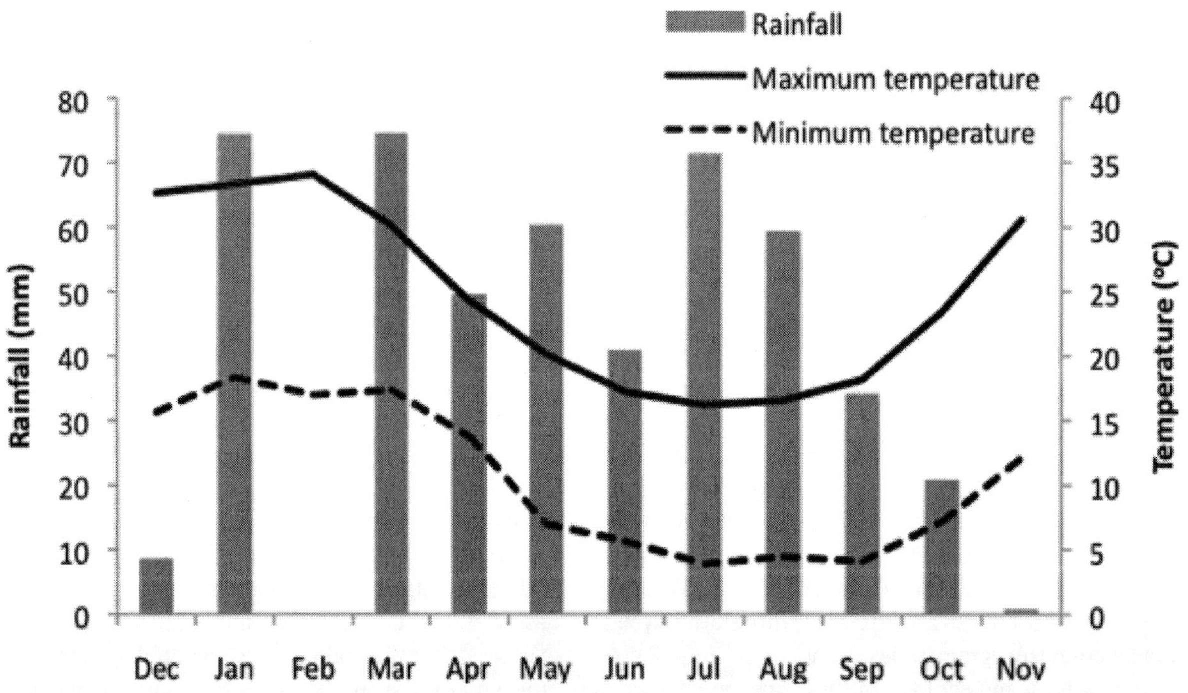

Figure 1. The total monthly rainfall and average maximum and minimum temperature from December 2015 to November 2016, from the Northam weather station 10111 (Department of Science Information Technology and Innovation 2018).

the seed from the roadside population (2 and 31 May 2016). Likewise, the population from York WA had the crop seed reach 50% emergence significantly later than the roadside seed (2 and 19 May 2016). There was no significant difference in emergence of seed harvested from high or low rainfall zones.

Great brome grass also reached 50% emergence on 2 or 3 May 2016 (Figure 3). There was only one population (Grass Valley WA) where the seed harvested from the crop had a significant delay in reaching 50% emergence (on 8 May 2016) compared to the seed from the roadside (2 May 2018).

Great brome grass populations reached 90–100% emergence 1–2 weeks after seedlings first emerged in May (Figure 3). By comparison, many barley grass populations took four months to reach 100% emergence (Figure 2).

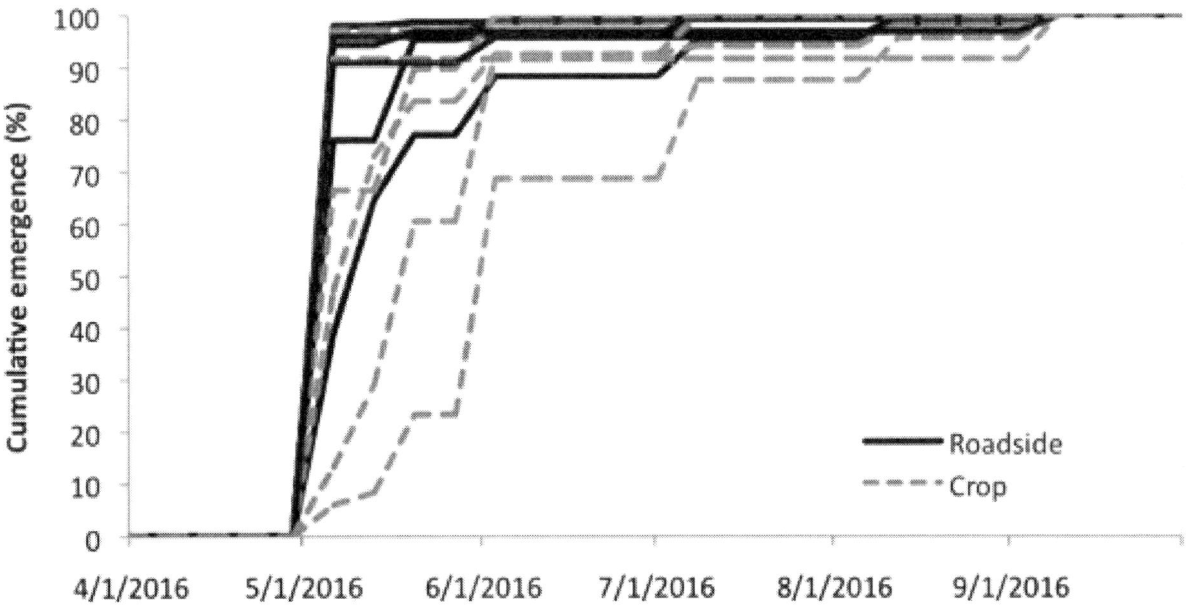

Figure 2. The cumulative emergence of seed from barley grass populations collected from six locations, with seed taken from the cropped field or adjoining roadside in each location.

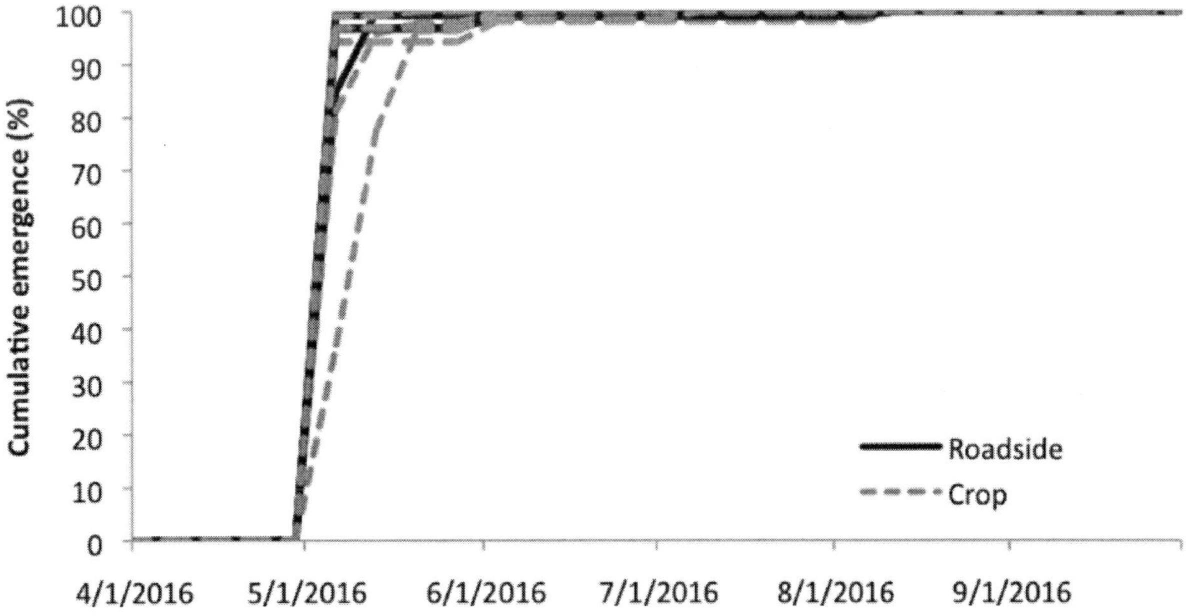

Figure 3. The cumulative emergence of seed from great brome grass populations collected from six locations, with seed taken from the cropped field or adjoining roadside in each location.

DISCUSSION

Barley grass populations in WA have the same capacity to develop delayed emergence as was observed in SA with smooth barley grass populations. This will affect the management of barley grass. Where emergence is delayed, non-selective herbicides will provide less effective control (Gill and Fleet 2012). However, the crop will be more competitive against the late emerging weeds, and early (dry) seeding will ensure the crop is as large as possible when the weeds emerge (Gill and Fleet 2012, Storrie 2014, Borger and Hashem 2018). However, there are few effective herbicides for selective control of barley grass in cereal crops, and growers will need to rotate crops to control barley grass (Storrie 2014). Development of delayed emergence was not consistent; only evident in two populations. Growers need to assess the emergence of barley grass on their own property, to determine an effective management strategy. However, most barley grass populations had staggered emergence, and so growers will need to plan for late weed control to prevent seed set.

Great brome grass did not develop delayed emergence, in contrast to the SA populations (Kleemann and Gill 2013). This is possibly because brome grass has not been managed as intensively in WA as in SA. The severity of brome grass as a weed has been increasing in Australia over the last 15 years (Llewellyn *et al.* 2016). Therefore, as great brome grass becomes a more problematic weed in WA, increased intensity of management may lead to evolution of delayed emergence to avoid weed control tactics early in the season. However, there is increasingly widespread use of harvest weed seed destruction in WA (Walsh *et al.* 2017). Both great brome grass and barley grass shed seed prior to and during crop harvest (Walsh and Powles 2014, Borger and Hashem 2018). If these species emerge late in the season and avoid a non-selective herbicide, then they are likely to retain a greater proportion of seed at harvest, to be destroyed by harvest weed seed destruction. The evolutionary pressure from harvest weed seed destruction may prevent these weeds evolving late emergence strategies. Delayed emergence is a costly strategy to avoid weed control techniques, as late emergence in a competitive crop reduces weed growth and seed production (Zimdahl 2004, Borger and Hashem 2018). Therefore, it is reasonable to assume that not all populations of brome and barley grass develop this strategy.

ACKNOWLEDGMENTS

This research was funded by the Grains Research and Development Corporation project 'Seed bank biology of emerging weeds' UA00156. Thanks are due to the growers who donated the weed seed, and to Nerys Wilkins and Dave Nicholson for assisting with this trial.

REFERENCES

Borger, C. and Hashem, A. (2018). What's the worst weed in Wongan Hills? Proceedings of the Grains Research Updates. (Grains Industry Association of Western Australia, Perth, Western Australia).

Department of Science Information Technology and Innovation (2018). Daily climatic data: Patched point dataset & DAFWA weather stations. http://srig–web/ppd/ (accessed 27 January 2016).

Gill, G.S. and Fleet, B. (2012). Seed dormancy and seedling recruitment in smooth barley (*Hordeum murinum* ssp. *glaucum*) populations in southern Australia. *Weed Science* 60, 394-400.

Kleemann, S.G.L. and Gill, G.S. (2013). Seed dormancy and seedling emergence in ripgut brome (*Bromus diandrus*) populations in southern Australia. *Weed Science* 61, 222-9.

Llewellyn, R., Ronning, D., Clarke, M., Mayfield, A., Walker, S. and Ouzman, J. (2016). Impact of weeds on Australian grain production: the cost of weeds to Australian grain growers and the adoption of weed management and tillage practices. https://grdc.com.au/Resources/Publications/2016/03/Impact–of–weeds–on–Australian–grain–production (accessed 20 March 2018).

Storrie, A.E. (2014). 'Integrated weed management in Australian cropping systems' (Grains Research and Development Corporation, Australia).

VSN International (2017). 'GenStat for Windows 15th edition.' (VSN International, Hemel Hempstead, United Kingdom).

Walsh, M., Ouzman, J., Newman, P., Powles, S. and Llewellyn, R. (2017). High levels of adoption indicate that harvest weed seed control is now an established weed control practice in Australian cropping. *Weed Technology* 31, 341-7.

Walsh, M.J. and Powles, S.B. (2014). High seed retention at maturity of annual weeds infesting crop fields highlights the potential for harvest weed seed control. *Weed Technology* 28, 486-93.

Zimdahl, R.L. (2004). 'Weed-crop competition: a review' (Blackwell Publishing, Ames, Iowa).

Management of *Chromolaena odorata* in the Douglas Shire

Simon J. Brooks[1] and Peter Logan[2]
[1] Tropical Weeds Research Centre, Department of Agriculture and Fisheries, PO Box 187, Charters Towers, Queensland 4820, Australia
[2] Douglas Shire Council, PO Box 723, Mossman, Queensland, 4873, Australia
(simon.brooks@daf.qld.gov.au)

Summary *Chromolaena odorata* (L.) King and Robinson (Siam Weed) is a serious weed across tropical and subtropical areas of Africa, Asia and multiple Pacific Islands, and was the target of a nationally cost-shared weed eradication program after it was discovered in Australia in 1994. *Chromolaena odorata* was discovered in the Douglas Shire near Mossman in 2003. At the conclusion of the broader eradication program in 2012, greater progress was being made towards its removal in this shire compared with larger infestations in other areas. Staff from the Douglas Shire, and other local stakeholders, have continued surveying and controlling known and new infestations. Ongoing localised management, mapping and record keeping provides a case study for the continuing control of this former eradication target. This discrete incursion is seen as a high priority by the shire and a threat to valuable agricultural, environmental and tourism assets. Further dispersal from infestations is likely via various recreational and agricultural thoroughfares.

Keywords Tropics, Siam weed, Daintree.

INTRODUCTION

Chromolaena odorata (L.) King and Robinson (Siam weed) is a large, multi-stemmed, perennial shrub in the Asteraceae family. It has long been recognised as one of the most invasive plants of tropical and subtropical regions of Africa, Asia and Oceania (Zachariades *et al.* 2009). It was first discovered on mainland Australia in 1994 near the towns of Mission Beach and Tully on the wet tropical coast of north-east Queensland. *Chromolaena odorata* has since been recorded across a variety of agricultural and environmental habitats and rainfall zones (Figure 1) in north Queensland, including the drier inland tropical areas of the upper Herbert River catchment, and catchments to the west of Townsville (Figure 1). In recognition of the serious threat *C. odorata* posed to tropical and sub-tropical Australia, a national, cost-shared, weed eradication program commenced in 1994.

The eradication program concluded in 2012, as more infestations were being discovered each year (Jeffery 2012), making the management of the entire incursion increasingly difficult. There were varying degrees of infestation and progress towards eradication across catchment and local government areas across the incursion. This paper illustrates the progress that has been made on the most northern part of the incursion in the Douglas Shire around the town of Mossman on the far north Queensland coast. Information on the local post eradication activities is also presented along with the risks posed by the potential spread of *C. odorata* in the Douglas Shire and further north.

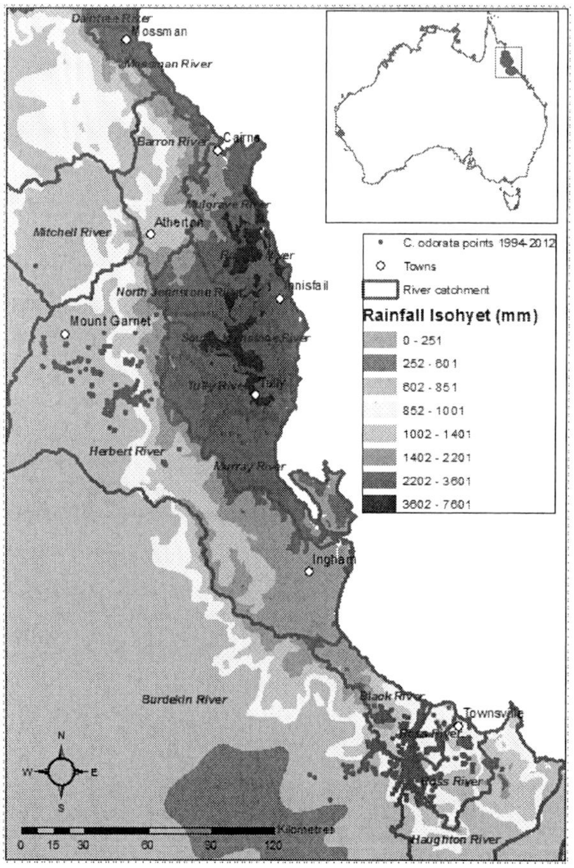

Figure 1. Overview of all recorded *Chromolaena odorata* locations 1994–2012 across river catchment boundaries and rainfall zones.

MANAGEMENT 2003–2012

Field crews searched buffer areas twice a year extending for at least 150 m around points where at least one *C. odorata* has been recorded. Additional surveys were conducted along creeks and rivers. Presence or absence data were derived from field records for every known or new waypoint. Waypoint records were aggregated into static one hectare cells (100 × 100 m). These cells are used in program reporting as 'management areas'. They were not survey units. A management area was added to the database if a new waypoint was detected beyond the boundaries of a previously known management area.

Reporting of management areas commenced in 2009. Previously, a buffered infestation-based system was used. Under the buffer system, infestations could expand and merge to create uneven sizes so all historical data had to be recalculated each year (Brooks 2017). The management area system provided a consistent spatial and temporal scale for reporting on field activities. Data collected prior to 2009 was re-analysed using this new scale. However, its implementation required a finer scale collection of location data, which was reflected in recording of more management areas in 2010 surveys (Figure 2). Every 12 months, point records are summarised to allocate a status of 'control' = plants present, or 'monitoring' = plants absent for each management area. Management areas only entered a monitoring phase when absence data was recorded in the last 12 month period. Management areas relapsed to the control phase if plants were found.

By 2012, there were 134 management areas, with at least one plant present once. However, 46% were in a monitoring phase with no plants present for between one and seven years (Figure 2).

MANAGEMENT 2013–2016

Douglas Shire Council control activities have included controlled burns, foliar herbicide application and physical control of isolated plants, as was also the case between 2003 and 2012. Small C. *odorata* plants and seeds on the soil surface can be controlled by fire (S. Brooks, unpubl. data). Controlled burns have been conducted in 2013 and 2015 on the Mowbray Range (the eastern most points on Figure 3) to aid population suppression along with other control measures. Flowering plants are readily detectible from the air in late May and June. Aerial surveys have been conducted annually targeting *Hiptage benghalensis* (L.) Kurz. and *C. odorata* in cooperation with Queensland Parks and Wildlife Service. Ground activities have also been assisted by Jabalbina Rangers, landholders, Terrain NRM and Biosecurity Queensland.

The Douglas Shire Council has been using an internal electronic database to record dates, resources and multiple categories of control activities in a 'reason(s) to visit' field from 2013 to 2016. These records are on a broad property scale, including multiple 'lot on plans'. Where properties were visited and 'survey' or 'check-up' was recorded as the 'reason to visit' and there was no treated area or method, these were considered to be *C. odorata* absence records. Where 'fire-burn' was a reason to visit, plants were assumed to be present. The broader scale of control records were not directly comparable to the pre-2012 records and were analysed separately. Control and monitoring phases were allocated annually to each property, using the same process described for management areas. The broader scale of data collection meant only a quarter of properties have progressed to a monitoring phase in 2016 (Table 1). Large and

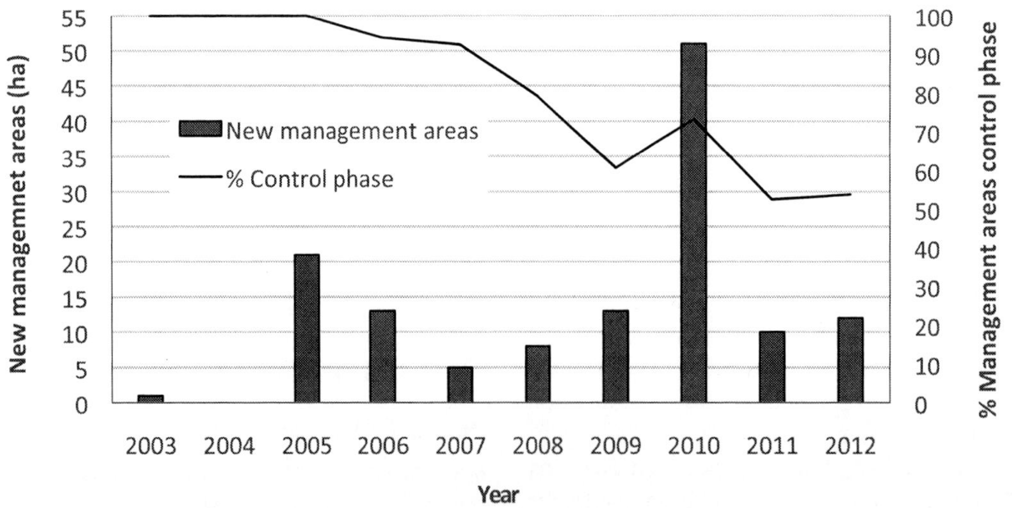

Figure 2. Discovery of new *Chromolaena odorata* management areas (1 ha each) and the percentage of management areas with plants (control phase) at yearly intervals in the Douglas Shire.

unevenly sized infestations were one of the reasons for moving to a management area reporting system during the initial eradication program (Brooks 2017).

Annual summaries of information from the Douglas Shire Council database including the number of property visits, total area treated (spray and manual control) and hours on sites are shown (Table 1). There were 355 visits recorded, where 83 recorded no treatment, 10 were associated with fires and half of the remaining 263 control visits treated between 1 and 50 m^2 of *C. odorata*. The small treated areas, absence records and total work hours show that the Shire has continued to invest in survey, as well as control activities. Visits are concentrated in May, June and July, but also occur between January and October.

Additional waypoints have been recorded through ground and aerial surveys. Ground surveys and points can also be instigated by verifying a public report of *C. odorata*, resulting from local awareness activities including display specimens and social media. The waypoints are a comparable scale to the eradication program data and could be used to extract management area locations, aggregated between 2013 and 2016.

For a species that is considered to be primarily wind-dispersed, *C. odorata* has repeatedly been observed spreading along creek lines (Brooks *et al.* 2017). In the Douglas Shire, plants have also been found near farm tracks, cane headlands, grazing land, and in forested areas (Figure 3), underlying the need for effective chemical control measures for a range of situations (Vitelli *et al.* 2018, and references there-in).

Table 1. Annual *Chromolaena odorata* data from Douglas Shire field records.

	2013	2014	2015	2016
Properties	29	33	38	39
Monitor %	20.7	21.2	34.2	25.6
Total visits	90	85	115	65
Work hours	717	1309	1580	1286
Treated area (ha)	2.13	9.57	8.06	8.87

Figure 3. Location of *Chromolaena odorata* waypoints and management areas prior to 2012 (red) and after 2012 (yellow) near the town of Mossman in northern Queensland. Map also shows water courses (blue lines) endangered regional ecosystems (purple hash), 'of concern' regional ecosystems (orange hash) (see text) and national park boundaries (green line). Ecosystems are classified by their biodiversity status (Queensland Herbarium 2016).

MANAGING FUTURE RISKS

Since 2003, there has been substantial investment in on-ground control of *C. odorata* by a range of agencies in the Douglas Shire. The difficult task of removing this disparate incursion is highlighted by the recording of new points and active areas of seedling recruitment annually to 2016. While the eradication program concluded in 2012, the risk posed by this species remained and Douglas Shire Council saw the need to coordinate and continue survey and control activities. *C. odorata* is included in the shire's biosecurity plan, raising awareness of the highest priority weeds in the area, as it remains a threat to valuable local agricultural, environmental and tourism assets.

Endangered regional ecosystems (Queensland Herbarium 2016) in relation to *C. odorata* infestations have been illustrated (Figure 3). There are records of *C. odorata* within areas of endangered regional ecosystems 7.3.10a (a mesophyll vine forest) and 100–200 m away from 7.3.23a (simple-complex semi-deciduous notophyll to mesophyll vine forest) and 7.3.17 (a complex mesophyll vine forest). Endangered lowland rainforest ecosystems such as these occur as narrow fragments of creek-line vegetation, outside of reserves and amongst agricultural land. The edges of these thin strips of vegetation are susceptible to damage by a range of invasive plants including *C. odorata*. The spread of *C. odorata* along creek line vegetation, headlands and farm tracks (Figure 3) could facilitate on-going invasion into sugar cane and other horticultural crops. Isolated points show the potential for water and vehicles to spread this weed (Figure 3).

Many different vegetation types are present on reserved and unreserved vegetation in this area. South of Mossman, *C. odorata* has been recorded in endangered regional ecosystem 7.3.12b (a *Eucalpytus*, *Corymbia* and *Melaleuca* open forest with a well-developed vine forest understory) (Figure 3). In, and to the north of, the Mowbray National Park, *C. odorata* has been recorded within 'of concern' regional ecosystem 7.11.49 (a *Eucalyptus* and *Corymbia* open forest to woodland on metamorphic foothills) and endangered 7.11.16a (a *Eucalyptus portuensis* K.D.Hill and *Corymbia intermedia* (R.T.Baker) K.D.Hill & L.A.S Johnson open forest to woodland). Within 100 m of the southeastern most *C. odorata* points (Figure 3) is a small area of endangered regional ecosystems 7.11.39a (grassland dominated by *Themeda triandra* Forssk.). These examples demonstrate the different types of vegetation where *C. odorata* has been (or soon could be) recorded, and where it will continue to invade if not controlled. These open forest vegetation types may be suitable areas to periodically utilise fire, in conjunction with conservation and other land management objectives.

Elsewhere, *C. odorata* has spread into open, partly shaded and riparian vegetation along the North Johnstone River in the Wet Tropics World Heritage Area (C. Roach, pers. comm.). Currently, small infestations (the western most points on Figure 3) are approximately two km east of the entrance to the Mossman Gorge in the Daintree National Park, an essential component of the local tourist infrastructure. Local, state and indigenous groups are also controlling the spread of the wind-dispersed invasive vine *H. benghalensis* into the Mossman Gorge World Heritage Area from the east of the National park.

In addition to wind and water dispersal, *C. odorata* can also be spread by recreational and agricultural vehicles and equipment, locally and between catchments (Brooks *et al.* 2017). An isolated infestation was found near Mount Lewis, 16 km west of Mossman in 2014 and is being surveyed and controlled by Queensland Parks and Wildlife Service. Other isolated plants in Mareeba and Tableland Shires were found since 2013, with larger infestations found around Lake Tinaroo in 2017. These discoveries indicate an expanding presence on agricultural and recreational land uses, with the further spread of *C. odorata* from local infestations also possible via recreational and agricultural activities north to Cape York Peninsula.

Regardless of the scale, progressing infestations to an absence phase requires both effective control and evidence of absence. The shire's record system provides a basis for tracking treatments and human resources. By inferring absence, it provides a coarse guide to the progress in removal of *C. odorata* from the Douglas Shire and reducing the many risks this weed poses.

ACKNOWLEDGMENTS

Data presented was diligently collected by field staff from a range of local and state agencies.

REFERENCES

Brooks, S.J. (2017). What can I do with weed control data? Proceedings of the 14th Queensland Weed Symposium, ed. T. Sydes, pp. 119-25. (Queensland Weed Society, Brisbane).

Brooks, S.J., Setter, S.D. and Gough, K.L. (2017). Siam weed dispersal mechanisms? Proceedings of the 14th Queensland Weed Symposium, ed. T. Sydes, pp. 153-8. (Queensland Weed Society, Brisbane).

Jeffery, M. (2012). Eradication: lessons learnt from 17 years of the National Siam weed eradication program. Proceedings of the 18th Australian Weeds Conference, ed. V. Eldershaw, pp. 92-5. (Weed Society of Victoria, Melbourne).

Queensland Herbarium (2016). Regional Ecosystem Description Database (REDD). Version 10.0 (December 2016) (Dept of Science, Information Technology and Innovation, Brisbane).

Vitelli, J.S., Panetta, F.D., Madigan, B.A. and Van Haaren, P.E. (2018). Chemical options for the control of Siam Weed (*Chromolaena odorata*) in natural ecosystems. *Ecological Management and Restoration* 19, 88-93.

Zachariades, C., Day, M., Muniappan, R. and Reddy. G.V.P. (2009). *Chromolaena odorata.* In 'Biological Control of Tropical Weeds using Arthropods'. eds R. Muniappan, G.V.P. Reddy, and A. Raman. pp. 130-62. (Cambridge University Press, New York).

Progress in the eradication of *Mikania micrantha* from Australia

Simon J. Brooks[1] and Mick Jeffery[2]
[1] Tropical Weeds Research Centre, Biosecurity Queensland, Department of Agriculture and Fisheries, PO Box 187, Charters Towers, Queensland 4820, Australia
[2] Biosecurity Queensland, Department of Agriculture and Fisheries, PO Box 20, South Johnstone, Queensland 4859, Australia.
(simon.brooks@daf.qld.gov.au)

Summary *Mikania micrantha* Kunth (Mikania vine) was first discovered near Mission Beach in north Queensland in 1998 and this species was included in the nationally cost-shared 'National Tropical Weeds Eradication Program' when it commenced in late 2003. The progress that is being made towards the eradication of this serious tropical weed is prompting decisions about the type and duration of program resources deployed to survey areas with continuous records of plant absence.

Issues still remain in detecting this vine amongst a backdrop of green vines and rainforest trees, the longevity of the seed soil seed bank and preventing the occasional reproductive relapse. Research and field observations have identified that control prior to July each year should prevent seed production, but canopy disturbance from cyclones increases emergence and growth rates.

Keywords Cyclones, tropics, declaring eradication.

INTRODUCTION

Mikania micrantha Kunth (Mikania vine) is a rampant, smothering tropical weed, readily capable of vegetative dispersal and seed dispersal by wind, water or machinery. It is one of the most serious weeds across tropical Asia, the Indian sub-continent and Pacific regions (Day *et al.* 2016). *Mikania micrantha* was first discovered near Mission Beach in north Queensland in 1998 and this species was included in the nationally cost-shared 'National Tropical Weeds Eradication Program' (NTWEP) when it commenced in late 2003. *Mikania micrantha* was also discovered near Forrest Beach, Ingham and Speewah, (Brooks *et al.* 2008) but most active infestations are near Mission Beach and Bingil Bay. This area also includes several of the multiple *Limnocharis flava* (L.) Buchenau and *Miconia calvescens* DC. infestations that are also targeted by this program. Single locations of *Miconia racemosa* (Aubl.) DC. and *Miconia nervosa* (Sm.) Triana are not located near any *M. micrantha* infestations. Program resources are managed across all the target species, so new discoveries and changes in search area and frequency of any target species can influence the resources available for all the species.

There has been a substantial period since details on the progress of the eradication of *M. micrantha* were published (Brooks *et al.* 2008). The eradication program is constantly changing (Jeffery and Brooks 2016) and many changes are only documented in internal reports or presentations. Therefore, an update on the progress of eradication of *M. micrantha* and the methods used to measure this progress are presented below.

ERADICATION REPORTING METHODS

Four times a year, field crews search buffer (polygon) areas at least 200 m around points known to have had at least one *M. micrantha* recorded. Additional surveys are conducted along creeks and rivers (Jeffery and Brooks 2016). Presence or absence is one of the parameters derived from field records for every known or new Global Positioning System (GPS) waypoint. Additional waypoints are added if plants are found more than 30 m away from previous points. For the purposes of program reporting waypoint records are aggregated into static one hectare cells (100 m × 100 m), generated as a 'grid layer' across the entire incursion. These cells are used in program reporting as 'management areas', they are not survey units. A new management area is added to the database if a new waypoint falls beyond the boundaries of a previously known management area.

Reporting of management areas commenced in 2010, previously a system based on infestation buffers was used (Brooks *et al.* 2008). Under the buffer system, infestations were different sizes and could expand and merge. To account for the merging of infestations, past annual status data had to be recalculated (Brooks 2017). The management area system means that eradication reporting is across a scale that is spatially consistent over time. Every six months point records are summarised to allocate a status of 'control' = plants present or 'monitoring' = plants absent for each management area. Management areas only enter a monitoring phase when absence data is recorded in the last 2 × 6 month periods, this progression is via evidence of absence. The amount of time management areas have been in the monitoring phase is calculated as the time since last detection, which is rounded down to an annual value in years. If plants are recorded in a management area which is in the monitoring phase it relapses to a control phase, for at least 2 × 6 month periods.

In recent years, at the suggestion of external reviewers (Jeffery and Brooks 2016), the program has also started using the time since last reproduction as a measure of eradication progress. In cases where no seed production has been observed, the discovery date is used as a 'defacto' time since last seed production. The time since last seed production (or discovery) accrues annually unless there is a seed production event (reproductive escape) causing the management area to suffer a reproductive relapse. The last reproduction data is determined at the end of each financial year from a single date for each management area.

Both the last detection and last reproduction or discovery data are plotted as histograms for annual program reporting at the end of each financial year. While this data has the same sample size and appears similar (Tables 1 and 2) it was created via different processes and at different intervals, as described above.

ERADICATION PROGRESS

By December 2017 there were 89 management areas with at least one *M. micrantha* vine present once. Seventy five percent (67 of 89) of the management areas had progressed to a monitoring stage (plant absence for more than a year) (Figure 1, Table 1). All management areas near the towns of Ingham and Forrest Beach have been in a monitoring phase for three or more years. Since 2012, there has been a consistent decline in the percentage of management areas in the control phase (plants present) (Figure 1). When plants are encountered, they are effectively controlled with fluroxypyr or triclopyr based herbicides (Brooks and Setter 2014). Occasionally isolated vines or any in riparian areas may be physically controlled by field crews with care taken to remove and destroy all vegetative fragments.

Sixty seven of the 89 management areas have absence records for between 1 and 19 years (Table 1). The NTWEP considers management areas to be 'provisionally eradicated' after a monitoring phase (absence) of 5 years. No decline in the percentage of 'provisionally eradicated' management areas is a current milestone the program must meet and report against each financial year. With the exception of one management area where seed was kept dry by nursery plastic and germinated when it was disturbed, all relapses from the monitoring phase have been from one, two or three years since last detection.

Discovery The 2010 transition to reporting management areas allowed eradication progress to be documented for portions of larger infestations. Across the program this transition also caused a spike in the discovery of new management areas with finer scales of recording and reporting. However, this is less evident in the *M. micrantha* data (no new areas in 2010, Figure 1) where each infestation was reasonably well documented by waypoints prior to 2010, including discovery and control information dating back to 1998.

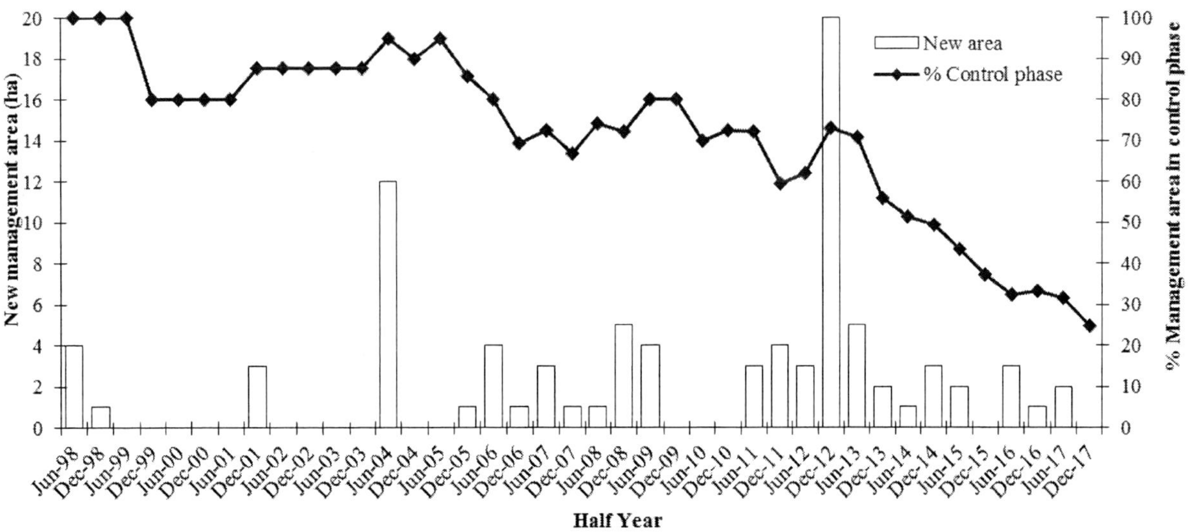

Figure 1. Discovery of new *M. micrantha* management areas (1 ha each) and the percentage of management areas with plants (control phase) at 6 monthly intervals.

Most of the new management areas recorded in August and September 2012 were in the Mission Beach area and are thought to have been the result from increased search effort north of currently known infestations. The passage of severe Tropical Cyclone Yasi over the area in early February 2011 may have also contributed as it created a much lighter environment for *M. micrantha* germination and seedling growth (see Brooks and Jeffery 2018, this proceedings).

The 14 new management areas discovered between September 2013 and May 2017 were all adjacent to previously known management areas. They are likely to be the result of local dispersal and emergence from the soil seed bank. There were no new management areas recorded in the six months to the end of December 2017. The discovery of new *M. micrantha* areas are more likely to be via on ground searches by trained weed officers, than as a result of public information and display specimens (Jeffery and Brooks 2016).

Reproductive relapses Only 45 of the 89 management areas have seeding plants recorded either on discovery or via reproductive relapses. Hence, the data in Table 2 is influenced by the discovery date of each management area. The five management areas in the zero years since last reproduction or discovery (Table 2) comprise of three newly discovered areas and two reproductive relapses in the 2016–2017 financial year. There were no reproductive relapses in 2015–2016, and the three areas shown in the one year line in Table 2 were new discoveries. Discoveries or reproductive relapses in 2012–2013 have resulted in a spike in the 4 year line (Table 2) after new management areas were following a crew scale up and cyclone Yasi. Research trials and field observations have identified that control prior to July each year should prevent seed production. However, there are still some reproductive escapes, or relapses; particularly leafless stems reaching several metres into lower parts of the tree canopy. Detecting this vine amongst a backdrop of green vines and rainforest trees and preventing the occasional reproductive relapse is a difficult task but has been aided by having a designated crew for the Mission Beach/Bingil Bay area and Ingham within the program. Across all infestations, field crews surveyed or resurveyed a gross total of 935 hectares in 2016–2017. Field crews have no choice but to record plants that have developed fractionally beyond flowering as seeding. Although undesirable, newly mature vines with intact seed heads have limited opportunities for the dispersal of viable seed, or for incorporation into the soil seed bank. The last local collection of mature seed suitable for seed research was in 2012.

Table 1. Summary of years since last plant was detected for 89 *M. micrantha* management areas as of December 2017, note years with 0 management areas are not shown (n = 89).

Years since last detection	Management areas (ha)
0	22
1	14
2	11
3	6
4	8
5	11
6	3
7	6
8	2
10	1
11	1
12	3
19	1

Table 2. Summary of years since last seeding plant was detected or the management area was discovered as of June 2017, note years with 0 management areas are not shown (n = 89).

Years since last seeding or discovery	Management areas (ha)
0	5
1	3
2	10
3	7
4	25
5	9
6	3
7	2
8	9
9	3
13	8
15	3
19	2

When the categories of years since last reproduction are split by whether the management areas are in control or monitoring phases (July 2017, data not shown), all of the control phase cells are in the six years since last reproduction category or less. Once management areas pass seven years since last reproduction, and multiple years continuous of absence are recorded, then confidence that the seed bank is exhausted increases.

FUTURE ISSUES

The positive trends in decline of the active areas, discovery and reproductive relapses show the overall progress made towards the eradication of this serious tropical weed. This progress is prompting discussion about the type and duration of program resources deployed to survey areas with continuous records of plant absence. Specifically considering the need to maintain a quarterly visit schedule, the possibility for transitional arrangements to local management for some areas and when to stop searching and declare areas eradicated. Decisions about these topics are likely to be based on multiple criteria including: the time since last reproduction; length of the recent monitoring phase; total recorded population or infested area; number of monitoring phase or reproductive relapses; and the data from the surrounding management areas. Importantly these criteria will include a spatial component as management areas rarely exist in isolation and are arbitrary divisions. Mature discoveries or reproductive relapses could influence the surrounding management areas within the dispersal buffer. As there is always uncertainty about the status of the soil seed bank, the use of multiple criteria may help manage the uncertainty and overcome any limitations of applying a single value for seed bank longevity in all situations. There are some indications from an ongoing buried seed packet trial (Brooks and Setter 2012) that seven years since last reproduction may be insufficient as one of criteria to declare management areas eradicated (S. Brooks unpubl. data). The 25 *M. micrantha* management areas with over seven years since last reproduction illustrated in Table 2 provide a sample to develop, test and refine these criteria.

Although plants are becoming increasingly rarer, the occasional reproductive relapse will suppress future progress. Eradication progress took a step back 15 months after cyclone Yasi, so there remains the potential for increased emergence and growth rates as a result of future cyclone(s).

The decreasing proportion of management areas in the control phase demonstrates that plants are effectively treated when found. Even allowing for new areas, there have been fewer active (control phase) areas each year for the past five years. Field crews should eventually face the possibility of searching for years and never seeing a naturalised vine. Despite the difficult working environment, substantial progress is being made towards the eradication of one of the most serious tropical weeds.

ACKNOWLEDGMENTS

The data presented has been collected by numerous dedicated field workers in the Four Tropical Weeds Program and stakeholder agencies (particularly Hinchinbrook Shire Council and Queensland Parks and Wildlife), who repeatedly and diligently survey tropical forests, disturbed vegetation margins and crocodile infested creeks for single green vines amongst a continuum of green vegetation.

REFERENCES

Brooks, S.J. (2017). What can I do with weed control data? Proceedings of the 14th Queensland Weed Symposium, ed. T. Sydes. pp. 119-25. (Queensland Weed Society, Brisbane).

Brooks, S.J., Panetta, F.D. and Galway, K.E. (2008). Progress towards the eradication of mikania vine (*Mikania micrantha*) and limnocharis (*Limnocharis flava*) in northern Australia. *Invasive Plant Science and Management* 1, 296-303.

Brooks, S.J. and Jeffery, M. (2018). The effects of cyclones on a tropical weed eradication program. Proceedings of the 21st Australasian Weeds Conference, 9–13 September, Sydney, New South Wales, eds S. Johnson, L. Weston, H. Wu and B. Auld, pp. 119-23. (Weed Society of NSW).

Brooks, S.J. and Setter, S.D. (2012). Soil seed bank longevity information for weed eradication target species. *Pakistan Journal of Weed Science Research* 18, 73-83.

Brooks, S.J. and Setter, S.D. (2014). Increased options for controlling mikania vine (*Mikania micrantha*) with foliar herbicides. Proceedings of the 19th Australasian Weeds Conference, ed. M. Baker, pp. 409-12. (Tasmanian Weed Society, Hobart).

Day, M.D., Clements, D.R., Gile, C., Senaratne, K.A.D., Shen, S., Weston, L.A. and Zhang, F. (2016). Biology and impacts of Pacific Islands invasive species. 13. *Mikania micrantha* Kunth (Asteraceae). *Pacific Science* 70, 257-85.

Jeffery, M. and Brooks, S.J. (2016). Eradication in the tropics: constantly changing and adapting. Proceedings of the 20th Australasian Weeds Conference, eds R. Randall, S. Lloyd and C. Borger, pp. 23-7. (Weeds Society of Western Australia, Perth).

The effects of burial depth and water stress on Melastome weed seeds

Simon J. Brooks[1], Rose K. Easton[2] and Kirsty L. Gough[1]
[1] Tropical Weeds Research Centre, Department of Agriculture and Fisheries, PO Box 187, Charters Towers, Queensland 4820, Australia
[2] School of Earth and Environmental Sciences, University of Queensland, St Lucia, Queensland 4702, Australia
(simon.brooks@daf.qld.gov.au)

Summary Two seed trials were conducted to provide insight into the conditions under which high priority tropical rainforest weeds from the family Melastomataceae can establish. Seeds of *Clidemia hirta, Miconia calvescens, Miconia nervosa* and *Miconia racemosa* were placed at depths of 0, 2, 5, 9, 14 and 20 mm in small containers and incubated. Emergence declined with depth, with no emergence from 9 mm depth or greater. Seeds of the same species were also subjected to moisture stress in incubated Petri dishes, via exposure to water potentials (stress) between 0 and −1 MPa. Despite their different native and invasive home ranges seeds of the *Miconia* species germinated at relatively low water potentials between 0 and −0.33 MPa. In comparison, *Clidemia hirta* seed germinated under higher moisture stress at −0.5 and −0.67 MPa. These weeds form persistent soil seed banks which may need managing for decades. Management actions and potential impacts can be informed by understanding the factors influencing their establishment from seed.

Keywords Eradication, polyethylene glycol, rainforest.

INTRODUCTION

Eradication programs are seeking to manage the soil seed bank by preventing additions and running down the viable seed to extinction. Yet the processes governing seed germination and seedling emergence are rarely documented for species invading tropical rainforest, though some are implied from native and invasive geographical extents. Estimates of the potential distribution and impacts of these species can also be based on their native and invasive ranges. However, the Melastomataceae species used in these trials have different native and introduced geographical ranges (Table 1). The infestations of *Miconia* species are the targets of a nationally cost-shared 'National Tropical Weeds Eradication Program' (NTWEP) (Jeffery and Brooks 2016). *Clidemia hirta* was included in the NTWEP until the discovery of the second local infestation in 2015 (Brooks *et al.* 2016).

Table 1. Local infestations, broad native and introduced ranges and seed lot sources of species tested.

Species and life form	Australian extent and discovery*	Other invasive range	Native range**	Seed sources used in trials
Clidemia hirta (L.) D. Don (shrub)	2 infestations Julatten (2001), Misty Mountain (2015)	Many tropical and subtropical countries and islands	Widespread, Central and South America and the Caribbean, 23° N to 27° S	Fresh glasshouse May 2016 (originally Julatten), field Misty Mountain July 2015
Miconia calvescens DC. (small tree)	Multiple infestations in north Qld and NSW (1997–2014)	Widespread mostly on tropical islands	Common across Central and eastern South America, 17° N to 25° S	Field, Mossman area (July 2009)
Miconia nervosa (J.E Smith) Triana (shrub)	1 infestation Whyanbeel (2004)	None known outside South America	Widespread Central and South America. 18° N and 18° S	Field, Whyanbeel (December 2012)
Miconia racemosa (Aubl.) (shrub)	1 infestation Kuranda (2001)	None known	Limited, Caribbean and north west coast of South America 19° N to 2° S	Shade house plants from Kuranda (February 2013)

*Information from Jeffery and Brooks (2016), Brooks *et al.* (2016), references there-in and unpublished.
**Native range information is inferred from maps of herbarium specimens held by the Missouri and New York Botanical Gardens.

Past seed research Seeds of all species are less than 250 µm long and experiments on the same Melastome seed lots (Table 1) have shown that seeds only start to germinate after two or three weeks, and may take eight weeks to cease germinating (S. Brooks unpubl. data). Although slow, 95 to 100 % of the viable seed germinates, so there is little evidence of physiological dormancy under ideal incubator conditions. Buried seed packet trials on these four species are now three to seven years old, both the buried packet trials and local field control records indicate that buried seeds can persist in the soil for many years (Brooks and Setter 2014).

To explain the occurrence of persistent but non-dormant seeds, studies have suggested that small seeded tropical pioneer and early successional species have 'risk averse' seeds (Daws *et al.* 2008, Silveira *et al.* 2013). Where seeds only germinate when conditions are suitable for an extended period and the chance of survival is high. Such species have small seed reserves to draw on and with superficial surface roots seedlings are also thought to be susceptible to short periods of drought (Pearson *et al.* 2002 (citing Engelbrecht *et al.* 2001)). The *Miconia* species are targeted for eradication, and progress towards their eradication is made when absence records accrue annually under climatically suitable conditions. The local tropical climate is presumably suitable for germination during wetter periods of most years, however the germination requirements of each of these species have not been documented. As moisture levels have an influence on the germination and survival of seedlings, a trial to investigate the effects of declining water potentials (increasing moisture stress) on seed germination was conducted.

Seed burial depth could also influence the establishment and control of Melastome seedlings. The emergence of *Miconia argentea* (Sw.) DC seedlings dropped dramatically from 0 to 2.5 and 5 mm and was not recorded at soil depths of 10 mm or deeper; further, *M. argentea* did not germinate under dark conditions (Pearson *et al.* 2002). A second trial was conducted to determine the effect of burial depth on the emergence of seedlings.

METHODS

Seed germination Seed sources and ages are listed (Table 1). The irregular availability of field seed collections is a limitation of researching eradication target species (Brooks and Setter 2016). Since collection, the seed has was air dried, removed from fruit and stored in a refrigerator at 3°C. Each experimental unit (Petri dish or aluminium tray) was labelled, placed on larger trays which were enclosed in a clear plastic bag and placed in a Thermoline® incubator running a 30/20°C, 12h/12h, day/night temperature regime for the duration of the trials. Trials commenced in September 2016.

Water stress experimental treatments A laboratory trial was conducted on the same seed lots (Table 1), except the *C. hirta* seed was sourced from Misty Mountains. This trial investigated the effect of moisture stress on seed germination. Petri dishes with 50 seeds of each seed lot were treated with a poly-ethylene glycol (PEG) solution. The equation of Michel (1983) was used to determine the amount of PEG 8000 (Fisher Scientific) needed to create solutions with water potentials of −0.08, −0.17, −0.33, −0.5, −0.67, −0.83 and −1 MPa. Each treatment × seed lot combination was replicated three times. Each dish contained 2 × Whatman No 1, 90 mm filter papers and was placed in plastic bags and incubated for eight weeks. Germination (production of a radicle) was recorded every 2 to 7 days for 12 weeks with the germinated seeds removed and counted, dishes and topped up with water as required. Un-germinated seeds were subjected to physical tests of viability.

Depth experimental treatments Melastome seeds were placed at soil depths of 0, 2, 5, 9, 14 and 20 mm and seedling emergence recorded. In preparation for the trial, several kilograms of a locally obtained (commercial) predominantly clay (+ organic/sand) soil mix was passed through a 1 mm sieve and baked for 120 hours at 70°C. Three 250 mL samples of this soil were weighed, placed in nursery tube pots with a small base lining of weed mat, weighed again and immersed in water overnight. After immersion the tubes were removed from the water, covered with a plastic bag and allowed to drain for three days, and reweighed. The weights were used to calculate the soils bulk density (dry weight of a known volume), the field capacity and the wet weight of a known volume. The volume and weight of soil when saturated was used to maintain the weights of the experimental units (aluminium trays) near field capacity for the duration of the experiment and to prevent waterlogging the different volumes of soil.

The trial was conducted in small round flat-bottomed aluminium trays (weigh boats, generic brand from Proscitech Townsville), 75 mm diameter and 32 mm deep. Each depth and seed lot treatment was replicated three times in separate trays. The base of each tray was covered with 10 mm of soil. A 50 mm diameter filter paper was placed in the middle on the base layer of soil and one lot of 50 seeds were placed on the middle of the filter paper. To create the 2, 5, 9, 14 and 20 mm depth treatments, prepared dry soil was carefully added on top of each filter paper and seed layer. Each tray was weighed with dry soil and bought up to the field capacity weight with distilled water. During the trial the trays were watered to avoid disturbing the soil above the seed and to keep trays close to the target wet weight. The trial ran for 58 weeks during which the trays were allowed to dry out between weeks 13 and 18, before being re-wet. Seedling emergence was recorded every 2 to 7 days and seedlings removed.

Statistics Regression equations were fitted to the untransformed germination or total emergence data for each treatment using Genstat® 16th edition VSN International Ltd. A Poisson distribution with a log link function fitted the emergence depth data, however the mean data is shown in the results. Logistic non-linear regression curves were fitted to mean germination data for each seed lot in the water stress trial.

RESULTS

Water stress experiment The most marked comparison between the two genera represented is at −0.5 MPa where 0% of the *M. calvescens* and *M. nervosa* germinated, 1.3% (2 seeds) of the *M. racemosa* germinated and 81.3% of the *C. hirta* germinated (Figure 1). With 41.3% germination at −0.67 MPa *C. hirta* seed is more tolerant of moisture stress conditions than the three *Miconia* species. Despite their different native and invasive ranges the three *Miconia* species showed a similar moisture stress threshold with no germination lower than −0.5 MPa. The viability of the *M. calvescens* seed may have been influenced by the age of the seed.

Depth experiment Seedling emergence declined from unburied seed to a depth of 5 mm and no seedlings emerged from the 9, 14 or 20 mm treatments (Figure 2). There was some variation in the 2 and 5 mm data due to emergence near cracks that appeared in the clay soil. The variability of emergence of the

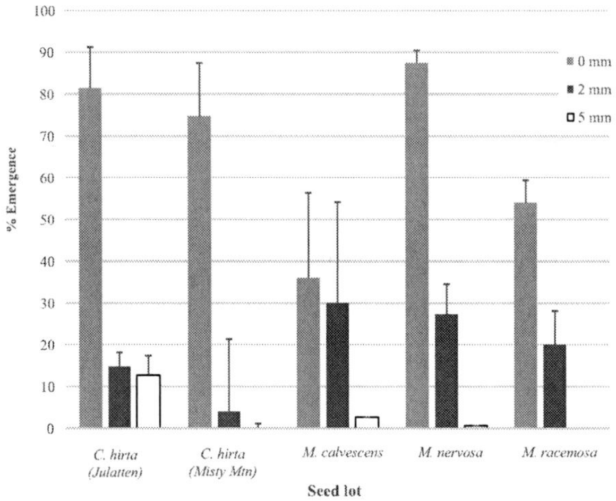

Figure 2. Mean emergence of different Melastome seed lots (+ SE) from burial under 0, 2 and 5 mm of soil. 9, 14 and 20 mm results not shown, all = 0.

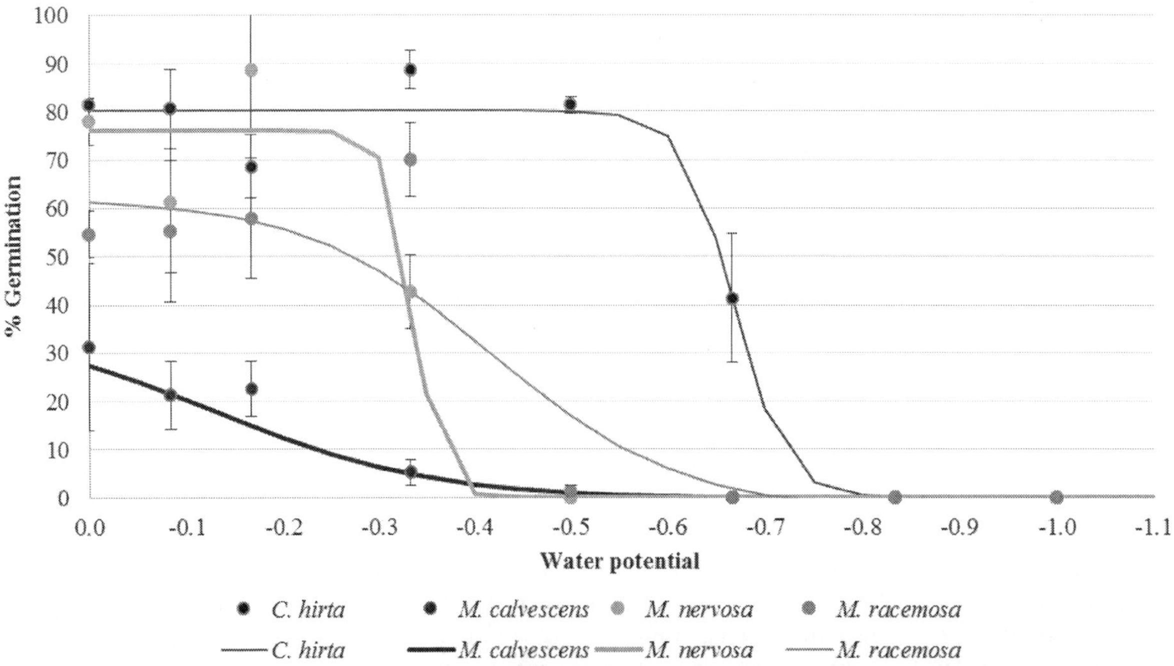

Figure 1. Mean germination of Melastome seed lots (± SE) exposed to levels of decreasing water potential. Points are species means of three replicates for each treatment. Lines are logistic curves plotted from parameters (not shown) estimated in Genstat.

M. calvescens seed was thought to be as a result of the age of the seed.

DISCUSSION

The depth experiment had similar results to those of Pearson *et al.* (2002) and showed that small seeded Melastomes will only emerge if seed is near the soil surface. This is consistent with studies (Pearson *et al.* 2002, Silveira *et al.* 2013) that have identified litter gaps, a light requirement and shallow depth as conditions under which species with small seed reserves will establish. Similarly, seed that is incorporated into the soil profile is unlikely to germinate but some will remain viable and create a persistent buried soil seed bank. If there are events that erode or expose soils with buried seed, such as uprooted trees or flood water, then seedlings could emerge and require control. Following these events the emergence of *Miconia* seedlings after a long period of plant absence could influence the progress of the eradication program.

As the species appear to lack any physiological dormancy, they are unlikely to respond to germination stimulants. However, there may be physical or chemical methods that deplete the emergent portion of the seed bank depth in the top centimetre of the soil. Seed bank depletion treatments applied to the top soil layer may assist in reducing the length and cost of long term control operations and the risk of plants escaping detection and producing fruit. Methods such as flame weeding could be applied to small dense areas of recruitment or around isolated mature plants expediting the transition to a state where plants are continually absent. Further research on the depth profile in the soil seed banks of the *Miconia* species is being planned. A test of light and dark germination requirements may assist in explaining the results of the depth trial.

Eco-physiological data can be used to refine or corroborate climatic parameters used to predict the potential distributions of invasive species (Kriticos *et al.* 2005). The three melastome shrub species have limited local distributions on which to base estimates of potential distribution, impacts and cost-share apportionment. Information to predict the invasive potential of *M. racemosa* is limited by narrow climatic ranges and such information could greatly under-estimate its potential distribution. Despite their different native and invasive ranges, the three *Miconia* species showed a similar threshold of moisture stress and higher moisture requirements for germination than *C. hirta*. A greater tolerance of lower moisture levels could assist in explaining the wider native range and indicate a greater potential invasive extent of *C. hirta*. The moisture trial was an important step in defining the germination requirements of each species. This trial will be combined with current temperature gradient studies to create a fuller picture of each species germination requirements.

The combination of these two trials with age to maturity trials (Brooks and Setter 2014) should provide a deeper understanding of field emergence patterns under local conditions. This information helps to build the biological profile, including potential impacts, for those species with limited known invasive ranges and local distributions while improving the management of these high priority invasive tropical Melastomes.

ACKNOWLEDGMENTS

Shane Campbell and Wayne Vogler assisted with trial plans during an industrial placement subject for the School of Earth and Environmental Sciences (University of Queensland). Wayne Vogler and Kelsey Hosking are thanked for comments on the draft.

REFERENCES

Brooks, S.J. and Setter, S.D. (2014). Issues and solutions for researching weed eradication target species. Proceedings of the 19th Australasian Weeds Conference, ed. M. Baker, pp. 255-8. (Tasmanian Weed Society, Hobart).

Brooks, S., Setter, S.D, Gough, K.L and Setter, M.L (2016). Increasing foliar herbicide options for controlling *Clidemia hirta*. Proceedings of the 20th Australasian Weeds Conference, eds R. Randall, S. Lloyd and C. Borger, pp. 321-5 (Weeds Society of Western Australia, Perth).

Daws, M.I., Crabtree L.M., Dalling J.W., Mullins C.E. *and* Burslem D.F.R.P. *(*2008*).* Germination responses to water potential in neotropical pioneers suggest large-seeded species take more risks. *Annals of Botany* 102, 945-51.

Jeffery, M. and Brooks, S.J. (2016). Eradication in the tropics: constantly changing and adapting. Proceedings of the 20th Australasian Weeds Conference. eds R. Randall, S. Lloyd and C. Borger. pp. 23-7. (Weeds Society of Western Australia, Perth)

Kriticos, D.J., Yonow, T. and McFadyen R.E. (2005). The potential distribution of *Chromolaena odorata* (Siam weed) in relation to climate. *Weed Research* 45, 246-54.

Michel, B.E. (1983). Evaluation of the water potentials of solutions of polyethylene glycol 8000 both in the absence and presence of other solutes. *Plant Physiology* 72, 66-70.

Pearson, T.R.H., Burslem, D.F.R.P., Mullins, C. E. and Dalling, J.W. (2002). Germination ecology of neotropical pioneers: Interacting effects of environmental conditions and seed size. *Ecology* 83, 2798-807.

Silveira, A.O., Fernandes, W.G. and Lemos-Filho, J.P. (2013). Seed and seedling ecophysiology of neotropical Melastomataceae: Implications for conservation and restoration of savannas and rainforests. *Annals of the Missouri Botanical Garden* 99, 82-99.

The NSW Biological Control Taskforce – collaborating to maintain a functional biocontrol agent delivery pipeline

Troy R.F. Brown
NSW Department of Primary Industries, Grafton Primary Industries Institute, Trenayr Road, Junction Hill, New South Wales 2460, Australia
(troy.brown@dpi.nsw.gov.au)

Summary There is an increasing need to apply self-sustaining weed management technologies. History has shown that biocontrol is one of the most cost-effective, self-sustaining management technologies for the control of weeds. Biocontrol is continually under scientific refinement for not only maximising returns on investment, but also maximising safety and impact.

The NSW Biological Control Taskforce (the taskforce) funds and promotes biocontrol research alongside rearing, release, and monitoring programs (implementation activities) through a shared investment commitment. Target weeds are reviewed and selected annually by the Taskforce Steering Committee following a prioritisation framework developed in conjunction with Commonwealth Scientific and Industrial Research Organisation (CSIRO), NSW Department of Primary Industries (NSW DPI), NSW Office of Environment and Heritage (NSW OEH) and the NSW Wales Environmental Trust (The Trust).

Biocontrol agents included in the implementation activities have received approval for release from the Australian Federal Government and are reared, released and monitored strategically to achieve the desired result. Current weed species targeted for implementation include salvinia (*Salvinia molesta* D.S.Mitch.), cat's claw creeper (*Dolichandra unguis-cati* (L.) Lohmann), madeira vine (*Anredera cordifolia* (Ten.) Steenis), crofton weed (*Ageratina adenophora* (Spreng.) R.M.King & H.Rob), water hyacinth (*Eichhornia crassipes* (C.Mart.) Solms), smooth tree pear (*Opuntia monacantha* (Willd.) Haw), water lettuce (*Pistia strateoites* L.), boxing glove cactus (*Cylindropuntia fulgida* F.M.Knuth var. *mamillata* (Engelm.) Backeb) and Hudson pear (*Cylindropuntia rosea* (DC.) Backeb).

New members are welcomed encouraged to engage as soon as possible with the NSW Biocontrol Taskforce. Collaboratively, the Taskforce aims to develop cost effective biocontrol solutions to combat weeds.

Keywords Weed biological control, NSW Biocontrol Taskforce.

Geospatial assessment to predict potential African olive (*Olea europaea* spp. *cuspidata*) infestation sites in remnant vegetation communities

Doug Campbell
Upper Hunter Weeds Authority, PO Box 122, Muswellbrook, New South Wales, Australia
(uhwa@muswellbrook.nsw.gov.au)

Summary African olive (*Olea europaea* ssp. *cuspidata*) is a highly invasive evergreen exotic tree whose African native range is similar climactically to the New South Wales' (NSW) Upper Hunter Valley. Its ability to form dense monocultures has led to it being listed as key threatening process for native vegetation communities in NSW. Presently, infestations are widespread in the lower Hunter Valley with the upper Hunter Valley recording a relevantly small number of isolated infestations. Changing land use, notably the cessation in grazing, has been linked to an increase African olive range, with avian seed dispersal considered a key factor in its spread. This study used existing information on infestation locations, avian dispersal range, land use and vegetation type to develop maps via geospatial analysis. All new infestations identified during this study were in areas where grazing had been excluded, with 73% of these previously unknown infestations recorded on roadsides. The link between grazing and seedling recruitment can be used to inform both surveillance and control programs, resulting in a more effective and efficient control program on African olive, which will assist in maintaining the biological integrity of remnant vegetation in Upper Hunter.

Keywords African olive, avian seed dispersal, biodiversity, geospatial analysis, Hunter Valley, *Olea europaea* ssp. *cuspidata*.

Causing a stink: *Paederia foetida* found on mainland Australia

Joel Daniels[1] and Stephen B. Johnson[2]
[1] Liverpool City Council, 33 Moore Street, Liverpool, New South Wales 2170, Australia
[2] Department of Primary Industries, Locked Bag 6006, Orange, New South Wales 2800, Australia
(stephen.johnson@dpi.nsw.gov.au)

Summary *Paederia foetida* (skunk or stink vine) was found in riparian vegetation near Liverpool (south western Sydney) in December 2015. It was later positively identified by the National Herbarium of NSW. Previously skunk vine has only been found in Australian territory on Christmas Island. The species is on the target list of weeds of concern to northern Australia as part of the Northern Australia Quarantine Strategy (NAQS). A native of southern and eastern Asia, skunk vine has invaded both Hawaii and four other continental USA states, becoming regionally problematic by invading a range of natural habitats. It has also invaded the Mascarene Island group in the Indian Ocean and New Caledonia in the Pacific.

Skunk vine adapts to a wide range of conditions. The leaves of this perennial vine emit a sulphurous smell when injured. It is also a weed of sugarcane, in the nursery trade and is commonly found in lawns. It is spread by wind or birds, in landscaping material and garden waste, but is also spread intentionally by people, particularly as a traditional medicinal crop grown for its anti-inflammatory properties. The Liverpool infestation has been delimited at <0.5 ha and is undergoing eradication. The response to skunk vine demonstrates a number of key lessons, that is the importance of: identifying potential threats; regular routine surveillance in high risk areas to detect these threats early; positive taxonomic determination/s; professional networks in identifying the broader threat posed by species; and communicating these threats to others should they eventuate.

Effect of nitrogen rate and weed control treatments on quantitative and qualitative yield of sugar beet

Mohammad A. Dastorani and Mohammad Armin
Department of Agriculture, Syadati Bol. Sabzevar Branch, Islamic Azad University, Sabzevar, Iran
(armin@iaus.ac.ir)

Summary The effect of nitrogen rate and weed control treatments on quantitative and qualitative yield of sugar beet was studied in 2015. This experiment was carried out following a split plot design with three replications. Factors were nitrogen rates as one and two times the recommended dose (200 kg N ha^{-1}) and Weed control treatments consisted of (1) a weedy check, (2) hand-hoeing twice, at 3 and 6 weeks after sugar beet emergence (WAE), (3) hand-hoeing three times, at 3, 6 and 9 WAE, (4) postemergence application of Phenmedipham + Ethofumesate + Desmedipham (PED) at 3 WAE + one hoeing at 6 WAE and (5) one hoeing at 3 WAE + PED application at 6 WAE. Increasing the amount of nitrogen reduced weed density, impure sugar and pure sugar content whereas it increased the root yield, sugar yield, α-amino and sodium content. Weed dry weight, sugar content in molasses, potassium content and alkality were all not affected by nitrogen amount. Results showed that the weed management treatments had a significant effect on all traits except sodium content and alkality. In the 'weedy' treatment, weed density and weed dry weight were 41% and 60% higher than in the other weed control treatments, respectively. Weed interference decreased root yield (62%), pure sugar content (35%), impure sugar content (31%) and sugar yield (76%). Overall, all weed control treatments improved root yield compared with the non-weeded check. However, highest root and sugar yields were obtained by hand hoeing three times at 3, 6 and 9 weeks after sugarbeet emergence.

Keywords Integrated weed management, herbicides, nitrogen, sugar beet, weeding.

The impact of two biological control agents, *Passalora ageratinae* and *Baeodromus eupatorii*, on *Ageratina adenophora* (crofton weed): synergism or antagonism?

Juliette Fontaine[1], Gavin C. Hunter[2] and Louise Morin[2]
[1]AgroSup Dijon, 26 Boulevard Petitjean, BP 87999, 21079 Dijon cedex, France
[2]CSIRO Health and Biosecurity, Black Mountain Science and Innovation Park, Clunies Ross Street, Acton, Australian Capital Territory 2601, Australia
(gavin.hunter@csiro.au)

Summary *Ageratina adenophora* (crofton weed) is an environmental weed in many countries due to its high reproductive capacity and ability to thrive in many environments. In Australia, crofton weed is a garden escapee that has become invasive along the eastern coast. This prompted the search for and successful release, in 2014, of the fungal biological control agent *Baeodromus eupatorii* (crofton weed rust) to aid in the management of crofton weed. An additional fungal pathogen of crofton weed, *Passalora ageratinae*, was accidently introduced in Australia in the 1950s. This fungus causes leaf lesions leading to defoliation of the weed. It has been suggested that these two pathogens in combination can reduce the growth of crofton weed but this has never been empirically tested. The aim of this study was to investigate the pathogenicity of these two fungal species on crofton weed and determine if there is evidence of an interaction between these two pathogens on the weed. A manipulative experiment was undertaken where crofton weed plants were exposed to treatments of either pathogen alone and the two pathogens in combination. Results from these experiments indicate that *B. eupatorii* alone is most effective at reducing the overall biomass of crofton weed while *P. ageratinae* alone is less effective than the two pathogens in combination at reducing the overall biomass. Interestingly, in the treatment with both fungal pathogens it appears that *P. ageratinae* may delay the symptom development of *B. eupatorii* suggesting a possible antagonistic effect of *P. ageratinae* on *B. eupatorii*.

Keywords Crofton weed, *Passalora ageratinae*, *Baeodromus eupatorii*, agent interactions.

Seed-set reduction of milk thistle (*Sonchus oleraceus* L.), wild turnip (*Rapistrum rugosum* L.) and flaxleaf fleabane (*Conyza bonariensis* L.) following a crop topping application of saflufenacil (Sharpen® WG Herbicide) in cereals

Ian Francis, Marco Montagna, Russell Ison and Adam Perfrement
BASF Australia Ltd, Level 12, 28 Freshwater Place, Southbank, Victoria 3006, Australia
(ian.francis@basf.com)

Summary Crop topping for seed set reduction is the application of selective and non-selective herbicides to crops at late stage of development to reduce seed set and thus the future weed population. Saflufenacil (Sharpen® WG Herbicide) is a broadleaf herbicide of the Protoporphyrinogen Oxidase Inhibitors (PPOi) mode of action group. In 2017, registration was granted for saflufenacil to be used in wheat, barley and triticale for the reduction in seed set of wild radish (*Raphanus raphanistrum* L.). This has become an important resistance management tool and further work has been conducted to similarly assess the ability of saflufenacil to reduce the seed-set from mature plants of additional troublesome weeds in the cereal system. Trial work generating efficacy data has targeted full flowering milk thistle (*Sonchus oleraceus* L.), wild turnip (*Rapistrum rugosum* L.) and flaxleaf fleabane (*Conyza bonariensis* L.) in cereal crops. Saflufenacil was applied at watery ripe growth stage of wheat and barley (BBCH 71) compared against 2,4-D, glyphosate, and diquat all applied at their respective registered crop timings. Saflufenacil significantly reduced the ability of targeted weeds to set seeds when compared to the untreated control and other alternate treatments. Moreover, saflufenacil showed excellent rapid desiccation of large weeds, meaning that crop harvestability was improved. The combination of reduced number of seeds and high levels of desiccation of mature weeds provides a significant reduction in weeds seed that is able to contribute to the weed seed population in a field for following seasons. In conclusion, saflufenacil can be an excellent tool in managing weed seed populations in cropping systems with obvious herbicide resistance management implications.

Keywords Crop topping, seed-set, milk thistle, wild turnip, flaxleaf fleabane, saflufenacil, resistance.

Integrated weed management cotton (*Gossypium hirsutum* L.) in conventional and ultra narrow row spacing system

Ali R. Ghavi[1] and Mohammad Armin[2]
[1] Former Ms.e. student, Sabzevar Branch, Islamic Azad University, Sabzevar, Iran
[2] Department of Agriculture, Syadati Bol. Sabzevar Branch, Islamic Azad University, Sabzevar, Iran
(armin@iaus.ac.ir)

Summary The effect of cultivation system, weed control treatments, and their interaction on cotton yield and weed growth were studied as a factorial experiment in a randomised complete block design with three replications in Sabzevar in 2015–2016. Factors were cultivation system (conventional (75 cm row spacing)) and ultra narrow row (UNR, 20 cm row spacing) and five weed control treatments, e.g., hand hoeing three times, Ethalfluralin (at recommended dose), Ethalfluralin (50% recommended dose) + hoeing once, Ethalfluralin (50% recommended dose) + hand hoeing twice and non-weeded check. The results showed that UNR system had more height (15.5%), leaf area index (14.5%), the percentage of light absorbed (15.5%), yield (22.45%), lint yield (16.7%), cotton yield (25.6%), biological yield (18.5%) and less lint weight per boll (9.02%), weed dry weight (16.5%), lint index (7.75%) than in conventional system. For all of the traits in this experiment, three times hand weeding was most useful in increasing the characteristics over other control methods which were not significant, with Ethalfluralin (50% recommended dose) + hand hoeing twice at 45 and 60 day after emergence treatment. In UNR system, reducing herbicide dose had more yield and yield components compared to conventional system. Overall results showed that planting system in UNR cultivation in the presence of weed competition, had more cotton yield compared to conventional systems.

Keywords Cotton, low herbicide dose, ultra narrow row, weed competition.

A useful *in vitro* bioassay for evaluation of weed suppression provided by cover crop residues

Saliya Gurusinghe[1], Sajid Latif[1,2], William Brown[1,3], Paul Weston[1,3] and Leslie A Weston[1,3]
[1] Graham Centre for Agricultural Innovation, Charles Sturt University, Wagga Wagga, New South Wales 2678, Australia
[2] School of Animal and Veterinary Sciences, Charles Sturt University, Wagga Wagga, New South Wales 2678, Australia
[3] School of Agriculture and Wine Sciences, Charles Sturt University, Wagga Wagga, New South Wales 2678, Australia
(sgurusinghe@csu.edu.au)

Summary Legume cover crops provide rotational diversity resulting in reduced erosion and improved soil health, nitrogen availability, weed control, and moisture retention. Legume covers have also been shown to successfully suppress weeds due to their physical and/or chemical properties. We utilised a unique *in vitro* soil bioassay (Weston and Mueller 1989) to further evaluate the suppressive potential of selected legume residues, performed in combination with replicated field studies. Dried field-grown cover crop residues (including clover spp., French and yellow serradella, vetch, lupin, field and chick pea) were evaluated for weed suppression at various rates using a control treatment of inert cellulose. Experiments also evaluated soil type and the presence of soil microbiota on residue activity using selected indicator species [barnyard grass (*Echinochloa crus-galli*), annual ryegrass (*Lolium rigidum*), cress (*Lepidium sativum*), and radish (*Raphanus raphanistrum*)]. Higher rates of residues were associated with greater reductions in seed germination and growth, as compared to the cellulose control. Rates as low as 1.0 g residue 100 cm^{-2} (~1000 kg ha^{-1}) showed significant suppression of seed germination and seedling growth; biserrula, yellow serradella and subterranean clover were most suppressive. Some treatments provided selective control of some weeds i.e. chickpea and field pea suppressed dicot weeds most significantly. Soil type also impacted suppression, with the greatest reduction in growth observed in red sodosol soils vs. cracking clays or higher organic soils. Soil pasteurisation resulted in decreased phytotoxicity of all dried residues, suggesting that microbial activity was instrumental in causing increased suppression of weed growth.

Keywords Legumes, biserrula, French serradella, soil microbiota, phytotoxicity, mulch effect.

Egeria (*Egeria densa* Planch.) presents an increasing threat to Australian waterways

Peter Harper[1] and Nimal Chandrasena[2]
[1] Bettersafe Pest and Weed Management, 33 Lagonda Drive, Ingleburn, New South Wales 2565, Australia
[2] GHD Pty Ltd, 20 Smith Street, Parramatta, New South Wales 2650, Australia
(peter@bettersafe.com.au)

Summary Egeria (*Egeria densa* Planch.), a species native to South-East Brazil and Argentina, is now widely distributed in many countries across the world. Aquarium trade is the recognised pathway of introduction to other countries. Egeria is a globally important invasive species because of its capacity to establish, spread and displace native aquatic plants in waterways. Following introduction in the 1950s, Egeria is now considered as naturalised in Australia and is tolerated, as an environmental aquatic weed that can occasionally cause problems. In recent years, we have observed a significant increase in its distribution and abundance in major rivers (Hawkesbury-Nepean; Georges Rivers), and a large number of urban creeks, lakes, ponds and other minor waterways across the Sydney basin. Egeria is also present in other parts of NSW and in Queensland, particularly, around urban centres (Brisbane, etc.). Nutrient enrichment and variable flows are common in some infested waterways, which are likely contributory factors of increasing abundance.

Our view is that proactive management of new infestations of Egeria may reduce its rate of spread. Mechanical control efforts are usually ineffective under good growing conditions, unless water drawdown is achievable (as in small ponds and lakes). Diquat-Hydrogel® controls Egeria effectively as it the species is quite sensitive to the herbicide. Given that turbidity in water often makes Diquat ineffective, mixing with Hydrogel, as a carrier, allows the herbicide to sink onto plant leaves with much reduced dissipation in the vertical water column. We present the results of several case studies demonstrating cost-effective control using Diquat-Hydrogel.

Keywords Egeria, *Egeria densa*, aquatic weeds.

Vacuum weeding: the removal of wild oats (*Avena ludoviciana*) and annual ryegrass (*Lolium rigidum*) seed from wheat stubble

Linda M. Heuke, Shona L. Robilliard, Asad Shabbir and Michael J. Walsh
I.A. Watson Grains Research Centre, Sydney Institute of Agriculture, The University of Sydney, 12656 Newell Highway, Narrabri, New South Wales 2390, Australia
(linda.heuke@sydney.edu.au)

Summary Wild oats and annual ryegrass are weeds of significant economic importance in winter cropping regions throughout Australia. These weeds, especially wild oats, shed seed prior to harvest that subsequently cannot be targeted by harvest weed seed control techniques. A field trial was conducted to determine the weed seed collection potential of a vacuum system. The trial was established by spreading 100 seeds each, of wild oats and annual ryegrass evenly over the soil surface between wheat stubble rows in a 0.09 m^2 area. Three vacuum hose height treatments (1, 20 and 50 mm) were used to establish differing levels of suction. Vacuum removal close to the soil surface (1 mm) collected 82% of wild oats seeds and 77% of the annual ryegrass seeds. At the 20 mm height, 71% and 45% of wild oats and annual ryegrass seeds were collected, respectively. At 50 mm, very low levels of wild oat (11%) and annual ryegrass (0.4 %) seed was collected. These results clearly establish the potential for vacuum weeding, demonstrating that very high proportions (>75%) of shed wild oats and annual ryegrass seed can be effectively collected by vacuuming.

Keywords Wild oats, annual ryegrass, vacuum, weed, seed, removal.

Host testing and taxonomic clarification of a potential fungal biological control agent of sea spurge (*Euphorbia paralias*)

Gavin C. Hunter[1], John Lester[1], Isabel Zeil Rolfe[1], Mireille Jourdan[2], John K. Scott[3] and Louise Morin[1]
[1] CSIRO Health and Biosecurity, Black Mountain Science and Innovation Park, Clunies Ross Street, Acton, Australian Capital Territory 2601, Australia
[2] CSIRO European Laboratory, Campus International de Baillarguet, Montferrier-sur-lez, Montpellier 34980, France
[3] CSIRO Land and Water, 147 Underwood Avenue, Floreat, Western Australia 6014, Australia
(gavin.hunter@csiro.au)

Summary Sea spurge (*Euphorbia paralias*) is a fleshy perennial native to sand dunes of western and southern Europe. In Australia however the plant is an invader of coastal habitats and has become established on many beaches and dunes across southern Australia. Worryingly, sea spurge is now widespread and abundant on beaches along the New South Wales South Coast and is progressively moving northwards. An effective biological control agent against sea spurge would be a welcome tool in the fight against this invasive beach weed. In 2009 a cercosporoid hyphomycete fungus, *Passalora euphorbiae*, was isolated from sea spurge near La Salie on the French Atlantic coast where it was found to cause impressive leaf and stem disease on the plant. Initial host testing with this fungus in France was encouraging and the fungus was imported into quarantine in Australia for further host-specificity testing. A comprehensive host test list encompassing *Euphorbia* species from the four *Euphorbia* sub-genera has been established and this study will report on the initial host-testing of the potential fungal biocontrol agent on these *Euphorbia* species. In addition, although the fungus was originally classified within the Ascomycete fungal genus *Passalora* recent taxonomic revisions of the fungal family Mycosphaerellaceae have raised doubts on the correct identity of the fungus. We therefore employ DNA sequence data from several informative nuclear and protein-coding gene regions, in combination with a range of informative morphological characteristics, to critically examine and finalise the taxonomy of this potentially important fungal biological control agent of sea spurge.

Keywords *Euphorbia*, sea spurge, fungal biocontrol agent, Mycosphaerellaceae, *Passalora*.

Evaluation of herbicide mixtures and manual weed control methods in maize (*Zea mays* L.) production

E.O. Imoloame
Crop Production Department, College of Agriculture, Kwara State University, Malete, PMB 1530, Ilorin, Kwara State, Nigeria
(oyaimoloame@yahoo.com)

Summary Field trials were conducted in 2015 and 2016 cropping seasons at the Teaching and Research Farm of Kwara State University in the southern Guinea savanna of Nigeria to evaluate some herbicide mixtures and manual weed control method in the production of maize. The study was based on the hypothesis that herbicide mixture plus one supplementary hoe weeding (SHW) at 6 WAS will minimise weed infestation and give higher maize yield. The experiment consisted of 10 treatments as follows: metolachlor + atrazine at 1.0 + 2.0 kg ai ha^{-1}, metolachlor + atrazine at 2.0 + 2.5 kg ai ha^{-1}, metolachlor + atrazine at 3.0 + 3.0 kg ai ha^{-1}, pendimethalin + atrazine at 1.0 + 2.0 kg ai ha^{-1}, pendimethalin + atrazine at 2.0 + 2.5 kg ai ha^{-1}, pendimethalin+ atrazine at 3.0 + 3.0 kg ai ha^{-1}, metolachlor + atrazine at 1.0 + 2.0 kg ai ha^{-1} plus one supplementary hoe weeding (SHW) at 6 weeks after sowing (WAS), pendimethlin + atrazine at 1.0 + 2.0 kg ai ha^{-1} plus one SHW at 6 WAS, hand weeding at 3 and 6WAS and a weedy check. These treatments were laid out in randomised complete block design with three replicates. Results showed that metolachlor + atrazine and pendimethalin + atrazine at 1.0 + 2.0 kg ai ha^{-1} plus one SHW at 6 WAS significantly reduced weed dry matter and weed density ($P<0.05$), had taller plants and larger leaf area ($P<0.05$), increased maize yield ($P<0.05$) and gave higher economic returns in both years of the study. Therefore, the two herbicide mixtures plus one SHW at 6 WAS are recommended as alternatives to two hoe weeding for maize production.

Keywords Agro-chemical, hoe weeding, economic returns.

African boxthorn stakeholder survey: Impacts and management goals for biocontrol

Kylie B. Ireland[1], Michelle Rafter[2], Nagalingam Kumaran[2], S. Raghu[2] and Louise Morin[1]
[1] CSIRO Health and Biosecurity, Black Mountain Science and Innovation Park, Clunies Ross Street, Acton, Australian Capital Territory 2601, Australia
[2] CSIRO Health and Biosecurity, Ecosciences Precinct, 41 Boggo Road, Dutton Park, Queensland 4102, Australia
(kylie.ireland@csiro.au)

Summary African boxthorn (*Lycium ferocissimum*) is a widespread and significant environmental and agricultural weed in regional Australia. An online survey was used to engage key stakeholders affected by African boxthorn, and ask them about the impacts and desired management goals for this weed. More than 200 people participated in the survey, with approximately 80% of respondents actively involved in natural resources management. A range of reasons were provided by respondents as to why this weed was problematic, with 75% of them identifying preventing stock access to water as the main economic impact and 95% reporting that it was displacing native vegetation. Close to 80% of respondents indicated that biological control solutions would be useful in managing this weed. In terms of management outcomes, respondents ranked reduction in new infestations, reduction in management costs and reduction in negative impacts on native flora and fauna to be the top three management objectives to which biological control needs to make a significant contribution. The management objectives gleaned from this survey will guide the selection of candidate biological control agents, and the subsequent evaluation of their performance if they are deemed safe for release into the Australian environment.

Keywords Biological control, stakeholder engagement.

Early invader Opuntioid cacti in Victoria

Bec James[1] and Kate Blood[2]
[1] Department of Environment, Land, Water and Planning, Box 3100, Bendigo Delivery Centre, Victoria 3554, Australia
[2] Department of Environment, Land, Water and Planning, PO Box 7, Beaufort, Victoria 3373, Australia

Summary Opuntioid cacti are a group of invasive plants that form one of the sub-families in the Cactaceae family. They pose a serious risk to the biodiversity of Victoria and appear to be spreading with more species and infestations popping up in the landscape. In 2017 the Victorian Government declared all *Cylindropuntia* and *Opuntia* species (with a few exceptions) under the *Catchment and Land Protection Act 1994*, prohibiting the trade of these species and their propagules. In 2017, new species of Opuntioid cacti have been discovered in Victoria. Westwood Pear (*Opuntia streptacantha* Lem.) was identified at Wallpolla Island at Murray Sunset National Park in north-west Victoria and jumping cholla (*Cylindropuntia prolifera* (Engelm.) F.M.Knuth) located in Long Gully, a suburb of Bendigo, central Victoria. A third yet to be determined specimen has been collected at Wargan Bushland Reserve near Mildura. Specimens of all three have been submitted to the National Herbarium of Victoria. As awareness of invasive cacti grows, more species and sites are expected to be reported. Public land managers now have improved tools to make better investment decisions.

Keywords Opuntioid cacti, early invader.

Using detection dogs as part of an integrated weed management strategy – A local government perspective

Melanie Kelly[1] and Nicole Gill[2]
[1] Natural Resource Department, Glamorgan Spring Bay Council, PO Box 9, Triabunna, Tasmania 7190, Australia
[2] Independent Environmental Consultant, Hobart, Tasmania, Australia
(mel@freycinet.tas.gov.au)

Summary Serrated tussock (*Nassella trichotoma*) is a Weed of National Significance that impacts on both native grassy ecosystems and agricultural grazing land. Serrated tussock was inadvertently introduced to Tasmania in 1956, and has been the target of many government-funded eradication and management campaigns for many years. Situated on Tasmania's east coast, Glamorgan Spring Bay (GSB) is a rural municipality with scattered infestations of serrated tussock spreading over a relatively small number of properties, many of which are active grazing properties, and others which are rural residential land, some with absentee landowners. Since 2005 the GSB Council has undertaken an integrated, cross land tenure approach to weed management, in line with the municipal strategic weed management plan. Landholders throughout the municipality have been supported by the GSB Council, NRM South (the regional NRM organisation), and State and Federal Governments to manage their serrated tussock since it was first detected in the municipality. In some locations, populations of serrated tussock have been reduced to a point where it is difficult to detect any remaining plants. There are also other locations (such as north-facing rocky banks) where it is highly likely outlying plants may be found. Glamorgan

Spring Bay Council is now trialling the use of a conservation detection dog to seek out remaining tussock plants in difficult country, where it may be harder for human eyes to detect, to enable both mapping and control. This activity is embedded within the broader strategic management framework. The dog being trialled is a working line German Shepherd from Garsova Kennels; his primary trainer is veteran detection dog trainer Steve Austin. The dog handlers are all people with extensive serrated tussock management experience.

Triticale competitive ability with annual ryegrass (*Lolium rigidum* Gaudin) as affected by variety, sowing rate and row spacing

Husam S.M. Khalaf, Brian M. Sindel, Paul E. Kristiansen, Robin S. Jessop and Craig Birchall
Department of Agronomy and Soil Science, School of Environmental and Rural Science, University of New England, Armidale NSW 2351, Australia
(hkhalaf@myune.edu.au)

Summary One strategy for dealing with herbicide resistant weeds is to identify strongly competitive crop cultivars that suppress weed growth and/or withstand the competitive effects of weeds. Triticale (X *Triticosecale* Wittmack) is often ranked as more tolerant to environmental stresses than wheat (*Triticum aestivum* L.), however, little is known about varietal differences in its competitive ability with weeds. Annual ryegrass (*Lolium rigidum* Gaudin) is one of the most serious weed species in Australian cereal cropping largely because it has evolved herbicide resistance, is widespread and competes strongly with grain crops. A field trial was conducted to examine the competitive ability of four triticale cultivars and one wheat cultivar against annual ryegrass as influenced by row spacing (15 or 30 cm) and sowing rate (50 or 100 kg ha^{-1}). Cultivar Bogong had the highest grain and biomass yield of the triticale cultivars in the ryegrass-free treatment. In contrast, cultivars KM10 and Tuckerbox demonstrated greater competitive ability than the other triticale cultivars by sustaining smaller growth reductions when grown with ryegrass (10.4 and 8.5% biomass reduction, respectively). No interactions were observed with row spacing and sowing rate. The comparative wheat cultivar Spitfire had a lower grain yield and biomass than the triticale cultivars in both ryegrass treatments. There appears to be scope for utilising triticale competition in weed management, however, the optimal agronomy may vary with the cultivar.

Keywords Grain yield, biomass, herbicide resistance, wheat, cultivar.

INTRODUCTION

Annual (or rigid) ryegrass (*Lolium rigidum* Gaudin) is a problematic weed in Australian winter cereal cropping systems (McGillon and Storrie 2006). Previous studies have shown that many populations have developed resistance to herbicides such as glyphosate and diclofop-methyl (Christopher *et al.* 1991, Powles *et al.* 1998, Owen *et al.* 2007). In addition, it strongly competes with grain crops such as wheat (*Triticum aestivum* L.), causing significant yield losses (Wu *et al.* 1998). Annual ryegrass is one of at least 142 weed species worldwide that are resistant to more than one mode-of-action (MOA) group including 7 MOA groups which are no longer effective against many weed species (Heap 2018). Herbicide resistance development in weed species, particularly in Australia (Lemerle *et al.* 2004), has encouraged many researchers and farmers to examine alternatives to herbicides in weed control and improving cereal crop production.

Increasing crop competitive ability against weeds through manipulations of either sowing rates or row spacing is suggested for reducing yield losses due to weeds in wheat cropping systems (Lemerle *et al.* 2004, Mason *et al.* 2007). A wheat density of 120–240 plants m^{-2} can reduce ryegrass competition and achieve high grain yield (Tanji *et al.* 1997). Koscelny *et al.* (1990) stated that decreasing wheat crop row spacing to less than 18 cm could result in higher wheat yield by 5–7% due to increased efficiency of resource utilisation (Holliday 1960, Auld *et al.* 1983). Crop varietal selection is another factor that may change crop competitiveness through giving yield improvement and weed suppression; however, the cultivar competitive benefit can be influenced by the interaction of environment (including management) and genotype (Lemerle *et al.* 2001).

The use of vigorous crop varieties and optimised planting configurations could help improve crop competitiveness and thus manage herbicide resistance, potentially reducing costs and increasing yield. While some studies have shown that manipulation of sowing rates, row spacing and crop cultivars is effective for improving wheat competitive ability with weeds like annual ryegrass (Lemerle *et al.* 2004, Olsen *et al.* 2006, Beres *et al.* 2010a), and some studies with other weeds have included triticale (X *Triticosecale* Wittmack) (e.g. (Beres *et al.* 2010b), no studies exist in Australia or globally that focus on triticale competitiveness against annual ryegrass. Therefore, this experiment aimed to investigate the influence of cultivar, crop sowing rate and row spacing on triticale competitiveness with annual ryegrass.

MATERIALS AND METHODS

A field experiment was carried out during winter 2015 at the Laureldale Research Station, University of New England, Armidale NSW (latitude 30.515° S and longitude 151.665° E). The weather in 2015 was relatively warm (average 21.4°C) and the average monthly rainfall was 54.6 mm from August-December (Burr 2016).

Four triticale lines (cultivars Tuckerbox, Tobruk, Bogong and KM10) and one comparative wheat cultivar (Spitfire) were planted in early August 2015. Two sowing rates (50 and 100 kg ha^{-1}), two row spacing treatments (15 and 30 cm) and two ryegrass treatments (with and without) were used in this experiment with four replicates. The plots were 1.8 m by 5 m and 1.65 m by 5 m for 30 and 15 cm row spacings respectively. A randomised complete block design was used. Annual ryegrass seeds were harvested from a commercially grown crop in Victoria, Australia, and sown at 1.8 g m^{-2} to produce approximately 300 plants m^{-2} (Lemerle *et al.* 2013). The seeds were mixed with sand (Lemerle *et al.* 2013) and were distributed on each plot by hand immediately prior to crop planting on 7 August 2015. The broadleaf weed species which grew in the plots were; *Polygonum aviculare* L., *Sonchus oleraceus* L., *Amaranthus* sp., *Vicia* sp., *Trifolium* sp. and *Daucus carota* L.

Measurements The samples, taken at crop maturity, included four 1 m length rows for the 30 cm row spacing and eight 1 m length rows for the 15 cm row spacing treatments respectively. The crop biomass (t ha^{-1}) and grain yield (t ha^{-1}) were recorded.

Data analysis The total biomass and grain yield data were analysed with generalised linear models and lsmeans in R version 3.1.2 (R Core Team 2015). Model structures were simplified using a reduction in Akaike Information Criteria values as the criteria, while residuals were checked after analysis to validate the models used.

RESULTS

There was a highly significant interaction between crop cultivar, sowing rate and row spacing ($P<0.01$) in terms of the crop grain yield. Bogong had the greatest grain yield at high sowing rate (100 kg ha^{-1}) and wide row spacing (30 cm), about 4.21 t ha^{-1}, followed by Tobruk at about 4.10 t ha^{-1} in the same treatment. In contrast, the wheat cultivar Spitfire had its greatest grain yield at the low sowing rate (50 kg ha^{-1}) and narrow row spacing (15 cm), about 3.39 t ha^{-1} and had the lowest grain yield at high sowing rate with either narrow or wide row spacing in comparison with triticale cultivars (Figure 1).

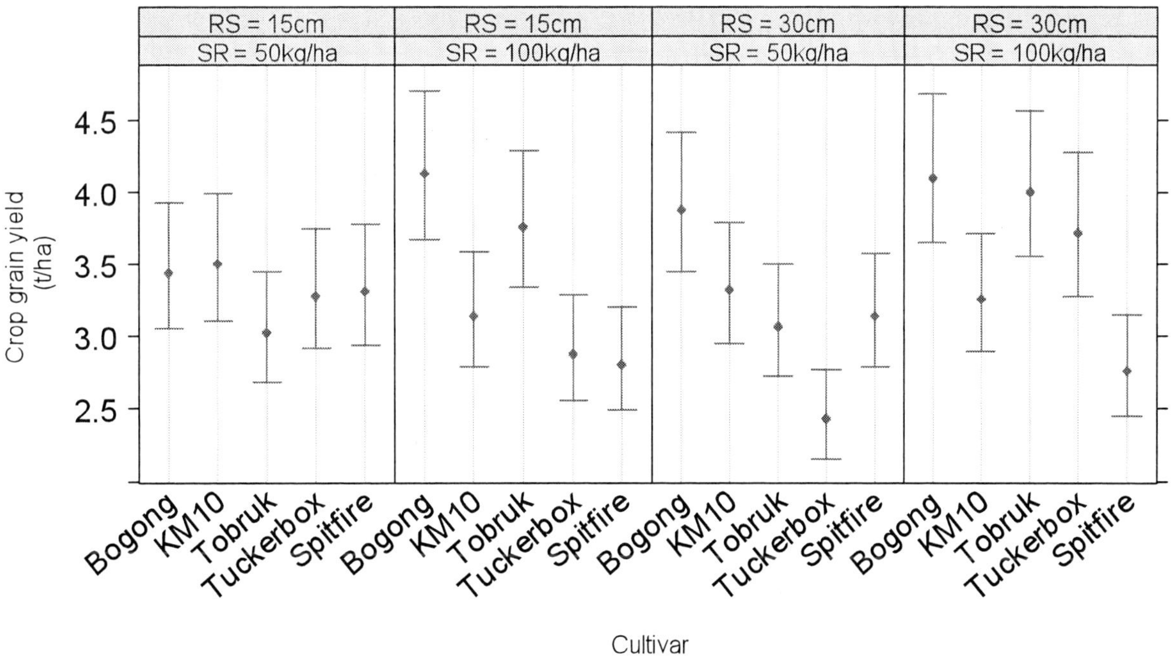

Figure 1. The effect of crop cultivar, row spacing (RS = 15 cm and 30 cm) and sowing rate (SR = 50 kg ha^{-1} and 100 kg ha^{-1}) on the crop grain yield. The error bars show the 95% confidence intervals.

Tuckerbox grain yield declined as row spacing increased at the 50 kg ha^{-1} sowing rate but increased as row spacing increased at 100 kg ha^{-1}. Other varieties were not significantly affected by row spacing (Figure 1). Tobruk appeared to respond positively in grain yield to increasing sowing rate from 50 kg ha^{-1} to 100 kg ha^{-1} at both row spacings. Likewise, Tuckerbox responded to increasing sowing rate at the 30 cm spacing. Other cultivars were not affected (Figure 1).

There was a significant interaction between crop cultivar and ryegrass treatment for crop biomass ($P<0.05$). Bogong had the greatest biomass with no ryegrass at about 10.8 t ha^{-1}. Biomass for Bogong and Tobruk was significantly lower under ryegrass competition than with no ryegrass while there was no significant difference in biomass between triticale cultivars, with or without ryegrass present. In both ryegrass treatments, the wheat cultivar Spitfire had less biomass than the triticale cultivars and was highly susceptible to ryegrass competition (Figure 2).

DISCUSSION

Grain yield varied depending on row spacing and sowing rate, but not ryegrass presence. For crop biomass, Bogong, Tobruk and Spitfire were more sensitive to annual ryegrass pressure than KM10 and Tuckerbox. The ability to resist weed competition pressure in KM10 and Tuckerbox (10.5 and 8.5% biomass reductions) may be due to greater plant height and number of tillers (data not shown). Previous research has found that the tall short-season triticale produced the greatest crop biomass and demonstrated a high competitive ability with weeds (Beres *et al.* 2010b). Triticale cultivars performed well in production and competition with ryegrass in comparison with the tested wheat cultivar, Spitfire. There appears to be scope for utilising triticale competition in the management of weeds due to greater biomass and grain yield under crop/weed/competition conditions compared with wheat, however, the optimal agronomy may vary with the cultivar used. Lemerle *et al.* (2004) found that increasing crop sowing rate, which means crop plant population, can increase weed suppression and reduce grain yield losses due to crop competitive ability depending on environmental conditions, season, and wheat cultivars.

This experiment forms one part of a larger study. Other experiments in this study will examine the effect of crop seed size, nutrient levels and water levels on the competitive ability of triticale crop cultivars with annual ryegrass.

ACKNOWLEDGMENTS

We thank Anthony McKinnon, David Edmonds and Dr Karl Anderson for their help in the field trial. HSMK acknowledges the Higher Committee for Education Development in Iraq for their financial support and sponsorship of his PhD study.

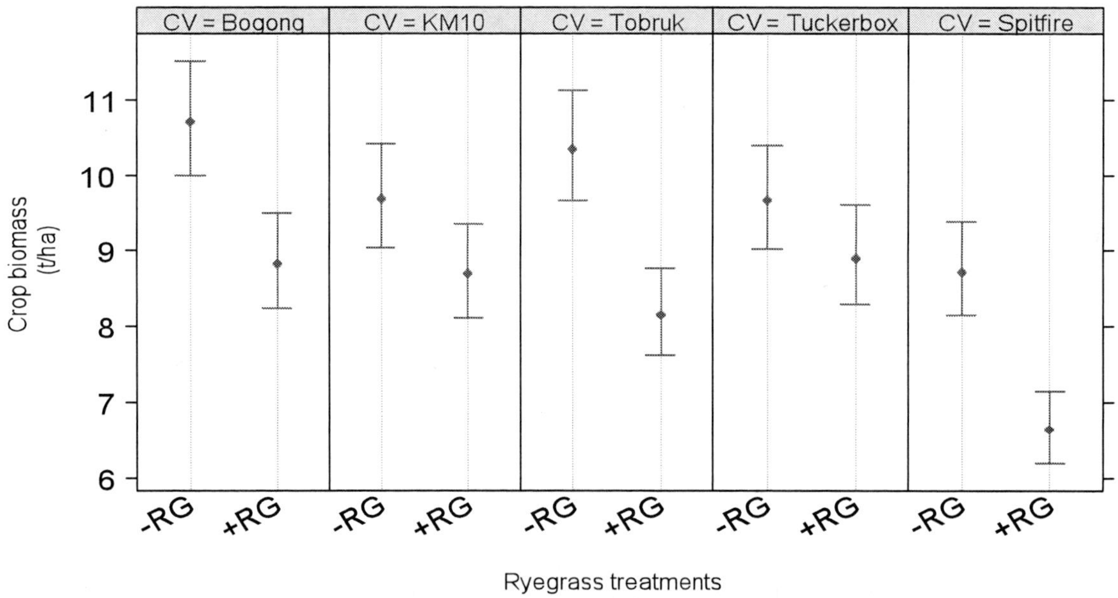

Figure 2. The effect of crop cultivar and ryegrass treatments (−RG and +RG) on the total crop biomass. The error bars show the 95% confidence intervals.

REFERENCES

Auld, B., Kemp, D. and Medd, R. (1983). The influence of spatial arrangement on grain yield of wheat. *Australian Journal of Agricultural Research* 34, 99-108.

Beres, B.L., Clayton, G.W., Harker, K.N., Stevenson, F.C., Blackshaw, R.E. and Graf, R. J. (2010a). A sustainable management package to improve winter wheat production and competition with weeds. *Agronomy Journal* 102, 649-57.

Beres, B.L., Harker, K.N., Clayton, G.W., Bremer, E., Blackshaw, R.E. and Graf, R.J. (2010b). Weed-competitive ability of spring and winter cereals in the Northern Great Plains. *Weed Technology* 24, 108-16.

Burr, P. (2016). Annual weather review Armidale NSW. http://www.weatherarmidale.com/review2015.html (accessed 1 August 2018).

Christopher, J.T., Powles, S.B., Liljegren, D.R. and Holtum, J.A.M. (1991). Cross-resistance to herbicides in annual ryegrass (*Lolium rigidum*) II. Chlorsulfuron resistance involves a wheat-like detoxification system. *Plant Physiology* 95, 1036-43.

Heap, I. (2018). The international survey of herbicide resistant weeds, http://www.weedscience.org/Summary/Country.aspx?CountryID=1 (accessed 1 August 2018).

Holliday, R. (1960). Plant population and crop yield. *Nature* 186, 22-4.

Koscelny, J.A., Peeper, T.F., Solie, J.B. and Solomon Jr, S.G. (1990). Effect of wheat (*Triticum aestivum*) row spacing, seeding rate, and cultivar on yield loss from cheat (*Bromus secalinus*). *Weed Technology* 4, 487-92.

Lemerle, D., Cousens, R.D., Gill, G.S., Peltzer, S. J., Moerkerk, M., Murphy, C.E., Collins, D. and Cullis, B.R. (2004). Reliability of higher seeding rates of wheat for increased competitiveness with weeds in low rainfall environments. *Journal of Agricultural Science* 142, 395-9.

Lemerle, D., Lockley, P., Koetz, E. and Diffey, S. (2013). Herbicide efficacy for control of annual ryegrass (*Lolium rigidum* Gaud.) is influenced more by wheat seeding rate than row spacing. *Crop and Pasture Science* 64, 708-15.

Lemerle, D., Verbeek, B. and Orchard, B. (2001). Ranking the ability of wheat varieties to compete with *Lolium rigidum*. *Weed Research* 41, 197-209.

Mason, H., Navabi, A., Frick, B., O'Donovan, J. and Spaner, D. (2007). Cultivar and seeding rate effects on the competitive ability of spring cereals grown under organic production in Northern Canada. *Agronomy Journal* 99, 1199-207.

McGillon, T. and Storrie, A. (2006). Integrated weed management in Australian cropping systems – A training resource for farm advisors. (GRDC, Adelaide, South Australia).

Olsen, J., Kristensen, L. and Weiner, J. (2006). Influence of sowing density and spatial pattern of spring wheat (*Triticum aestivum*) on the suppression of different weed species. *Weed Biology and Management* 6, 165-73.

Owen, M., Walsh, M., Llewellyn, R. and Powles, S. (2007). Widespread occurrence of multiple herbicide resistance in Western Australian annual ryegrass (*Lolium rigidum*) populations. *Crop and Pasture Science* 58, 711-8.

Powles, S.B., Lorraine-Colwill, D.F., Dellow, J.J. and Preston, C. (1998). Evolved resistance to glyphosate in rigid ryegrass (*Lolium rigidum*) in Australia. *Weed Science* 46, 604-7.

R Core Team. (2015). 'R: A Language and Environment for Statistical Computing'. Vienna.

Tanji, A., Zimdahl, R. and Westra, P. (1997). The competitive ability of wheat (*Triticum aestivum*) compared to rigid ryegrass (*Lolium rigidum*) and cowcockle (*Vaccaria hispanica*). *Weed Science* 45, 481-7.

Wu, H., Pratley, J., Lemerle, D., Haig, T. and Verbeek, B. (1998). Differential allelopathic potential among wheat accessions to annual ryegrass. Proceedings of the 9th Australian Agronomy Conference, eds D.L. Michalk and J.E. Pratley. http://www.regional.org.au/au/asa/1998/6/139wu.htm?print=1 (accessed 1 August 2018).

Effect of moisture stress on morphological and physiological attributes of *Amaranthus viridis* and *A. retroflexus*

Asad M. Khan[1], James P. Hereward[2], Jeff A. Werth[3], Gimme H. Walter[2] and Bhagirath Singh Chauhan[1]
[1] Queensland Alliance of Agriculture and Food Innovation, University of Queensland, Australia
[2] School of Biological Sciences, University of Queensland, Australia
[3] Department of Agriculture and Fisheries, Toowoomba, Queensland, Australia
(asad.khan@uqconnect.edu.au)

Summary Cotton growers in Australia face significant challenges eradicating weeds from the cotton cropping system. Poor weed management can cause up to 90% yield loss in cotton due to competition with the crop for nutrients and water. *Amaranthus retroflexus* is one of the most common, widely distributed and troublesome weeds of cotton growing areas of Queensland and New South Wales. Numerous factors have enabled this species to become dominant and difficult-to-control. These factors include rapid growth rate, high fecundity, genetic diversity, and ability to tolerate adverse conditions. Crop growth and yield is known to be affected by many factors especially water stress; however, the effect of water stress on growth and propagation of weed is not deeply understood. A water stress experiment was conducted at University of Queensland wire house to determine the effect of duration and degree of water stress on growth and seed production of two *A. retroflexus* (Gatton and Goondiwindi) biotypes. The experiment included four water stress treatments of 100, 75, 50, 25 of soil water content applied at 7 days interval. No significant difference was observed in growth, development and seed production of 2 different biotypes. Highest plant height and number of leaves and branches were observed at 75 % of soil water content. While the lowest number of leaves, branches were observing in the water soil water content of 25%. The lowest level of seed production was also observed in 25% soil water content treatment. The study determined that the *A. retroflexus* can survive and produce seed under high degree of water stress.

Keywords Weeds, *Amaranthus*, water stress.

Weed suppressive potential of selected pasture legumes against annual weeds in southeastern Australia

Sajid Latif[1,2], Paul A. Weston[1,3], John W. Piltz[1,4], Saliya Gurusinghe[1], William B. Brown[1,3], Jane C. Quinn[1,2] and Leslie A Weston[1,3]
[1] Graham Centre for Agricultural Innovation
[2] School of Animal and Veterinary Sciences
[3] School of Agriculture and Wine Sciences
Charles Sturt University, Wagga Wagga, New South Wales 2678, Australia
[4] NSW Department of Primary Industries, Agriculture Institute, Wagga Wagga New South Wales 2650
(slatif@csu.edu.au)

Summary Poor adaptation of numerous annual and perennial legumes across south eastern Australia due to acidic soils, insufficient rainfall resulting in false breaks, high input costs, soil acidification and the emergence of herbicide resistant weeds have resulted in limited options for producers seeking to establish productive pastures in the Riverina region of New South Wales (NSW). To overcome these challenges, numerous novel pasture species have been selected and recently released for establishment. However, limited knowledge exists regarding their ability to suppress weeds during establishment and in subsequent years when they regenerate. Field trials were conducted at Wagga Wagga, NSW over a two year period to evaluate: (a) the suppressive potential of eight selected pasture legumes against annual weeds; and (b) the impact of autumn (March) and winter (June) sowing dates on stand establishment. Weed and crop cover and biomass were assessed each year in replicated trials in 2016–2017. Results suggested that species mixtures with more than one pasture species were not significantly different ($P<0.05$) in terms of establishment and subsequent weed infestation. However, autumn sowing of arrowleaf clover, yellow serradella cv. Avila and French serradella/bladder clover generally resulted in increased ($P<0.05$) pasture crop cover over two growing seasons. Arrowleaf clover and biserrula cv. Casbah were strong performers and produced significantly greater crop biomass and also less weed biomass. However, weed suppression and subsequent biomass was not always related to competition for resources and production of total crop biomass. This was the case for yellow serradella cv. Santorini in both 2016 and 2017 where weed biomass was significantly reduced but total crop biomass produced was limited. This suggested that factors other than resource competition, such as allelopathy, were associated with weed suppression. Arrowleaf clover, biserrula cv. Casbah and yellow serradella cv. Santorini were generally the best and most reliable annual legumes with respect to yearly regeneration and pasture weed suppression in the Riverina.

Keywords Hard-seeded legume, weed suppression, competition, establishment.

INTRODUCTION

Pasture legumes have been an integral part of mixed farming systems in southeastern Australia as they provide high-quality nutrition for livestock and improve soil fertility due to their ability to fix atmospheric nitrogen. The selective use of pasture legumes has also been instrumental in effectively managing weeds in such rotational cropping systems (Mengel *et al.* 2001, Loi *et al.* 2005a, Loi *et al.* 2005b). Traditionally, subterranean clover has been the most widely used pasture in southern Australia but unpredictable weather patterns resulting in lower and inconsistent spring rainfall, high cost of legume establishment and low persistence mainly due to low hardseed levels have impacted on the suitability of subterranean clover in the region (Loi *et al.* 2001, Loi *et al.* 2005a).

To overcome these challenges, numerous novel self-regenerating annual pasture legumes including *Biserrula pelecinus* L. (biserrula), *Ornithopus sativus* Brot. (French serradella), *Ornithopus compressus* L. (yellow serradella), *Trifolium glanduliferum* Boiss. (gland clover), *Trifolium spumosum* L. (bladder clover) and *Trifolium vesiculosum* Savi. (arrowleaf clover) have been recently released in New South Wales (NSW). These pasture legumes are well adapted to the south eastern region, due to their successful establishment in diverse soils, drought tolerance, higher herbage production and nutritional quality relative to traditional pasture species. However, limited knowledge exists regarding their establishment following initial seeding and in subsequent years of regeneration. To maximise the adoption of introduced species/cultivars under suitable soil types and rainfall zones, further research is required in NSW to assess the ability of selected pasture legumes to establish and produce improved yields while suppressing annual weeds to determine their suitability within sustainable cropping systems. Therefore, the objectives of this study were to evaluate: (a) the suppressive potential of selected annual pasture legumes against dominant annual weeds; and (b) the impact of sowing season; autumn (March) or winter (June) on establishment and regeneration of annual legumes in southeastern Australia.

MATERIALS AND METHODS

Field experiments were established in 2016 and 2017 at the Graham Centre field site and the Agricultural Research Institute field site in Wagga Wagga, NSW (35°S, 147°E). All experiments were arranged in a randomised complete block design with five replicates per treatment. The soil type was characterised as a fine red sodosol at both trial sites. Treatments were planted at standard sowing rates using a precision cone planter in 20 × 3 metre with 15 cm row spacing. All pasture treatments were sown on two different sowing dates either in the first week of March or first week of June and thereafter referred to as autumn and winter sowing dates respectively, except in the case of those not suitable agronomically for autumn sowing. All plots were fertilised with 100 kg ha^{-1} of fertiliser (Croplift® 12, 1:2:1, Incitec Pivot, Victoria) at the time of sowing. The second trial site was established similarly in winter 2017 at the Wagga Wagga Agricultural Research Institute using only monoculture pasture treatments in 9 × 2 m plots with similar row spacing and seeding rates.

Percentage stand establishment of pasture crops and weeds was recorded using visual ratings on a scale of 0–100 performed at physiological maturity of the pasture crops when crops were at the flowering stage. In addition to qualitative data, two representative pasture samples were collected from each plot at a height of approximately 2 cm the soil surface (to simulate grazing by sheep) using 0.25 m^2 quadrants from monoculture plots to determine the total biomass of crop and weeds in 2016 and 2017. Plant matter was sorted into pasture legume and weed species before being dried at 90°C for 96 hours and weighed. All weeds were collected and bulked irrespective of species and dry weights were recorded. Above ground crop competitive traits including normalised vegetative index (NDVI), leaf area index (LAI), and light interception (LI) were assessed for each pasture legume species at physiological maturity in 2017 to determine the relative effect of each trait on total weed suppression. An analysis of variance (ANOVA) was performed to determine the effect of sowing season and pasture crop on ground cover and biomass of crops and weeds using the GenStat statistical software package (18th Edition, VSN International Ltd, Hertfordshire, United Kingdom). The fixed model included year, sowing date and pasture crops while the random effect was block number. A multivariate analysis was performed using IBM SPSS statistical package (IBM SPSS® Inc. 2001, version 24.0 for Windows, United States of America) using a partial least square regression (PLSR) analysis model to determine the relative effect of competitive traits of weed suppression (Reiss *et al.* 2018).

RESULTS

Results of visual ratings performed for pasture crops at physiological maturity demonstrated that percentage ground cover of crops and weeds varied significantly ($P<0.05$) across pasture species. Arrowleaf clover produced the highest ground cover of all pasture species followed by biserrula cv. Casbah and gland clover, while cultivars of yellow serradella produced the least pasture ground cover. Arrowleaf clover initially exhibited the lowest percentage of weed infestation followed by biserrula cv. Casbah and yellow serradella cv. Santorini (Figure 1).

Generally, mean percentage crop cover was slightly to moderately higher for the autumn sowing date compared to winter sowing but the overall effect of sowing date was not significant ($P>0.05$) except for arrowleaf clover and both cultivars of yellow serradella.

Biomass of pasture crops and weed species was also obtained in at crop maturity in 2016 and 2017 and these varied significantly ($P<0.05$) across pasture species. Overall, crop biomass of all treatments ranged from 44–90 g m^{-2} while weed biomass ranged from 5–30 g m^{-2} (Figure 1). Arrowleaf clover produced the highest crop biomass followed by biserrula cv. Casbah and gland clover. In contrast, yellow serradella cv. Santorini and subterranean clover produced the lowest crop biomass. Interestingly, arrowleaf clover had the lowest weed biomass followed by yellow serradella cv. Santorini. Remarkably, yellow serradella cv. Santorini produced the lowest biomass but also maintained low weed infestation.

A PLSR analysis was conducted to determine the relative effect of competitive traits on weed biomass. Competitive traits explained up to 30% of the variance in two latent factors when dataset from all species was analysed in one model suggesting a separate analysis for each species separately is appropriate. Overall, total weed biomass at physiological maturity was strongly correlated with crop biomass followed by LI.

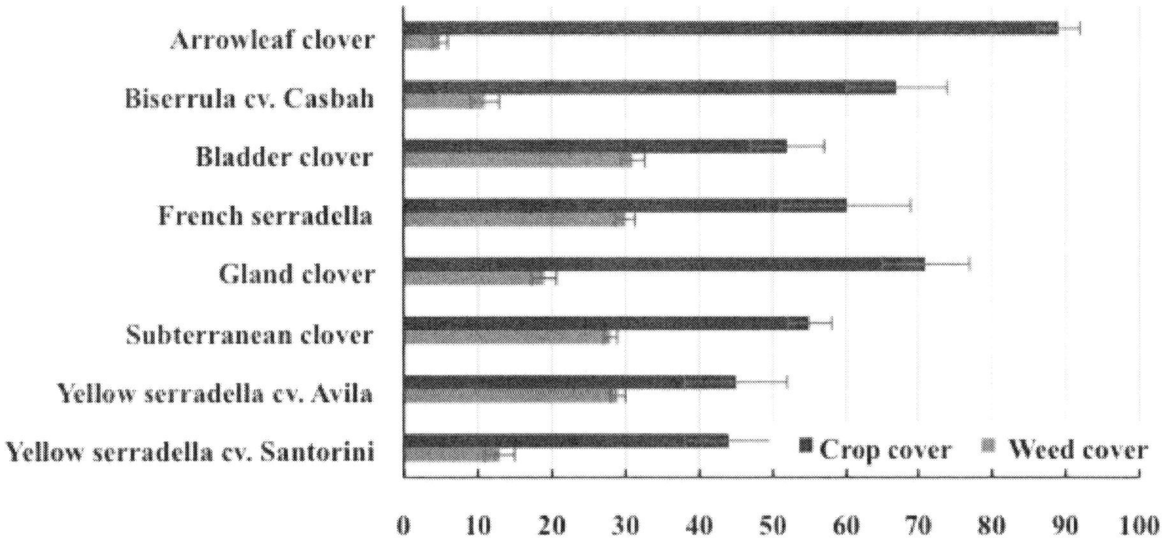

Figure 1. Visual ratings performed on a scale of 0 to 100 where zero represents no crop cover and 100 represents complete and uniform ground cover established in Wagga Wagga, NSW. Ratings were taken at physiological maturity of the pasture crops which was achieved in the third week of October and November, respectively, and averaged over both years. Error bars indicate the standard error of means.

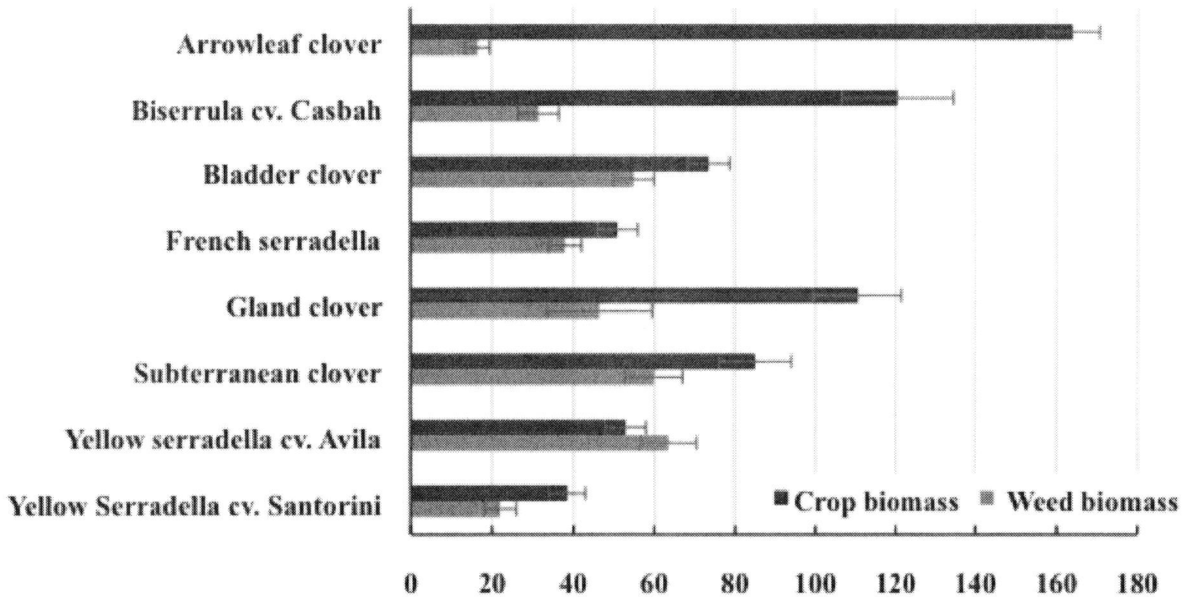

Figure 2. Biomass of crop and weeds (g m^{-2}) assessed at physiological maturity of the pasture crops averaged across 2016 and 2017 Wagga Wagga, NSW. Weed biomass represents the composite biomass of all weeds. Error bars indicate the standard error of means.

Crop biomass appeared to have the strongest inverse effect on total weed biomass in the case of biserrula cv. Casbah, bladder clover, French serradella, subterranean clover and yellow serradella cv. Avila. In addition, LAI was observed to have a strong inverse effect on total weed biomass in the case of arrowleaf clover and gland clover. Clearly, crop biomass, NDVI and LAI were not associated with weed suppression in case of yellow serradella cv. Santorini.

DISCUSSION

Results of two years of several field trials in the Riverina region of NSW clearly demonstrated significant variation in the competitive ability of individual pasture legumes species and cultivars, resulting in variable weed suppression against annual weeds. Arrowleaf clover, biserrula and yellow serradella were generally the most reliable and competitive annual legumes with respect to stand establishment, yearly regeneration and consistent suppression of pasture weed species in two years of experiments performed at multiple field sites in Wagga Wagga. Sowing of pasture legumes in autumn or winter is feasible but crops performed more consistently when sown in June. In this study autumn sowing did not offer significant advantages compared to winter sowing in suppressing annual weeds. Arrowleaf clover and biserrula have been reported to have greater ability to successfully regenerate in Mediterranean-like climatic conditions. In general, these two legumes produced significantly greater crop biomass, and as a result, considerably less weed biomass due to low light penetration of the dense crop canopy and competition for resources. Results of PLSR analysis demonstrated that, in general, crop biomass had the strongest inverse effect on total weed biomass for many pasture species and likely also impacted weed seed germination and seedling growth. Several mechanisms could be associated with the ability of these pasture legumes to suppress weeds including their ability to rapidly develop dense canopies limiting light interception at the soil surface. In addition, yellow serradella may exhibit the ability to suppress weeds through allelopathic interference. Additional information related to the mechanisms involved in weed suppression, including the potential for some species to release allelochemicals over time, will be useful in the selection of new pasture species/cultivars for sustainable weed management solutions, particularly when considering the increasing importance of herbicide resistance in pasture weeds.

ACKNOWLEDGMENTS

The authors acknowledge funding support for this project from Meat and Livestock Australia (MLA) Project B. WEE 0146 and the Australian Centre for International Agricultural Research (ACIAR) for providing a PhD scholarship for SL. The author also acknowledges technical support provided by Graeme Heath and Simon Flynn in collecting field data.

REFERENCES

Loi, A., Howieson, J.G., Nutt, B.J. and Carr, S.J. (2005a). A second generation of annual pasture legumes and their potential for inclusion in Mediterranean-type farming systems. *Australian Journal of Experimental Agriculture* 45, 289-99.

Loi, A., Howieson, J.H. and Carr, S.J. (2001). Register of Australian herbage plant cultivars. *Biserrula pelecinus L.* (biserrula) cv. Casbah. *Australian Journal of Experimental Agriculture* 41, 841-2.

Loi, A., Revell, C. and Nutt, B.J. (2005b). Domestication of new forage legumes improves the productivity and persistence of pastures in Mediterranean environments. Proceedings of the 1st COST 852 Workshop, eds B.E. Frankow-Lindberg, R.P. Collins, A. Luscher, M.T. Sebasria, A. Helgadottir, pp 165-75. (Swedish University of Agricultural Sciences, Sweden).

Mengel, K., Kirkby, E.A., Kosegarten, H. and Appel, T. (2001) 'Principles of Plant Nutrition'. (Springer, Dordrecht).

Reiss, A., Fomsgaard, I.S., Mathiassen, S.K. and Kudsk, P. (2018). Weed suppressive traits of winter cereals: Allelopathy and competition. *Biochemical Systematics and Ecology* 76, 35-41.

Angled onion, *Allium triquetrum* L., a beautiful flower with a bitter taste

V. Lesieur[1,2], M. Jourdan[2], T. Thomann[2], A. Sheppard[2,3], J.F. Martin[1], M.S. Tixier[1], L. Morin[3], S. Raghu[4] and B. Gooden[3]

[1] CBGP, Montpellier SupAgro, INRA, CIRAD, IRD, Univ. Montpellier, Montpellier, France

[2] CSIRO Health and Biosecurity, European Laboratory, Montferrier sur Lez, France

[3] CSIRO Health and Biosecurity, GPO Box 1700, Canberra, Australian Capital Territory 2601, Australia

[4] CSIRO Health and Biosecurity, Brisbane, Australia

Summary It is now well recognised that many invasive plants have escaped from gardens and landscapes where they were originally planted for ornamental purposes. Angled onion, *Allium triquetrum* L. (Amaryllidaceae), is another example of such an escapee. This bulb-forming herbaceous perennial plant, native to the west and central Mediterranean, is now widespread and invasive in different parts of the world. In Australia, the weed is declared noxious in southern parts of the country and its uncontrolled spread threatens biodiversity in natural environments. In addition, *A. triquetrum* is also problematic in agriculture, causing milk and meat taint with strong onion odour. Because of the lack of effective control methods, a classical biological control approach has been initiated by CSIRO and Montpellier SupAgro (France). The first phase of the program includes native range surveys for characterising the natural enemies community (arthropods and fungi) associated with the target plant, which seems to be poorly understood across its native range. Here, we summarise the relevant information from a biological control perspective based on literature review and present the results of the first field surveys carried out in the native range.

Keywords Weed, Amaryllidaceae, natural enemies' community, arthropods, fungi.

Diversity and extent of mutations endowing resistance to AHAS-inhibiting herbicides in *Sisymbrium orientale* in Australia

Weihua Long[1,2], Peter Boutsalis[2], Mahima Krishnan[2], Jenna M. Malone[2] and Christopher Preston[2]
[1] The Institute of Industrial Crops, Jiangsu Academy of Agriculture Science; 50 Zhongling St, Nanjing, Jiangsu 210014, China
[2] School of Agriculture, Food and Wine, the University of Adelaide; PMB 1, Glen Osmond, Adelaide, South Australia 5001, Australia
(christopher.preston@adelaide.edu.au)

Summary Indian hedge mustard (*Sisymbrium orientale*) (IHM) is an important broadleaf weed of crops in Australia. Resistance to sulfonylurea (SU) herbicides that inhibit acetohydroxyacid synthase (AHAS) is extensive in IHM populations in Australia, but resistance to imidazolinone (IMI) herbicides has only recently been a problem. The AHAS-mutation profile of 65 IHM populations collected randomly from crop fields was examined to better understand the extent and types of resistance present. Screening with the herbicides chlorsulfuron and imazamox showed that 40.4% of the random populations were resistant to SU herbicides and 10.8% were resistant to IMI herbicides. Various mutations were identified in SoAHAS by sequence analysis, which included not only the reported amino-acid substitutions on the Pro-197 and Trp-574, but also the new substitutions on Pro-197 and the new mutation on Asp-376 for this species. One population with possible non-target-site resistance was identified. Furthermore, germination studies with fresh seed demonstrated that the mutations in SoAHAS resulted in lower and slower germination; however, other factors had a larger effect on germination ability in *S. orientale*. The molecular basis of AHAS-inhibiting herbicide-resistant populations of *S. orientale* was endowed by mutations in the SoAHAS in all but one population. Mutations at Pro-197 endowing resistance to SU herbicides were most common; however, mutations at Trp-574 that provide resistance to IMI herbicides are becoming more common.

Keywords Acetohydroxyacid synthase (AHAS), *Sisymbrium orientale*, germination test.

Growth responses of two Australian biotypes of common sowthistle (*Sonchus oleraceus* L.) to varied soil moisture regimes

Gulshan Mahajan[1], Michael Walsh[2] and Bhagirath S. Chauhan[1]
[1] Queensland Alliance for Agriculture and Food Innovation (QAAFI), The University of Queensland, Gatton, Queensland 4343, Australia
[2] Sydney Institute of Agriculture, School of Life and Environmental Sciences, The University of Sydney, Narrabri, New South Wales 2390, Australia
(g.mahajan@uq.edu.au)

Summary Common sowthistle is one of the most troublesome weeds in the northern grain region of Australia. It is well known fact that water stress played a great role in influencing crop growth and yield. However, effects of water stress on weed growth and seed production are poorly understood. In this experiment, growth behaviour of two population of sowthistle weed (Dalby and Gatton biotype) were studied at 100, 75, 50 and 25% of soil water holding capacity (WHC). At 100% WHC, Dalby biotype had higher plant height than Gatton biotype; while at other soil moisture levels, no such differences were observed The highest plant height in Gatton biotype was observed at 100% WHC; while Gatton attained highest plant height at 50% WHC. The seed production in Dalby biotype did not vary with varied soil moisture levels. The Gatton biotype had similar seed production at 75 and 50% WHC; but significantly higher than 25% WHC. It was interesting that Gatton biotype had higher seed production than Dalby biotype at 50%WHC; while at 100%WHC, seed production in Dalby biotype was higher than Gatton biotype. Averaged over soil moisture levels, Dalby biotype had greater biomass than Gatton biotype. The root weight, proline and phenolics contents were also higher in Dalby biotype than Gatton biotype. This study shows that sowthistle has capacity to survive and reproduce even under a higher degree of water stress; but response varied with the biotypes.

Keywords Water stress, weed ecology, rainfall, proline, phenolics, seed production.

Phalaris minor resistance is increasing and solutions urgently needed to farmers of Punjab and Haryana

Malwinder Singh Malhi[1] and Pavitarpal Singh Pangli[2]
[1] Syngenta India Ltd
[2] Borlaug Farmer Association for South Asia, 499- Phase-2 Urban-Estate Dugri Ludhiana-141013 Punjab, India
(malwinder.malhi@syngenta.com; msmalhi2000@yahoo.com)

Summary Wheat is a major cereal crop in north India (Punjab and Haryana). It is grown on an area of 60 million hectare with production of around 160 metric tonnes and average yield of 4.6 metric tonne per hectare. *Phalaris minor* is the major problem weed in wheat which leads to heavy losses in wheat grain yield. Therefore, timely control measures are needed to mitigate the yield loss. The resistance in *P. minor* to herbicides of different modes of action is increasing in almost all wheat growing areas, including the rice-wheat rotation areas. A recent farmers' survey in more than 50 villages in Punjab indicated that some farmers (15%) used sulfosulfuron alone or tank-mix of sulfosulfuron and clodinafop before first irrigation. Some farmers used tank-mix of clodinafop plus metribuzin (30%), pinoxaden plus metribuzin (10%), mesosulfuron plus iodosulfuron (10%), and metribuzin alone (5%) after first irrigation. The surveyed farmers reported that the available recommended chemicals are not effective now, which they attribute to poor quality of the herbicides or development of resistance in *P. minor* against herbicides.

Alternative solutions for the effective management of *P. minor* in the region include direct-drill of wheat into standing stubbles of rice (to reduce the weed density), improved spray technology (to enhance herbicides efficacy), rotation of different herbicidal modes of action (to extend life of herbicides) and use of new chemicals (to provide immediate relief to wheat growers). The adoption of these alternative strategies will effectively manage *P. minor* and will reduce wheat yield loss in north India.

Keywords India Haryana Punjab farmers, herbicides, *Phalaris minor*.

Allelopathic activity of *Hosta alismifolia* against *Lactuca sativa* L.

Mohammad Shamim Hasan Mandal
Graduate School for International Development and Cooperation, Hiroshima University, 1-5-1 Kagamiyama, Higashihiroshima, Japan 739-0044
(mshmandal@gmail.com)

Summary Allelopathy of common plants is being investigated by researchers around the world over the years. However, there is little research done on allelopathy of *Hosta* spp. In this study, the inhibitory properties of *Hosta alismifolia* were investigated. Aerial parts of the plants were collected from Kochi prefecture, Japan, and air dried. Water extractions of the dried materials were performed to produce four (0.5%, 1%, 2% and 4%) concentrations. A laboratory bioassay was performed where *Lactuca sativa* L. was used as the receiving plant. The results show that, water extract had significant inhibition against lettuce germination and growth. The water extracts induced physiological changes on lettuce seedlings, reducing amounts of chlorophylls, and carotenoids. Shoot length inhibition was 13.85% and 93.24% over control at 2% and 4% concentrations, respectively. In case of root length all concentrations had significant inhibition over control (51.33%, 69.42%, 89.46% and 99.28% inhibition over control at the 0.5, 1, 2 and 4% concentrations, respectively). High-performance liquid chromatography analysis revealed presence five major phenolic acids namely protocatechuic acid, *p*-hydroxybenzoic acid, ferulic acid, *p*-coumaric acid, benzoic acid and ellagic acid. Among the five detected phenolic acids, ferulic acid content was the highest (2.07 mg g^{-1} dry weight) in amount followed by benzoic acid (0.88 mg g^{-1} dry weight). Further bioassays using the detected phenolics to reveal their allelopathic effects against lettuce are to be performed. The results indicate *Hosta alismifolia* plant contains plant growth inhibitors and can be potentially used for biological weed control.

Keywords Allelopathy, phenolics, *Hosta alismifolia*.

The role of biochar pyrolysis temperature on hexazinone sorption-desorption in soil using the batch-equilibrium method

Kassio F. Mendes, Gustavo M. Chitolina, Mayara C. Arruda, Leonardo V. Junqueira, Cássio S. Almeida and Valdemar L. Tonisielo
Ecotoxicology Lab, Center of Nuclear Energy in Agriculture, University of São Paulo, Piracicaba, SP, Brazil
(kassio_mendes_06@hotmail.com)

Summary Pyrolysis of organic waste involves the thermal transformation of biomass under partial or no oxygen supply to solid (biochar), liquid (bio-oil) and gas phases. Pyrolysis temperature controls the physico-chemical properties (e.g. composition, particle and pore size distribution) of the biochar, which in turn influence herbicide retention in the soil, mainly herbicides applied in pre-emergence, as hexazinone. The aim of this research was to assess the role of biochar pyrolysis temperature on hexazinone sorption-desorption in a Brazilian soil. Triazine-6-^{14}C-hexazinone (radiochemical purity 99.7%, specific activity 3.14 MBq mg^{-1}) sorption-desorption was evaluated using the batch-equilibrium method from Freundlich model and five solutions (0.63, 1.25, 1.88, 2.50, and 3.13 mg L^{-1}) were used. Biochar was prepared from eucalyptus (*Eucalyptus grandis*) at 450, 550, 650, 750, 850 and 950°C. The biochar was added to soil at 1% (w w^{-1}) ratio. The K_f values calculated for the hexazinone sorption on the unamended and all biochar-amended soils was 0.18 (50.49%) and ranged from 0.12 to 0.24 μmol$^{(1-1/n)}$ L$^{1/n}$ kg^{-1} (57.57–70.33%), respectively. Desorption on the unamended and all biochar-amended soil was 26.17% and ranged from 27.23 to 19.80%, respectively. The $1/n_{(desorption)}$ are greater than $1/n_{(sorption)}$, suggesting that the hexazinone sorption by all treatments was reversible. H values ranged from 1.25 to 1.95 on all treatments. As biochar pyrolysis temperature increased, hexazinone sorption also increased and desorption increased. The results showed that biochar pyrolysis temperature can influence on hexazinone sorption-desorption in the soil, which this material reduces the availability of herbicide in the soil solution for weed control.

Keywords Residual herbicide, retention process, soil amendments.

Could *Phytophtora* species associated with declining populations of invasive European blackberry be used for biological control?

Louise Morin[1], John Lester[1], Patrick Gleeson[1], Treena I. Burgess[2], Giles E. St. J. Hardy[2] and John K. Scott[3]
[1] Commonwealth Scientific and Industrial Research Organisation (CSIRO), Health and Biosecurity, Black Mountain Science and Innovation Park, GPO Box 1700, Canberra, Australian Capital Territory 2601, Australia
[2] Centre for Phytophthora Science and Management, School of Veterinary and Life Sciences, Murdoch University, 90 South Street, Murdoch, Western Australia 6150, Australia
[3] Commonwealth Scientific and Industrial Research Organisation (CSIRO), Land and Water, Private Bag 5, PO, Wembley, Western Australia 6913, Australia

Summary European blackberry (*Rubus fruticosus* agg.) is a thorny invasive shrub that grows primarily in southern areas of Australia where annual rainfall exceeds 700 mm. It readily invades land along watercourses, competing against native plants and pasture, and preventing access to the water by native fauna and livestock. *Phytophtora* species have been found to be associated with severe blackberry dieback in Western Australia. The objective of this study was to determine the biological control potential of two *Phytophthora* species by comparing their pathogenicity on blackberry and testing the specificity of the most promising species on a range of non-target plants. The two species, *P. bilorbang* and *P. pseudocryptogea*, have been found to occur naturally in soil in other states. Under glasshouse conditions, *P. pseudocryptogea*, grown on solid substrate and applied to the soil, killed or significantly reduced biomass of blackberry plants when exposed to simulated flooding events. In contrast, plants treated with *P. bilorbang* did not differ from untreated, control plants. In a series of subsequent experiments, *P. pseudocryptogea* was found to adversely affect a range of non-target species, including some *Acacia* and *Eucalyptus* species. On the basis on these results, it was decided not to proceed with field trials.

Keywords *Rubus fruticosus*, Weed of National Significance, inundative biocontrol, soil-borne disease.

Germination ecology of resistant and susceptible populations of barnyard grass (*Echinochloa colona* (L.) Link.) collected from Queensland

Navneet Kaur Mutti, Gulshan Mahajan and Bhagirath S. Chauhan
Queensland Alliance for Agriculture and Food Innovation (QAAFI), The University of Queensland, Gatton, Queensland 4343, Australia
(n.mutti@uqconnect.edu.au)

Summary Barnyard grass is the most prevalent summer weed in the northern grain region of Australia occupying the regional ranking of 1 in terms of yield loss and revenue. Barnyard grass is highly competitive, prolific in nature and difficult to control by commonly used herbicides in Australia. Here, barnyard grass has developed resistance to glyphosate and other groups of herbicides. For effective weed management, it is important to understand the seed germination ecology of herbicide-resistant and susceptible populations of barnyard grass. Therefore, laboratory experiments were conducted to evaluate the effect of temperature, light, salinity, pH, soil burial depth and the amount of sorghum crop residue on germination and emergence of glyphosate-resistant and susceptible populations of barnyard grass. The research findings from this study may help growers in developing appropriate weed management strategies for barnyard grass.

Keywords Weed biology, seed germination, burial depth, osmotic potential, integrated weed management.

Field evaluation of selected canola competitive cultivars for suppression of natural weed populations

James M. Mwendwa[1], William B. Brown[1], Paul A. Weston[1], Shamsul K.M. Haque[2], Jeffrey D. Weidenhamer[3] and Leslie A. Weston[1]
[1] Graham Centre for Agricultural Innovation, Locked Bag 588, Wagga Wagga, New South Wales 2678, Australia
[2] Eurofins Agroscience Services Pty Ltd, 50 Leewood Drive, Orange, New South Wales 2800, Australia
[3] Department of Chemistry, Geology and Physics, Ashland University, Ashland, Ohio 44805, USA
(jmwendwa@csu.edu.au)

Summary In 2014–2016, replicated field trials were performed to evaluate mechanisms of weed suppression in Australian canola genotypes in moderate to low rainfall zones at Wagga Wagga (572 mm) and Condobolin (449 mm) NSW, respectively. In 2015–2016, a split plot design with and without trifluralin as the main plot and cultivar as the sub plot was employed for trials; in this experiment 8 cultivars including hybrid and open-pollinated cultivars were assessed. At each site, crop and weed growth were monitored at various phenological stages including early season, vegetative, grain-filling, harvest and post-harvest. Certain cultivars exhibiting early vigour and also the ability to intercept light due to leaf canopy structure were associated with increased suppression of in-crop weed growth in canola trials; in addition, improved post-harvest weed suppression was associated with the presence of remaining crop residues after harvest. Cultivars GT-50, Hyola 600RR and Hyola 50 were the most weed suppressive and consistently higher yielding in each year at both locations. CB Taurus and GT-50 provided higher weed suppression only after harvest when residues remained in plots for 150 days post-harvest. Pre-emergence trifluralin treatment resulted in improved crop yields in contrast to untreated plots for most, but not all, cultivars. In this case, the cultivars that possessed rapid early growth and vigour and significantly reduced light availability at the soil surface limited weed growth in the absence of trifluralin. Our results indicated that establishment of certain canola cultivars effectively resulted in enhanced in-crop and post-harvest weed suppression, with or without the use of post-emergent herbicides during the growing season, particularly for common spring and summer annual weeds which were problematic post-harvest.

Keywords Weed suppression, canopy architecture, phenology, crop residue, annual weeds.

An inventory of Queensland prioritised invasive plant species for management and research

Olusegun O. Osunkoya[1], Christine Perrett[1], Kerri Moore[2], Sam Nicol[2], Jens Froese[2], Kristy Gooding[3] and Shane Campbell[4]
[1] Invasive Plant and Animal Science Unit, Biosecurity Sciences, Department of Agriculture and Fisheries, Eco-sciences Precinct, Brisbane, Queensland 4102, Australia
[2] CSIRO, Eco-sciences Precinct, Brisbane, Queensland 4102, Australia
[3] Local Government Association of Queensland, PO Box 230 Fortitude Valley, Brisbane, Queensland 4006, Australia
[4] Invasive Plant and Animal Science Unit, Biosecurity Sciences, Department of Agriculture and Fisheries, Tropical Weeds Research Centre, Charters Towers, Queensland 4820, Australia
(Olusegun.osunkoya@daf.qld.gov.au)

Summary There are currently ~300 declared weeds in Queensland (QLD), Australia, but this list requires a review as the last known exercise dealing with weed risk assessment and prioritisation was undertaken about two decades ago. In this paper we propose an updated list of invasive alien (plant) species of significant concern in QLD, based on information derived from pest management plans of local governments, herbarium records and published/grey literature, expert knowledge, and opinion from diverse groups of stakeholders. Weed diversity varies significantly between local government areas and regions. Regions on the mainland eastern seaboard of the State share similar weed communities, unlike western regions and the Torres Strait islands, which share fewer, weed species. Stakeholders identified the top research and management priorities for the weed list. These were: studies involving biological control options (34.8% of respondents), public awareness and education (18.5%), effective herbicide-use and application techniques (15.2%), ecology, taxonomy and risk analysis (11.5%), and adaptive pasture management (9.3%). Based on occurrence and distribution across local government areas/regional jurisdictions and on-ground stakeholders' perceived weed severity, a weed priority list of high-medium and-low impact scores for policy, research and management was compiled for each region and State-wide.

Keywords Biological invasions, stakeholder elicitation, weed risk assessment, weed impact, weed prioritisation, weed spread.

INTRODUCTION

Biological invasions, a key component of global change, have significant environmental and economic impacts worldwide (McGeoch *et al.* 2016). Due to increasing globalisation, most geographical regions or ecosystems have numerous candidate invasive alien species (IAS) in need of attention. Given the large number of problematic introduced pest plants and animals and limited resources for their management, it is necessary to prioritise them for adaptive research and strategic control options. For example, in Australia there are currently 3100+ alien plants with a broad range of environments and industries affected (AG 2018). Prioritisation of IAS requires information on abundance (density), extent of spread (current and potential distribution), socio-economic-ecological impact, and management achievability of candidate species (Parker *et al.* 1999, McGeoch *et al.* 2016, Booy *et al.* 2017).

We define pest species prioritisation as the process of ranking environmental IAS (including those impacting the grazing industries, but excluding those exclusive to cropping ecosystems) to guide policy, research and management. Here we focus on prioritisation of established invasive plants of Queensland (QLD), Australia using data obtained from the grey literature, herbarium records and stakeholders elicitation. Bebawi *et al.* (2002) did a similar exercise, but restricted the exercise to plants of the tropics. Our aims are to:

1. describe variations in QLD invasive flora that are of great concern across all regions of the State; and
2. develop a prioritised IAS list at regional and whole State levels based on spread and perceived impact.

MATERIALS AND METHODS

Pests are managed at a regional level in QLD by local government (LG) authorities, and consequently each LG has a different list of IAS. Each LG belongs to one of 10 Regional Organisations of Councils (hereafter 'regions'), which are administrative groupings of nearby LG councils that share relatively similar geographic and climatic features. To examine the full list of IAS at a statewide level, we accessed QLD LG pest management plans (PMPs) during February to June 2016 by *e*-downloading or requesting via email from the appropriate authority (74 LGs in all). For each LG PMP, IAS lists were extracted and entered into a relational database. Each IAS was recorded together with information about its priority listing for perceived threats and management need (rated: high, medium or low). Across LGs, The IAS and associated ecological data (e.g. life form, habitat invaded, date of first record in the State [i.e. time since invasion/residency time] based on QLD herbarium records (HERBRECS) etc.) were aggregated into regional lists based on their regional affiliations.

From October 2016 through to March 2017, we convened a round of regional stakeholders' workshops around the State for each of the 10 regions using the platform of Local Government Association of QLD. Participants ($n \geq 15$) were made up of LG officials tasked with pest control and management, farmers, elected council officials, and State Government biosecurity officers. At each workshop, the region's established weed list (extracted and aggregated from LG PMPs), together with identified State emerging IAS, was presented to the group. We made initial presentations and discussions around the scoring methods (see below) and terminology to reduce the potential for language-biased misunderstanding, followed by open, facilitated discussions. Finally, through deliberation and consensus-building, each species was assigned a single priority opinion (high, medium or low) on 'perceived' pest impact and management expediency. Research and management needs for each species were also recorded. In the process, some species were added, removed from the regional list, or changed rank/rating compared to what was in the LG PMPs.

For each region, LG pest lists were aggregated by summing the number of LG that listed and/or rated a given IAS as a significant concern and therefore requires management. The IAS qualitative ratings and distribution derived from LG PMPs and from the elicitation workshops were converted to numerical (given arbitrary values of 8, 5, and 2 for high, medium and low ratings, respectively) or absence-presence values (1 for presence of a given rating, e.g. high, and 0 for its absence) for further analyses (see below).

To explore similarities between IAS lists at a regional level, the data were subjected to multivariate analyses of non-metric multidimensional scaling (NMDS) in two dimensions using Bray-Curtis dissimilarity measure. The Primer software (ver. 6) was used with region as the main factors. Responses to open-ended questions (e.g. research needs for management) were summarised by grouping major opinions or answer types, and recording the number of respondents (i.e. region) that mentioned each opinion.

A plethora of indices exist for estimating pest spread (invasiveness) and impact (see Barney *et al.* 2013). We used the framework of Parker *et al.* (1999) to quantify regional spread and impacts of invaders based on the linear equation $I = A \times R \times E$, where I is the invader's impact, A is the invader's mean local abundance (density), R is the range or area occupied by the invader (i.e. its distribution), and E is the invader's per unit effect. We used the proportion of LGs within each region that listed and rated a given invader as our estimate of A and proportion of regions within QLD with the weed as the estimate of R. Note that we assume that if the weed is listed then it will be abundant, now or in the near future, so listing by LG serves as a crude proxy for mean spread (see Kumschick *et al.* 2012, Panetta 2016 for a similar approach). As in many IAS assessment reports (e.g., McGeoch *et al.* 2016, Pearson *et al.* 2016), quantifying E (i.e. per unit weed effect) is untenable due to the large number of species screened for. Thus we used, as a proxy, the mean of the consensus ratings generated at the regional stakeholders' meetings as a measure of effect (E) of each listed weed. This involved conversion of priority ratings generated from qualitative, three point state [high-medium-low] to quantitative values- assigned 8, 5 and 2, respectively (see Kumschick *et al.* 2012, McGeoch *et al.* 2016) with the average regional priority rating values as input to derive a final impact (I) score.

RESULTS

In all, a total of 295 species were identified from the LG PMPs of which only 107 species (36.3%, range 13.6–90.9%) were of significant concern at the regional stakeholders' workshops (Figure1). Queensland invasive flora is taxonomically diverse, representing 161 genera and 64 families. The five most important families based on number of species assessed and prioritised were: Fabaceae (16 species), Asteraceae (13 species), Poaceae (12 species), Cactaceae (10 species), and Verbenaceae/Amaranthaceae (5 species each).

NMDS ordination of QLD IAS based on presence-absence data indicated that range of pest plant species varied widely across the inland regions and the Torres Strait islands, while those of coastal regions appeared homogenous (Figure 2). These variations (of 25.0–90.0% dissimilarity) reflect roughly the geographical continuum of the State, consisting of (a) northern, (b) southern, and (c) western regions.

Using the invasiveness index in Parker *et al.* (1999) (a product of $A*R$), the twenty most highly ranked pest plant species in QLD in decreasing order

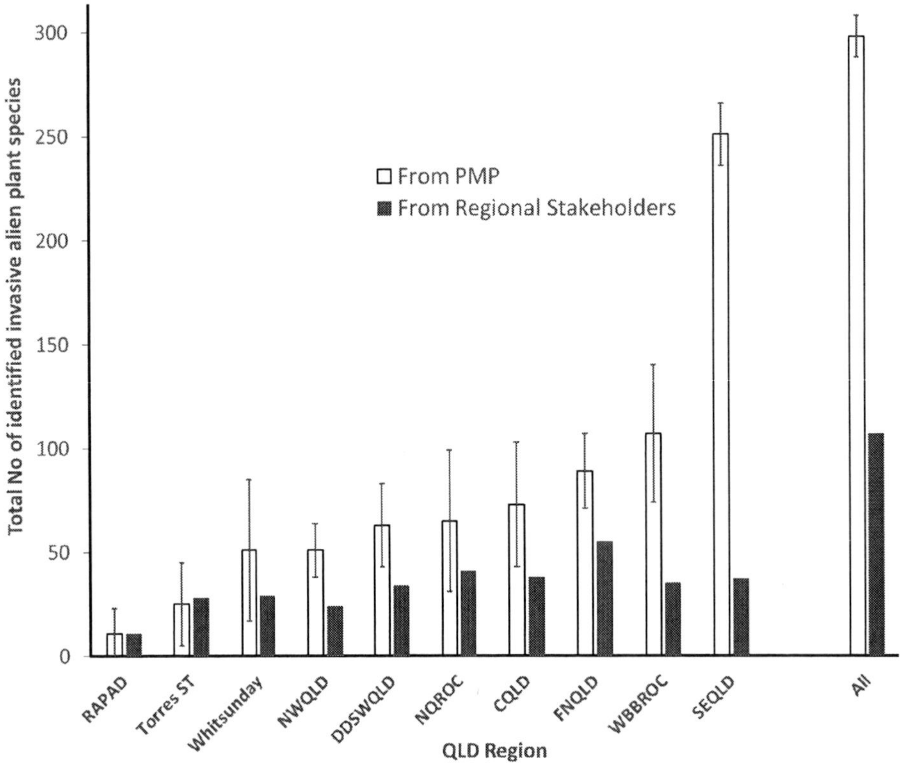

Figure 1. Number of IAS (±SE) across regions of QLD identified from scoping of LG pest management plans (PMP) and by stakeholders at regional workshops. Regions are: RAPAD – Remote Area and Planning Authority Development [Central West]; Torres ST – Torres Straight Islands; NWQLD – North West QLD; DDSWQLD – Darling Downs QLD [South West]; NQROC – North QLD; CQLD – Central QLD; FNQLD – Far-North QLD; WBBROC – Wide Bay Burnett region; and SEQLD – South East QLD.

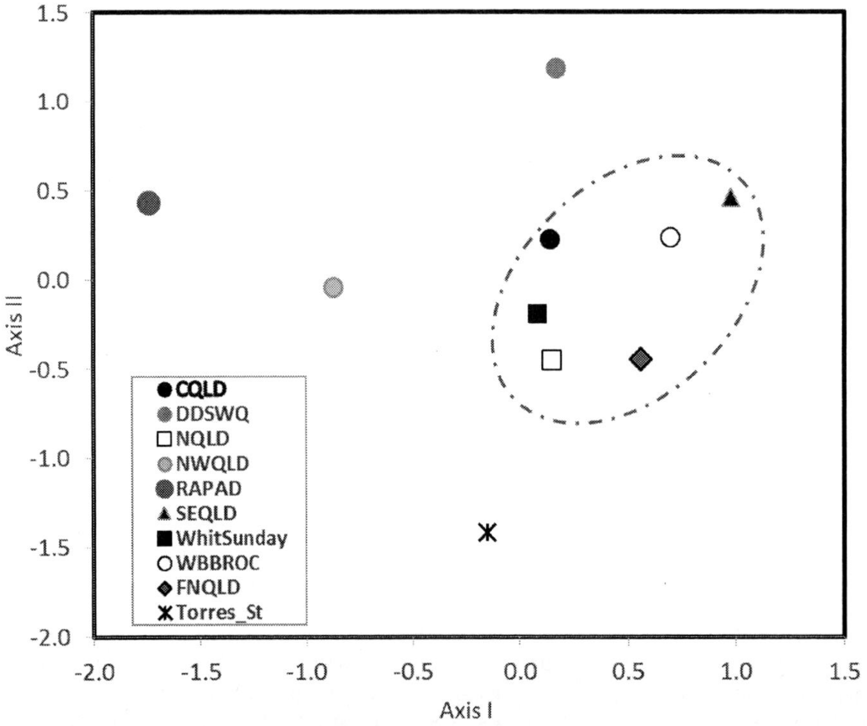

Figure 2. NMDS ordination of ten regions of QLD, Australia based on presence-absence data of IAS. Dashed circle indicates similarity in IAS for the eastern regions. Region abbreviations as in Figure 1.

with estimated residency time (years.) based on the date of first appearance of the IAS in the QLD herbarium records are: *Parthenium hysterophorus* L. (parthenium – 62 yr); *Bryophyllum delagoense* (Ecklon and Zeher) Schinz (mother of millions – 104 yr); *Cryptostegia grandifilora* (Roxb.) R.Br. (rubber vine – 157 yr); *Parkinsonia aculeata* L. (parkinsonia – 157 yr); *Vachellia nilotica* (Benth.) Brenan (prickly acacia – 117 yr); *Sporobolus pyramidalis* P. Beauv., *S. natalensis* (Steud.) Dur and Schinz (giant rat's tail grass – 57 yr); *Salvinia molesta* D.S. Mitchell (Salvinia – 64 yr); *Jatropha gossypifolia* L. (bellyache bush – 129 yr); *Eichhornia crassipes* (Mart.) Solms (water hyacinth – 127 yr); *Harrisia martini* (Labour.) Britton (harrisia cactus – 59 yr); *Lantana camara* L. (lantana – 176 yr); *Prosopis glandulosa* Torr. (mesquite – 117 yr); *Ziziphus mauritiana* Lam. (chinee apple – 101 yr); *Hymenachne amlexicaulis* P. Beauv. (Hymenachne – 44 yr); *Pistia stratiotes* L. (water lettuce – 71 yr); *Dolichandra unguis-cati* (L.) L.G. Lohmann (cat's claw creeper vine – 70 yr); *Leucaena leucocephala* (Lam.) de Wit. (leucaena – 96 yr); *Opuntia* and *Cylindropuntia* spp. (e.g., *Cylindropuntia rosea* (DC.) Backeb. Hudson pear – 44 yr); *Xanthium strumarium* L. (Noogoora burr – 90 yr); and *Senna obtusifolia* (L) Irwin and Barneby (sicklepod – 146 yr). If regional impact score ($I=A*R*E$) rather that mean regional invasiveness (spread) score is used, the order of ranking of pest species changed somewhat but not significantly, with *Senecio madagascariensis* Poir. (fire weed – 99 yr) and *Tecoma stans* (L.) Kunth (yellow bells/tecoma – 93 yr) coming up on the list. Other pest species making up the prioritised list and their scores can be requested from the senior author.

At the stakeholders' workshops, research and management needs for prioritised IAS were identified, and they include needs on: (i) biological control options (34.8% of respondents); (ii) public awareness and education (18.5%); (iii) effective herbicide use and application techniques (15.2%); (iv) ecology, taxonomy and risk assessment (11.5%): (v) pasture and integrated management (9.3%); and (vi) eradication of outliers (4.4%).

DISCUSSION

The weed flora of QLD is diverse within and between regions, with greater variation between regions in western QLD and that of coastal eastern part of the State (Figure 2). The coastal similarities in IAS composition suggest that QLD eastern regions have shared challenges and may benefit from working cooperatively as a single unit (with the exception of Torres Strait, a top end region, which tends to have its own distinctive weed assemblage). In contrast, weeds in the western regions and the Torres Strait are largely disparate; while there are still benefits from cooperating on individual shared species, it may be necessary to develop region-specific management strategies.

The use of biological control agents, including their redistribution and efficacy, was the most frequent requirement of stakeholders' research and management needs. This suggests that there is strong community support for the QLD and Australian governments in continuing investment in biocontrol ventures, including fund provision to prospect for agents to combat prominent emerging invaders, including grasses such as *Sporobolus pyramidalis* and *S. natalensis* (giant rat's tail grass).

The quest to find the best procedure for pest prioritisation continues. Currently there are over 30 procedures published and recommended (see the comprehensive list in Nentwig *et al.* 2016). However, many have argued that there is no objectively correct way to carry out pest species prioritisation, as long as certain guidelines are met. These guidelines include: the provision of a logical scientific basis for the approach, transparency; generality of questions that are applicable to a range of situations; minimisation of subjective views; relatively few questions; and ability to aggregate all available data into a single index (Kumschick *et al.* 2012, McGeoch *et al.* 2016). In this study we have fulfilled these criteria. We initially extracted weed lists from PMPs of LGs, aggregated the LG data into a regional list and presented the weed lists to stakeholders, followed by the use of a single stakeholders' consensus rating per candidate weed to ensure that subjective views are minimised. Finally the qualitative ratings were converted to quantitative data that serves as input in the estimation of the Parker *et al.* (1999) weed invasiveness and impact indices. These are steps that are transparent, traceable and repeatable. The next phase is to explore how the prioritised target species list can be refined further to account for management achievability. This step, the pest risk management scenario, calls for another round of stakeholder elicitation to examine the probability of IAS eradication and/or control. Combining risk assessment and risk management creates a matrix that defines the final IAS list – thus ultimately influencing policy in terms of adaptive weed research and management options.

ACKNOWLEDGMENTS

We thank many stakeholders across the state of QLD who participated in the regional workshops.

REFERENCES

AG, Australian Government (2018). Weeds in Australia. http://environment.gov.au/biodiversity/invasive/weeds (accessed 4 April 2018).

Barney, J.N., Tekiela, D.R., Dollete, E.S. and Tomasek, B.J. (2013). What is the "real" impact of invasive plant species? *Frontiers in Ecology and the Environment* 11, 322-9.

Bebawi, F., Campbell, S. and Stanley, T. (2002). Priority lists for weed research in the wet-and dry-tropics of north Queensland. *Plant Protection Quarterly* 17, 67-73.

Booy, O., Mill, A.C., Roy, H.E., Hiley, A., Moore, N., Robertson, P., Baker, S., Brazier, M., Bue, M. and Bullock, R. (2017). Risk management to prioritise the eradication of new and emerging invasive non-native species. *Biological Invasions* 19, 2401-17.

Kumschick, S., Bacher, S., Dawson, W., Heikkilä, J., Sendek, A., Pluess, T., Robinson, T. and Kühn, I. (2012). A conceptual framework for prioritization of invasive alien species for management according to their impact. *NeoBiota* 15, 69-100.

McGeoch, M.A., Genovesi, P., Bellingham, P.J., Costello, M.J., McGrannachan, C. and Sheppard, A. (2016) Prioritizing species, pathways, and sites to achieve conservation targets for biological invasion. *Biological Invasions* 18, 299-314.

Nentwig, W., Bacher, S., Pyšek, P., Vilà, M. and Kumschick, S. (2016). The generic impact scoring system (GISS): A standardized tool to quantify the impacts of alien species. *Environemtal Monitoring and Assessement* 188, 315-28.

Panetta, F.D. (2016). Dealing with 'new' alien plants: Risk assessment and risk management. Proceedings of the 20th Australasian Weeds Conference, eds R. Randall, S Lloyd, and C. Borger, pp. 183-9. (Weeds Society of Western Perth, Australia).

Parker, I.M., Simberloff, D., Lonsdale, W., Goodell, K., Wonham, M., Kareiva, P., Williamson, M., Von Holle, B., Moyle, P. and Byers, J. (1999). Impact: Toward a framework for understanding the ecological effects of invaders. *Biological Invasions* 1, 3-19.

Pearson D.E., Ortega Y.K., Eren Ö. and Hierro J.L. (2016) Quantifying "apparent" impact and distinguishing impact from invasiveness in multispecies plant invasions. *Ecological Applications* 26, 162-73.

Flupropanate resistant serrated tussock (*Nassella trichotoma*) identified at multiple locations in the Monaro, New South Wales

Jo Powells
South East Local Land Services, PO Box 26, Cooma, New South Wales 2630, Australia
(Jo.Powells@lls.nsw.gov.au)

Summary The continual use of group J herbicides (e.g. flupropanate) for more than 15–20 years has resulted in the development of herbicide resistant serrated tussock (*Nassella trichotoma*) biotypes. Local archived reports indicate that NSW Agriculture commenced serrated tussock control trials using flupropanate in the early 1980s near Dalgety on the Monaro and the herbicide remains in common use today. Jointly funded by South East Local Land Services, Snowy Monaro Regional Council and the Australian Government's National Landcare Program, a multi-year serrated tussock herbicide resistance testing project commenced in 2016. In the first year of testing, 13 samples of serrated tussock seed were collected from across the Monaro region and tested for flupropanate resistance at the application rates of 1.25, 2.0 and 3.0 L ha^{-1}. Five of the thirteen samples were found to have a high level of resistance to flupropanate with these samples originating from two different localities (Bredbo and Dalgety). Increasing the rate of application of flupropanate from 1.25 to 2 and 3 L ha^{-1} did not improve the control of these resistant biotypes which confirmed the high-level of resistance. The results from the second year of the project where over 40 seed samples have been collected at strategic locations around the region are pending. These results have significant impacts for how serrated tussock is managed on the Monaro in the future and reinforce the need for integrated approaches to be taken for the control of serrated tussock and the importance of herbicide resistance testing.

Keywords Herbicide resistance, serrated tussock, Monaro.

The effect of weed interference with cotton in conventional and ultra narrow row spacing condition

Ahmad Rafeizade[1], Mohammad Armin[2] and Matin Jamimoeini[2]
[1] Former Ms.e. student, Sabzevar Branch, Islamic Azad University, Sabzevar, Iran
[2] Department of Agriculture, Syadati Bol. Sabzevar Branch, Islamic Azad University, Sabzevar, Iran
(armin@iaus.ac.ir)

Summary To investigate the effect of weed interference duration on yield and yield components of cotton (*Gossypium hirsutum* L.) in conventional and ultra narrow row (UNR) condition, a field experiment was conducted at Sabzevar in 2014–2015. Experiment was conducted as factorial arrangement in a randomised complete block design with three replications. Experimental factors were conventional and UNR and duration of weed interference (0, 30, 40, 50 and 60 days after emergence. Row spacing in conventional and UNR were 20 and 70 cm, respectively. The result showed that that UNR had 13.24% and 59.47 lower weed density and weed dry matter compared with conventional, respectively. In ultra narrow row system, Cotton had more plant height, seed cotton yield, lint yield, biological yield and seed yield and less number of branches, boll number and boll weight than conventional system. Increasing of interference duration was decreased plant height (18.31%), lateral branches (25.78%), boll number (37.04%), boll weight (44.86%), biological yield (44.25%) seed cotton yield (61.22%), lint yield (61.52%) and seed yield (60.80%). Fitted logistic regression function coefficients showed that the onset of yield loss was earlier (34.5 days after emergence) in conventional condition while in UNR condition the onset of yield loss was postpone to 41.64 days after emergence. In conclusion, results showed in both conventional and ultra narrow row condition, weed interference reduces seed cotton yield that this reduction was more in conventional than ultra narrow row condition.

Keywords Cotton, competition, ultra narrow row, interference, yield loss.

Studies on tolerance of chickpea (*Cicer arietinum* L.) to some pre- and post-emergence herbicides

K. Rahamdad[1], I.A. Khan[2] and S.S. Shah[1]
[1] Department of Agriculture, Bacha Khan University Charsadda-Pakistan
[2]Department of Weed Science, The University of Agriculture, Peshawar-Pakistan
(drrahamdad@bkuc.edu.pk)

Summary Herbicides application is an integral part of farmers' cultural practice in modern agricultural systems. Weeds always cause serious losses in legumes due to poor competition with weeds. In order to quantify the tolerance of chickpea against pre-emergence herbicides (pendimethalin, s-metolachlor) and post-emergence herbicides (Clodinafop-propargyl, fenoxaprop-p-ethyl, di-methyl urea) a pot studies were undertaken during 2014–2015 and 2015–2016. Chickpea variety Karak-I was used for the purpose under the CR Design in the department of Weed Science, The University of Agriculture Peshawar. Data were recorded at plant height, number of branch per plant, number of seed per plant, number of pods per plant, number of nodule per plant, nodule fresh and dry biomass and seed protein contents during the investigations. The results revealed that the treatment had a significant effect on all the examined parameters of chickpea except nodule fresh and dry biomasses. Moreover, the toxic effect of pre-emergence herbicide on chickpea was significantly higher as compared to post-emergence herbicides. Minimum chickpea plant height (50.50 cm), number of nodule per plant (17.83) and seed protein (14.13 %) was recorded in pendimethalin, while highest seed protein content (21.75 and 21.15 %) was recorded for control and fenoxaprop-p-ethyl. Thus, it is concluded from the present data that chickpea has less tolerance for pre-emergence herbicides pendimethalin and s-metolachlor as compared to post-emergence herbicides so farmers are advised to give preference to post-emergence herbicides instead of a pre emergence while controlling weeds in chickpea through chemicals.

Keywords Chickpea, herbicides, protein, pendimethalin, weed.

Competitive wheat (*Triticum aestivum*) cultivars: A potential tool for managing herbicide resistant weeds

Abhimanyu Rinwa[1] and Manpreet Singh[2]
[1] Department of Agronomy, Punjab Agricultural University, Ludhiana 141004, Punjab, India
[2] Regional Research Station, Punjab Agricultural University, Abohar 152116, Punjab, India
(abhimanyu-agr@pau.edu)

Summary Competitive wheat cultivars have strong potential to reduce yield losses due to weeds; however, yield has always been preferred over competitiveness by breeders or farmers while choosing cultivars. The evolution and rapid expansion of herbicide-resistant weeds, and limited new herbicide chemistry availability, has renewed interest in harnessing competitive potential of wheat cultivars for managing such weeds. We discussed the recent understanding of the crop traits to explain variation in competitive wheat cultivars and the concept of suppression and tolerance related to weeds associated with wheat. The cumulative effects of allelopathy and competition determine the weed suppression potential of a given cultivar. Weak competitive ability of resistant weeds over susceptible weeds has been reported in previous findings; which provides opportunity to reduce the menace of resistant weeds through use of competitive cultivars. Weed research in the last decade although provide great foundation for breeding of competitive cultivars, the genetic variability in wheat for competitiveness is still limited. Further research is required to define the nature of relationships between cultivar traits and agronomic practices (like planting density, sowing rate, row orientation and sowing pattern) for managing diverse weed flora specific to a region or seasons. Such studies may help in reducing herbicide doses or managing herbicide resistance through use of herbicide in conjunction with competitive cultivars. This would be a step forward for sustained wheat productivity as well as human and ecosystem health.

Keywords Suppression, allelopathy, weed management, herbicide resistance, wheat, competitive cultivars.

Efficacy of harvest weed seed control systems on wild oats

Shona L. Robilliard, Linda M. Heuke, Asad Shabbir and Michael J. Walsh
I.A. Watson Grains Research Centre, Sydney Institute of Agriculture, The University of Sydney, 12656 Newell Highway, Narrabri, New South Wales 2390, Australia
(shona.robilliard@sydney.edu.au)

Summary Harvest weed seed control (HWSC) is an important tool for weed management in Australian cropping systems, leading to more sustainable crop production with reduced herbicide reliance. The efficacy of this weed control approach has now resulted in their widespread use with 40% of Australian growers using some form of HWSC. However, HWSC efficacy is directly related to the proportion of weed seed production that is collected during harvest. Weed species, such as wild oats, that are prone to seed shedding may not be effectively targeted by HWSC systems. A preliminary trial conducted at Narrabri, NSW during the 2016 harvest determined that HWSC treatments reduced wild oat populations by only 5–20%. However, the very high wild oat plant density (>600 plants m^{-2}) at this site indicates a large seedbank that would have masked HWSC treatment effects. Further trials were established during the 2017 harvest comparing chaff tramlining, chaff cart and narrow windrow burning treatments on wild oat populations in wheat and chickpea crops. Wild oat emergence counts in autumn 2018 will be used to assess the efficacy of these treatments on wild oat populations. The results from these trials will confirm if HWSC systems can be relied on to target wild oat populations in northern region cropping systems.

Keywords HWSC, seed shed, wild oats.

Interventional weed management methods in direct seeded rice

Pasoubady Saravanane
Department of Agronomy, Pandit Jawaharlal Nehru College of Agriculture and Research Institute, Karaikal, Puducherry Union Territory 609 603, India
(psaravanane@rediffmail.com)

Summary Rice (*Oryza sativa* L.) is the most important staple food crop of India. A shift in crop establishment from traditional manual transplanting of seedlings to direct-seeding has occurred in many Asian countries, including in India. Direct seeded rice has many advantages over transplanted rice. It consumes less water and matures earlier than the transplanted crop. However, weeds are the most important biotic constraint in direct seeded rice production. A field experiment was carried out during 2016–2017 at Pandit Jawaharlal Nehru College of Agriculture and Research Institute, Karaikal, Puducherry Union Territory, India to study the effect of interventional weed management strategies on the diverse weed flora, rice yield and benefit:cost ratio in direct seeded rice under unpuddled conditions.

Grassy weeds dominated the weed flora, with jungle rice (*Echinochloa colona* (L). Link) the major weed. A higher jungle rice population was noticed when weed management was done through manual weeding with a rotary weeder. Integration of herbicide and manual (hand) weed control resulted in a lower population of jungle rice, which resulted in better weed control, higher rice yield and a superior benefit:cost ratio. Uncontrolled weeds lead to 71% yield loss in direct seeded rice.

Keywords Herbicides, manual weeding, benefit:cost ratio.

INTRODUCTION

Rice is the most important food crop for half of the world population. The global grain provides 35–80% of total calorie uptake to more than 2.7 billion people. It is the most important staple food crop of India, grown over an area of 42.2 million ha with a production of 84.74 million t. A number of factors including: inadequate availability of water supplies; along with increased labour costs; labour shortages at the peak of planting activities; and declining profitability of rice production, have encouraged many rice farmers in south east Asian countries, including India, to shift from transplanting to direct-seeding in irrigated areas. Relatively inexpensive and cost-efficient herbicides and early maturing modern varieties have helped this change. Direct-seeding rice production systems are subject to greater weed pressure than conventionally transplanted systems. This is because weeds are suppressed by flooding and transplanted rice seedlings have a 'head start' over germinating weed seedlings (Moody 1983). Direct seeding is practised in two forms: wet-seeding; and dry-seeding. In wet-seeded rice, pre-germinated seeds are broadcast onto the puddled soil. In dry-seeding rice, non-pre-germinated seeds are sown onto dry-ploughed, unpuddled soil that could be dry or moist.

The most important problem in direct seeded rice is weed competition. Failure to manage the weeds could result in 50–90% rice yield loss (Chauhan and Johnson 2011). Considering this, we conducted a field experiment to study the effect of interventional weed management methods in direct seeded rice under unpuddled situation at Karaikal, Puducherry Union Territory, India.

MATERIALS AND METHODS

A field experiment was conducted at Pandit Jawaharlal Nehru College of Agriculture and Research Institute, Karaikal, Puducherry (11°56′ N latitude, 79°53′ E longitude, altitude 8 m), Puducherry Union Territory between September 2016 – January 2017. The soil was sandy clay loam in texture, at pH: 6.94, low in available nitrogen (119 kg ha^{-1}) and high in available phosphorus (24 kg ha^{-1}) and potassium (366 kg ha^{-1}). The experiment was laid out in randomised block design with ten treatments. The treatments were: mechanical weeding using a rotary weeder at 20, 40 and 60 days after sowing (DAS) (T_1); 3 manual (hand) weeding events occurring at 20, 40 and 60 days after sowing (DAS) (T_2); post-emergence application of bispyribac sodium 25 g ai ha^{-1} (T_3); or along with chlorimuron + metsulfuron 4 g ai ha^{-1} (T_4); pre-emergence application of oxadiargyl 100 g ai ha^{-1} followed by (*fb*) post-emergence application of bispyribac sodium 25 g ai ha^{-1} (T_5); or pyrazosulfuron ethyl 20 g ai ha^{-1} *fb* post-emergence application of bispyribac sodium 25 g ai ha^{-1}(T_6); pre-emergence application of pendimethalin 1000 g ai ha^{-1} *fb* post-emergence application of bispyribac sodium 25 g ai ha^{-1} (T_7); pre-emergence application of pendimethalin 1000 g ai ha^{-1} *fb* one manual (hand) weeding at 40 DAS (T_8);

pre-emergence application of pendimethalin 1000 g ai ha^{-1} *fb* one manual (hand) weeding at 40 DAS (T_9); and an unweeded control (T_{10}).

The relative density of jungle rice for each treatment was recorded with a quadrat (0.25 m × 0.25 m) placed randomly at four spots in each plot at 80 DAS, and computed using the procedure suggested by Raju and Reddy (1998). Weed control efficiency (WCE, %) was calculated using the formula WCE = (weed biomass in non-treated plot-weed biomass in treated plot/ weed biomass in non-treated plot) × 100. The rice cultivar ADT 46 was sown with the spacing of 25 cm × 10 cm, and all the recommended package of practices except weed control were adopted during the period of experimentation. The data on rice grain yield were subjected to statistical scrutiny as per the lsd procedure in Panse and Sukhatme (1967). The Benefit:Cost (B:C) ratio was calculated using the existing market rates.

RESULTS

The major weed flora observed in the experimental field were jungle rice (*Echinochloa colona* (L.) Link), Chinese sprangletop (*Leptochloa chinensis* (L.) Nees), eclipta (*Eclipta prostrata* (L.) L.), small flowered nutsedge (*Cyperus difformis* (L.)), rice flatsedge (*Cyperus iria* L.) and lesser fimbristylis (*Fimbristylis littoralis* Gaud). Analyses of the relative weed densities revealed that jungle rice was the major grassy weed infesting the experimental field (Table 1). Sequential pre- and post-emergence application of pendimethalin and bispyribac sodium herbicides, respectively, integrated with hand weeding, resulted in lower jungle rice weed density (46.4%) and higher weed control efficiency (94.1%). However, a single application of post-emergence application of bispyribac sodium alone resulted in poor weed control efficiency (51.5%).

A higher rice yield and better B:C ratio was observed with sequential applications of pre- and post-emergence application of pendimethalin and bispyribac sodium herbicides alone when integrated with manual (hand) weeding (3.64–3.84 t ha^{-1} and 1.65–1.72, respectively).

DISCUSSION

Sequential applications of pre- and post-emergence application of pendimethalin and bispyribac sodium herbicides integrated with manual weeding effectively reduced jungle rice density by 49.2% when compared to the unweeded control. Das and Duary (1998) indicated that pendimethalin, which belongs to dinitroanaline group, prevents germination and arrests root and shoot development of susceptible weeds by acting as mitotic poison (microtubule assembly inhibitor). In contrast, large numbers of jungle rice plants were recorded with mechanical weeding using rotary weeder. This was due to the escape of the target weed in the intra-row area and the availability of light for the germination of weed seeds. Benvenuti and Macchia (1995) indicated that light plays an important role in the germination of some weed species to overcome the dormancy. Chauhan (2012) observed that light stimulates the germination of jungle rice.

Table 1. The effect of weed management interventions on weed, crop and economics in direct seeded rice. The text '*fb*' denotes 'followed by'.

Treatment	Jungle rice relative density (%)	Weed control efficiency (%)	Rice yield (t ha^{-1})	B:C ratio
Mechanical weeding with rotary weeder thrice	87.8	48.6	1.84	0.50
Manual weeding alone thrice	69.2	86.5	3.52	1.24
Bispyribac-sodium	60.5	51.5	2.60	1.20
Bispyribac + (chlorimuron + metsulfuron)	83.3	66.7	2.80	1.27
Oxadiargyl *fb* bispyribac	91.8	71.5	3.20	1.47
Pyrazosulfuron *fb* bispyribac	91.3	80.7	3.04	1.49
Pendimethalin *fb* bispyribac	78.1	77.0	2.86	1.17
Pendimethalin *fb* bispyribac *fb* manual weeding	46.4	94.1	3.84	1.65
Pendimethalin *fb* manual weeding	68.5	88.2	3.64	1.72
Unweeded control	91.3	–	1.12	0.05
LSD ($P<0.05$)	–	–	0.78	v

Uncontrolled weeds resulted in a 71% yield loss compared to the sequential application of pre- and post-emergence applications of pendimethalin and bispyribac sodium integrated with hand weeding. Economically, the application of pre-emergence pendimethalin integrated with a manual weeding at 40 DAS was superior to other weed management interventions. Therefore, results of the present study clearly indicate the supremacy of interventional methods in tackling weed problem in direct seeded rice.

ACKNOWLEDGMENTS

All India Coordinated Research Project on Weed Management (AICRP-WM), Directorate of Weed Research, Jabalpur, India provided financial support and Department of Agronomy, PAJANCOA&RI, Karaikal, India provided institutional support to carry out this project.

REFERENCES

Benvenuti, S. and Macchia, M. (1995). Effect of hypoxia on buried weed seed germination. *Weed Research* 35, 343-51.

Chauhan, B.S. (2012). Weed ecology and weed management strategies for dry seeded rice in Asia. *Weed Technology* 26, 1-13.

Chauhan, B.S. and Johnson, D.E. (2011). Growth response of direct seeded rice to oxadiazon and bispyribac sodium in aerobic and saturated soils. *Weed Science* 59, 119-22.

Das, T.K. and Duary, B. (1998). Pendimethalin. An excellent herbicide for weed control in crops. *Intensive Agriculture* 36, 18-20.

Moody, K. (1983). The status of weed control in Asia. *FAO Plant Protection Bulletin* 30, 119-20.

Panse, V.G. and Sukhatme, P.U. (1967). Statistical methods for agricultural workers. (ICAR, New Delhi).

Raju, R.A. and Reddy, M.N. (1998). Phytosociological studies of rainy season weeds with special reference to *Imperata cylinderica* (L.) Raeuhl in Godavari Delta. *Indian Journal of Weed Science*, 30, 182-8.

Bionomics and effectiveness of Aak fruit fly: A potential biocontrol candidate for calotrope in Australia

Asad Shabbir[1,2], Shahid Ali[2] and Kunjithapatham Dhileepan[3]
[1] Sydney Institute of Agriculture, School of Life and Environmental Sciences, University of Sydney, Narrabri, New South Wales 2390, Australia
[2] Ecology and Environment Management Research Laboratory, Department of Botany, University of Punjab, Quaid-e-Azam Campus, Lahore 54590, Pakistan
[3] Biosecurity Queensland, Department of Agriculture and Fisheries, Ecosciences Precinct, Boggo Road, Dutton Park, Queensland 4102, Australia
(asad.shabbir@sydney.edu.au)

Summary Calotrope, *Calotropis procera* (Aiton) W.T.Aiton (Apocynaceae) is a serious weed of rangelands in northern Australia. Calotrope is a target weed approved for biological control in Australia. The native range of this weed is south Asia, the Middle East and North Africa, where several natural enemies of the weed have been reported. Among them, pre-dispersal seed feeding Aak fruit fly *Dacus persicus* Hendel (Diptera: Tephritidae) has been identified as a prospective biological control agent for calotrope in Australia. Pre-dispersal seed predators are regarded as effective biological control agents that could potentially reduce the future recruitment and spread of the weed. Life history and damage potential of the Aak fruit fly was studied under both laboratory and field conditions in its native range in Lahore, Pakistan. The duration of life cycle of *D. persicus* (egg to adult) was 42.2 ± 0.38 days (mean ± se) with a range of 38 to 50 days. The females start laying eggs 5.2 ± 0.2 days after emergence and a typical female infests on average 4.5 ± 0.2 fruits during its life-span. The average life span of an adult fly was 16.4 ± 0.68 days. The Aak fruit fly larvae destroyed (100%) all immature seeds and internal tissue of infested pods of the host plant. The field host specificity, high reproductive capacity and damage potential of the fruit fly indicate that this agent hold promise to be considered as a potential candidate agent for biological control of *C. procera* in Australia.

Keywords *Dacus persicus*, native range, effectiveness, classical biological control, rubber bush, co-evolved natural enemy, Pakistan.

Managing barnyard grass through crop competition: The role of row spacing and crop density

Asad Shabbir and Michael J. Walsh
Sydney Institute of Agriculture, School of Life and Environmental Sciences, University of Sydney, Narrabri, New South Wales 2390, Australia
(asad.shabbir@sydney.edu.au)

Summary Barnyard grass (*Echinochloa* spp.) are the most problematic weeds in summer cropping systems in Australia. Increased reliance on herbicides for the fallow management of barnyard grass has resulted in the evolution of glyphosate-resistant populations which are now widespread across the northern cropping region. While herbicides continue to be relied on, non-chemical control strategies are needed to complement these herbicide treatments and ensure the effective management of barnyard grass. Field trials at Narrabri during summer 2017–2018, quantified the effect of mung bean and sorghum plant densities and row spacings on the biomass and seed production of barnyard grass. Once crop plant seedlings had emerged, young barnyard grass seedlings were transplanted at a density of 5 plants m^{-2} into designated areas within each plot. An equivalent sized plot area was maintained weed-free. The biomass of barnyard grass was reduced by up to 88% with narrow row spacing in both sorghum (50 cm) and mung bean (25 cm) crops. The better weed suppression observed in narrow row spacings was a result of the early crop canopy closure and shading of barnyard grass plants. When planted at narrow row spacing (25 cm), the increased density (30 plants m^{-2}) of mung bean resulted in the maximum biomass reduction (24%) of barnyard grass. This study has highlighted that crop competition through reducing row spacing and altering the crop density is an effective control strategy to control barnyard grass that should be considered in integrated weed management in summer crops.

Keywords *Echinochloa colona*, *Echinochloa crus-galli*, crop competition, sowing rate, cultural weed control, mung beans, sorghum.

Successful biocontrol of *Echium* spp.

Paul R. Sullivan
Weed Research Unit, NSW Department of Primary Industries, Tamworth Agricultural Institute,
4 Marsden Park Road, Calala, New South Wales 2340, Australia

Summary The exotic species *Echium plantagineum* L. (Paterson's curse) and *Echium vulgare* L. (viper's bugloss) were significant noxious weeds of pastures and natural habitats in southern Australia. The more prolific *E. plantagineum* infests 33 million hectares of pastures and, in 2002, was estimated to cost the wool and meat industries A$125 million pa. The *Echium* biocontrol program started in 1972. Between 1988 and 1996, seven biocontrol agents were released across southern Australia. This study examined the impact of three biocontrol agents on *E. plantagineum* and *E. vulgare*: *Mogulones larvatus*, *M. geographicus* and *Dialectica scalariella*. A study site was established near Lithgow (NSW) with the three agents being released in 1995/6. All three agents established. Over a nine year period (1995–2003), impact on flowering by all agents and plant damage by *D. scalariella* and *M. larvatus* were measured monthly on both *Echium* spp. Seed bank data was monitored annually (1997–2003). Increasing populations corresponded to increased plant attack. The two *Mogulones* spp. killed up to 84% of *E. plantagineum* plants in one year and feeding by *D. scalariella* caused 4% to 60% leaf damage. Moreover, by 2003, all three agents contributed to a decrease in plants that flowered by 31% and a decrease in the *Echium* seed bank by 85%. This study shows that all three agents contribute to biocontrol of *Echium* weeds. By 2010, *Echium* weeds across Australia had been successfully controlled by a combination of these three agents, two new agents *Longitarsus echii* and *Meligethes planiusculus* and a more drought tolerant biotype of *M. larvatus*.

Keywords *Echium plantagineum*, *Echium vulgare*, Paterson's curse, viper's bugloss, *Mogulones larvatus*, crown weevil, *Mogulones geographicus*, root weevil, *Dialectica scalariella*, biological control, biocontrol.

Influence of resource conservation techniques on weed control in green manure maize-pulse cropping system

R. Thirumalaikumar, N.S. Venkataraman, K. Balakrishnan, R. Babu and A. Rathinasamy
Agricultural College and Research Institute, Tamil Nadu Agricultural University, Madurai- 625 104, Tamil Nadu, India
(thiruhid@gmail.com)

Summary Field experiments were conducted at Agricultural College and Research Institute, Madurai during *Kharif* 2016, *Rabi* 2016–2017 and summer 2017. The experiments were laid out in split-split plot design with three replications. The treatments viz., tillage (T), weed (W) and nutrient (N) management practices were assigned as main, sub and sub-sub plots respectively. The experimental results revealed that minimum tillage (T_2) with PE herbicide (Atrazine) followed by one hand weeding (W_2) + 75% NPK + Vermicompost + Bio-fertiliser (N_2) recorded lesser weed density, higher weed control efficiency and enhanced productivity than the conventional practices in maize. Similarly in black gram, zero tillage (T_2) with crop residue mulch (W_1) + 75% NPK + Vermicompost + Bio-fertiliser (N_2) effectively suppressed the weed growth and achieved the higher weed control efficiency and maximise the economic return when compared to farmers practice. From the present investigation, it can be concluded that, conservation tillage with PE herbicide in maize followed by crop residue mulch in black gram, with 75% NPK + Vermicompost + Bio-fertiliser provides an appropriate management option for the farmers to control weeds effectively for increased crop growth and higher system productivity besides improving soil health.

Keywords Minimum tillage, zero tillage, crop residue mulch, bio-fertiliser, maize, black gram.

Comparison of site-specific, whole field and seedbank physical weed control treatments

Michael J. Walsh[1], Salah Sukkarieh[2] and Guy R. Coleman[1]
[1] Sydney Institute of Agriculture, School of Life and Environmental Sciences, University of Sydney, Narrabri, New South Wales 2390, Australia
[2]Australian Centre for Field Robotics, University of Sydney, Camperdown, New South Wales 2006, Australia
(m.j.walsh@sydney.edu.au)

Summary Herbicides are the dominant weed control method in grain production systems due to their high efficacy, reliability and low cost. Yet their ongoing use is threatened by herbicide-resistance, fewer herbicide discovery programs, and increasing regulation. Thus, there is a need and considerable interest in the use of physical methods for weed control. There are a wide range of highly effective physical weed control approaches that can potentially be used in cropping systems. The many physical weed control technologies developed over the last 50 years have been reviewed and compared according to their energy requirements for effective weed control. Physical weed control methods have been divided into four groups, broadcast (whole paddock), intra-row, site-specific and below ground (seed) control, in accordance with application method. In general, site-specific weed control approaches (0.005–11.5 MJ ha^{-1}) were orders of magnitude more energy efficient than broadcast thermal (1000–75,000 MJ ha^{-1}), broadcast tillage (2.5–75 MJ ha^{-1}) and below ground treatments (24–280 GJ ha^{-1}). The most energy effcient site-specific approaches included forms of electrocution, laser pyrolysis and hoeing. For broadcast weed control, cultivation techniques are the most energy efficient, over broadcast thermal systems. However, for no-till farming operations, steaming, infrared, and inter-row flame weeding were found to be the most efficient. Seedbank targeting, while very energy expensive, creates an opportunity for maximum weed control with microwave treatment on smoothed soil the most suitable technique for this approach.

Keywords Site-specific weed control, physical weed control, energy requirements.

Environmental weed advisory list update, Victoria

Matt White[1], Kate Blood[2] and Bec James[3]
[1] Department of Environment, Land, Water and Planning, Arthur Rylah Institute, 123 Brown Street, Heidelberg, Victoria 3084, Australia
[2] Department of Environment, Land, Water and Planning, PO Box 7, Beaufort, Victoria 3373, Australia
[3] Department of Environment, Land, Water and Planning, Box 3100, Bendigo Delivery Centre, Victoria 3554, Australia
(Kate.Blood@delwp.vic.gov.au)

Summary Invasive plants present a serious and ongoing threat the Victoria's biodiversity. There is an emerging need to usefully discriminate between the temporal risk posed by individual species, such that, in combination with other asset based data, resources allocated to pest management can be deployed to best advantage. In 2008–2009 five advisory lists of environmental weeds were published providing risk scores for about 680 species across Victoria. They have been very useful tools for land managers to help prioritise weed management activities. The Weeds at the Early Stage of Invasion (WESI) project commissioned an update of the lists in 2017. The update has now incorporated all known naturalised Victorian plant species including potential and early invader weeds, extending the list to 1780 plant taxa, of which, 1227 are currently considered to be environmental weeds. All environmental weeds have been ranked with respect to management urgency, by means of an objective 'expert system'. The advisory list is intended to provide 'non-statutory' advice to conservation managers.

Keywords Environmental weed, risk rating, early invaders, advisory list.

Physiological and biochemical resistance cost of multiple herbicide resistance in *Echinochloa crus-galli*

Xia Yang, Yongfeng Li, Zichang Zhang and Tao Gu
Institute of Plant Protection, Jiangsu Academy of Agricultural Sciences, Nanjing 210014, China
(xiayang_njau@hotmail.com)

Summary Barnyard grass is one of the major malignant weeds infesting rice fields in China and elsewhere in the world. The expansion of herbicide-resistant barnyard grass is increasingly threatening rice production systems. Here, a multiple-resistant *Echinochloa crus-galli* population to quinclorac, penoxsulam and bispyribac-sodium were determined by the whole-plant bioassay. To explore the difference in fitness adaption between the multi-herbicide resistant (R) and susceptible (S) biotypes, a comparative proteomics analysis using iTRAQ were performed. Ten down-accumulated proteins including PSI and PSII, cytochrome b6 and F-type H^+-transporting ATPase, and seven light-harvesting complex I chlorophyll *a/b* binding proteins were identified in the R versus S biotypes. In addition, various xenobiotic metabolising enzymes which play crucial roles in abiotic stress response (such as catalase, Cu/Zn-SOD, aldehyde dehydrogenase, P450, and NADPH-cytochrome reductase) were also down-accumulated in the R biotype. Meanwhile, photosynthetic parameters and physiological indexes of the six-leaf stage seedlings were measured in the R and S biotypes. The R biotypes showed a reduced Chl *a* and carotenoid content by 4.72% and 17.06%, respectively, and a reduced specific leaf area by 47.10% when the first fully expanded leaf was measured. The net photosynthetic rate shows a slight decrease by 18.41% in the R biotypes (11.70 ± 0.57 μmol CO_2 $m^{-2}s^{-1}$) relative to that of the S biotypes (14.34 ± 0.53 μmol CO_2 $m^{-2}s^{-1}$). Therefore, proteomic and physiological analysis demonstrate that the physiological and biochemical resistance costs are associated with multiple herbicide resistance in barnyard grass.

Keywords *Echinochloa crus-galli*, herbicide resistance, proteomics, photosynthesis, resistance cost.

Efficacy of pyrazosulfuron-ethyl to manage weeds of rainy season rice under non-puddled field condition and its effect on the subsequent wheat crop

Taslima Zahan[1], Abul Hashem[2] and Md. Moshiur Rahman[3]
[1] Bangladesh Agricultural Research Institute, Gazipur-1701, Bangladesh
[2] Department of Primary Industries and Regional Development, PO Box 483, Northam, Western Australia 6401, Australia
[3] Bangladesh Agricultural University, Mymensingh-2200, Bangladesh
(taslimazahan_tzp@yahoo.com)

Summary Non-puddled transplanting is a promising establishment technique for rainy season rice. A study was carried out at Mymensingh, Bangladesh in transplanted rainy season rice to evaluate weed control efficiency of pyrazosulfuron-ethyl alone or in sequential application with post-emergence herbicide under non-puddled strip-tilled field condition. Residual effects of those herbicides were also examined on germination and growth of the subsequently grown crop in rice-wheat cropping pattern. Seven treatment combinations comprised of four herbicides, 1 pre-emergent (PEM): pyrazosulfuron-ethyl, 1 early post-emergent (EPOST): orthosulfamuron, and 2 late post-emergent (LPOST): butachlor plus propanil and 2,4-D amine, along with one weedy check and one weed-free check were tested. Grasses and sedges were the major weeds in non-puddled strip-tilled rice with *Cynodon dactylon* (L.) Pers. the most dominant species. Herbicide treatments provided a wide range of control (above 50–95%) on all types of weeds. Sequential application of PEM with EPOST or LPOST or both offered higher weed control than sole application of PEM by 27–65%. Pyrazosulfuron-ethyl (PEM) followed by (fb) orthosulfamuron (EPOST) fb butachlor plus propanil (LPOST) gave the most effective weed control as well as maximum rice grain yield. Moreover, bioassay study showed that tested herbicides had no adverse effect on germination and growth of the next season wheat crop. Therefore, the study suggests sequential application of pyrazosulfuron-ethyl with post-emergence herbicide(s) in any of the tested combinations to attain the optimum rice yield, however rotational use of herbicides along with maintaining appropriate rates and timing needs to be strictly followed to obtain maximum benefit.

Keywords Non-puddled rice, herbicide, herbicide persistence, weed control.

INTRODUCTION

Rice-wheat is the main cropping system of Bangladesh (Sarker *et al.* 2014) and traditionally puddling is done to transplant rainy season rice. But, the strip-tilled non-puddled transplanting technique is gaining popularity over the conventional puddled system because of less cultivation cost involvement (Islam *et al.* 2014), conservation of soil health (Lindwall and Sonntag 2010) and production of more yield than the conventional system (Haque *et al.* 2016).

Weed management always plays a vital role to achieving rice yield. But high labour costs are causing farmers to shift from manual weeding to chemical weeding. Pyrazosulfuron-ethyl is a very common and effective pre-emergence herbicide, usually used to control grass and broadleaf weeds of puddled transplanted rice (Naveen *et al.* 2012). However, the efficacy of this selective herbicide in the diversified weed community of strip-tilled non-puddled transplanted rainy season rice is not well known. Moreover, the persistence of this herbicide in the soil and the residual effect of the herbicide on the subsequent crop is a matter of great concern. Therefore, the study was done to evaluate performance of pyrazosulfuron-ethyl alone, or in a sequential application with post-emergence herbicide(s) to obtain most effective weed control in non-puddled transplanted rainy season rice, and to examine the residual effect of herbicides on the subsequent wheat crop.

MATERIALS AND METHODS

The study was initiated in a sandy clay loam soil (pH 7.2) at Agronomy Field Laboratory, Bangladesh Agricultural University, Mymensingh from July 2014 to February 2015. The monthly average of maximum air temperature (T_{max}) ranged from 23.9–33.5°C and average minimum temperature (T_{min}) was 13.5–26.7°C. The highest T_{max} was in April and the lowest T_{min} was in January. The experimental site received 1951.4 mm of rainfall during the study period. The highest rainfall was recorded in August and no rain occurred in November and December.

Main experiment Efficacy of pyrazosulfuron-ethyl was tested using seven treatments:

- T_1: Pre-emergent (PEM) Pyrazosulfuron-ethyl (15 g ai ha^{-1}).
- T_2: PEM Pyrazosulfuron-ethyl (15 g ai ha^{-1}) followed by (fb) late post-emergent (LPOST) butachlor plus propanil (0.7 kg ai ha^{-1}).
- T_3: PEM Pyrazosulfuron-ethyl (15 g ai ha^{-1}) fb LPOST 2,4-D amine (1.6 kg ai ha^{-1}).
- T_4: PEM Pyrazosulfuron-ethyl (15 g ai ha^{-1}) fb early post-emergent (EPOST) orthosulfamuron (75 g ai ha^{-1}) fb LPOST butachlor plus propanil (0.7 kg ai ha^{-1}).
- T_5: PEM Pyrazosulfuron-ethyl (15 g ai ha^{-1}) fb EPOST orthosulfamuron (75 g ai ha^{-1}) fb LPOST 2,4-D amine (1.6 kg ai ha^{-1}).
- C_1: Weed-free check (4 × manual weeding).
- C_2: Weedy check.

The study was laid out in a randomised complete block design with three replications. Pre-planting Roundup® (glyphosate) was applied at 1.54 kg ai ha^{-1} in the experimental site prior to transplanting of rice seedlings. After one week of pre-planting spray, the land was fertilised and strip-tilled. The land was then inundated for 48 hours and after two days, rice seedlings (cultivar BINA dhan-7) were transplanted within the strips maintain 20 cm line to line distance and 15 cm hill to hill distance. Herbicides were applied by hand operated knapsack sprayer fitted with flat-fan nozzle at a spray volume of 300 L ha^{-1}. PEM, EPOST and LPOST herbicides were applied at 3 days after transplanting (DAT), at 15 DAT and at 22 DAT, respectively. Weed related data were recorded from randomly selected three quadrats (50 cm × 50 cm) in each plot. Grain and straw yields were recorded from central area of the plots at crop maturity.

Residual effect study This study had received the same treatments and design of the main rice experiment by following a micro-plot bio-assay technique. Micro-plots of 1 m^2 were established within each plot with 100 wheat seeds. Data on emergence, shoot length and dry matter of wheat were recorded up to 25 days after sowing.

Recorded data were analysed by using statistical package program 'R' (version 3.3).

RESULTS

The rice field was infested by seven weed species namely *Cynodon dactylon* (L.) Pers., *Echinochloa colona* (L.) Link, *Cyperus rotundus* L., *Cyperus difformis* L., *Fimbristylis miliaceae* (L.) Vahl, *Ludwigia decurrens* Walter and *Cynotis auxillaris* (L.) Sweet.

The highest densities of all the species were found in the weedy plots and the herbicide treatments offered a significant reduction in the densities of all species by 40–100% during 35 days after transplanting (Figure 1). Except for *C. dactylon*, the densities of all weed species were lower in those plots that received pyrazosulfuron-ethyl and one early post- or both early and late post-emergence herbicides in a sequential application in comparison to the sole PEM pyrazosulfuron-ethyl. Application of PEM pyrazosulfuron-ethyl followed by EPOST orthosulfamuron fb LPOST butachlor plus propanil gave complete reduction of all the species except *C. dactylon*; however this grass weed was reduced by 98% in terms of weed density and by 95% in terms of weed biomass compared to the weedy check (Table 1). On the other hand, the application of PEM followed by LPOST butachlor plus propanil or PEM followed by EPOST orthosulfamuron fb LPOST 2,4-D amine also offered 85–90% and 82–86% reductions of total weed density and biomass, respectively. Among the herbicide treatments, pyrazosulfuron-ethyl alone gave the lowest reduction in total weed density and biomass.

Rice grain yields also varied with herbicide treatments (Table 1). The lowest yield was recorded from the weedy check and the highest yield was obtained from PEM pyrazosulfuron-ethyl followed by EPOST orthosulfamuron fb LPOST butachlor plus propanil

Table 1. Effect of herbicide treatments on total weed density (TWD, plants m^{-2}) and biomass (TWB, g m^{-2}) at 35 days after transplanting and grain yield (GY, t ha^{-1}) of non-puddled rainy season rice during 2014.

Treatments	TWD	TWB	GY
T_1	36.7 b	10.11 b	3.58 d
T_2	10.7 c	4.19 d	5.30 b
T_3	33.3 b	6.40 c	4.68 c
T_4	2.3 d	1.24 e	6.18 a
T_5	15.7 c	3.33 d	6.01 a
C_1	105.3 a	23.82 a	2.69 e
C_2	–	–	5.99 a
Level of significance	***	***	***
CV (%)	8.68	10.28	3.18

In a column, figure(s) with different letters are significantly different at 5% level of significance (as per Tukey's Honest Significant Difference Test).
*** = 0.1% level of significance.

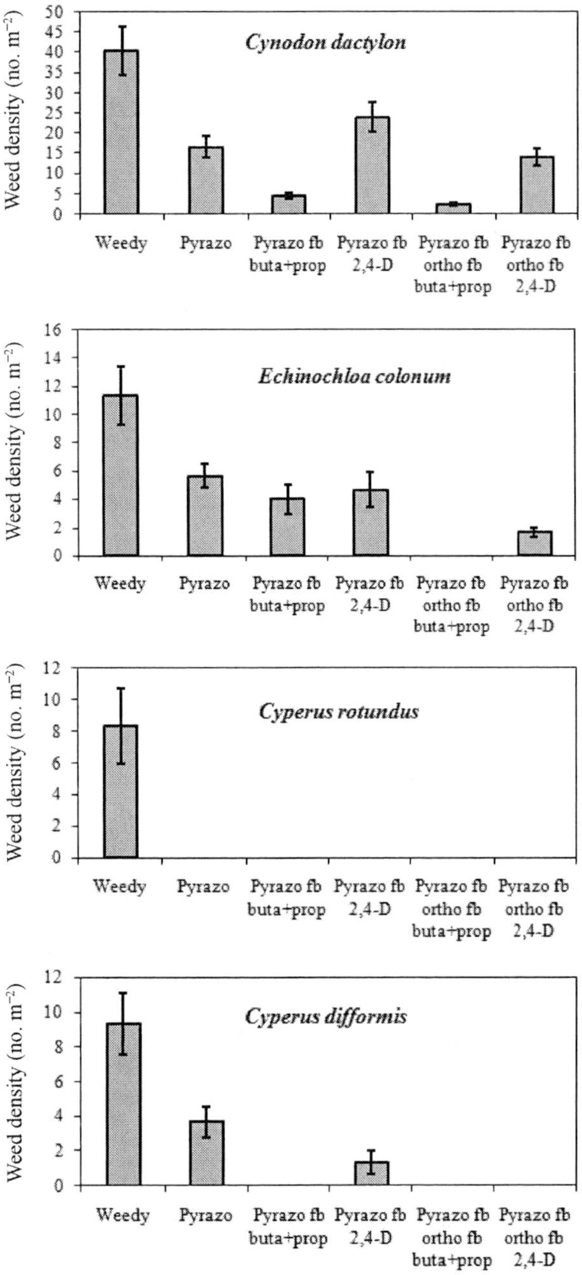

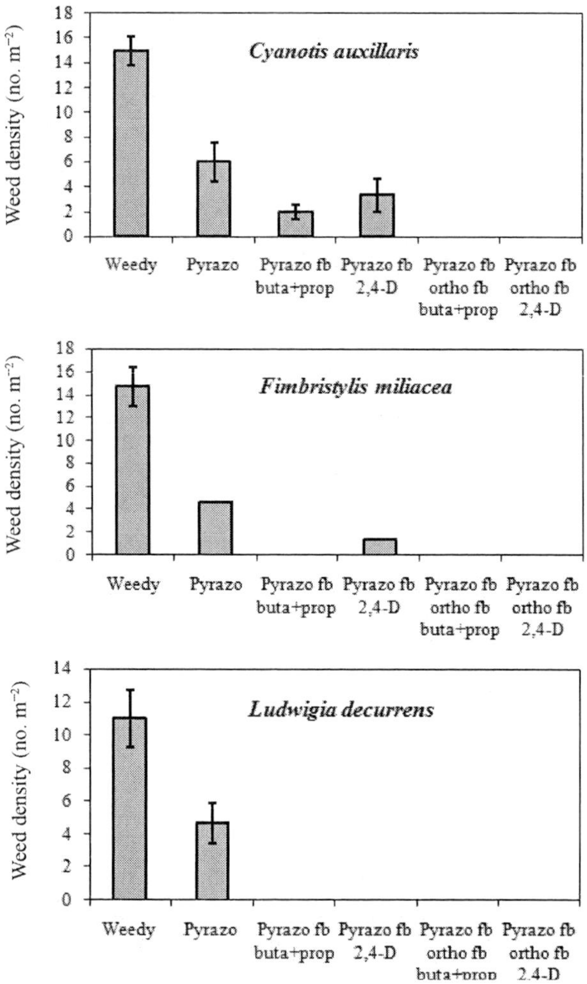

Figure 1. Effect of herbicide treatments on densities of different weed species at 35 days after transplanting of rainy season rice in strip-tilled non-puddled field, Mymensingh, Bangladesh during 2014.

treatment that had a rice yield more than double that of the weed-free check.

Results of the residual effect study showed that emergence, shoot length and biomass of the succeeding wheat were not significantly affected by the herbicides applied in the preceding rice crop (Table 2).

DISCUSSION

The strip-tilled non-puddled rice field was infested by seven weed species, and, except for *Cynodon dactylon*, all species were successfully controlled by herbicide treatments. This might be related to the regenerative ability of *C. dactylon* from its stolons. However, the pyrazosulfuron-ethyl followed by butachlor plus

Table 2. Residual effect of herbicide treatments on emergence (EMER, %) and shoot length (SL, cm) and biomass (WDM, g m^{-2}) of wheat at 25 days after sowing during 2014.

Treatments	EMER	SL	WDM
T_1	87.7	26.0	12.20
T_2	89.0	28.7	10.97
T_3	84.0	24.3	9.33
T_4	90.7	27.4	10.33
T_5	84.3	25.4	11.83
C_1	86.0	24.8	8.84
C_2	87.7	26.6	10.80
Level of significance	ns	ns	ns
CV (%)	4.47	9.79	15.7

ns = non-significant.

propanil was effective on *C. dactylon* both with and without addition of orthosulfamuron into the sequence.

The weed control of PEM followed by LPOST 2,4-D amine was similar to the sole application of PEM pyrazosulfuron-ethyl and the performance of both these treatments was poorer than any other sequential herbicide treatments. Only PEM is not sufficient to provide control of all types of weeds for the whole crop growing period. Application of PEM and POST will control all types of weeds effectively for a long period and therefore proper crop growth can be ensured. But, an application of 2,4-D amine as LPOST with PEM did not offer any additional benefit in weed control because 2,4-D amine had no effect on grass weeds and the experimental field was mostly dominated by *C. dactylon* (Figure 1).

Treatments with PEM pyrazosulfuron-ethyl followed by EPOST orthosulfamuron followed by LPOST butachlor plus propanil or LPOST 2,4-D amine provided similar rice grain yield as the weed-free check. Moreover, the emergence of the succeeding wheat crop was not significantly affected by any of the herbicide sequences applied in the rice. Shoot length and biomass of wheat were also not adversely affected by the residue of rice herbicides. Vaghasia and Nadiyadhara (2013) reported similar findings with other herbicides: that emergence of wheat was not affected by imazethapyr, imazamox, quizalofop-p-ethyl and fenoxaprop-p-ethyl applied in the preceding groundnut crop.

The persistence of herbicide in the soil is highly dependent on several soil and climatic factors and also on herbicidal properties. However, the half-lives of the tested herbicides are not more than three months and the half-life of pyrazosulfuron-ethyl ranges between 16–21 days if applied at 25–50 g ai ha^{-1} (Naveen *et al.* 2012). Curran (2001) indentified some factors that largely influence herbicide persistence in soil such as rainfall, organic matter content and pH of soil. As such, the residual effect of herbicides could vary from soil to soil and location to location.

ACKNOWLEDGMENTS

The authors would like to acknowledge Australian Centre for International Agricultural Research for providing funding.

REFERENCES

Curran, W.S. (2001). Persistence of herbicides in soil. Agronomy Facts 36, Penn State Extension, Ag Communications and Marketing, Pennsylvania, United States of America. pp. 1-4.

Haque, M.E., Bell, R.W., Islam, M.A. and Rahman, M.A. (2016). Minimum tillage unpuddled transplanting: An alternative crop establishment strategy for rice in conservation agriculture cropping systems. *Field Crops Research* 185, 31-9.

Islam, A.K.M.S., Hossain, M.M. and Saleque, M.A. (2014). Effect of unpuddled transplanting on the growth and yield of dry season rice (*Oryza sativa* L.) in high Barind Tract. *The Agriculturists* 12, 91-7.

Lindwall, C.W. and Sonntag, B. (2010). 'Landscape transformed: The history of conservation tillage and direct seeding.' University of Saskatchewan, Canada.

Naveen, D.V., Gowda, R.C. and Mamatha, B. (2012). Field studies on persistence of pyrazosulfuron-ethyl in soil, ground water and residues in transplanted rice. *An Asian Journal of Soil Science* 7, 8-12.

Sarker, M.A.Z., Alam, M.A., Hossain, A. and Mannaf, M.A. (2014). Agro-economic performance of crop diversification in rice based cropping systems of northwest Bangladesh. *Agriculture, Forestry and Fisheries* 3, 264-70.

Vaghasia, P.M. and Nadiyadhara, M.V. (2013). Effect of post-emergence herbicides in groundnut and its residual effect on succeeding crops. *International Journal of Forestry and Crop Improvement* 4, 54-8.

Alterations in the 5 'untranslated region of the *EPSPS* gene influence *EPSPS* over-expression in glyphosate-resistant *Eleusine indica*

Chun Zhang, Li Feng and Xing-shan Tian
Guangdong Provincial Key Laboratory of High Technology for Plant Protection, Institute of Plant Protection, Guangdong Academy of Agricultural Sciences, Guangzhou 510640, P.R. China
(zhangchun_0726@163.com)

Summary The herbicide glyphosate inhibits the enzyme 5-enolpyruvylshikimate-3-phosphate synthase (EPSPS). Over-expression of the *EPSPS* gene is one of the molecular mechanisms conferring glyphosate resistance in weeds, but the transcriptional regulation of this gene is poorly understood. The *EPSPS* gene was found to be significantly up-regulated following glyphosate treatment in a glyphosate- resistant *Eleusine indica* population from South China. To further investigate the regulation of *EPSPS* over-expression, the promoter of the *EPSPS* gene from this *E. indica* population was cloned and analysed. Two upstream regulatory sequences, Epro-S (862 bp) and Epro-R (877 bp) of *EPSPS* were obtained respectively from glyphosate-susceptible and -resistant *E. indica* plants by HiTAIL-PCR. The Epro-S and Epro-R sequences were 99% homologous, except for the two insertions (3 bp and12 bp) in the Epro-R sequence. The 12-base insertion of the Epro-R sequence was located in the 5′-UTR-Py-rich stretch element. The promoter activity tests showed that the 12-base insertion resulted in significant enhancement of the Epro-R promoter activity, whereas the 3-base insertion had little effect on Epro-R promoter activity.

In conclusion, alterations in the 5′-UTR-Py-rich stretch element of *EPSPS* are responsible for glyphosate induced *EPSPS* over-expression. Therefore, *EPSPS* transcriptional regulation conferred glyphosate resistance in this *E. indica* population.

Keywords *Eleusine indica*, EPSPS, promoter, 5′-UTR-Py-rich stretch element, insertion.

Investigation of root exudation and phytotoxic potential of diverse *Sorghum* spp. through regulation of sorgoleone

Xiaocheng Zhu[1], Paul A. Weston[1], Ian D. Godwin[2], Sally Norton[3] and Leslie A. Weston[1]
[1] Graham Centre for Agricultural Innovation (Charles Sturt University and NSW Department of Primary Industries), Charles Sturt University, Wagga Wagga, New South Wales 2678, Australia
[2] School of Agriculture and Food Sciences, The University of Queensland, Brisbane, Queensland 4072, Australia
[3] Australian Grains Genebank, Agriculture Victoria, Horsham, Victoria 3400, Australia
(leweston@csu.edu.au)

Summary Sorgoleones are a group of long chain benzoquinones specifically exuded in large quantities by the root hairs of *Sorghum* spp. Sorgoleones inhibit electron transport of higher plants and can completely inhibit both photosynthesis and respiration in sensitive seedlings. However, research on sorgoleone has focused solely on African *Sorghum* spp., including *S. bicolor* and *S. halepense*. Australia, by contrast, has 17 native *Sorghum* species which have not yet been examined with respect to their photosynthetic system, root architecture or metabolism. Eighteen *Sorghum* species were assembled and evaluated for sorgoleone biosynthesis, with an emphasis on pathway flux and gene regulation. This collection represents the majority of species in the genus, including 1, 4 and 13 species that are native to Asia, Africa and Australia, respectively. On average, *Sorghum* species of African, Asian or Australian origin produced approximately 1.79%, 0.68% and 1.66% root exudate of the fresh weight of living roots, respectively. Metabolic profiling of extracts enriched in exuded sorgoleones was performed using methanol root dips of 10-day-old seedling roots using UPLC coupled to time of flight mass spectrometry (UPLC/MS ESI QToF). Sorgoleones were positively identified by comparison to known standards. Exudates differed qualitatively with Asian and African *Sorghum* species yielding over 50-fold greater abundance of sorgoleone compared to Australian *Sorghum*. At this time, flux analysis of the biosynthetic pathways present in African, Asian and Australian genotypes is underway with specific focus on genes regulating biosynthesis and root architecture. Such studies will contribute to crop improvement and potentially weed suppression.

Keywords Australian native *Sorghum*, sorgoleone, allelopathy.

Authors' index

Page numbers in bold represent primary author.

Acharya, Paras **176**
Adams, Vanessa 71
Adkins, Steve W. 47, 318, 334, 338
Adu-Yeboah, Patricia **328**
Ajamian, Chad 175
Albrecht, David E. 312
Ali, Shahid 405
Almeida, Cássio S. 388
Alqaderi, Husham A. **329**
Armel, G.R. **236**
Armin, Mohammad 362, 365, 398
Arruda, Mayara C. 388
Asad, Muhammad **333**
Asaduzzaman, Md **35**
Asaeda, Takashi 280
Awan, Tahir Hussain **300**

Babu, R. 408
Bai, Lianyang 26
Bailey, Kay 70
Baillie, Craig P. 125
Bajwa, Ali A. **318**, **334**, **338**
Balakrishnan, K. 408
Bangma, Brendon 64
Beames, Louise 261
Beautrais, Josef 77
Beckie, Hugh J. **1**
Bell, Kerry 133
Belson, Sue 147
Birchall, Craig 374
Blayney, Andrew 55
Blood, Kate 70, **197**, 372, 410
Boorman, Des 96
Borger, Catherine P.D. 153, **158**, **341**
Bourdôt, Graeme W. **82**
Boutsalis, Peter **10**, 12, 16, 384
Bower, Hank 187
Bower, Sue 187
Bowman, Troy A.J. 91, **267**
Brawata, Renee 181
Brodie, Chris **322**
Brodie, Graham I. 140
Brooks, Simon J. **119**, **345**, **350**, **354**
Broster, John 139, 144, **145**, 163, 295
Brown, Troy R.F. **359**
Brown, William B. 146, 295, 366, 379, 391
Brunton, David J. **12**
Buckmaster, Tony 86
Buddenhagen, Christopher E. 36
Burgess, Lester 64
Burgess, Treena I. 389
Burley, John 64
Burton, Tegan 68
Busch, Marco **132**
Busi, Roberto **23**
Byrne, Margaret 72

Cacho, Oscar **187**
Caldwell, Jo 126
Caldwell, Pete **103**
Callander, Jason 215
Cameron, John 169
Campbell, Doug **360**
Campbell, Shane 392
Campe, Ruth **238**
Cao, Jingjing 21
Chambers, Alison 144, 145, 163
Champion, Paul D. **108**
Chandrasena, Nimal 273, 367
Chang, Hsing-Chung **175**
Chauhan, Bhagirath S. 11, 20, 32, 47, 300, 301, 318, 334, 338, 378, 385, 390
Chen, Jinyi **24**
Chen, Yuchi **312**
Cherry, Hillary **96**, 175, 187, **327**
Chitolina, Gustavo M. 388
Chynoweth, Richard J. 36
Cirocco, Robert M. **225**
Cizdziel, James V. 42
Clark, Judy 211
Coleman, Guy R. 409
Coleman, Michael **49**
Conaty, Meredith 54
Congreve, Mark **169**
Conolly, Jenny 181
Constantine, Angela B. 109, 193
Cook, Tony 15, 133
Cripps, Michael G. **216**
Csurhes, Steve **115**
Cullen, Jim 223
Daniels, Joel **361**
Dastorani, Mohammad A. **362**
Dawson, P. 131
Day, Michael 226
Dhileepan, Kunjithapatham 215, 281, 405
Dodd, Michael B. 77, 82, 216
Dong, Mingchao 21
Dorjee 86
Downey, Paul O. **102**, 231
Dowsett, Claire A. **36**
Duke, Stephen O. 42

Easton, Rose K. 354
Elson, Phil 107
Evans, Jessica 4, 192

Facelli, José M. 225
Fagg, Jamie 230
Farebrother, Will 175
Feng, Li 416
Ferris, Centaine 211
Fleet, Benjamin 328
Fomsgaard, Inge 146
Fontaine, Juliette **363**
Francis, Ian 236, **237**, **243**, **244**, **364**
French, Kristine 230, **285**
Froese, Jens 392

Fry, Scott **271**
Fyfe, Christine 49

Galea, Victor J. 247
Gannaway, Dennis J. **104**
Ghavi, Ali R. **365**
Gibney, Wendy 60
Gill, Gurjeet 12, 16, 137, 148, 328
Gill, Nicholas **266**
Gill, Nicole 373
Glanznig, Andreas **235**
Gleeson, Patrick 389
Godwin, Ian D. 417
Goggin, Danica 24
Goode, Ben **230**
Gooden, B. 383
Gooding, Kristy 392
Gopurenko, David **302**, 307
Gough, Kirsty L. 354
Goulter, Ken C. **247**
Graham, Victoria 71
Gu, Tao 21, 411
Gupta, Dorin 140
Gurusinghe, Saliya **366**, 379

Haegi, Laurence A.R. 307
Hain, Sarah 64
Hamilton, Mark **126**, 175
Han, Heping 24, 25, 26
Hancock, Nola 71
Haque, Shamsul K.M. 391
Hardy, Giles E. St. J. 389
Harms, Nathan E. 204
Harper, Peter **273**, **367**
Harvey, Greg 138
Harvey, Kerinne 69, **203**
Hashem, Abul 133, **153**, 158, 341, 412
Hassan, Marwan El 4
Havell, David C. **101**
Healey, Madaline 64
Heard, Gavin 237, 244
Heim, Rene 175
Heinson, Michaela 91, 107
Hendriks, Janneke 238
Hereward, James P. 11, **20**, 378
Heuke, Linda M. 138, 163, **368**, 401
Hignell, Kim **251**
Hodges, David **48**
Höffken, Wolfgang 238
Hofstra, Deborah E. 108
Hollenbach, Eva 238
Hona, Shane R. **55**
Hopwood, Michael 290
Humphries, A.W. 43
Hunter, Gavin C. 33, 157, 222, **223**, 224, 363, **369**
Hutzler, Johannes 238

Imoloame, E.O. **370**
Inkson, Terry 202
Iqbal, Nadeem **47**
Ireland, Kylie B. 33, 64, 157, 222, 223, **224**, **371**
Ison, Russell 237, 243, 244, 364
Ivory, Susan 107

Jaca, Thulisile P. **201**
Jackman, Sarah D. 216
Jalaludin, Adam **139**, 163
James, Bec 197, **372**, 410
James, Trevor K. **27**, 36, 48, 55
Jamimoeini, Matin 398
Jeffery, Mick 119, 350
Jessop, Robin S. 374
Johansen, Chris 133
Johnson, Stephen B. **86**, 96, 102, 361
Jones, Peter 131, 226
Jourdan, Mireille 157, 220, 221, 223, 369, 383
Junqueira, Leonardo V. 388

Kämmerer, Lara 238
Keenan, Michelle 11, 20
Kelly, Jane E. **295**
Kelly, Melanie **373**
Khalaf, Husam S.M. **374**
Khan, Asad M. **378**
Khan, I.A. 399
Khan, Muhammad J. **140**
Kleemann, Samuel G.L. **137**, **148**, 328
Koetz, Eric 35
Kraus, Helmut 238
Kreuz, Klaus 238
Krishnan, Mahima 15, **22**, 384
Kristiansen, Paul E. 49, 265, 329, 374
Kriticos, Darren J. **77**, 82
Kumaran, Nagalingam **33**, 222, 371
Kwong, Raelene M. 69, **204**

Lalla, Reshnee 201
Lang, Kate 54
Lang, Peter 322
Latif, Sajid 366, **379**
Laursen, Bente B. 146
Lee, Caroline 147
Lefoe, Greg **69**
Leishman, Michelle **71**, **234**
Lemerle, Deirdre 34, **64**, 133
Lerchl, Jens 238
Lesieur, V. 157, 220, **221**, **383**
Lester, John 369, 389
Li, Yongfeng **21**, 411
Liebl, R. 236
Llewellyn, Rick **147**, 164
Logan, Peter 345
Long, Weihua **384**
Loukopoulos, Panayiotis 312
Lu, Huan **25**
Lukitsch, Bert 200

Macintyre, Jillian **272**
Magnussen, Craig **100**
Mahajan, Gulshan **32**, **385**, 390
Maher, Tim 71
Mahmood, Zahid 333
Malhi, Malwinder Singh **386**
Malone, Jenna M. 10, **15**, 16, 328, 384
Maloney, Luke 192
Manalil, Sudheesh 47, **301**
Mandal, Mohammad Shamim Hasan **387**
Manea, Anthony 71
March, Nathan **257**
Marini, Danila 147
Marshall, Graham 49

Martin, J.F. 157, 220, 221, 383
Matloob, Amar 32
Matthews, Robert 126
Maul, Jude E. 42
McCarthy, Cheryl L. **125**
McConnachie, A.J. **131**, 203, 232
McInerney, Catherine 193
McKiernan, Adam B. **138**
Meffin, Ross **91**
Melland, Rachel **286**
Mendes, Kassio F. **388**
Merriam, Alicia Blyth **16**, 22
Mietzner, Thomas 238
Mifsud, Charlie **202**
Miller, Jessica **261**
Milling, Philip **177**
Mitterdorfer, Belinda 86
Mnqeta, Zezethu 231
Monjardino, Marta 147, **164**
Montagna, Marco 237, 243, 244, 364
Moore, Geoff 72
Moore, Kerri 222, 281, 392
Mooy, Rebecca **68**
Moran, Nicholas P. **109**
Morin, Louise 33, **157**, 220, 221 222, 223, 224, 230, **232**, 363, 369, 371, 383, **389**
Moroni, Sergio 34
Mudassar, Muhammad 333
Muellner, Petra 48
Munakamwe, Zachariah **193**
Munday, Christine J. **72**
Mutti, Navneet Kaur **390**
Mwendwa, James M. **146**, **391**

Nachtrieb, Julie G. 204
Nagalingam, K. 157
Neilsen, James **54**
Ng, Katherina 192
Nicholls, Joshua 211
Nicol, Sam 392
Nielsen, S.G. 43
Nielson, R. 236
Norton, Sally 417
Nxumalo, Menzi 201

Odgers, Tim 104
Ollivier, M. 157, **220**, 221
Oosterhout, Elissa van 60
Opden, Roni 200
Opena, Jhoana **34**
Osten, Vikki 133
Osunkoya, Olusegun O. **281**, **392**
Ouzman, Jackie 147
Owen, Mechelle J. 25

Pain, Nicholas 64
Panetta, Dane 187
Pangli, Pavitarpal Singh 386
Parsons, Sandra **4**, 192
Passeretto, Kellie **87**
Paterson, Iain D. **231**
Pene, Heidi M. 27
Peng, Qiong **26**
Perfrement, Adam 364
Perrett, Christine 392
Petrovic, Tijana 22
Phantavong, Sengphet 64
Picard, Laurent 238
Piltz, John W. 379
Piper, Melissa 230
Plater, Colin 239, 245
Pokhrel, Sita Tiwari 49
Potter, Shauna **70**
Powells, Jo **397**
Powles, Stephen B. 23, 24, 25, 26
Pratley, James E. 5, 34, 43
Preston, Aaron L. **5**
Preston, Christopher 10, 12, 15, 16, 22, 137, 328, 384
Price, A. 43
Price, Richard J. 235
Pukallus, Kelli **211**

Quinn, Jane C. 146, 295, 312, 379

Rafeizade, Ahmad **398**
Rafter, Michelle A. 33, 157, **222**, 371
Raghu, S. 33, 157, 220, 221, 222, 371, 383
Rahamdad, K. **399**
Rahman, Md. Moshiur 412
Raphael, Bo 4, **192**
Rasheed, Muhammad 333
Rathinasamy, A. 408
Rayner, Annie **144**, 163
Read, John L. 91
Reddy, Krishna N. 42
Rees, Steven R. 125
Revell, Clinton 72
Riding, Natasha 226
Riikonen, Peter 247
Rimando, Agnes M. 42
Rinwa, Abhimanyu **400**
Robilliard, Shona L. 163, 368, **401**
Robinson, Randall W. 280
Robinson, Wayne 295
Rowles, Alexei D. 109
Rutherford, Henry **107**
Ryan-Colton, Ellen 91

Sagliocco, Jean Louis 204
Sandral, G.A. 43
Saravanane, Pasoubady **402**
Schooler, Shon 230
Scott, Angela L. 101
Scott, John K. 369, 389
Shabbir, Asad 368, 401, **405**, **406**
Shah, S.S. 399
Shaw, Justine D. 265
Sheehan, Matthew R. 70
Shephard, Adam 290
Sheppard, A. 157, 220, 221, 383
Shi, Boyang 215, 281
Shilpakar, R. 131
Sindel, Brian M. 49, **265**, 329, 374
Singh, Manpreet 400
Snow, Elizabeth L. **226**
Storrie, Andrew 133, **323**
Sukkarieh, Salah 409
Sullivan, Paul R. **233**, **407**

Tana, Tamado 334, 338
Taylor, Cameron 133
Taylor, Dianne B.J. **215**
Taylor, Malcolm 239

Taylor, Steve **181**
Thirumalaikumar, R. **408**
Thomann, T. 157, 220, 221, 383
Thornby, David F. 11, 20
Tian, Xing-shan 416
Tixier, M.S. 157, 220, 221, 383
Tomkins, Kerrie 175
Tonisielo, Valdemar L. 388
Tresch, Stefan **242**
Turner, Peter 232

Uddin, Md N. **280**

van der Meulen, Annie **163**
van Koten, Chikako 216
Velappan, Geetha 22
Velvin, Frances G. **124**
Venkataraman, N.S. 408
Verbeek, Birgitte **60**
Vink, Cor J. 216
Vitorino, M.D. 157
Vogler, Wayne 257

Walsh, Michael J. 32, 138, 139, 145, 144, 163, 368, 385, 401, 406, **409**
Walter, Gimme H. 20, 378
Wang, Aisuo 302
Watling, Jennifer R. 225
Waycot, Michelle 322
Weidenhamer, Jeffrey D. 146, 391
Wells, Gregory S. **239**, **245**
Werth, Jeff A. **11**, 20, 378
Westaway, John O. **200**
Weston, Leslie A 146, 295, 312, 318, 366, 379, 391, 417
Weston, Paul A. 146, 295, 366, 379, 391, 417
Whalley, Ralph D.B. 329
White, Andrew 230
White, Matt **410**
Whitsed, Rachel 72
Widderick, Michael J. **133**, 138, 139, 145, 163
Williams II, Martin M. **42**
Williams, Laura K. 265
Wilson, Susan C. 265
Witschel, Matthias 236, 238
Wood, Alan 224
Wu, Hanwen 34, 146, **290**, 302, 307

Yang, Xia 21, **411**
Yeates, Alice 230
Yu, Qin 21, 24, 25, 26

Zahan, Taslima **412**
Zahid, Waqas 265
Zeil Rolfe, Isabel 369
Zhang, Chun **416**
Zhang, Zichang 21, 411
Zhu, Xiaocheng **307**, 312, 318, **417**
Zubair, H.M. **43**

Keyword index

α-tubulin mutation 24
1-aminocyclopropane-1-carboxylate oxidase (ACO) 21
2,4-D 22
2,4-D amine 139
5′-UTR-Py-rich stretch element 416

Acetohydroxyacid synthase (AHAS) 384
Adelaide 107
Adventure volunteers 68
Advisory list 410
African lovegrass 181
African olive 360
Agent interactions 363
Agent prioritisation 222
Agixa 239
Agricultural weed 102
Agriculture Victoria 193
Agro-chemical 370
Allelopathy 34, 43, 387, 400, 417
Alligator weed 273, 327
ALS inhibitor resistance 16
Amaranthus 378
Amaryllidaceae 383
Amino acid 42
Amitrole 323
Annual ryegrass 5, 43, 137, 144, 153, 163, 237, 244, 368
Annual weeds 391
Aquatic weed management 271
Aquatic weeds 367
Arthropods 157, 221, 383
Arylex active 245
Assembly 20
Assessment 96
Atlas of Living Australia 69
Australia 281
Australian native *Solanum* 307
Australian native *Sorghum* 417
Automatisation 242
Avian seed dispersal 360

Baeodromus eupatorii 363
Balloon vine 232
Barley grass 328
Barley Grass RIM 164
Barnyard grass 34, 239
Base temperature 36
Behaviour change 60
Bellyache bush 215
Benefit:cost ratio 402
Best practice management 70
Binding coefficient 169
Bio-economic modelling 164
Bio-fertiliser 408
Biocontrol. *See* Biological control
Biodiversity 281, 360
Biological control 33, 215, 216, 220, 222, 224, 230, 232, 233, 371, 407
Biological control agents 104
Biological invasions 281, 392
Biological weed control 226
Biology 301
Biomass 374
Biosecurity 48, 82, 109
Biserrula 366
Bitou 272
Black gram 408
Boats 68
Boerhavia 200
Boneseed 272
Bowling Green Bay 271
Breakdown pathway 169
Broad leaved weeds 333
Broadleaf weed 236, 237, 243
Broadleaved pepper tree 232
Brome grass 163
Brome RIM 164
Bromus diandrus 158, 341
Burial depth 32, 47, 390
Bush-regeneration 273
Californian thistle 216
Canola 245
Canopy architecture 391
Capsules 247
Carcass contamination 295
Cardiospermum grandiflorum 232
Case studies 286
Cassytha pubescens 225
Cat's claw creeper 104
CBSM 60
Cerastium fontanum 265
Cereals 237, 243, 244
Chaff lining 163
Chaff tramlining 163
Chickpea 138, 399
Chilean needle grass 181
Chromolaena odorata 226
Cinmethylin 238, 244
Citizen scientists 193
Classical biological control 157, 405
Clethodim resistance 137
Climate change 234, 301
CLIMEX 77, 82
Co-dependent partnership funding model 203
Co-evolved natural enemy 405
CO_2 assimilation 33
Coastal ecosystems 271
Collaboration 104, 176, 286
Colour cameras 125
Common sowthistle 22, 138, 157, 163, 221
Communication 233
Community 286
Community engagement 104
Community-based social marketing 60
Competition 146, 285, 379, 398
Competitive cultivars 400
Compliance 100
Conflict species 91
Containment 107

Control 27
Control tools 108
Conyza bonariensis 222
Corn 42
Cost benefit analysis 82
Cotton 35, 54, 365, 398
Cover crops 49
cp4 epsp 42
Crofton weed 363
Crop competition 133, 137, 138, 164, 406
Crop residue 391
Crop residue mulch 408
Crop topping 364
Cross-resistance 16
Crown weevil 407
Cultivar 374
Cultural weed control 406
Cyclones 350
Cylindropuntia spp. 226
CYP450 26

D1 protein 25
Dacus persicus 405
Daintree 345
Data 192
Data analytics 242
Decision analysis 187
Decision support tools 70, 96
Declaring eradication 350
Defoliation 33
Delimitation 126
Delta arrowhead 204
Detection 187
Detection dogs 373
Dialectica scalariella 407
Dicamba 54
Diclofop 26
Dinitroaniline 24
Disaster 119
Distribution 238
DNA barcodes 302
DNA barcoding 312
DNA sequencing 318
Dormancy 341
Drill-sown rice 239
Drones 125
DT50 169

Early detection 193
Early invader 372
Early invader pilot projects 197
Early invaders 410
Echinochloa colona 11, 21 406, 411
Echium plantagineum 407
Echium vulgare 407
Ecological interactions 220
Ecology 55, 148
Economic returns 370
Economic threshold 300, 338
Ecosystem monitoring 230
Effectiveness 405
Efficacy 333
Egeria 273, 367
Egeria densa 367
Eleusine indica 416
Emex australis 158
Energy requirements 409
Engagement 235
Environmental pest plants 101
Environmental Trust 232
Environmental weeds 4, 72, 102, 124, 265, 410
Environmentally-safe 96
EPSPS 416
Eradication 11, 115, 124, 126, 187, 200, 202, 265, 327, 354
Establishment 231, 379
Ethylene biosynthesis 21
Euphorbia 369
Euphorbia paralias 232
Exotic 4
Extension 49, 233

Faba bean 138
Fatty acid thioesterases 238
Flaxleaf fleabane 157, 364
Fleabane 20, 139
Floating Landcare 68
ForageMax 245
French serradella 366
Frogbit 202
Fungal biocontrol agent 369
Fungal genomics 223
Fungal pathogen 230
Fungi 157, 221, 383

Genetic assay 302
Genome sequencing 20
Geospatial analysis 360
Germination 35, 36, 158, 329, 341
Germination test 384
Gical control 231
GIS 176
Glufosinate 323
Glyphosate 15, 35, 42, 139, 177, 247, 323, 328
Glyphosate resistance 11, 54
Glyphosate-resistant crops 42
Golf 10
Gorse 225
Government regulation 64
Grain yield 374
Granule-bound starch synthase 1 307
Grass weeds 236
Graze and grain 245
Great Barrier Reef (GBR) Ramsar 271
Great Victoria Desert 267
Green life industry 96
Green thistle beetle 216
Group I herbicides 22

Hair grass 36
Hand weeding 49
Hard-seeded legume 379
Harvest weed seed control 163
Harvester front 144
Hawkesbury estuary 68
Hawkweed 175, 327
Heat waves 301
Herbicide 412
Herbicide metabolism 169
Herbicide persistence 412
Herbicide resistance 12, 139, 145, 169, 374, 397, 400, 411
Herbicide resistance monitoring surveys 16
Herbicide technology 261
Herbicide trial 55
Herbicide-tolerant 323
Herbicides 251, 267, 290, 333, 362, 386, 399, 402

Heritage plants 101
Hieracium aurantiacum 126
Historical perspective 102
Hoe weeding 370
Hordeum leporinum 158, 341
Hordeum spp. 295, 328
Horticulture 49, 71
Host specificity 224
Hosta alismifolia 387
Hunter Valley 360
HWSC 144, 401
Hybrid canola 137
Hydrothermal time 36

Image analysis 125
Impact assessment 108
India Haryana Punjab farmers 386
Indian mallow 27
Indigenous communities 267
Industry 286
Inheritance 22
Inhibition 43
Insertion 416
Integrated weed management 1, 32, 64, 157, 295, 300, 362, 390
Integrated weed management strategy 373
Interference 398
Inundative biocontrol 389
Invasion mechanisms 234, 285
Invasive 71, 201
Invasive alien species 318
Invasive grasses 181
Invasive pests 100
Invasive plants 77, 96, 115, 226, 235
Invasive species 271
Invasive species impacts 230
Invasive weed 251
Investment frameworks 235
IPM 300
Iridaceae 201
Iris pseudacorus 201
Irrigation channel 323
Islands 103
ITS 312
IWM 164

Kalanchoe delagoense 226
Kosciuszko National Park 175

Land zones 271
Leaf area 33
Leaf cactus 232
Leaf entry 169
Legumes 366
Levy 100
Life cycle 55
Light 47
Lime 153
Lineal reserve 323
Lipid biosynthesis 238
Listronotus appendiculatus 204
Livestock 147
Local government 373
Lolium rigidum 5, 24, 158
Loop-mediated isothermal amplification 302
Low herbicide dose 365
Lucerne 43

Madeira vine 104
Maize 408
Management 12, 102
Management effectiveness 87
Manual weeding 402
matK 312
Meta analysis 133
Metabolic resistance 26
Metabolomics 146
Methodology 233
Metrics 286
Metsulfuron-methyl 251
Miconia 119
Microwave soil heating 140
Mikania 119
Mikania micrantha 226
Milk thistle 364
Mineral 42
Minimum tillage 408
Mismatches 102
Mitchell grasslands 257
Mode of action 238, 242
Mogulones geographicus 407
Mogulones larvatus 407

Moisture deficiency 333
Molecular analysis 220
Monaro 397
Monitoring 108
Mowing 295
Mulch effect 366
Multidisciplinary 286
Multiple herbicide resistance 10
Multiple resistance 1
Multispectral imagery 131
Mung beans 406
Mycosphaerellaceae 369

Nassella neesiana 302
Nassella tenuissima 109
Nassella trichotoma 302, 373
National Interest Pest Response Programme 124
Native bio-control 225
Native range 405
Natural enemies' community 221, 383
Natural resources 267
Natural systems 285
Neem 261
New incursion 251
Nitrogen 225, 362
Nodulation 43
Non-crop 323
Non-indigenous plants 72
Non-puddled rice 412
Northern Australia 261
Northern Beaches 272
NSW Biocontrol Taskforce 359
Nutrient enrichment 280

Olea europaea ssp. *cuspidata* 360
Opuntioid cacti 70, 372
Orange hawkweed 126
Ornamental plants 96
Osmotic potential 32, 390

Paederia foetida 361
Pakistan 405
Panicum hillmanii Chase 312
Panicum spp. 312
Paraquat 323

Parasitic plant-host association 225
Parkinsonia 211
Parthenium hysterophorus 334, 338
Parthenium weed 318
Participatory training 193
Partnerships 68
Passalora 369
Passalora ageratinae 363
Passive surveillance 193
Pasture legumes 34
Paterson's curse 407
Pathogenicity phenotype 223
Pendimethalin 399
Penoxsulam 21
Pereskia aculeata 232
Phalaris minor 386
Phenolics 385, 387
Phenology 391
Phenrica guerini 231
Photosynthesis 33, 225, 411
Phragmites australis 280
Phylogenetic tree 312
Physical weed control 409
Phytotoxicity 366
Pine tree 103
Plant pathogen 224
Poa annua 10, 265
Policy 102, 200
Polyethylene glycol 354
Post-release evaluations 231
Pre-dispersal seed predation 204
Pre-emergence 244
Pre-emergent 12
Precision agriculture 147
Prediction 86
Prescription map 125
Prevention 4
Prickly acacia 215, 257
Primary production 102
Prioritisation 48, 86, 87
Prioritisation framework 232
Prioritisation tool 77
Priority 4
Priority weeds 192
Prohibited matter 202
Proline 385
Promoter 416
Prosulfocarb 12
Protein 399
Proteomics 411
psbA gene mutation 25
PSII herbicide 25
Pucciniales 223

Queensland Department of Agriculture and Fisheries 211
Quinclorac 21

Railway 323
Rainfall 385
Rainforest 354
Rangelands 72
Ranking 48
Rapid response 202
Recommendations 133
Remote sensing 175
Removal 368
Research 235
Research impact 286
Research protocol 133
Residual herbicide 388
Residue 146
Resistance 236, 237, 243, 328, 364
Resistance cost 411
Resistance mechanism 15
Resource allocation 146
Resource manipulations 280
Respiration 33
Restoration 280
Retention process 388
Review 133
Rhizomes 329
Rhizosphere 42
Rice 140, 239
Rinskor active 239
Risk assessment 48
Risk rating 410
RNA-sequencing 26
Roadside 323
Root weevil 407
Rubber bush 405
Rubus fruticosus 389
Rumex hypogaeus 158
Ryegrass 145

Safety 247
Saflufenacil 364
Sagittaria platyphylla 204
Salt stress 47
'Scattergun' 257
Schinus terebinthifolius 232
Screening tool 71
Sea spurge 232, 369
Seasonality 271
Seed 368
Seed germination 32, 47, 390
Seed production 158, 385
Seed shed 401
Seed-set 364
Seedbanks 49, 140
Serrated tussock 181, 373, 397
Shedding 158
Siam weed 345
Silverleaf nightshade 307
Silvertop grass 239
Simulated herbivory 222
Sisymbrium orientale 384
Site-specific weed control 409
Skeleton weed 223
Skunk vine 361
Small-holder farmers 64
Soil acidity 153
Soil amendments 388
Soil chemistry 281
Soil enzymes activity 281
Soil microbes 280
Soil microbiota 366
Soil-borne disease 389
Sonchus oleraceus 15, 138, 158
Sorghum 406
Sorgoleone 417
South Africa 201
Sowing rate 406
Sowthistle 20, 139, 220
Soybean 42
Spatial grazing 147
Spatial occurrence 82
Sporobolus spp. 226
Spot spray 125
Spot treatment 177

Spread 82
Stakeholder elicitation 392
Stakeholder engagement 91, 222, 371
State prohibited weeds 193
Stellaria media 265
Strategic weed management 77, 87
Sugar beet 362
Suppression 400
Surveillance 119
Survey 192

Tebuthiuron 257
Technology 267
Tecoma stans 232
Temperature 15, 47
Tolerant 35
Topramezone 243
Training 60
Translocation 169
Tree weeds 247
Triallate 12
Trifludimoxazin 236, 237
Trifluralin metabolism 24
Trifluralin resistance 24
trnL intron-*trnF* 312
Tropics 345, 350
Turf 10
Turnip weed 163

UAV 126, 177
Ulex europaeus 225
Ultra narrow row 365, 398
University 286
Uptake 238
UU 211

Vachellia nilotica 257
Vacuum 368
Vegetation community 230
Velvetleaf 27
Viper's bugloss 407
Virtual fencing 147
Volunteer support 64
Vulpia 295

Water availability 225
Water quality 104
Water solubility 169
Water stress 378, 385
Weed 82, 200, 221, 368, 383, 399
Weed biological control 203, 204, 359
Weed biology 32, 290, 390
Weed competition 365
Weed control 147, 153, 239, 261, 412
Weed detection dog 327
Weed ecology 300, 385
Weed futures 234
Weed impact 281, 392
Weed lists 102
Weed management 176, 233, 272, 286, 290, 329, 334, 338, 400
Weed management tools 242
Weed mapping application 87
Weed prioritisation 392
Weed risk 96
Weed risk assessment 77, 82, 86, 108, 392
Weed risk management 77, 86
Weed seed collection 144
Weed spread 392
Weed suppression 146, 379, 391
Weed-crop interference 300, 334
Weeding 362
Weeds 48, 60, 68, 96, 235, 245, 290, 378
Weeds at the Early Stage of Invasion 197
Weeds of National Significance 70, 104, 107, 225, 389
See also WoNS
Western Cape bridal creeper 107
Wetting/drying regimes 271
Wheat 140, 374, 400
Wild oats 26, 145, 163, 243, 368, 401
Wild radish 25, 153, 163, 237
Wild turnip 364
Wilding conifers 103
Willow primrose 273
Winter grass 10
WoNS 192, 307

XtendFlex 54

Yellow bells 232
Yield losses 334, 338, 398

Zea mays 334, 338
Zero tillage 408
Zoned management 91

21AWC
SYDNEY 2018 WEED BIOSECURITY ~ PROTECTING OUR FUTURE

Council of Australasian Weed Societies Inc. (CAWS)
Massey University, Private Bag
11-222 Palmerston North 4442
New Zealand

ISBN 978-1-5108-8847-0

CURRAN ASSOCIATES INC.
proceedings.com